# Lecture Notes in Networks and Systems 1224

The series "Lecture Notes in Networks and Systems" publishes the latest developments in Networks and Systems—quickly, informally and with high quality. Original research reported in proceedings and post-proceedings represents the core of LNNS.

Volumes published in LNNS embrace all aspects and subfields of, as well as new challenges in, Networks and Systems.

The series contains proceedings and edited volumes in systems and networks, spanning the areas of Cyber-Physical Systems, Autonomous Systems, Sensor Networks, Control Systems, Energy Systems, Automotive Systems, Biological Systems, Vehicular Networking and Connected Vehicles, Aerospace Systems, Automation, Manufacturing, Smart Grids, Nonlinear Systems, Power Systems, Robotics, Social Systems, Economic Systems and other. Of particular value to both the contributors and the readership are the short publication timeframe and the world-wide distribution and exposure which enable both a wide and rapid dissemination of research output.

The series covers the theory, applications, and perspectives on the state of the art and future developments relevant to systems and networks, decision making, control, complex processes and related areas, as embedded in the fields of interdisciplinary and applied sciences, engineering, computer science, physics, economics, social, and life sciences, as well as the paradigms and methodologies behind them.

Indexed by SCOPUS, EI Compendex, INSPEC, WTI Frankfurt eG, zbMATH, SCImago.

All books published in the series are submitted for consideration in Web of Science.

For proposals from Asia please contact Aninda Bose (aninda.bose@springer.com).

Anu Bajaj · Ana Maria Madureira ·
Ajith Abraham
Editors

# Hybrid Intelligent Systems

## 23rd International Conference on Hybrid Intelligent Systems (HIS 2023), December 11–13, 2023, Volume 2: Large Language Models and Machine Learning

Springer

*Editors*
Anu Bajaj
Department of Computer Science and Engineering
Thapar Institute of Engineering and Technology
Patiala, Punjab, India

Ana Maria Madureira
Departamento de Engenharia Informática
Instituto Superior de Engenharia
Porto, Portugal

Ajith Abraham
School of Artificial Intelligence
Sai University
Chennai, India

ISSN 2367-3370 ISSN 2367-3389 (electronic)
Lecture Notes in Networks and Systems
ISBN 978-3-031-78924-3 ISBN 978-3-031-78925-0 (eBook)
https://doi.org/10.1007/978-3-031-78925-0



This Springer imprint is published by the registered company Springer Nature Switzerland AG
The registered company address is: Gewerbestrasse 11, 6330 Cham, Switzerland

# Preface

Welcome to the 23rd International Conference on Hybrid Intelligent Systems (HIS 2023) and the 19th International Conference on Information Assurance and Security (IAS 2023). HIS-IAS 2023 was held in 5 different cities, namely Olten, Switzerland; Porto, Portugal; Kaunas, Lithuania; Greater Noida, India; Kochi, India, and in online mode. HIS-IAS 2023 is hosted and sponsored by the Machine Intelligence Research Labs (MIR Labs), USA, and supported by Bennett University, Greater Noida, India.

23rd International Conference on Hybrid Intelligent Systems (HIS 2023) was focusing on synergistic combinations of multiple approaches to develop the next generation of intelligent systems. A fundamental stimulus to the investigations of HIS 2023 is the awareness that combined approaches will be necessary if the remaining tough problems in Artificial Intelligence are to be solved. The objectives were to increase the awareness of the research community of the broad spectrum of hybrid techniques, to bring together AI researchers from around the world to present their cutting-edge results, to discuss the current trends in HIS research, to develop a collective vision of future opportunities, to establish international collaborative opportunities, and as a result to advance the state of the art of the field.

Information assurance and security has become an important research issue in the networked and distributed information sharing environments. Finding effective ways to protect information systems, networks and sensitive data within the critical information infrastructure is challenging even with the most advanced technology and trained professionals. The 19th International Conference on Information Assurance and Security (IAS 2023) aims to bring together researchers, practitioners, developers, and policy makers involved in multiple disciplines of information security and assurance to exchange ideas and to learn the latest development in this important field.

HIS - IAS 2023 received 540 papers from 44 countries, and based on the review process, we have accepted 240 papers. Each paper was reviewed by at least five or more reviewers and based on the outcome of the review papers were accepted for inclusion in the conference proceedings. Proceedings were divided into Five Volumes. This volume is focused on *Large Language Models and Machine Learning.* Large language models represent a breakthrough in natural language processing technology, using deep learning techniques and massive amounts of data to generate human-like text. Deep learning plays a pivotal role in advancing artificial intelligence by enabling machines to learn from data, model complex relationships, and make intelligent decisions across a wide range of applications. Its versatility, scalability, and performance make it a cornerstone of modern AI research and development. This volume which comprises of 48 articles is focused on feature learning; scalability; flexibility, state-of-the-art performance in applications, unstructured data handling; continuous improvement; etc.

First, we would like to thank all the authors for submitting their papers to the conference, for their presentations and discussions during the conference. Our thanks go to program committee members and reviewers, who carried out the most difficult work by carefully evaluating the submitted papers. Our special thanks to the following plenary speakers, for their exciting plenary talks:

- Christine Zarges, Aberystwyth University, UK
- Theresa Schmiedel, University of Applied Sciences and Arts Northwestern Switzerland, Basel, Switzerland
- Ke Feng, Singapore-ETH Centre, The National University of Singapore, Singapore
- Stefka Fidanova, Bulgarian Academy of Sciences, Bulgaria
- Diego Oliva, Universidad de Guadalajara, Mexico
- Sebastian Ventura, University of Cordoba, Spain
- João Pedrosa, INESCTEC, Portugal
- Nuno Bettencourt, Instituto Superior de Engenharia do Porto, Portugal
- Milan Tuba, Singidunum University, Serbia
- Dalia Kriksciuniene, Vilnius University, Lithuania
- Kusum Deep, Indian Institute of Technology Roorkee, India

We express our sincere thanks to the organizing committee chairs for helping us to formulate a rich technical program. Enjoy reading the articles!

Ana Maria Madureira
Cengiz Kahraman
Ajith Abraham
General Chairs

Yuchuan Sun
Milan Tuba
Program Chairs

Anu Bajaj
Publication Chair

# Contents

# Performance Evaluation of ChatGPT on BITSAT: Engineering Entrance Examination

Syed Ehtesham, Harpreet Singh Bedi, and Ravindra Giriraj Bhardwaj(✉)

Department of Mechanical Engineering, Birla Institute of Technology and Science Pilani, Dubai, United Arab Emirates
ravindra@dubai.bits-pilani.ac.in

**Abstract.** ChatGPT, a powerful natural language processing tool, is an Artificial Intelligence (AI) language model designed to communicate with humans. The education field has achieved a new perspective with AI playing a very important role. AI has opened avenues to comprehend subjects, get answers to questions, and receive recommendations and problem-solving solutions. The present study aims to assess the performance outcome of ChatGPT in solving Birla Institute of Technology and Science Admission Test (BITSAT). BITSAT is an engineering entrance examination conducted in India. An adjudication criterion based on Accuracy, Concordance, and Insight is implemented for performance evaluation of ChatGPT responses. A dataset of questions and answers for subjects including physics, chemistry, English proficiency, logical reasoning, and mathematics is generated from the previous two years' exams. ChatGPT responses pertaining to English have the highest accuracy (72%), concordance (52%), and insight (68%), while least with physics at 30%, 28%, and 30%, respectively. The variation in the outcomes can be attributed to factors such as lack of domain expertise, and limited ability to understand the context and generate original insights. It is therefore imperative for academia to be abreast of the strengths and weaknesses of ChatGPT, so as to make an effective utilization of ChatGPT as a tool in the field of academics and research.

**Keywords:** ChatGPT · Large Language Model · BITSAT · OpenAI · Education Technology

## 1 Introduction

### 1.1 Introduction to ChatGPT

Currently, there has been an increasing trend to improve the overall experience of higher education by incorporating new technology and techniques. For example, implementation of learning management systems (LMS), video assisted learning, gamification, virtual and augmented reality in education and technology showed increased student engagement and educational planning [1, 2]. Despite the numerous benefits that technology has brought to education, there are still concerns about how it may affect higher education institutions. AI is currently being applied in education in a variety of ways,


A. Bajaj et al. (Eds.): HIS 2023, LNNS 1224, pp. 1–9, 2025.
https://doi.org/10.1007/978-3-031-78925-0_1

ranging from chatbots that give 24/7 student support to personalized learning algorithms that adapt to the needs of each student. AI is used to automate administrative activities, deliver computer-based tutorials called as intelligent tutoring systems (ITS) and evaluate massive volumes of data to design new educational techniques and policies [3–5]. AI can bring numerous educational benefits such as customized learning experience and student specific performance evaluation. This can increase student involvement and motivation, leading to higher academic outcomes [3].

A recently developed AI tool called ChatGPT is an OpenAI Natural Language Processing (NLP) chatbot based on Generative Pretrained Transformer (GPT). ChatGPT interacts with the user in a conversational discourse and produces responses that appear human alike. It is intended to participate in natural and intelligent conversations with humans by utilizing modern NLP techniques to interpret and reply to user inquiries. ChatGPT has the capacity to learn and adapt, and can also recall the previous inputs from the user. ChatGPT may provide knowledge on a variety of topics or subjects, answer the asked questions, make recommendations, and engage in casual chat, among other things. Because of its ability to learn from massive volumes of data, it can provide increasingly accurate and relevant responses over time [6]. Sok et al. have discussed the various advantages and drawbacks of using ChatGPT in research and education [3]. According to the report, ChatGPT offers five primary advantages, including creating learning assessments, improving pedagogical practice, providing virtual one-on-one coaching, developing outlines, and brainstorming ideas. However, there are dangers associated with questions of academic integrity, unequal learning assessments, incorrect information, and excessive dependence on AI tools [7].

It is found that ChatGPT is a substantial advancement in natural language processing models for medical question answering [8]. The performance of ChatGPT on the United States Medical Licensing Exam (USMLE), which consists of three exams showed that without any professional instruction or support, ChatGPT passed all three examinations or came very close to completing them [9]. Giannos et al. evaluated the performance of ChatGPT on standardized admission tests in the United Kingdom based on a dataset of 509 questions from different examinations [10]. Thus, ChatGPT holds promise as a supplement to subjects and tests that assess aptitude, problem-solving and critical thinking, and reading comprehension. However, limitations of ChatGPT in scientific and mathematical knowledge underline the importance of further research [10]. Moreover, it is important to evaluate the ChatGPT responses not only quantitatively based on the accuracy of responses, but also qualitatively based on logical explanation, question comprehensibility, and interactive description. To this end, the current study is designed to evaluate the performance of ChatGPT to solve an engineering entrance examination (BITSAT) based on accuracy, concordance, and insights of the obtained responses.

### 1.2 Background of BITSAT

The Ministry of Education of the Government of India has designated the Birla Institute of Technology and Science (BITS) Pilani as an Institution of Eminence [11], one of the top institutions of higher learning in India and a deemed to be university. BITSAT is a computer-based online exam for applicants to BITS Pilani's integrated undergraduate programs at its various campuses. In this work, the response of ChatGPT for BITSAT-2022 and BITSAT-2021 examinations is evaluated. Each BITSAT session includes 130 questions on physics, chemistry, English, logical reasoning, and mathematics (or biology for those seeking admission in Bachelor of Pharmacy). The structure of a question paper is given in Table 1. There are four options given for each objective type (multiple choice) question, but only one is correct. Each correct response earns three points, each wrong response results in a deduction of one point, while unattempt questions have no points. The cutoff marks for different science and engineering disciplines differs, therefore the criterion of cutoff is not studied in this work.

**Table 1.** Structure of a BITSAT examination and the number of questions in four different areas.

| | Part-I | Part-II | Part-III | | Part-IV |
|---|---|---|---|---|---|
| Subject | Physics | Chemistry | English | Logical reasoning | Mathematics |
| Number of questions | 30 | 30 | 10 | 20 | 40 |

## 2 Methodology

The methodology of evaluating BITSAT exams using ChatGPT is illustrated in Fig. 1. A total of 234 multiple-choice questions (MCQ) out of 300 questions are encoded into ChatGPT and the response for each question is recorded. The remaining 66 were image-based questions which are not encoded as the current version of ChatGPT does not supports image-based input. Each response of ChatGPT is assessed using the accuracy, concordance, and insight (ACI) adjudication criteria. For any wrong or erroneous response, training and feedback are given to ChatGPT and another attempt is made. The second response is further assessed as per ACI-criteria, and if it is incorrect, no further training is provided to ChatGPT. The ACI criteria provides a framework for evaluating the rigor, quality, and the relevance of ChatGPT responses. The responses generated by ChatGPT can fall into three categories: accurate, inaccurate, or indeterminate. Additionally, the explanations provided with each response are also evaluated for their concordance (in agreement) or discordance (in disagreement) with respect to the obtained response. It has been observed that sometimes ChatGPT provides the correct response either without an explanation or with an explanation that contradicts itself. In such cases, the response is considered discordant. Moreover, the level of insight provided in the explanations is evaluated based on five parameters: non-definitional, unique, non-obvious, valid, and compendious. These parameters help gauge the depth and clarity of understanding in the explanations. To quantify the level of insight, a metric called the "density of insight" (DOI) is computed for each explanation. The DOI is calculated as the ratio of the number of insights present in the explanation to the total number of insight criteria. A DOI value of 1 indicates a fully insightful explanation, while any other value suggests a non-insightful explanation.

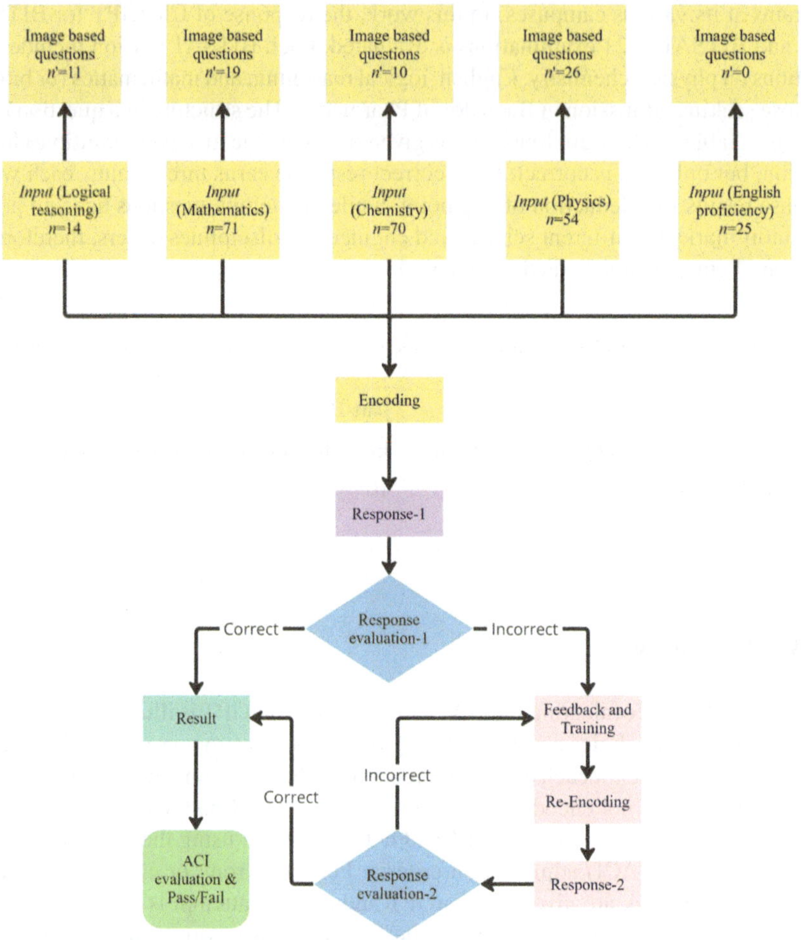


**Fig. 1.** Flowchart showing sourcing, encoding, training, and adjudicating the responses of ChatGPT. *Note:* $n$ and $n'$ refer to the number of questions evaluated, and the number of questions exempted, respectively.

## 3 Results and Discussion

Evaluations of ChatGPT's performance based on ACI criteria is discussed in this section. Accuracy occurs when ChatGPT provide a choice that matches with official examination answer key wheras inaccuracy occurs when ChatGPT selects incorret answer choice. Indeterminancy occur when ChatGPT provides an output which is not an answer choice or determines that not enough information is available or returns wrong answer after training. The accuracy of ChatGPT responses before and after training for physics, chemistry, English, logical reasoning, and mathematics subjects is shown in Fig. 2. Before training, the English responses have highest accuracy of 72%, whereas Physics responses have the lowest accuracy of 30%. Additionally, it is observed that physics and mathematics questions tend to get more erroneous and ambiguous responses from ChatGPT possibly because of the inability of the AI model to properly analyse the provided input, a lack of training, and complexity of the problem. Following the training regime, the responses of logical reasoning questions have the highest improvement in accuracy from 36% to 56%. As seen, after training ChatGPT provides more indeterminate responses as compared to before training. This shows that ChatGPT is unable to comprehend the questions and requires rigorous training. In the case of mathematics, there is not much improvement observed even after training, and the level of indeterminate responses increases from 21% to 49%. These indeterminate responses could be frustrating for students seeking clear answers and explanations. Overall, the inaccuracy and indeterminacy of ChatGPT implies that it is unable to understand problems with higher complexity and possibly requires more training.

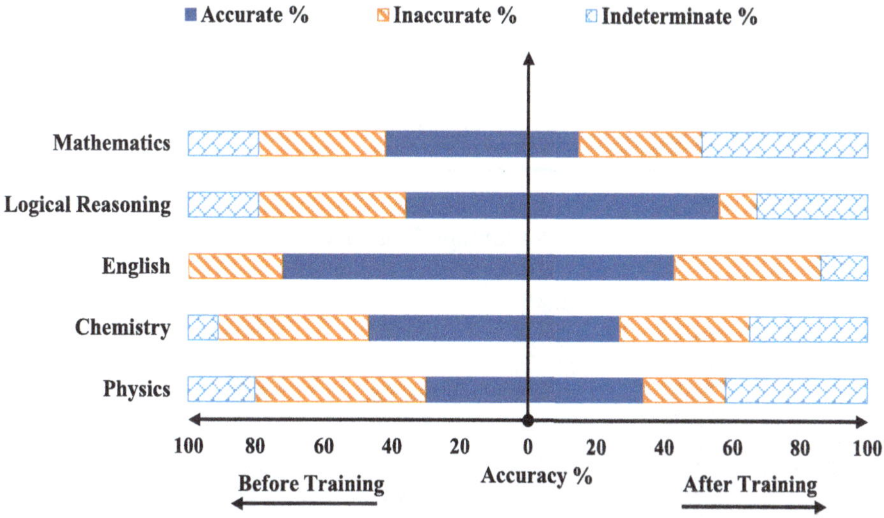


**Fig. 2.** Accuracy percentage of the ChatGPT responses before and after training for various subjects.

Next, the level of concordance of responses from each subject are evaluated as shown in Fig. 3. ChatGPT response is considered concordant when its explanation affirms the answer and negates all remaining choices. Contradictory, if any part of the explanation contradicts itself or there is no explanation for the answer then the response is considered as discordant. English has the highest overall concordance of 52% compared to the least concordance of 28% in Physics (Fig. 3a). This shows that ChatGPT is more adaptable to solve linguistic based problems such as English language considered in this study. It would be a breakthrough if ChatGPT functions in many different languages and solves a variety of linguistics problems. Although, most of the time ChatGPT gives explanation to each question, there are few questions that either require straight forward answers or the AI determines that no explanation is required. Accordingly, apart from the overall concordance, another important factor to consider here is the percentage of accurate discordant responses. Such a scenario occurs when the answer is true but there is no explanation or argument to back it up. It is observed that English has the highest percentage of accurate discordance 28%, followed by Mathematics and Physics with 7% and 6%, respectively as shown in Fig. 3(b). Chemistry and Logical Reasoning have 100% accurate concordance as shown in Fig. 3(b). This discordance between responses and explanations can be misleading and negatively affect the AI tool credibility.

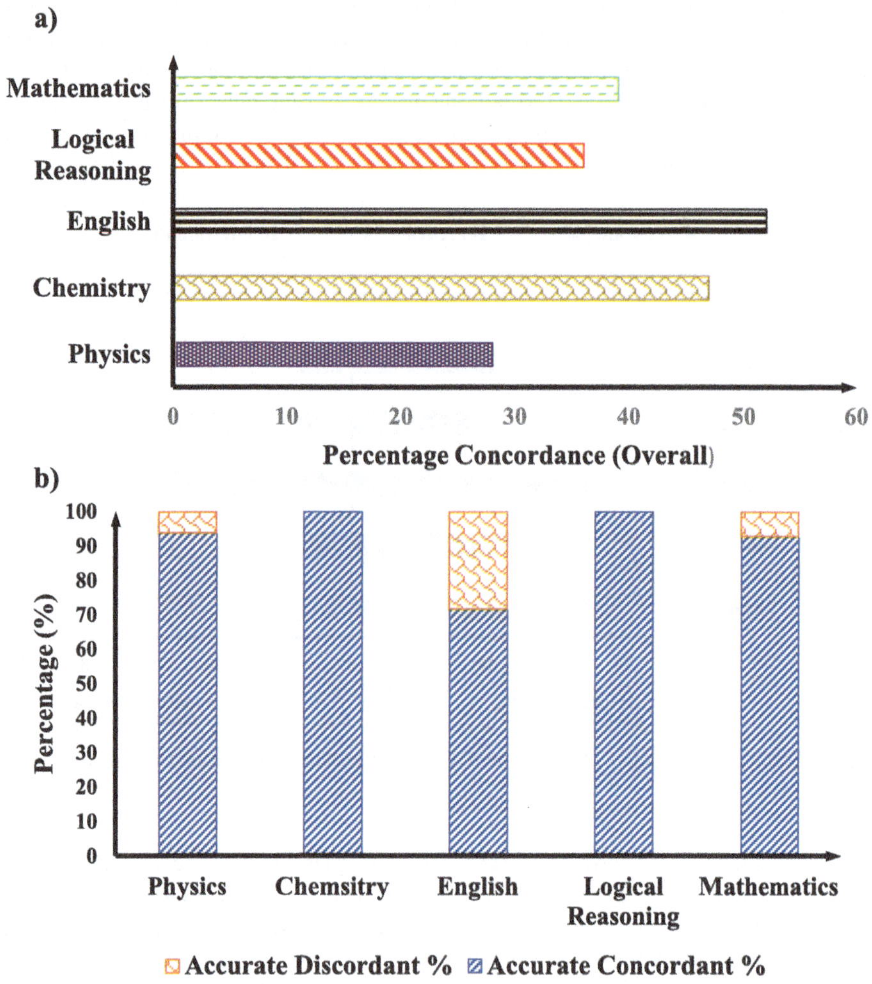


**Fig. 3.** (a) Overall percentage concordance and (b) accurate concordance and discordant responses for different subjects.

Apart from the accuracy and concordance of the responses, another factor that is evaluated here is the insightfulness of the ChatGPT responses. Insightful responses are those whose responses have a DOI = 1, meaning they satisfy all the parameters (non-definitional, unique, nonobvious, valid, compendious). Figure 4 presents the percentage insightfulness of responses obtained from each subject. While, English has shown an insightfulness of 68%, physics responses are least insightful of 29.6%. Altogether, ChatGPT showed an average of 43.1% insightfulness in the explanation. The higher the insightfulness, deeper the explanation and understanding are present in ChatGPT responses, helping to gauge the effectiveness of ChatGPT. It is evident that ChatGPT is powerful in giving insightful explanations for linguistic problems. DOI acts as an indicator that more training is needed to improvise the ChatGPT response for questions with non-insightful and zero-DOI responses. This will make the AI tool more beneficial and trustworthy, especially for students preparing for competitive exams.

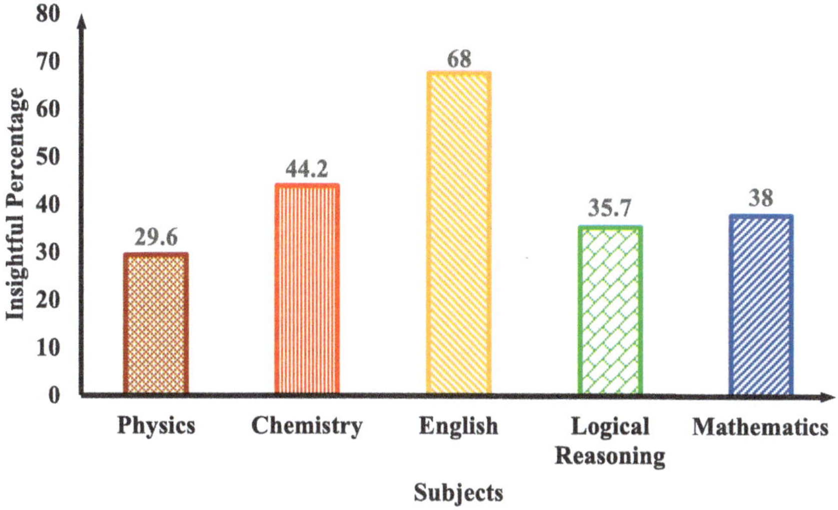


**Fig. 4.** Percentage insightfulness of ChatGPT responses for various subjects.

A comparison of insightful, non-insightful, and zero-insightful percentage for all the subjects is shown in Fig. 5. These characteristics are determined based on the density of insight (DOI). The response is considered insightful if DOI = 1, non-insightful if DOI < 1, and zero insight if DOI = 0. It can be seen that English has the most insightful responses of 68%, least non-insightful responses about 32% and no zero-insightful responses. Physics has the least insightful responses of 29.62%, most non-insightful responses of 51.87% and zero-insight responses of about 18.51%. This observation signifies that ChatGPT provides the most insightful responses for linguistic problems as compared to scientific problems. Overall, the average insightful, non-insightful and zero-insight responses by ChatGPT are 43.126%, 45.264%, and 11.626% respectively. This shows that ChatGPT produces more non-insightful responses as compared to insightful responses. The points awarded to each response for insight are done in accordance with the ACI adjudication criterion. It can be observed that the subjects involving mathematical equations have more non-insightful and zero-insight responses possibly because of the inability of the present version of GPT to understand complex problems and equations.

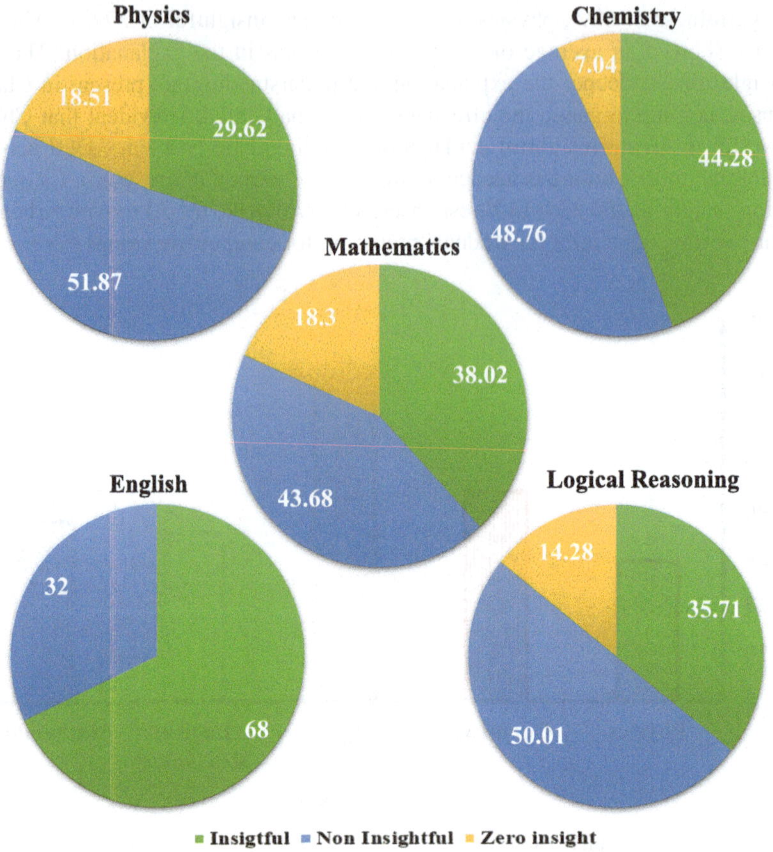


**Fig. 5.** Density of Insight (in percentage) of ChatGPT responses for various subjects.

## 4 Conclusion

The purpose of this study is to assess the performance of ChatGPT in solving the BITSAT entrance examination to predict its performance outcome pertaining to ACI (Accuracy, Concordance, and Insight) adjudication criteria. The findings reveal a noteworthy trend in the responses pertaining to different subjects characterized by a prevailing occurrence of most inaccurate answers for physics before training (50%), and the most indeterminant answers for mathematics after training (49%), and again physics having the most non-insightful and zero insight responses (51.87% and 18.51% respectively). The variation observed in the ChatGPT performance can be attributed to factors such as a lack of domain expertise. This limitation could be a significant concern, especially when students rely on AI models for accurate explanations and insights in specific subjects. These results can be linked to the limitations of ChatGPT in limited ability to comprehend the concepts and lack of discipline competency. Nevertheless, ChatGPT's performance in English must be acknowledged as it has projected the highest accuracy, concordance, and insight in the BITSAT examination, which can be attributed to a high degree of proficiency in understanding and answering explicit questions. The scope of improving the models' performance in other domains such as physics, chemistry and mathematics should be explored by enhanced focus on target programming and training which can prove beneficial for potential students preparing for examinations. However, it is critical to recognize the limits of ChatGPT such as answers being inaccurate, indeterminate, discordant, and non-insightful as seen in this study.

## References


1. Pasha, S.A., Youssef, E., Sharif, H.: Role of virtual reality in improving students' LMS experiences: structural equation modelling based study. In: 2021 International Conference of Modern Trends in Information and Communication Technology Industry (MTICTI) (2021)
2. Huang, W.H.-Y., Soman, D.: Gamification of education. Rep. Ser. Behav. Econ. Act. **29**(4), 37 (2013)
3. Perez, J.: Artificial intelligence (AI) in education: Impact & examples. 2023 [cited 2023 12/07/2023]. https://www.questionpro.com/blog/ai-in-education/
4. Srinivasa, K.G., Kurni, M., Saritha, K.: Harnessing the power of AI to education. In: Learning, Teaching, and Assessment Methods for Contemporary Learners: Pedagogy for the Digital Generation, pp. 311–342. Springer Nature Singapore, Singapore (2022)
5. Holmes, W., Tuomi, I.: State of the art and practice in AI in education. Eur. J. Educ. **57**(4), 542–570 (2022)
6. Kumar, S.: ChatGPT introduction. (2023). https://www.testingdocs.com/chatgpt-introduction/
7. Sarin Sok, K.H.: ChatGPT for education and research: a review of benefits and risks. SSRN.:12 (2023)
8. Gilson, A., et al.: How does ChatGPT perform on the United States Medical Licensing Examination? The implications of large language models for medical education and knowledge assessment. JMIR Med. Educ. **9**, e45312 (2023)
9. Mbakwe, A.B., et al.: ChatGPT Passing USMLE Shines a Spotlight on the Flaws of Medical Education, p. e0000205. Public Library of Science, San Francisco, CA (2023)
10. Giannos, P., Delardas, O.: Performance of ChatGPT on UK standardized admission tests: insights from the BMAT, TMUA, LNAT, and TSA examinations. JMIR Med. Educ. **9**, e47737 (2023)
11. Pilani, B.: Bitsat Brochure (2023)

# A Review on Weather Prediction Based on Deep Learning Model

Mohit Kumar[1], Vinay Saroya[2], Kamal Kumar Gola[1](✉), and Sumit Kumar[3]

[1] COER University, Roorkee 247667, Uttarakhand, India
mgagiyan@gmail.com, kkgolaa1503@gmail.com
[2] DIT University, Roorkee 248009, Uttarakhand, India
saroyavinay18@gmail.com
[3] Haridwar University, Roorkee 247667, Uttarakhand, India
mail2dr.sumit@gmail.com

**Abstract.** Many industries rely on weather forecasting, including agriculture, transportation, and disaster management. Traditional weather forecasting models rely on complex numerical calculations and empirical approaches, which typically produce low accuracy and reliability. Deep learning developments, on the other hand, have opened up new opportunities for improving weather forecast accuracy. This paper looks at the use of deep learning algorithms in weather forecasting and presents a unique approach to enhancing prediction accuracy and reliability. Machine learning has achieved incredible success in a wide range of programmes, including predicting rain, a complex meteorological variable with significant societal ramifications. This study employs a novel form of gradient-boosted trees to forecast seasonal precipitation in the region of South America. To determine the appropriate hyperparameters to use with the gradient-boosting approach, the Optima framework employs Bayesian optimization. Seasonal precipitation predictions based on data collected from observation are juxtaposed with predictions produced by the Brazilian National Institute of Space Research (INPE, Portugal), the gradient-based boosting method, and deep neural network algorithms. The results reveal that the enhancing method outperforms other approaches in a variety of situations.

**Keywords:** Machine Learning · Weather Forecasting · Bayesian Technique · Optima Framework · Numerical atmospheric models

## 1 Introduction

Weather prediction has always been challenging due to the dynamic nature of atmospheric processes. The Global Forecasting System (GFS) & the European Centre to feed Medium-Range Weather Forecasting (ECMWF) are two traditional numerical weather prediction models), use physical equations and meteorological measurements to simulate atmospheric processes. These models, however, have trouble capturing localized weather patterns and dealing with complicated nonlinear conversations. Deep learning has emerged as an effective approach for dealing with complex problems in a wide range


A. Bajaj et al. (Eds.): HIS 2023, LNNS 1224, pp. 10–20, 2025.
https://doi.org/10.1007/978-3-031-78925-0_2

of domains, particularly artificial intelligence, natural language processing, and speech recognition. Convolutional neural networks (CNNs) are deep learning algorithms that excel in learning patterns and extracting meaningful features from huge data sets [1]. As a result, deep learning is a realistic weather forecast approach. The capability of deep learning models to learn representations and features directly from raw data eliminates needed for human feature engineering. This is especially valuable in weather forecasting since the complex interactions of numerous atmospheric elements are not fully understood and may change over time. Deep learning models can capture these complicated interconnections and adapt to changing circumstances, resulting in more accurate predictions. Deep learning has transformed artificial intelligence, with remarkable results in domains such as image recognition, natural language processing, and speech synthesis. By acquiring structured representations from data, deep learning has the potential to enhance weather prediction systems. Deep learning algorithms may extract important properties from huge amounts of atmospheric data, perhaps revealing deep links and patterns that would be difficult to uncover using traditional methods. This study intends to investigate the potential of deep learning algorithms in weather prediction by utilizing their capacity to learn from large datasets of historical weather observations, satellite images, and atmospheric simulations.

### 1.1 Application of Deep Learning for Weather Prediction

Deep learning has emerged as an effective technique for a variety of applications, including weather forecasting. Weather forecasting entails evaluating complex patterns and dynamics in atmospheric data to properly anticipate future weather conditions.

**Image Recognition**. Algorithm based on deep learning can recognize and categories various weather patterns such as cloud formations, storm systems, and fronts by analyzing satellite images and radar data.
**Numerical Weather Prediction (NWP)**. By solving complicated mathematical equations, NWP models replicate the Earth's atmosphere. Deep learning can improve NWP models by understanding the correlations between different atmospheric variables and enhancing model output accuracy.
**Extreme Weather Event Prediction**. Deep learning models can be trained to detect patterns associated with extreme weather events, such as hurricanes, tornadoes, or heavy rainfall. By analyzing historical data, including atmospheric conditions and previous instances of extreme events, these models can provide early warnings and forecasts, helping in disaster preparedness and mitigation efforts.
**Short-term Weather Forecasting**. Deep neural networks (recurrent intelligent neural networks) also known as & CNN algorithms. (Convolutional neural networks) are machine learning algorithms that can analyses enormous quantities of real-time climate information like wind speed, humidity, temperature, and pressure to generate temporary weather forecasts. By learning correlations between time and geographical patterns in the data, these models can make accurate predictions spanning a few hours to several days ahead.
**Data Assimilation**. Deep learning can play a crucial role in data assimilation, which involves integrating real-time observations from weather sensors, satellites, and other sources into weather models. Deep learning models can learn to correct biases and

errors in observations, optimize the blending of data sources, and improve the accuracy of model outputs.

**Climate Modelling**. Deep learning techniques can also be applied to climate modelling, which involves simulating long-term climate patterns and trends. By analyzing historical climate data and learning the relationships between different variables, deep learning models can help predict future climate scenarios and assess the impact of climate change (Fig. 1).

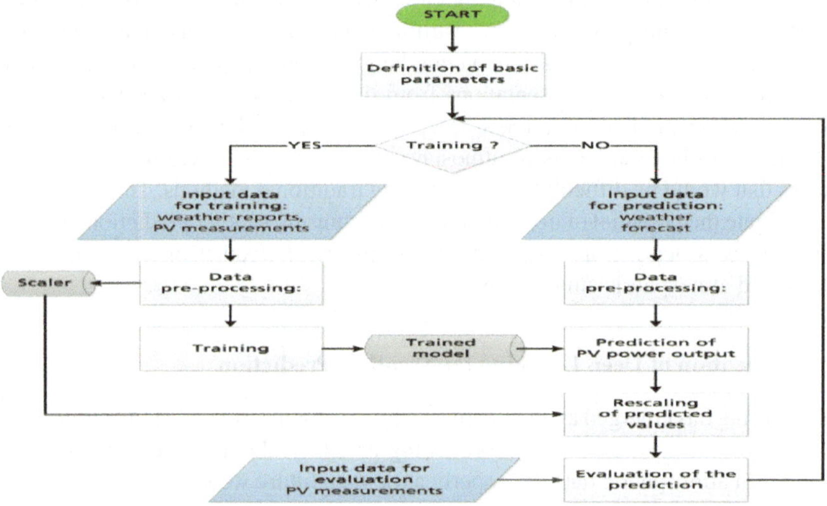


**Fig. 1.** Weather prediction based on deep learning model processing model

## 2 Literature Review

In recent years, [2] There has been an enormous increase in research focusing on using algorithms based on deep learning for forecasting the weather in recent years. Several research have shown that alternative deep learning architectures, such as CNNs (convolutional neural networks) and retinal neural networks (RNNs), can improve the accuracy of predictions for the weather. CNNs have showed potential in simulating the temporal relationships seen in meteorological data, whereas deep neural networks have shown interest in capturing geographical patterns via satellite images. Researchers have also looked at including additional learning components such as long-short-term memory, or LSTM, units and concentration mechanisms to capture persistent dependencies and identify significant atmospheric variables.

In this resulted in the publication of 500 publications since 2018, all of which focused on the use of methods of machine learning in numerical prediction of the weather and climate analysis. We identified the key subjects covered in the present research, as well as other features such as the meteorology domains studied, the methodologies used, and the nations commonly referenced in the introductory sections of these publications, using this dataset [3]. It is critical to recognize that this strategy has some limitations.

The search engine may favor articles that directly mention "ML" in the title or intangible potentially ignoring important papers that simply address it in the main text. Furthermore, the manual gathering of articles yielded just 500 papers, falling short of the standards specified by previous text mining research that investigate patterns in scientific literature. This research frequently relies on pre-prepared datasets, such as COVID-19. In our investigation, however, each publication was carefully reviewed by an area of expertise specialist in order to quickly reject irrelevant studies from further analysis.

Open-pit mines have a unique topographic characteristic that has a considerable influence on weather conditions within and around the mines. Understanding these trends is critical because of the environmental effects of mining activities. Such operations frequently produce significant amounts of fugitive dust, unpleasant chemicals, and atmospheric greenhouse gases (GHGs) [4]. Because of features like as travel circulations, shear layers, & plume meandering generated by mine pits, quantifying emissions from mines with open pits is a difficult method. Existing monitoring methodologies for open-pit coal emissions have geographical and temporal coverage constraints. While diagnostic methods are computationally efficient, predicting emission flow properly remains difficult. Although more accurate, prognostic models are more difficult to build up and computationally costly for practical usage.

According to the study findings, the emission flow from the freshly constructed mine, in order, which is more productive than the old my very own, was greater and displayed substantial diurnal changes. Wind speed at 10 m above the ocean's surface (S10) correlated strongly with emission flux, but other meteorological factors demonstrated smaller or negative associations. When compared to the previous mine, machine learning approaches produced superior estimations of emission flow for the water body and new mine. In estimating emission flux, the lagoon and new mine had reduced average biases and a smaller root-mean-square error (RMSE), but the old mine has the highest biases and RMSE [5].

Representing precipitation effectively in weather and climate predictions is a difficulty for biophysical parametrizations in mathematical models. Our findings are consistent with previous research comparing numerical representations and machine learning prediction methods. In particular, when built with the Optuna optimizer, the XGBoost model outperformed the TensorFlow (TF) deep neural networks in recognising precipitation patterns over the summer, wintertime, and spring seasons. However, during hind casting experiments, the TF model exhibited better performance in autumn. Further investigation is necessary to understand the specific meteorological pattern that the XGBoost model failed to capture during the autumn season. It is possible that incorporating an additional attribute as a new input for XGBoost could reveal a hidden pattern that the deep neural network can detect better. A database spanning from 1980 to 2020 was used to train and test the supervised XGBoost estimator [1]. Overall, the XGBoost model showed superior results compared to the TF model.

In We modified the DLWP-CS approach to establish relationships between three variables: total cloud cover, two-meter temperature, and total incident solar radiation, with precipitation using the ERA-5 reanalysis product. The DLWP-CS approach utilizes a residual UNET-based deep learning model to learn temporal relationships among these variables. It improves upon previous studies by incorporating cubed sphere mapping to reduce errors caused by the spherical nature of the datasets. However, the DLWP-CS approach requires an equal number of inputs and outputs for its model. This becomes computationally expensive for variables like precipitation, which depend on other atmospheric, oceanic, and land fields [6].

In Forecasts of sea surface temperatures (SSTs) tailored to specific locations have diverse applications across various fields. However, the commonly used numerical forecasting method for SSTs often exhibits significant deviations when applied to site-specific studies. Additionally, the accuracy of numerical estimations decreases considerably as the forecast lead time increases. To overcome these limitations, a hybrid approach combining numerical methods with data-driven techniques, such as neural networks (NNs), has been implemented, resulting in improved outcomes. Although this approach utilizing standard and traditional NNs has shown significant improvement, there is still potential for further enhancements. Which is resilient to data gaps and possesses a higher capacity to extract information from input spaces, offers an appealing alternative for time series prediction studies [7].

In Infectious illnesses have a tremendous influence on global public health and economic stability. The goal of this work was to create an improved infectious illness prediction model by combining statistics, machine learning, as well as deep learning approaches. However, there is little information available in India about dengue prediction using machine based deep learning methodologies. As a result, for dengue prediction, the study used predictive models based on machine learning (SVR, GBM), models trained with deep learning (LSTM), or a statistical framework (VAR). When compared to a lagged dataset, the LSTM model performed better (RMSE = 0.345, R = 0.9). The LSTM model effectively anticipated monthly outbreaks of dengue in Kerala state by combining meteorological factors and past dengue cases. These forecasts give essential information for public health officials, scholars, and planners to properly manage services and allocate medical resources. The suggested LSTM model for malaria prediction may be used in other Indian states [8].

Weather prediction, weather prediction utilizing a numerical interpretation of statistics that influence movement and climatic change. Many climate forecasting approaches are employed in research studies that address air pollution and the impact of greenhouse emissions on global climate change, in addition to short-term weather forecasts. Because the first live weather prediction system had just one layer, it could only depict transitory fluctuations in the layered architecture or the outside world. Computers now enable the creation of multilayer models (often tens to twenty) that they can solve vertical changes, temperature, and humidity. The basic climatic changes of enormous movement scales are predicted by these multi-mile species [9].

In we alter the DLWP-CS to construct connections between three variables: total cloud cover, two-meter temperature, and net incident ultraviolet (UV) rays to precipitation. DLWP-CS proposes the application of a residual UNET-based deep learning technique to analyses temporal correlations between the same/various variables as improves on earlier research by using cubed sphere projection to eliminate mistakes caused by the dataset's spherical form. However, their model would require the same number of inputs and outputs. DLWP-CS could prove computationally costly for variables like precipitation that are reliant on other components of land, sea, and atmosphere. Furthermore, their deep neural network model may drift at long lead periods without taking advantage of current state-of-the-art dynamical models. Which are utilized as precipitation precursors, will be adequately accurate with DLWP-CS [6].

In The rain-based PredRNN model beats the PredRNN/PredRNN_V2 models by requiring less than 30% of the training time required by both models. This benefit is especially useful when adding fresh training data and generating updated versions. Despite the reduction in computational demands, our improved model still produces high-quality, high-resolution anticipated pictures. The Rain PredRNN model, designed exclusively for precipitation now casting utilizing weather radar echo pictures, essentially merges the UNet plus PredRNN_v2 architectures. It successfully captures radar data's detailed spatial properties while minimizing testing and training time. The model handles both spatial and temporal information in time-series pictures effectively. Despite achieving equivalent outcomes with time spent training less than 30% that of the previous models and a 50% decrease in Mic value, there remain certain constraints that must be addressed. Our long-term objective is to improve the accuracy of the model for precipitation forecasting [10].

In The study proposes a new method for determining parameters linked via the inversion zone using radiometer characteristics and compares its performance to radiosonde observations. Although the approach correctly estimates the height of temperatures inversion, its accuracy in calculating temperature jumps connected to it is slightly lower. Wind data collected from wind profiler observations within the inversion layer, on the other hand, demonstrate high agreement of radiosonde data, especially in terms of parameters linked to maximum wind speed. The study makes use of scripts who automate the analysis of big datasets, especially data recorded at 20-minute intervals from January through April 2019, which included over 11,000 meteorological inputs covering both normal and extreme wind circumstances. Addressing this issue will need further research to improve the system's automated detection of these new severe climatic circumstances [2].

In the AGRI LPW product is automatically collocated with three forms of assessment data by our Quality Control System (QCS): radiosonde that was data, computer data, and satellite data. It provides a variety of assessment outcomes in near real-time and generates figures including statistical information obtained from these evaluations automatically. These numbers are constantly monitored to assure the quality of AGRI LPW goods. The current research includes monthly evaluation findings for November 2019, as well as more comprehensive information spanning the period of the 23rd of October, 2019 to November 15, 2020 [1].

In Based on the algorithm known as KNN, this work provides and evaluates two independent techniques for estimating power usage. The first approach, KNFTS, is concerned with time-related and date-related characteristics, whereas the second method, KNPTS, is concerned with sequences in the reply variable. Both systems shown suitable for estimating power usage and outscored existing artificial intelligence algorithms in experimental assessments. These approaches produce good results in forecasting future consumption values by extracting time and time-related information and using previous values of the outcome variable, thereby capturing the relationship across occurrences in the time series. The COVID-19 pandemic having a substantial influence on power consumption patterns, leading to concept drift and poor prediction model performance.

The models have to be retrained to solve this problem. The results revealed that retraining the algorithms using just data from the lockdown time frame resulted in an average increase of 6.67% in EDR [11] (Table 1).

**Table 1.** Comparative analysis of existing approaches

| Ref. & year | Methodology | Accuracy | Future enhancement |
|---|---|---|---|
| [2], 2022 | FLOWSTAR flow field model | The combined dataset gave shipping predictions for 70% of the period and land calculations for 84% of the time throughout the time period | This model would be enhanced to attain higher accuracy in real time |
| [14], 2021 | The design principles of Temporal Convolutional Networks (TCN) and Long Short-Term Memory (LSTM) are employed in numerical weather prediction (NWP) models. The Climate Research and Forecasting (WRF) model is recognized as an example of a NWP model | This study used ten surface atmospheric variables for predicting weather achieving an accuracy of 85% | The recommended lightweight deep model that focuses on 10 surface weather features is suitable for weather forecasting for a period of up to 12 hours. Surprisingly the model exceeded the latest-generation WRF model during the specified 12-hour period |
| [10], 2022 | It proposed RainPredRNN, a novel deep learning model for forecasting precipitation from weather radar echo pictures. This model employs UNet and PredRNN_v2 to shorten training and testing time | With an accuracy of 88.5%, we suggested a novel deep learning model, RainPredRNN, for moisture forecasting utilizing weather radar repeat pictures | The proposed approach produced equivalent results with a learning time that is less than 30% of the preceding versions' training time and 50% of the MAC value |
| [5], 2022 | There are three major approaches for calculating wind power. In the first phase, CityFed will employ computational fluid dynamics to compute local wind speed on building rooftops in a real city's region | The accuracy and efficacy of the model in predicting wind dispersion in a real-world metropolitan zone are investigated. When compared to weather station data, wind velocity has an RMSE of 0.265 m/s | The proposed approach was used to assess the wind potential of 25 building roofs. Wind potentials are the same for two different warming scenarios, RCP4.5 and RCP8.5 |

(*continued*)

**Table 1.** (*continued*)

| Ref. & year | Methodology | Accuracy | Future enhancement |
|---|---|---|---|
| [4], 2022 | There were two open-pit mines on the coal mining site, one old and one modern, as well as a tailings pond. The old mine, which had a depth of around 100 m and a span of approximately 2500 m, was mostly used for mining excavations | With 94.5% accuracy, the key modelling approaches were four ML techniques (multi-layer perceptron (MLP) artificial neural network (ML), gradient booster (GBR), XGBOOST (XGB), and support vector machines (SVM) | The PCC matrix was built using the old mine, new mine, and pond for the observed tracer (methane) mixing ratio, normalized emissions flow (estimated by WRF), and extra air state parameters (derived by WRF) |
| [7], 2020 | To forecast future sea surface temperature (SST) data, our speakers use the time series forecasting idea. For this prediction job, artificially formed neural networks (ANNs) are used, with the effect of previous SST values taken into account | A hybrid model with 92.5% accuracy | This study's findings indicate that there is still opportunity for expansion. However, the use of deep learning for all three periods, LSTM NN, effectively addresses this difficulty |
| [8], 2023 | The system is predictive performance (VAR, GBM, SVR, and LSTM) as measured by coefficient of determination (R2) and root mean square error (RMSE) | A 97.5% accurate predictive performance prediction model | The purpose of this work was to combine statistics, machine learning, and deep learning methodologies to produce a more accurate viral disease prediction model than previous models |
| [3], 2022 | The use of machine learning may have a huge impact on synoptic meteorology and climatology | NWP models with 98.5% accuracy with very high resolution | Techniques for machine learning will be included in many components of current NWP, and advances in novel computer clusters are tied to trends in climate models |

(*continued*)

**Table 1.** *(continued)*

| Ref. & year | Methodology | Accuracy | Future enhancement |
|---|---|---|---|
| [9], 2021 | To simulate energy generation using turbines built in Poland, four artificial intelligence approaches were used, including wind speed predictions using the NWP model and producing power time series | The ANN (Artificial Neuro Network) model was constructed in MATLAB R200b with a prior improvement of 92.50% | It proposed a wind energy system with day-ahead estimation based on several machine learning methods, a precise prohibited |
| [13], 2019 | There are two types of time-series models: non-linear and linear. Moving average models (also known as MA models) are used with AR models to develop auto-regressive motion average techniques (ARMA), which are employed in PV forecasting | The linear time-series equations used for PV forecasting have an accuracy of 89% and are a combination of auto regressive techniques (AR) and moving average techniques (MA) | Neither climate change nor the present energy crisis necessitates that renewable energies, notably solar PV electricity, be integrated into national power systems |

## 3 Challenges of Deep Learning for Weather Prediction

The use of deep learning for weather prediction confronts a number of issues that must be overcome before it can be successfully used. Here are a few of the major challenges:

**Data Limitations.** Effective training of Deep Learning models requires large volumes of high-quality data. However, weather data often suffers from limitations in terms of coverage, resolution, and historical duration. Obtaining comprehensive and diverse weather datasets can prove challenging, especially for specific regions or particular weather events. The availability of reliable ground-based observations, satellite data, and historical records is crucial for training models that can deliver accurate results.
**Interpretability and Explain ability**. Machine learning algorithms frequently lack interpretability, making it difficult to comprehend the underlying causes that influence their predictions. The use of transparent and explainable models in weather forecasting is critical for instilling confidence among meteorologists and stakeholders. Deep learning models' interpretability is being improved, and ways to provide light on the logic behind their predictions are being developed.
**Generalization to New Weather Conditions**. Weather patterns can exhibit significant variations, posing challenges for deep learning models trained on historical data to accurately predict unseen or extreme weather conditions. Adapting these models to

effectively handle rare events, novel climate regimes, or sudden changes in atmospheric dynamics is a complex task that requires careful consideration. It is crucial to ensure that these models possess robustness and generalization capabilities in order to achieve reliable weather predictions.

**Computational Resources and Efficiency**. Deep Learning models can necessitate a substantial amount of computer resources and time to train and inference, resulting in a high computational intensity. This is especially crucial in forecasting the weather, where real-time and near-real-time forecasts are required and time limits exist. To meet the needs of operational weather forecasting systems, it is critical to focus on building physically efficient models and utilizing parallel computing approaches.

## 4 Conclusion

The review study thoroughly examined the current literature on deep learning-based weather prediction, highlighting uncertainties related to the reliability of meteorological parameters. Spatial separation between input measurements introduces ambiguity and discrepancies in datasets, which could be addressed by considering the time it takes for air movement. Issues with the radiometer dataset's accuracy and air analyzer measurements scarcity contribute to uncertainty. The study suggests deploying additional measuring equipment, such as dropsondes from planes, for better characterization of the atmospheric boundary layer. The review also identifies research problems and future prospects, combining baseline, direct power prediction using an Artificial Neural Network (ANN), and hybrid model (ANN + fitted power curve) performance. It concludes that the paper is a valuable resource for researchers, users, and regulators in the field of deep learning-powered climate prediction, making forecasting simpler with fewer chances of failure. The approach has stabilized, but further growth is needed for global applicability and cross-disciplinary development.

## 5 Research Gaps

- To investigate the possibilities of several deep learning architectures for weather prediction and compare their performance to classic NWP models.
- To examine methods for preprocessing and extracting features to optimize the input data for deep learning models.
- To assess the impact of incorporating additional data sources, such as remote sensing data, on improving forecast accuracy and reliability.
- To analyze the computational requirements and scalability of deep learning models in operational weather forecasting systems.

## References

1. Monego, V.S., Anochi, J.A., de Campos Velho, H.F.: South America seasonal precipitation prediction by gradient-boosting machine-learning approach. Atmosphere (Basel). **13**(2), 243 (2022). https://doi.org/10.3390/atmos13020243
2. Stocker, J., et al.: Derivation of high-resolution meteorological parameters for use in airport wind shear now-casting applications. Atmosphere (Basel). **13**(2), 328 (2022). https://doi.org/10.3390/atmos13020328
3. Bochenek, B., Ustrnul, Z.: Machine learning in weather prediction and climate analyses—applications and perspectives. Atmosphere (Basel). **13**(2), 180 (2022). https://doi.org/10.3390/atmos13020180
4. Kia, S., Nambiar, M.K., Thé, J., Gharabaghi, B., Aliabadi, A.A.: Machine learning to predict area fugitive emission fluxes of GHGs from open-pit mines. Atmosphere (Basel). **13**(2), 210 (2022). https://doi.org/10.3390/atmos13020210
5. Mortezazadeh, M., Zou, J., Hosseini, M., Yang, S., Wang, L.: Estimating urban wind speeds and wind power potentials based on machine learning with city fast fluid dynamics training data. Atmosphere (Basel). **13**(2), 214 (2022). https://doi.org/10.3390/atmos13020214
6. Singh, M., et al.: Deep learning for improved global precipitation in numerical weather prediction systems
7. Sarkar, P.P., Janardhan, P., Roy, P.: Prediction of sea surface temperatures using deep learning neural networks. SN Appl. Sci. **2**(8), 1458 (2020). https://doi.org/10.1007/s42452-020-03239-3
8. Kakarla, S.G., et al.: Weather integrated multiple machine learning models for prediction of dengue prevalence in India. Int. J. Biometeorol. **67**(2), 285–297 (2023). https://doi.org/10.1007/s00484-022-02405-z
9. Bochenek, B., et al.: Day-ahead wind power forecasting in poland based on numerical weather prediction. Energies. **14**(8), 2164 (2021). https://doi.org/10.3390/en14082164
10. Tuyen, D.N., et al.: RainPredRNN: a new approach for precipitation nowcasting with weather radar echo images based on deep learning. Axioms. **11**(3), 107 (2022). https://doi.org/10.3390/axioms11030107
11. Gomez-Omella, M., Esnaola-Gonzalez, I., Ferreiro, S.: Short-term forecasting methodology for energy demand in residential buildings and the impact of the COVID-19 pandemic on forecasts. In: Lecture Notes in Computer Science (including subseries Lecture Notes in Artificial Intelligence. Lecture Notes in Bioinformatics), vol. 12498 LNAI, pp. 227–240 (2020) https://doi.org/10.1007/978-3-030-63799-6_18
12. Donadio, L., Fang, J., Porté-Agel, F.: Numerical weather prediction and artificial neural network coupling for wind energy forecast. Energies. **14**(2) (2021). https://doi.org/10.3390/en14020338
13. Mathe, J., Miolane, N., Sebastien, N., Lequeux, J.: PVNet: a LRCN architecture for spatio-temporal photovoltaic powerforecasting from numerical weather prediction, no. September, 2019, http://arxiv.org/abs/1902.01453 (2019)
14. Hewage, P., Trovati, M., Pereira, E., Behera, A.: Deep learning-based effective fine-grained weather forecasting model. Pattern. Anal. Applic. **24**(1), 343–366 (2021). https://doi.org/10.1007/s10044-020-00898-1

# Automated Detection and Mitigation of Toxic Comments Using XLNET Fine-Tuning Model

E. M. Roopa Devi[1](✉), R. Shanthakumari[1], R. Rajadevi[2], V. Sedhumadhavan[1], V. K. Sudharsan[1], D. Surendhar[1], and M. Rakesh[1]

[1] Department of Information Technology, Kongu Engineering College, Perundurai, Tamil Nadu, India
roopadevi.it@kongu.edu

[2] Department of Artificial Intelligence and Machine Learning, Kongu Engineering College, Perundurai, Tamil Nadu, India

**Abstract.** In the rapidly evolving realm of digital communication, the escalating prevalence of toxic comments poses a formidable challenge, encompassing offensive language that targets individuals or groups based on various attributes such as race, gender, religion, or beliefs. This research underscores the urgent and critical need for advanced tools capable of autonomously detecting and mitigating toxic comments across diverse digital platforms. Beyond serving as protective shields for individuals, these tools emerge as invaluable allies for content moderators, significantly enhancing their efficiency in managing user-generated content and contributing substantively to the ongoing battle against online toxicity. The study meticulously explores an array of methodologies, spanning traditional machine learning approaches to cutting-edge deep learning models like BERT and XLNet. Notably, XLNet emerges as a standout performer, showcasing exceptional capabilities and innovative self-attention mechanisms that enable the nuanced capture of intricate patterns within toxic comments, thereby contributing significantly to the success of this complex task. However, the research extends beyond the confines of academic exploration; it endeavors to make a tangible impact by fostering a safer, more inclusive online environment that benefits all users. Through a commitment to continuous research and innovation, the aim is to address the pervasive issue of toxic comments and uphold the integrity and safety of digital communication spaces. With XLNet as a powerful and efficient ally, significant strides are being made towards achieving this crucial and overarching objective, representing a progressive stance in the ongoing quest to create a digital landscape free from the harmful effects of toxic online interactions.

**Keywords:** Digital communication · Content filtering · Machine learning models · XLNet

## 1 Introduction

The phenomenal development of computer science and mobile communication has given one of the greatest inventions of the 21st century: the internet, where a person can communicate with anyone on this planet with only access to the Internet and a Smartphone.


A. Bajaj et al. (Eds.): HIS 2023, LNNS 1224, pp. 21–30, 2025.
https://doi.org/10.1007/978-3-031-78925-0_3

The rise of social media and blogging platforms has greatly transformed how people communicate. These platforms facilitate mass communication, with users expressing their thoughts and emotions through comments, feedback, and reviews, both positive and negative. Negative content, known as toxic, poses a challenge. With the internet's rapid growth, there are now 5.19 billion users. Twitter, for instance, enables users to follow, like, retweet, and reply to others, fostering discussions and information sharing globally. However, some individuals may use online anonymity to criticize others' beliefs. Social media supplies a common platform for the users to tell their opinions and express their ideas. Individuals can express their opinions without fear because there is no face-to-face contact between users. When conflicts arise, online arguments can escalate into uncomfortable and hostile exchanges, especially on social media. Comments can become discourteous, toxic, and even contain threats, rumours, and harassment. This can have severe consequences, leading to mental health issues. Therefore, it's crucial to identify and address toxic comments to combat problems like cyberbullying, online harassment, and offensive content. Law enforcement has even arrested people for posting offensive content on social media platforms. Cyberbullying can occur through various means, including text messages, forums, social media, and online games. It involves posting toxic content, spreading rumours, and targeting individuals, either privately or publicly. Criticism and body-shaming are common on platforms like Facebook and Instagram. Online harassment can also involve sexual abuse towards women. Despite content guidelines set by social media platforms, manually monitoring and addressing content violations is challenging due to the sheer volume of data. Many researchers have been working in this field, focusing on languages like English, Arabic, German, and Indonesian. This article extensively explores the critical issue of toxic comments on social media, addressing the challenges posed by cyberbullying, online harassment, and offensive content. Recognizing the severe consequences, including implications for mental health, the primary focus is on the identification and mitigation of toxic comments. Despite the implementation of content guidelines by social media platforms, the manual monitoring and addressing of violations prove to be challenging due to the sheer volume of data. Subsequent sections of this article delve into diverse methodologies employed for identifying and mitigating toxic comments. The exploration begins with an examination of traditional approaches, including Logistic Regression and Decision Trees, providing insights into their strengths and limitations. The narrative then transitions to the paradigm shift brought about by XLNet, a sophisticated model offering unique advantages in the field of toxic comment detection. This section elaborates on XLNet's bidirectional contextual comprehension, its exceptional performance in various natural language processing tasks, and its adaptability through fine-tuning. Detailed insights into XLNet's deep neural architecture and its capacity to extract intricate textual features are provided, highlighting its efficacy in identifying even the most subtle indicators of toxicity within comments. The emphasis on XLNet's adaptive nature is crucial, ensuring the model can be optimized to address evolving challenges and emerging forms of online toxicity.

## 2 Literature Survey

To identify the toxic comment analysis Furqan Rustam, Vaibhav Rupapara, Hina Fatima Shahzad, Imran Ashraf, Arif Mehmood, and Gyu Sang Choi [1] presented research that evaluates various machine learning models for toxic comment classification and introduces the RVVC ensemble. It examines the impact of dataset balancing methods (SMOTE, random under-sampling) and feature extraction (TF-IDF, BoW). Results highlight improved performance with balanced datasets, particularly with RVVC achieving high accuracy, but also emphasize computational complexity and potential influence of data balancing on reported accuracy. Future work includes multi-domain dataset exploration and more extensive experiments in toxic comment classification.

V. Varma, S. Gupta, P. Badjatiya and M. Gupta [2] conducted research centered on the detection of hate speech using deep neural network architectures.

The study discovered that these architectures surpassed current approaches in performance. It also emphasized the enhanced accuracy achieved by combining embeddings from deep neural network models with gradient boosted decision trees. Future investigations will explore the significance of user network features in this context, incorporating CNN, LSTM, and FastText embeddings as features for GBDTs.

R. Saia, S. Carta, R. Mulas, A. Corriga and D. Recupero [3] developed a machine learning approach for toxic comment classification, leveraging Apache Spark's Big Data capabilities, It introduces a machine learning approach harnessing Apache Spark's Big Data capabilities. Experiments emphasize the value of word embeddings, particularly domain-specific ones, for enhanced accuracy. The study advocates further exploration of contextual word embeddings and deep learning techniques to boost overall performance.

S. Zaheri D. Stroud and J. Leath [4] developed various approaches for toxic comment classification. The study highlights that combining these approaches into an ensemble enhances the F1-measure, particularly in cases of high data variance and classes with limited examples. First, a baseline classification solution was developed using a Naive Bayes [NB] technique. Being a result, NB, being a powerful and generally recognized classifier, served as a solid standard for subsequent work.

M. O. Olayiwola, P. A. Ozoh and A. A. Adigun [5] presented a study highlighting the importance of communication in the era of rising social media usage. The research aimed to create a multi-faceted logistic regression model for automated comment categorization, specifically targeting different forms of toxicity. The methodology involved data collection, preprocessing, and training using TF-IDF. Additionally, the study stressed the role of social media administrators in managing online conversations.

F. Rustam, A. Mehmood, I. Ashraf, S. Ullah, and G. Choi [6] proposing a voting classifier combining logistic regression and stochastic gradient descent for sentiment analysis of Twitter data. The study explored three feature extraction methods, favoring TF-IDF. The proposed classifier achieved high accuracy, outperforming non-ensemble models. However, a deep LSTM model did not perform well, suggesting potential for further deep learning experiments.

S. Ullah, M. Umer, A. Mehmood, I. Ashraf, S. Kumari, and G. S. Choi [7] presented a study introducing a unified deep neural network (DNN) architecture for Tweet classification, combining convolutional neural networks (CNN) with long short-term memory (LSTM) networks. Tested on three Twitter datasets with varying classes, the proposed model outperformed single LSTM and machine learning classifiers like VC, SVC, RF, and SGD. It also surpassed other proposed approaches. Future work includes using advanced convolutional networks and incorporating aspect-based sentiment analysis for improved performance.

L. Lusa and R. Blagus [8] research findings suggest that in high-dimensional scenarios, only k-NN classifiers using Euclidean distance benefit significantly from SMOTE, especially when variable selection precedes its application. The advantage is more pronounced with a greater number of neighbors. However, using SMOTE for k-NN without prior variable selection should be avoided, as it tends to overly favor the minority class and distorts classification outcomes.

V. Subramaniyaswamy, Logesh R., M. Abejith, Sunil Umasankar, and Arumugam Umamakeswari [10] introduce a Twitter sentiment analysis application. Combining lexicon-based scrutiny with deep data collection, the study proposes the "criticality" metric to assess threat levels for public events. This comprehensive approach provides nuanced insights valuable for security and intelligence applications, addressing the intricate relationship between social media sentiments and potential security concerns.

## 3 Proposed Methodology

This research focuses on multi-label toxic comment classification, using a mix of logistic regression, decision trees, and primarily XLNet models. The dataset includes annotated Wikipedia comments, classifying toxicity into categories like severe toxicity, toxic, identity hate, insult, threat and obscene. The main goal is to build a predictive model that assigns probability scores to each type of toxicity in individual comments. These model choices are based on their suitability for handling natural language processing complexities (see Fig. 1).

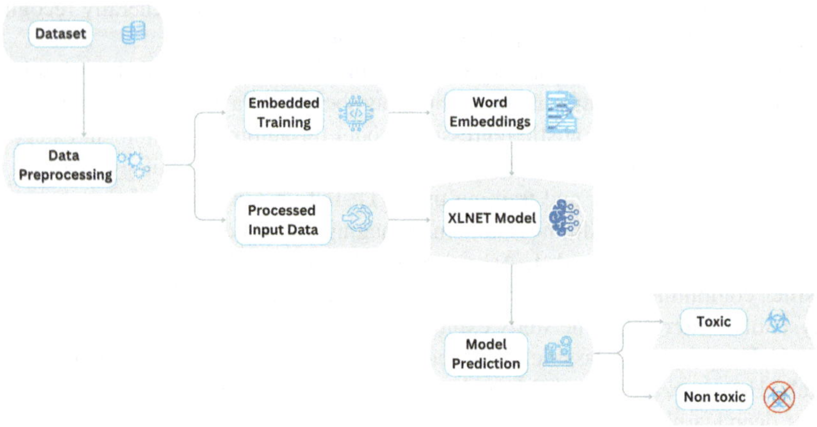


**Fig. 1.** Proposed framework for classifying toxic comments

## 3.1 Data Preprocessing

At the forefront of data management, the pivotal procedure entails the purification of noisy data by eliminating extraneous elements, which ultimately bolsters classification accuracy. This involves tokenization, padding, truncation, attention masks, number encoding, label encoding, and data loading.

### 3.1.1 Tokenization

Tokenization involves breaking a text document or sequence into discrete units, often words or subwords, which are referred to as tokens. These tokens serve as the fundamental elements in text analysis.

### 3.1.2 Padding and Truncation

Padding extends shorter sequences with tokens to match the length of longer ones, ensuring uniformity for text processing. Truncation involves shortening overly long sequences to fit within the model's memory capacity, allowing efficient computation and analysis of text data.

### 3.1.3 Attention Masks

Attention masks are binary matrices that determine which tokens within a sequence should be the focus of attention from other tokens. This is crucial for grasping context and the complex relationships between tokens, which is essential for a variety of natural language processing tasks.

### 3.1.4 Stopword Removal

Stopword removal entails eliminating common words, called "stopwords," from a text. These words, like "the," "and," "in," and "to" in English, are frequently used but carry little meaning in a sentence.

### 3.1.5 Text Cleaning and Lower-Casing

Lowercasing helps ensure uniformity by converting all text to lowercase, reducing case-related inconsistencies. Text cleaning further enhances data quality by removing special characters and punctuation, ensuring consistency in natural language processing tasks.

## 3.2 Classification and Transformer Models

In recent years, the transformer architecture, introduced in the 2017 paper 'Attention is All You Need,' has transformed machine learning for sequential data. Its self-attention mechanisms excel in various applications, notably natural language processing. Simultaneously, there's growing demand for robust classification models, and transformers have risen to prominence, revolutionizing classification across diverse domains beyond language.

### 3.2.1 Logistic Regression Model

Logistic regression serves as a model utilized for binary classification, and on occasion, it is extended to accommodate multiclass classification tasks. This technique is fundamental and extensively applied in both the realms of machine learning and statistics. In logistic regression, the relationship between the input data and the expected outcome (performance) is defined by the equation:

$$y = \frac{e^{(b0+b1*x)}}{1 + e^{(b0+b1*x)}} \tag{1}$$

In this equation:

'b0' represents the bias,
'y' is the anticipated performance or probability,
'b1' is the coefficient associated with the single input value 'x'.

### 3.2.2 Decision Tree

Decision trees are intuitive models used for data-driven decision-making. They create a hierarchical structure to classify or predict outcomes by asking questions about input features. At each node, a condition is evaluated on a feature, leading to one of two branches based on the condition's outcome. This process repeats until a leaf node, representing a class label, is reached.

### 3.2.3 BERT Model

BERT represents a significant breakthrough in natural language processing. It's the first bidirectional, unsupervised language model trained solely on text data, harnesses the power of the Transformer architecture to understand contextual relationships in text. Unlike sequential models, BERT uses a Transformer encoder to comprehend the entire word sequence simultaneously. While often referred to as bidirectional, it's more precise to describe it as non-directional. This unique characteristic empowers the model to comprehend a word's context by considering all of its neighboring words, both to the left and right. This is a distinct departure from context-free models like word2vec or GloVe, which generate a singular word embedding representation for each word in their vocabularies.

### 3.2.4 Transformer-XL Neural Network

XLNet, short for Transformer-XL, represents a notable progress in the realm of natural language processing (NLP). Introduced as a successor to BERT, XLNet leverages the transformer architecture to further enhance contextual understanding.

This section provides an in-depth exploration of XLNet's core mechanisms, such as permutation-based training and autoregressive modeling, highlighting its capacity to capture complex dependencies and context in textual data.

XLNet, when harnessed as a classifier, excels in understanding context, making it a potent tool for tasks like toxic comment analysis. Its ability to grasp the intricate contextual cues in language leads to more accurate and nuanced classifications. Unlike traditional classifiers, XLNet employs bidirectional learning, considering both preceding and subsequent words, which enhances its contextual awareness (see Fig. 2). Through transfer learning, it leverages pre-training on extensive text data, providing a strong foundation for classification tasks. In applications like content moderation, XLNet swiftly and effectively categorizes comments as toxic or non-toxic, ensuring responsible and ethical content management. Its prowess as a classifier heralds a new era of accuracy and context-awareness in NLP.

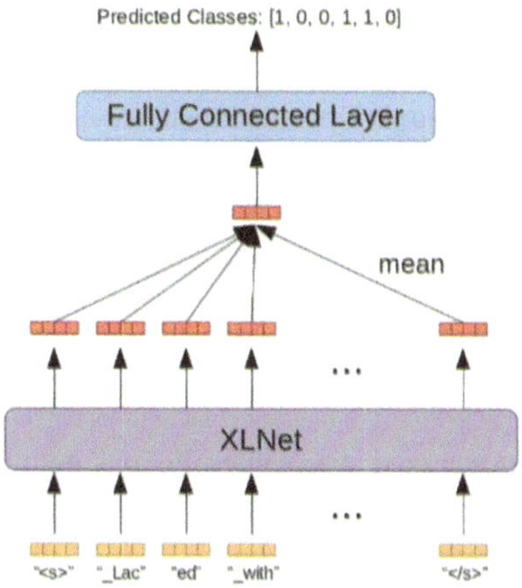


**Fig. 2.** Mean pooling the output vectors into a single vector

# 4 Experimental Result and Analysis

## 4.1 Evaluation Metrics

The effectiveness of machine learning models is gauged through the use of the following quantitative evaluation metrics. True Positives (TP), False Positives (FP), True Negatives (TN) and False Negatives (FN). TP signifies correct positive predictions, FP indicates incorrect positive predictions, TN represents correct negative predictions, and FN denotes incorrect negative predictions.

### 4.1.1 Accuracy

Accuracy assesses the number of correct predictions out of all predictions made by classifiers. A score of 1 signifies all predictions are correct, while 0 denotes no correct predictions. The calculation of accuracy is expressed as follows:

$$\text{Accuracy} = \frac{\text{TP} + \text{TN}}{\text{TP} + \text{TN} + \text{FP} + \text{FN}} \tag{2}$$

### 4.1.2 Precision

Precision, a measure of positive prediction accuracy, with a score of 1, indicates all predicted positives are true, but it doesn't account for false positives

$$\text{Precision} = \frac{\text{TP}}{\text{TP} + \text{FP}} \tag{3}$$

### 4.1.3 Recall

Recall, sometimes referred to as sensitivity, signifies the ratio of correctly identified positive instances to the total count of positive instances. Recall can be formally defined as follows:

$$\text{Recall} = \frac{\text{TP}}{\text{TP} + \text{FN}} \tag{4}$$

### 4.1.4 F1 Score

Precision and recall, when assessed independently, may not provide a complete view of classifier performance. The F1 score merges both metrics into a single score between 0 and 1 and it is determined as follows:

$$F1 = 2 \times \frac{Precision \times Recall}{Precision + Recall} \tag{5}$$

The dataset comprises comments from social media discussions, categorized by human raters to identify different types of toxic behavior. The dataset typically includes the text of the comments and the corresponding labels for each type of toxicity. These labels are binary, indicating whether a particular type of toxicity is present (1) or not present (0) in each comment.

Toxic comments cover a broad range of negative interactions and disrespect, while severe toxic comments represent a higher level of toxicity (see Fig. 3). Obscene comments contain offensive or sexually explicit content, threat comments express harm, and insult comments aim to offend. Identity hate comments target individuals or groups based on their identity. These categories help classify and address various online content and behavior. The study was conducted utilizing the Python programming language and a Python library commonly employed for training and building neural network models.

The comparative results of toxic comment classification among Decision Tree, Logistic Regression, BERT, and XLNet yields valuable insights. Decision Tree, achieving 86% accuracy, offers simplicity but might struggle with subtle toxic language nuances. Logistic Regression, with an 89% accuracy rate, shows better performance, indicating a deeper understanding of intricate data relationships. BERT, a pre-trained transformer model, stands out with an impressive 91% accuracy, highlighting its proficiency in comprehending complex linguistic contexts (see Table 1).

Figure 4 shows that the training loss for the XLNet model decreased steadily over epochs, while the validation loss also decreased, but at a slower rate. The marginal

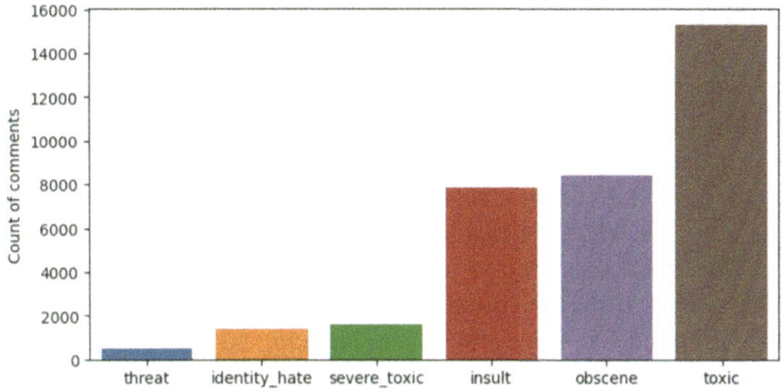


**Fig. 3.** Dataset distribution

**Table 1.** Performance comparison results of various models

| Models | Accuracy | Precision | Recall | F1 score |
|---|---|---|---|---|
| Decision Tree | 0.86 | 0.89 | 0.78 | 0.82 |
| Logistic Regression | 0.89 | 0.87 | 0.83 | 0.87 |
| BERT | 0.91 | 0.84 | 0.85 | 0.85 |
| XLNET | 0.95 | 0.96 | 0.92 | 0.93 |

difference between the training loss and validation loss suggests that the model did not exhibit overfitting to the training data. This suggests that the XLNet model is training well and is likely to perform well on new data. However, it's XLNet that stands as the star performer with an impressive accuracy of 95%, indicating its unparalleled ability to understand and classify toxic comments accurately.

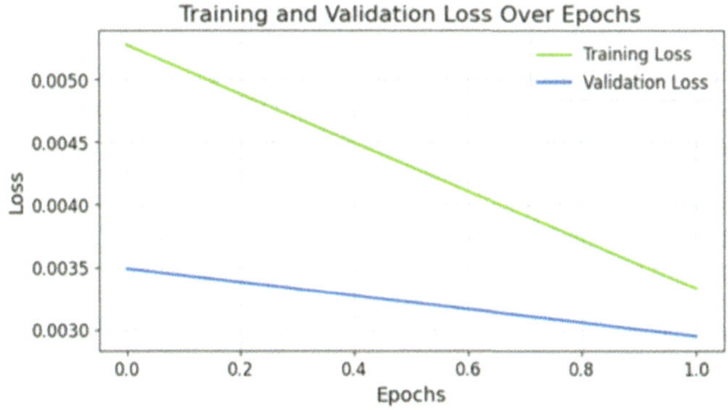


**Fig. 4.** Training and validation metrics over the number of epochs

## 5 Conclusion

Toxic comments are hurtful, harassing, intimidating, and they can discourage people from participating in online discussions. Toxic comments can also spread misinformation and hate speech. In this framework, the efficacy of four distinct machine learning models—decision tree, logistic regression, BERT, and XLNet—in classifying toxic comments is assessed. Findings highlight the crucial role of advanced NLP models, with XLNet leading, in enhancing toxic comment classification. The bidirectional context comprehension and permutation-based training streamline the identification and removal of toxic comments, suggesting that integrating XLNet models can create safer online spaces by mitigating the impact of toxic comments. The evolution of toxic comment classification involves advancing NLP models, embracing ethical AI practices, and enhancing real-time processing. Integrating explainability features and user feedback will refine models, while multimodal analysis aims to comprehensively understand toxic comments. These efforts aim to continually improve the accuracy and efficiency of toxic comment classification.

## References

1. Rupapara, V., Rustam, F., Shahzad, H.F., Mehmood, A., Ashraf, I., Choi, G.S.: Impact of SMOTE on imbalanced text features for toxic comments classification using RVVC model. IEEE Access. **9**, 78621–78634 (2021)
2. Badjatiya, P., Gupta, S., Gupta, M., Varma, V.: Deep learning for hate speech detection in tweets. In: Proceedings of the 26th International Conference on World Wide Web Companion, 2017, pp. 759–760
3. Carta, S., Corriga, A., Mulas, R., Recupero, D., Saia, R.: A supervised multi-class multi-label word embeddings approach for toxic comment classification. In: Proceedings of KDIR, 2019, pp. 105–112
4. Zaheri, S., Leath, J., Stroud, D.: Toxic comment classification. SMU Data Sci. Rev. **3**(1), 13 (2020)
5. Ozoh, P.A., Olayiwola, M.O., Adigun, A.A.: Identification and classification of toxic comments on social media using machine learning techniques. Int. J. Res. Innov. Appl. Sci. **4**(9), 1–6 (2019)
6. Rustam, F., Ashraf, I., Mehmood, A., Ullah, S., Choi, G.: Tweets classification on the base of sentiments for US airline companies. Entropy. **21**(11), 1078 (2019)
7. Umer, M., Ashraf, I., Mehmood, A., Kumari, S., Ullah, S., Choi, G.S.: Sentiment analysis of tweets using a unified convolutional neural network-long short-term memory network model. Comput. Intell. **37**(1), 409–434 (2021)
8. Blagus, R., Lusa, L.: Smote for high-dimensional class-imbalanced data. BMC Bioinf. **14**, 106 (2013)
9. Sankar, H., Subramaniyaswamy, V., Vijayakumar, V., Kumar, S.A.: Intelligent sentiment analysis approach using edge computing-based deep learning technique. Softw. Pract. Exp. **50**(5), 645–657 (2019)
10. Subramaniyaswamy, V., Logesh, R., Abejith, M., Umasankar, S., Umamakeswari, A.: Sentiment analysis of tweets for estimating criticality and security of events (2020)
11. Logesh, R., Subramaniyaswamy, V.: Learning recency and inferring associations in location based social network for emotion induced point-of-interest recommendation. J. Inf. Sci. Eng. **33**(6), 1629–1647 (2017)

# LSTM and ARIMA Comparison for Predicting Monitored Data from IoT Networks

Nguyen Xuan Dung and Ha Duyen Trung(✉)

School of Electronical and Electronic Engineering (SEEE),
Hanoi University of Science and Technology (HUST), Hanoi, Vietnam
trung.haduyen@hust.edu.vn

**Abstract.** Long Short-Term Memory (LSTM) is a form of neural network model, especially in the recurrent neural network (RNN). This paper uses Internet of Things (IoT) data aggregated from sensors measuring environmental parameters in the industrial zone into PostgreSQL database. The database includes temperature, humidity, particulate matter (PM) dust, soil temperature and moisture. The requirement is to predict time series with RNN regression neural network using the LSTM model in Python with Keras. The objectives of this work are to mine and analyze data retrieved from the server database and building models to make predictions of environmental parameters. Then, the model training was scheduled to automatically update every 24 hours by getting data automatically from the server's database and applying new data to the same old data and retrain the model with updated data in 24-hour intervals. The performance results of training data set on the LSTM model were compared with that on the AutoRegressive Integrate Moving Average (ARIMA) model in terms of environmental industrial parameters such as temperature, humidity, PM2.5, PM10 parameters.

**Keywords:** LSTM · ARIMA · Internet of Things Networks · Environmental Data

## 1 Introduction

Building an environmental monitoring network is one of the needs stemming from reality. Especially in the current era when the continuous development of industries serving the country's modernization process. When the Fourth Industrial Revolution (Industry 4.0) has emerged, it caused certain negative impacts on the environment. Environmental protection has become a key topic that is has become a major concern for society [1]. In the networks, IoT devices have been implemented to monitor environmental quality using different wireless connectivity. They also are integrated on the same Gateway. Obtained data can be visualized in real-time on the dashboard web user interface and app platforms. In context to the present standings of IoT, identification of the most prominent applications in the field of IoT networks have been highlighted and a scientometric review of the existing body of knowledge in the context of its application in the construction industry [2]. While several issues in the IoT domain are identified, investigate the source of those problems along with their remedial solutions.


A. Bajaj et al. (Eds.): HIS 2023, LNNS 1224, pp. 31–40, 2025.
https://doi.org/10.1007/978-3-031-78925-0_4

Data science is an interdisciplinary field that encompasses processes and systems for extracting knowledge within data in different structured and unstructured forms. Data science includes many areas of data analysis such as statistics, machine learning, data mining, predictive analysis. Data science uses a variety of techniques and theories such as operations science, information science, computer science, signal processing, probability modeling, and statistics studies, database, pattern recognition, visual description, fuzzy logic, data warehouse, artificial intelligence, decision support system, intelligent business. Methods that move towards big data are particular interest to data science. Data science has many applications in life such as machine translation, speech recognition, search engine, biomedical, healthcare, social sciences and economics, finance, insurance [3].

The advancement of machine learning has made a significant contribution to the growth of data science. Machine learning is a field of computer science that provides computers with the ability to learn through data and experience without requiring explicit programming. As an area of data analysis, machine learning is a method that allows complex models and algorithms building to find the knowledge hidden within the data that support decision-making and business processes [4]. Recently, Artificial Intelligence (AI), specifically machine learning, has emerged as a testament to the industrial revolution 4.0, from economics, education, medicine to housework, entertainment or even military [5]. Prominent applications in AI development come from many disciplines to solve a variety of problems. But most of the breakthroughs come from deep learning - a small array that is gradually expanding to each type of work, from simple to complex. Deep Learning has helped computers do things: categorize thousands of different objects in images, create annotations, communicate with people, even forecast weather. In the past few years, Deep learning has been mentioned a lot as new trend of the AI revolution. Deep learning exploits Big Data with much higher accuracy than other machine learning methods on the data set, especially for image processing [6].

In this paper, we firstly introduce a background of the Long Short-Term Memory architecture as a special type of RNN, and the AutoRegressive Integrate Moving Average (ARIMA) model in Sect. 2. Section 3 presents the case studies in IoT networks, formulates the problem, and presents the steps for conducting real-time series prediction using the LSTM model. Example applications of LSTM and ARIMA in IoT networks are also given in Sect. 3. Finally, this paper concludes in Sect. 4.

## 2 Backgrounds

### 2.1 Long Short-Term Memory (LSTM)

LSTM is a special type of RNN that is capable of learning distant dependencies. LSTM was introduced by Hochreiter & Schmidhuber [7] and has since been refined and popularized by many people in the industry. They work effectively across a wide range of problems on many different problems and gradually become more popular. LSTM is designed to avoid the problem of long-term dependency. Remembering information for a long time is their default property, LSTM cells are inherently capable of retaining long-term dependencies without specific training for memorization. Its internals can already be memorized without any intervention. Every recurrent network takes the form of a sequence of repeating modules of a neural network.

LSTM also has the same chained architecture, but the modules in it have a different structure from the standard RNN [8]. Instead of just one layer of neural networks, they have up to four layers that interact with each other in a very special way as illustrated in Fig. 1.

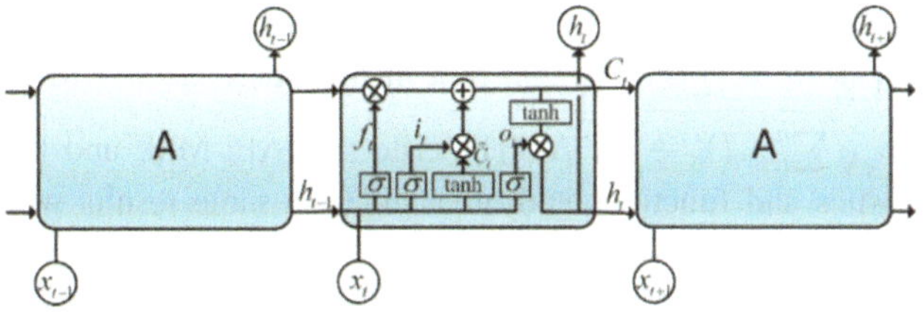


**Fig. 1.** The repeating module in an LSTM contains four interacting layers.

## 2.2 AutoRegressive Integrate Moving Average (ARIMA)

ARIMA model is another popular and widely used statistical method for time series forecasting. This method research factors independently (time series forecasting). The ARIMA model proposed by Box-Jenkins in 1976 [9–11], which is based on the AR autoregressive model and the MA moving average model. ARIMA is a quantitative forecasting model over time, the future value of the predictor variable will depend on the movement trend of that object in the past. The ARIMA model analyzes the correlation between observed data to provide a predictive model through the stages of model identification, estimating parameters from observed data, and checking estimated parameters to find the right model. The resulting model of the above process includes parameters that represent the degree of correlation on the data and is chosen to predict future values. The confidence limit of the forecast is calculated based on the variance of the forecast error.

The ARIMA class of models is quite broad. However, it does restrict us to considering stationary time series, or those series that exhibit stationarity after differencing. Furthermore, the models are linear, in the sense that predictions of future values are constrained to be linear functions of the observations. Now, very often, in practice, linearity and stationarity will be a sufficiently good approximation to reality. However, this will not invariably be so, in which case it would be sensible to contemplate a broader class of models. With the assumptions of linearity and stability, ARIMA models have the advantage of analytical parameterization. That is, they allow to represent a wide range of potentially useful predictive functions in models containing relatively few parameters. For example, instead of using ARIMA models, it is almost certainly faster and cheaper to develop linear predictions beyond the fitting of moderately high order autoregressive models [12].

Predictive errors are used to adjust the parameters of the forecasting model. Let $\varepsilon_t$ denote the forecast error at time $t$. $Y_t$ and $\hat{Y}_t$ are respectively the actual value and the predicted value at time $t$. The forecast error can be expressed as of $\varepsilon_t = Y_t - \hat{Y}_t$. The prediction model is evaluated well when the forecast error is small, in addition, the randomness of the error is also an important parameter to evaluate the accuracy of the forecast based on this assumption (random, normally distributed), so if the model is correct, the error should not be in any direction. There are four performance criteria for error evaluation commonly used in prediction can be expressed as of the mean absolute error (MAE) $MAE = \sum_{t=1}^{n}|\varepsilon_t|/n = \sum_{t=1}^{n}|Y_t - \hat{Y}_t|/n$, mean squared error (MSE) $MSE = \sum_{t=1}^{n}\varepsilon_t^2/n = \sum_{t=1}^{n}(Y_t - \hat{Y}_t)^2/n$, mean absolute percentage error (MAPE) $MAPE = \sum_{t=1}^{n}\frac{|\varepsilon_t|}{Y_t}/n = \sum_{t=1}^{n}\frac{Y_t - \hat{Y}_t}{Y_t}/n$, root mean square error (RMSE) $RMSE == \sqrt{\sum_{t=1}^{n}\varepsilon_t^2/n} = \sqrt{\sum_{t=1}^{n}(Y_t - \hat{Y}_t)^2/n}$. The criteria MAE, MSE and RMSE have the same characteristics and functions and often give the same results when evaluating. However, recommendation that if the error values of $\varepsilon_t$ are the same, the MSE criteria should be selected for evaluation. Conversely, if the error values of $\varepsilon_t$ are too different, the MAE criteria should be selected for evaluation. The RMSE criterion is the square root of the MSE criterion, so these two criteria are essentially one. The difference is that the value of the RMSE criterion is smaller. The MAPE criterion helps to evaluate the relative error. Assuming the mean error is 1 unit compared to the data value of 100, it is still small (1%). In opposition, an average error of 1 unit from the data value of 10 is considered large (10%). Therefore, when evaluating the forecast error with different data sets, the MAPE criterion should be used. In contrast, with the same data set but applying different forecasting methods, the MAPE criterion should not be applied because of the complexity in the calculation.

## 3 Case Studies in IoT Networks

### 3.1 An IoT Network

Figure 2 illustrates a horizontal architecture of the IoT network for industrial zone environmental monitoring system. In this architecture, air and wastewater sensors are embedded to IoT end devices for monitoring environmental parameters such as PM2.5, PM10 air dust, CO2, soil temperature, humidity, conductivity, etc. End devices communicate with IoT gateway via various wireless protocols to send monitored parameters to servers for data aggregation. Industrial environment management can be supported by exploiting the bigdata analysis.

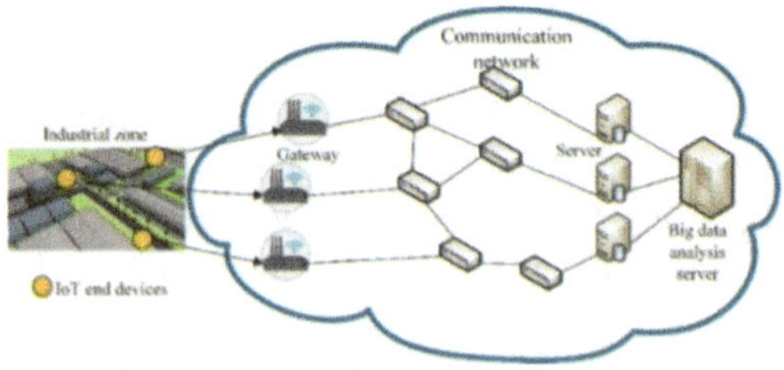


**Fig. 2.** An IoT network for industrial zone environmental monitoring system.

The system consists of many devices that are sensors that monitor environmental parameters and cameras. Each device uses a certain radio protocol. Data after measurement is sent back to Gateway. Here collected environmental data will be pushed to the cloud via MQTT (Message Queuing Telemetry Transport) protocol [13–19]. Cloud data is used for both model training and real-time display on user-friendly web and apps. The system is divided into each block to get a better understanding. As shown in the system overview, there are differential IoT devices with functions for monitoring different environments such as soil, wastewater, and air. There is also a security surveillance camera. Each set of devices uses a certain wireless communication protocol. Various sensors used for the IoT system, e.g., for the water environmental monitoring management, we used various types of sensors of PH, EC conductivity, and temperature. For the air environmental monitoring management, we use sensors of temperature, humidity, light intensity, $CO_2$, VOC, and dust PM2.5. Finally, we use sensors of temperature, humidity, and EC conductivity in for monitoring soil environmental management. Figure 3 shows the ZigBee and Bluetooth IoT devices used for deployment of IoT networks in an industrial zone. In the study, requirement is to predict time series with RNN regression neural network using LSTM and ARIMA models in Python with Keras.

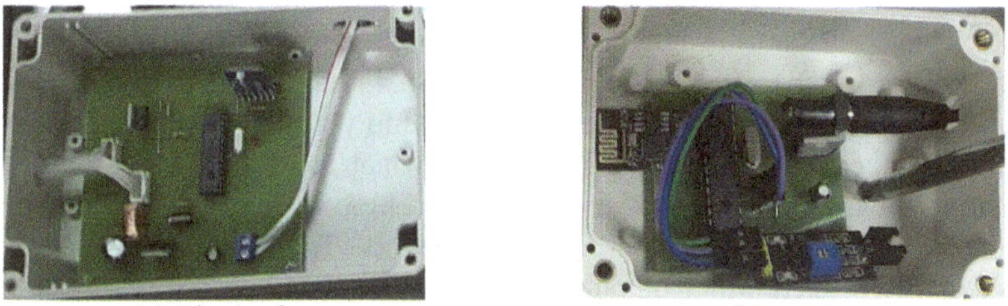

(a) Bluetooth (b) ZigBee

**Fig. 3.** Bluetooth and ZigBee IoT devices.

### 3.2 LSTM and ARIMA Implementation for Predicting Parameters

RNN models are based on the relationship between input and output data to learn past patterns, thereby making future predictions. In this paper, the open-source software library, Numpy, Pandas, Keras libraries together with Python 3.6 programming language were used to establish a short-term long-term neural network model LSTM, a form of neural network model, especially of the recurrent neural network RNN. We also use ARIMA to compare the results with LSTM.

Experiments using IoT data received from sensors measuring environmental parameters for PostgreSQL. The fields obtained by the database are from measurement of temperature, air humidity, soil, PM2.5, PM10. The requirement is to predict time series with recurrent neural network RNN using LSTM model in Python. Specifics of the problem include (i) mining and analyzing data retrieved from the server, (ii) building models to predict environmental parameters. The training of this model is scheduled to automatically update every 24 hours by automatically retrieving new data from the server's database, then applying the new data to the same old data, and then training the model again.

In the experiments, the data of the IoT devices of ZigBee and Bluetooth connectivity were acquired, processed, communicated, aggregated, and stored for environmental parameters such as temperature, humidity, soil temperature and moisture, air dust PM 10 and PM 2.5. These experimental data were accessed on server database that located at the research Lab from January 1, 2020, to January 1, 2022, to predict parameters for the next day. Data aggregation from the industrial park were stored into PostgreSQL database [20]. It then be retrieved by using psycopg2 to get data of environmental parameters by executing SELECT query in SQL. Then, save the data into a data frame as a csv file with the variables of "ts". In addition, data is always automatically fetched from the server every day by scheduled and appended to the variables in the csv file of the old dataset. Due to the data stored in the PostgreSQL database uses the Unix time system, it is necessary to convert the Unix time to Date time first (%Y-%m-%d %H:%M:%S.sss). The data from the IoT devices is forwarded to the server every 5 seconds, however the environmental parameters do not change so quickly. Therefore, the average will be calculated and aggregated hourly.

The data mining phase helps to understand patterns and biases in the data. Use `df.describe()` for data statistics. It is also possible to determine the number of samples, mean, standard deviation, max value, min value, which value accounts for 25%, 50%, 75% of the total values of each data field. From the data frame obtained, only time and variables are taken. Then, extract the NumPy array from the data frame and convert the integer values to floating point values, which are more suitable for modelling with neural networks. Similar to other neural networks, LSTM expects data to fall within the range of activation functions used by the network. The default activation function for LSTM is `sigmoid`, which outputs values from 0 to 1. This is the preferred range for time series data. Normalize the dataset using the `MinMaxScaler` pre-processor class from the scikit-learning library. Normalization requires that the data be provided in a matrix format with rows and columns.

Next, it is necessary to invert the scale on the forecasts to bring the values back to the original scale so that the results can be understood and a comparable error score can be calculated using `scaler.inverse_transform()`. With time series data, the temporal sequence of data is critical in time series modeling. A simple method is to split the ordered data set into training and test datasets. The LSTM model in Keras assumes that data is divided into input ($X$) and output ($Y$) components. For the time series problem, we can achieve this by using the observation of the temperature value from the time step ($t$) as input and the observation at the time step ($t + 1$) as the output.

However, multiple recent time steps can be used to make predictions for the next time step. This is called the window, and the size of the window is a parameter that can be adjusted for each problem. For example, given the current time ($t$), we want to predict the value at the next time in the series ($t + 1$), we can use the current time ($t$), as well as two times previous points ($t - 1$) and ($t - 2$) as input variables. When solved as a regression problem, the input variables are ($t - 2$), ($t - 1$), $t$, and the output variable is ($t + 1$). LSTM network model for regression in time series prediction problem with temperature variable. With the temperature value according to the hour of the day, it is necessary to determine the temperature value of the hours in the following days.

After converting the data into a form suitable for training the learning model, it will be represented as rows and columns. Each column will represent a feature with the model and may correspond to a separate lagged observation. Each row will represent a sample and will correspond to a new example with input and output components. Feature a column in a data set, such as a lag observation for a time series dataset. A sample is a row in a data set, such as an input and output series for a dataset time series. Given that the training dataset is defined as input X and output Y, it should be reformatted into the `[sample, time-step, feature]` format.

The shape of the input data should be specified in the LSTM layer using the "`input_shape`" parameter as a tuple specifying the expected number of observations to read per batch, the number of time steps, and the number of features. The batch size (`batch_size`) is usually much smaller than the total number of samples. It, together with the epoch number, determines the speed of the network in learning the data (how often the weights are updated). The final input parameter in determining the LSTM layer is the number of neurons, also known as the number of units or blocks of memory. The network requires a single neuron in the output layer with linear activation to predict the temperature value at the next time step. When compiling the network must specify a loss function (loss) and optimization algorithm (optimizer). In this work, we will use "`mean_squared_error`" as the loss function and the ADAM optimization algorithm. Use the Sequential Keras API to define the network. The network has a visible layer with one input, a hidden layer with four LSTM blocks, and an output layer that generates a single value prediction. The default sigmoid activation function is used for LSTM blocks. The network is trained for 100 epochs and a batch size of 1 is used.

It is necessary to invert the predictions using `scaler.inverse_transform` before calculating the error to ensure that the performance is reported in the same units as the original data (temperature values). When the LSTM model fits the training data, it can be used to make a prediction. It is possible to decide to fit the model once on all training data, then predict new time-steps at a time from the test data or re-adjustable model or update the model every step of the test data as new observations from the test data are provided. To make a prediction, call `predict()` function on the model. This requires a NumPy array input as an argument.

Finally, applying the built LSTM model, the above trained and tested LSTM model will be used to predict the 24-hour temperature value and other parameters outside the data set. We used a sample number of values in the test set as the input value. Then, convert these input values to a `list()` of values and use `model.predict()` with the input array argument above.

### 3.3 Results and Discussions

After going through the steps to build a regression LSTM model for the time series prediction problem, the results were able to generate predictions using the model. In the following Figures, the original raw dataset is illustrated in blue, the predictions on the dataset are presented in red. Specifically, Figs. 4 and 5 present the predictions of the environmental data of PM 10 and PM 2.5 air dust, as well as air temperature and humidity for in the industrial zone using the LSTM training model. It can be seen from the Figures that the predicted data always adapt with the trend line in the number of days. In addition, Fig. 6 illustrates the predictions of soil temperature and moisture using the LSTM model as well. Finally, Fig. 7 illustrates comparison of LSTM and ARIMA models for PM10 dust parameters. These results indicate that the LSTM model outperforms ARIMA in capturing daily trends that predictions of the LSTM model can be performed better than that of the ARIMA model in term of the number of days.

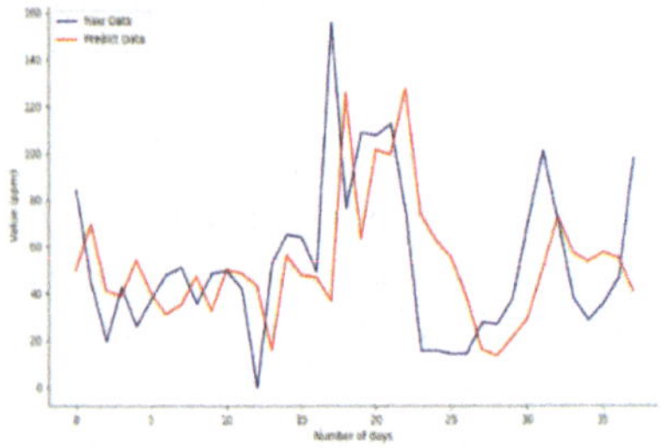

(a) PM 10 prediction.

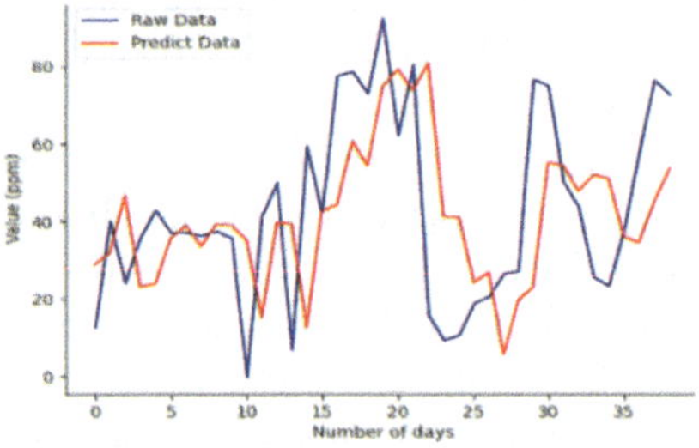


(b) PM 2.5 prediction.

**Fig. 4.** Prediction of air PM 10 and PM 2.5 and parameters.

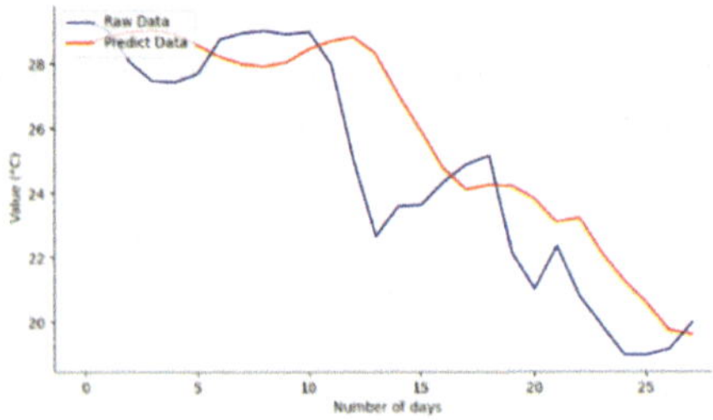


(a) Air temperature parameter prediction.

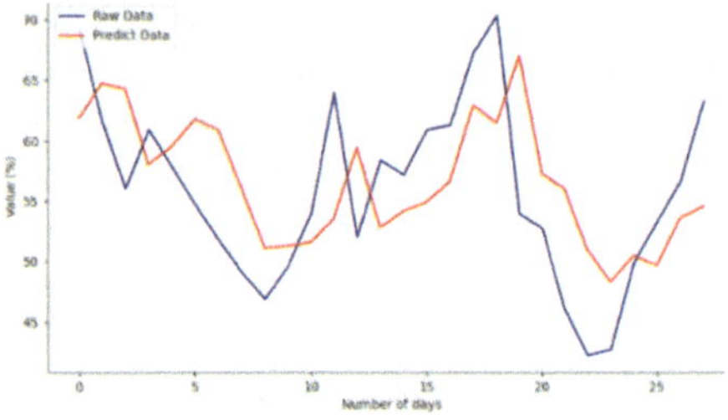

(b) Air humidity parameter prediction.

**Fig. 5.** Air temperature and humidity parameter predictions.

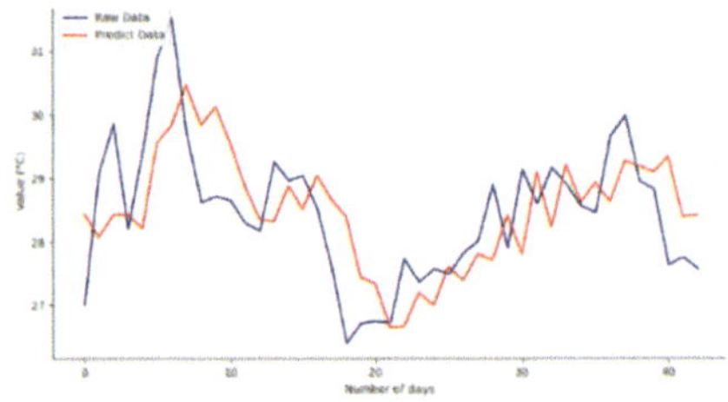

(a) Soil temperature parameter prediction

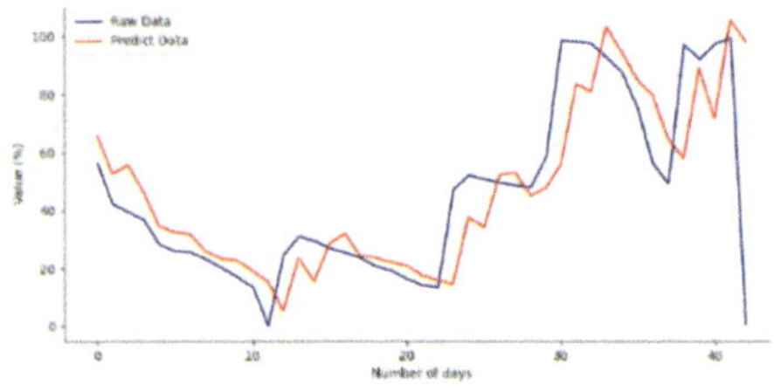

(b) Soil moisture parameters prediction.

**Fig. 6.** Predictions of soil temperature and moisture parameters using LSTM model.

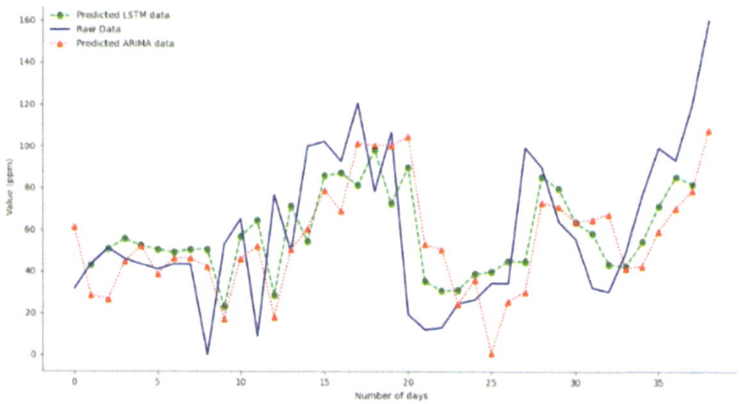


**Fig. 7.** Comparison of LSTM and ARIMA models for PM10 dust parameters.

## 4 Conclusions

This paper presented the theories of Long Short-Term Memory (LSTM) neural network model, the AutoRegressive Integrate Moving Average (ARIMA) model. The uses Internet of Things (IoT) data aggregated from sensors measuring environmental parameters in the industrial park into PostgreSQL database were also highlighted. The performance results of training data set on the LSTM model were compared with that on the ARIMA model in terms of environmental industrial parameters such as temperature, humidity, PM2.5, PM10 air dust, soil temperature and soil moisture parameters. LSTM is a big step in using RNN. It is to make it possible for all RNN steps to retrieve information from a larger set of information.

## References


1. Montori, F., Bedogni, L., Bononi, L.: A collaborative Internet of Things architecture for smart cities and environmental monitoring. IEEE Internet Things J. **5**(2), 592–605 (2018). https://doi.org/10.1109/JIOT.2017.2720855
2. Ghosh, A., Edwards, D.J., Hosseini, M.R.: Patterns and trends in Internet of Things (IoT) research: future applications in the construction industry. Eng. Constr. Archit. Manag. **28**(2), 457–481 (2021)
3. Sarker, I.H.: Data science and analytics: an overview from data-driven smart computing, decision-making and applications perspective. Preprints (www.preprints.org), 2021. https://doi.org/10.20944/preprints202104.0442.v1
4. Nguyen, G., Dlugolinsky, S., Bobák, M., et al.: Machine learning and deep learning frameworks and libraries for large-scale data mining: a survey. Artif. Intell. Rev. **52**, 77–124 (2019). https://doi.org/10.1007/s10462-018-09679-z
5. Sun, W., Liu, J., Yue, Y.: AI-enhanced offloading in edge computing: when machine learning meets industrial IoT. IEEE Netw. **33**(5), 68–74 (2019). https://doi.org/10.1109/MNET.001.1800510
6. Durga, S., Nag, R., Daniel, E. Survey on machine learning and deep learning algorithms used in Internet of Things (IoT) healthcare. In: 3rd International Conference on Computing Methodologies and Communication (ICCMC), 2019, pp. 1018–1022

7. Hochreiter, S., Schmidhuber, J.: Long short-term memory. Neural Comput. **9**(8), 1735–1780 (1997)
8. Bharati, S., Podder, P., Mondal, M.R.H.: Artificial neural network based breast cancer screening: a comprehensive review. Int. J. Comput. Inf. Syst. Ind. Manag. Appl. **12**, 125–137 (2020)
9. Box, G.E.P., Jenkins, G.M.: Time series analysis: forecasting and control, Revised Edition, Holden Day, San Francisco (1976). ISBN 978-1-118-61919-3
10. Ozaki, T.: On the order determination of ARIMA models. Appl. Stat. **26**, 290–301 (1977)
11. Li, W.K., McLeod, A.I.: Distribution of the residual autocorrelations in multivariate ARMA time series models. J. Roy. Stat. Soc. B. **43**, 231–239 (1981)
12. Newbold, P.: ARIMA model building and the time series analysis approach to forecasting. J. Forecast. **2**, 23–35 (1983)
13. Ngu, A.H., Gutierrez, M., et al.: IoT middleware: a survey on issues and enabling technologies. IEEE Internet Things J. **4**(1), 1–20 (2017)
14. Khanna, A., Kaur, S.: Internet of Things (IoT) applications and challenges: a comprehensive review. Wirel. Pers. Commun. **114**, 1687–1762 (2020)
15. Madakam, S., Ramaswamy, R., Tripathi, S.: Internet of Things (IoT): a literature review. J. Comput. Commun. **3**(3), 164–173 (2015)
16. Jiang, L., Xu, L.D., Cai, H., Jiang, Z., Bu, F., Xu, B.: An IoT-oriented data storage framework in cloud computing platform. IEEE Trans. Industr. Inform. **10**(2), 1443–1451 (2014)
17. Naik, N.: Choice of effective messaging protocols for IoT systems: MQTT, CoAP, AMQP and HTTP. In: Proceedings of 2017 IEEE International Systems Engineering Symposium (ISSE), 2017, pp. 1–7
18. Trung, H.D., Dung, N.X., Trung, N.H.: Building IoT analytics and machine learning with open source software for prediction of environmental data. Adv. Intell. Syst. Comput. **1375**, 134–143 (2020)
19. Trung, H.D., Hung, N.T., Trung, N.H.: Opensource based IoT platform and LoRa communications with edge device calibration for real-time monitoring systems. Adv. Intell. Syst. Comput. **1121**, 412–423 (2019)
20. Trung, H.D.: Database performance evaluation and applications of data science for IoT platform analysis. Int. J. Comput. Inf. Syst. Ind. Manag. Appl. **13**, 124–135 (2021) e-ISSN: 2150-7988

# Domain Adaptation with DIET-RASA and XLNet in Urgent Post Detection

Antonio Leandro Martins Candido[1,2](✉) and José Everardo Bessa Maia[1]

[1] Universidade Estadual do Ceará - UECE, 60714-903 Fortaleza, CE, Brazil
jose.maia@uece.br

[2] Instituto Federal Do Ceará - IFCE, 61609-090 Caucaia, CE, Brazil
leandro.candido@aluno.uece.br

**Abstract.** This paper presents an innovative solution for automated classification of urgent messages in online courses, using domain adaptation with DIET-RASA (Dual Intent and Entity Transformer) and XLNet (Extra-Long Transformer). Domain adaptation in this context is even more challenging because, in addition to the shift between distributions, the supports of the domain conditional distributions are also not coincident. We tackle this problem via generalization, mapping intent and lexicon from the source domain to the target domain using pre-trained models and embedding sequences of varying length. We provide in detail the main steps of the proposed pipeline and we also demonstrate the effectiveness of our approach in different scenarios, including cross-domain and low-resource environments. Our experiments show that our approach outperforms other state-of-the-art methods, being able to achieve an accuracy of up to 92% in some scenarios with a real-world dataset. Our approach can be used to offer personalized and agile support to students, regardless of the specificities of each domain.

**Keywords:** domain adaptation · online course forum · posts categorization · RASA · DIET · XLNet

## 1 Introduction

Among the various NPL applications, the automated classification of urgent messages written by students during their courses has become an essential task to provide fast and efficient support when immediate interventions are required. These messages can cover a variety of topics, from urgent academic questions to personal issues that affect student performance.

We soon came across a challenge known as the Unsupervised Domain Adaptation Problem (UDAP) [3]. In this context, we run the same classification task, denoted by $\mathcal{T}$, in two different environments: the source domain $\mathcal{X}_s$ and the target domain $\mathcal{X}_t$, the latter being somehow related to the former. In the $\mathcal{X}_s$ domain scenario, we have a set of data with labels, represented by $\mathcal{D}_s = \{x_i, y_i\}_{i=1}^{N}$, which allows us to train a classifier with an acceptable level of performance. However,


A. Bajaj et al. (Eds.): HIS 2023, LNNS 1224, pp. 41–51, 2025.
https://doi.org/10.1007/978-3-031-78925-0_5

when it comes to the domain $\mathcal{X}_t$, our only availability consists of unlabeled data, specifically $\mathcal{D}_t = \{x_i\}_{i=1}^N$. The main objective is to increase the performance of classifier when applied to the $\mathcal{X}_t$ domain.

Two distinct domains can exhibit disparities in both their feature spaces and their marginal probability distributions. These disparities may arise from the use of different languages or, in cases where the same language is used, from differences in word incorporation. When it comes to analyzing forum posts, these challenges become apparent when a model trained on one course forum is tasked with analyzing posts from a different course forum.

In essence, different courses have unique feature spaces and distribution patterns, causing certain words to appear frequently in one course but rarely in others. For example, consider words like "networks," "chains," and "limits," which can have distinct meanings in mathematics as opposed to literature courses.

In this context, the present work proposes an innovative and efficient approach for the automated classification of urgent messages, using a natural language processing pipeline based on the RASA framework. RASA[1] is an open source platform that offers advanced features for building chatbots and virtual assistants, allowing the integration of different natural language processing techniques to improve task performance.

To face the challenge of adapting the model to different domains and ensuring accurate classification of urgent messages, two fundamental components are incorporated into the pipeline: XLNet and DIET. XLNet is a pre-trained Transformer-based language model that uses the word permutation approach to better capture the bidirectional context of sentences. DIET, in turn, is a RASA intent and entity classifier, which uses deep learning to improve the understanding and interpretation of messages [8].

The great advantage of combining XLNet and DIET lies in the ability to perform effective domain adaptation, with the construction of a Common Sense Language Model (Fig. 1), allowing the model to be trained on different datasets, including those from different areas academic, and still maintain accuracy in classifying urgent messages.

The work presented here is expected to contribute significantly to the advancement of automated classification of urgent messages, facilitating rapid identification of students' needs and providing appropriate responses and support to ensure their progress and well-being throughout the course.

The main contributions of this work are summarized below:

- We developed a model capable of mapping intention and lexicon from the source domain to the target domain using pre-trained models and embedding sequences of varying length, aiming to reduce deviations in the distribution and support of the source and target domains.
- We demonstrate a possibility of reducing the problems faced in domain adaptation, in which classifications of posts from a target domain after training with posts from a different source domain do not usually provide good results.

[1] https://rasa.com/docs/rasa/.

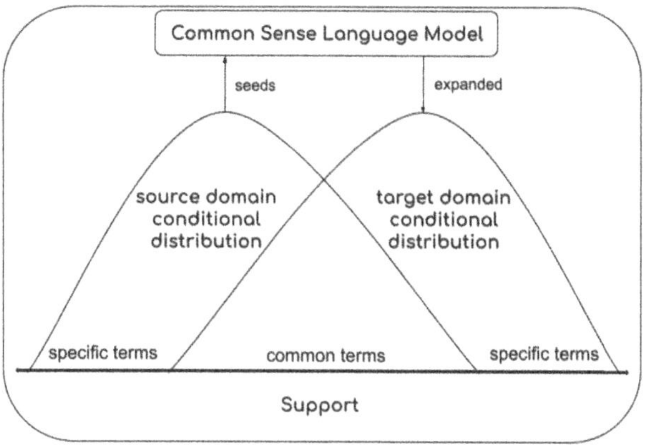


**Fig. 1.** Overview of the adopted framework. A set of initial semantic terms from the source domain is expanded, via the general language model, to be used by the task in the target domain.

- We investigated and encouraged the adoption of the XLNet language model after achieving better results compared to BERT, as it does not have token limits and performs permutations.

This article is structured as follows: Sect. 2 presents a literature review on work related to urgency message classification and domain adaptation; Sect. 3 details the main algorithms of the proposed approach; in Sect. 4 experiments and results are discussed, demonstrating the effectiveness of the approach in different scenarios; and, finally, Sect. 5 brings final considerations and perspectives for future work.

## 2 Related Works

This section presents a brief review of some of the work that influenced this research. A recent survey can be read at [9].

The study authors [1] present the dataset used in their research and establish a baseline for cross-domain classification performance using SVM and TFIDF representation without any knowledge of the target domain. They show that domain adaptation techniques are necessary due to the low accuracy achieved even by nonlinear classifiers like SVM.

The work of [5] presents a method called Co-Class that can identify posts from users with explicit intentions in online discussion forums. The proposed method uses transfer learning to train classifiers that can accurately identify intent posts even in domains where labeled data is not available. The method is designed to deal with the special difficulties of the problem, such as the limited set of features shared across different domains. Experimental results show that Co-Class outperforms several solid baselines, including supervised learning in the target domain and a transfer learning method.

Using the same dataset as [5], the authors of [11] evaluate a model that identifies intentions in discussion forum posts using multi-instance learning and multi-source transfer learning. The proposed method focuses on positive instances and transfers knowledge of unanimous positive instances across domains to minimize the effects of unintentional sentence bias. The results of their method outperform existing transfer learning methods. Overall, the work presents an innovative approach to identifying intentional posts in discussion forums that have potential applications beyond this specific domain.

Research emphasizes the importance of advances in domain adaptation and the potential benefits of proposed methods to increase the ability of classifiers to work effectively across different source and target domains [4,6,10,12].

In [2] the problem of the urgent need for instructor intervention in learning forums on MOOC platforms is discussed and a method is proposed to improve domain adaptation in post-classifiers trained in the source domain. The proposed method uses word embedding, seed words, and latent representation to address the challenge of cross-domain classification. The authors present results of experiments using the MLP classifier with (300, 64, 32) hidden neuron layers and the Adam optimizer. The work also compares the proposed method with published works and discusses potential practical applications of the research in the field of online education and MOOC platforms.

The authors [7] suggest directly reducing the discrepancy between the two domains by bringing the source and target subspaces together. They use subspaces composed of eigenvectors induced by a PCA, one for each domain. The optimization problem can be solved in a simple and closed way, leading to an extremely fast algorithm. Although the algorithm is fast, the accuracies of other approaches outweigh this advantage.

In [8], the authors explore Rasa for sentiment analysis, a text classification task that goes beyond the toolkit's original purpose of being conversational agents. Using pre-trained word embeddings, the DIET architecture achieved high accuracy (0.907, 0.816, and 0.858) on IMDb, MR, and SST2 datasets. Outperforming recent methods, this study suggests Rasa's potential as a versatile framework for text classification, enabling easy experimentation with various pipelines.

## 3 Algorithms

The main algorithms used in this work are described below. They were used in all scenarios, in the classification intra-domain and in the domain adaptation, comparing the results presented in the works of [1–3].

### 3.1 Transformer XLNet

XLNet is a variant of the Transformer architecture, which is a deep learning model initially designed for natural language understanding tasks such as language modeling, machine translation, and text generation. XLNet, short for

"Extra-Long Transformer", was introduced to resolve some limitations of previous models such as BERT (Bidirectional Encoder Representations of Transformers) and improve their performance on various NLP tasks. The Transformer architecture allows XLNet to capture contextual information from both directions, making it bidirectional. One of XLNet's key innovations is its permutation-based training strategy.

Unlike BERT, which uses a masked language model (MLM) objective, XLNet uses a permutation-based language model (PLM) objective. During training, it predicts the probability of a word given its context, but instead of considering just the left and right context like BERT, XLNet considers all possible permutations of the input sequence. This helps capture dependencies from both directions.

XLNet combines the strengths of autoregressive models (like GPT) and autocoding models (like BERT). This is done by factoring the probability of a sequence into a product of conditional probabilities. In other words, it considers context in both directions, but in an autoregressive way, similar to GPT models. Similar to BERT, XLNet can be tuned into specific downstream NLP tasks such as text classification, sentiment analysis, question answering, and more.

### 3.2 Dual Intent and Entity Transformer

DIET is a natural language understanding (NLU) component used in the Rasa framework to build conversational AI applications. DIET was designed to perform two essential tasks in dialogue systems: intent classification and entity recognition. DIET starts by encoding the input text using a pre-trained word embedding model. This converts the input words into numeric vectors, making them suitable for processing by neural networks.

The main task of DIET is to classify user intent, which determines the goal of user or request in the conversation. DIET uses a neural network architecture, typically a Transformer-based model, for intent classification. The Transformer model takes the encoded input text and produces an intent classification score for each possible intent in the predefined intent list, Fig. 2.

A softmax function (Eq. 1) is often applied to these scores to convert them into probability distributions over intentions, and the intention with the highest probability is selected as the predicted intention.

$$\sigma(z_i) = \frac{e^{z_i}}{\sum_{j=1}^{K} e^{z_j}} \quad for\ i = 1, 2, \ldots, K \tag{1}$$

DIET is trained using a combination of losses for intent classification and entity recognition. For intent classification, a categorical cross-entropy loss is commonly used. For entity recognition, a token-level cross-entropy loss is often applied (Eq. 2). During training, DIET learns to minimize these losses by adjusting model parameters (weights and biases) using backpropagation and gradient descent.

$$-\left(y\log(p)+(1-y)\log(1-p)\right) \tag{2}$$

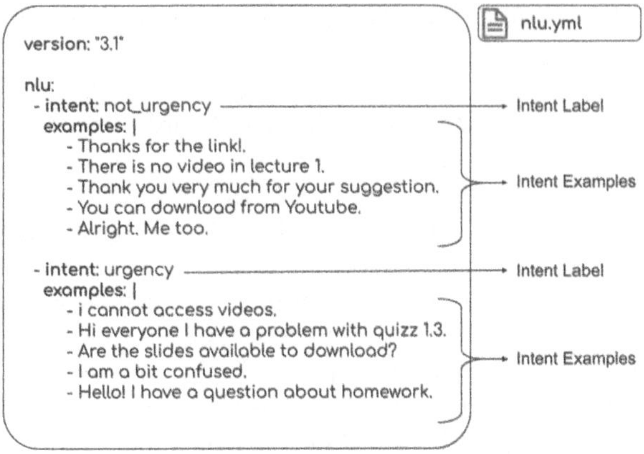


**Fig. 2.** Intent definition file for NLU: definition of all intentions that will be classified, including several examples of each one. Examples are mandatory.

After pre-training on a large corpus of text, DIET can be fine-tuned on domain-specific data to adapt it to specific tasks and vocabulary. Overall, DIET is a crucial component in building systems that understand user intentions and extract relevant information from their messages. It leverages neural network models such as Transformers and combines intent classification and entity recognition in a multi-task learning framework to achieve accurate NLU performance. In Fig. 3 we can see other information about the configurations used in building the model with DIET-RASA.

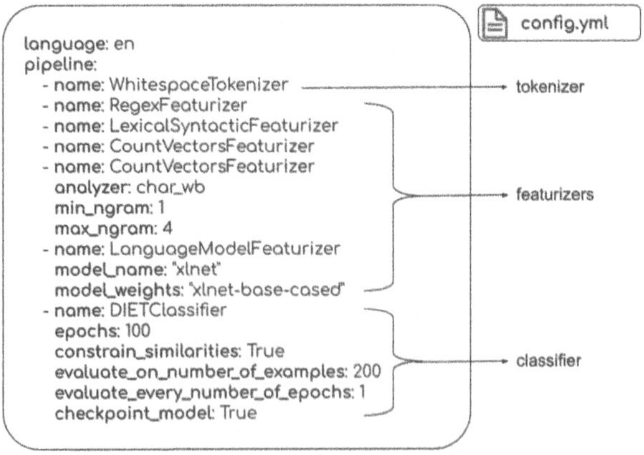


**Fig. 3.** DIET-RASA model pipeline file: choice of algorithms for pre-processing, vector representation, language model for embedding and classifier hyperparameters.

# 4 Results and Discussion

We fully use the structure provided by the RASA framework, including its API for sending and classifying each post contained in the dataset. All tests conducted used the DIET classifier with 100 epochs and the verification of similarity conflicts.

We present an intra-domain analysis (as discussed in Sect. 4.2) to establish a performance reference point, a kind of upper limit, taking into account the available data. In contrast, domain adaptation analysis (as explored in Sect. 4.3) offers a different perspective by providing a lower bound on performance since no specific knowledge of the target domain is used in this analysis.

## 4.1 Data Set

We use the Stanford MOOC forum posts dataset[2], comprising 29,604 anonymous contributions collected from 11 distinct courses. This dataset is accessible to academic researchers through a request process. These courses span three diverse domains: education, humanities/science, and medicine. To ensure quality, each post underwent meticulous manual labeling by three independent human coders, establishing it as a gold standard dataset. Each post was evaluated across six distinct categories, including opinion, question, answer, sentiment, confusion, and urgency. In the case of the opinion, question and answer categories, binary values were assigned. Regarding feeling, confusion and urgency, ratings were given on a scale of 1 to 7.

To provide a clearer perspective on the interpretation of the scale, we consider a non-urgent post rated 1, implying that it does not require immediate attention, while a post rated 7 indicates an extremely urgent matter that requires an immediate response from an instructor. To simplify this into a binary classification task, we transform the original scale ranging from 1 to 7 into binary values. Specifically, the post with a rating of less than 4 was assigned the value 0, meaning non-urgent, while the post with a rating of 4 or higher was assigned the value 1, meaning urgency. Additionally, we filtered posts containing only numerical content, resulting in a dataset consisting of 29,409 posts. Of these, 22,994 were classified as non-urgent and 6,415 were classified as urgent, as shown in Table 1.

## 4.2 Intra-Domain Analysis

The classification intra-domain is a stage of this study, aimed at classifying the urgent messages collected in three distinct areas of knowledge: Medicine (MED), Education (EDU) and Human and Social Sciences (HS) and, from the three courses DS1, DS2 and DS3. For all of the following scenarios, the values located on the diagonal represent results obtained when training the classifier using two of the data partitions, with subsequent evaluation occurring on the remaining third partition.

[2] https://datastage.stanford.edu/StanfordMoocPosts/.

**Table 1.** Grouped by area EDU, MED and H&S; by courses DS1 = Education EDUC115N How to Learn Math, DS2 = Course Medicine SciWrite Fall 2013 and DS3 = Course Humanities/Science Stat Learning Winter 2014. [2,3] use the first group and, [1,2] use the second.

| areas | no urgency | urgency | sets | no urgency | urgency |
|---|---|---|---|---|---|
| Education | 9386 | 461 | DS1 | 9386 | 461 |
| Humanities/Science | 5416 | 2112 | DS2 | 3252 | 1870 |
| Medicine | 8192 | 3842 | DS3 | 2034 | 977 |

**Table 2.** Precision results achieved in this work and in the works of [1–3] for intra-domain classification.

| method | — | —DS1— | —DS2— | —DS3— | —all— |
|---|---|---|---|---|---|
| RASA | mean | 0.98 | 0.87 | 0.90 | 0.92 |
| [1] | mean | 0.95 | 0.75 | 0.70 | 0.80 |
| [2] | mean | 0.95 | 0.75 | 0.74 | 0.81 |
| method | — | —EDU— | —MED— | —H&S— | —all— |
| RASA | mean | 0.98 | 0.88 | 0.91 | 0.92 |
| [2] | mean | 0.95 | 0.70 | 0.74 | 0.80 |
| [3] | mean | — | — | — | 0.87 |

**Discussion:** Analysis of the table reveals that, both for this study and for all comparative studies carried out, the best performances in intra-domain classification are consistently observed in the DS1 or EDU course. These courses represent the same set of data and stand out for having a significantly higher number of posts compared to the other domains. Furthermore, they have a greater number of posts labeled as urgent, which facilitates the process learning of classifier in identifying these posts. It is important to note that although the results on other data subsets are less impressive compared to DS1 or EDU, our approach, which uses DIET-RASA, outperforms the results achieved in other comparative studies.

### 4.3 Domain Adaptation Analysis

In this subsequent part, the advantages of domain adaptation obtained when employing the approaches mentioned in Sect. 3 will be highlighted. The results are recorded in Tables 3 and 4, referring to the two different configurations of the data set. As in the previous tables, the diagonal data were obtained by training the classifier with two of the partitions and subsequent evaluation on the remaining partition. On the other hand, the values in the last column represent the weighted averages and the macro average of the accuracy rates in each row of the table.

**Table 3.** Precision results achieved in this work and in the works of [1,2] for domain adaptation.

| data | method | —DS1— | —DS2— | —DS3— | —all— |
|---|---|---|---|---|---|
| DS1 | RASA | 0.78 | 0.70 | 0.71 | 0.73 |
| | [1] | — | 0.62 | 0.66 | 0.64 |
| | [2] | 0.78 | 0.68 | 0.69 | 0.72 |
| DS2 | RASA | 0.87 | 0.73 | 0.75 | 0.78 |
| | [1] | 0.76 | — | 0.57 | 0.66 |
| | [2] | 0.82 | 0.70 | 0.71 | 0.74 |
| DS3 | RASA | 0.83 | 0.75 | 0.73 | 0.77 |
| | [1] | 0.86 | 0.59 | — | 0.73 |
| | [2] | 0.82 | 0.69 | 0.73 | 0.75 |

**Discussion:** In Table 3, we can see that when a classifier is trained using data from the DS2 course and then tested on the DS1 course, it obtains an average accuracy of 0.87. However, when training is performed on data from the DS1 course and testing occurs on the DS2 course, the average accuracy decreases to 0.70. It is worth noting that the values on the main diagonal of the table represent the tests within the same domain. It is important to highlight that, even with this reduction in average precision, the results still surpass those of other comparable studies.

**Table 4.** Precision results achieved in this work and in the works of [2,3] for domain adaptation.

| data | method | —EDU— | —MED— | —H&S— | —all— |
|---|---|---|---|---|---|
| EDU | RASA | 0.77 | 0.71 | 0.74 | 0.74 |
| | [2] | 0.74 | 0.67 | 0.73 | 0.71 |
| | [3] | — | — | — | 0.74 |
| MED | RASA | 0.88 | 0.74 | 0.79 | 0.80 |
| | [2] | 0.74 | 0.70 | 0.72 | 0.72 |
| | [3] | — | — | — | 0.74 |
| H&S | RASA | 0.87 | 0.77 | 0.78 | 0.80 |
| | [2] | 0.84 | 0.70 | 0.76 | 0.76 |
| | [3] | — | — | — | 0.74 |

**Discussion:** Table 4 illustrates that when a classifier is trained using data from the MED domain and then tested in the EDU domain, it achieves an average accuracy of 0.88. However, when training occurs with data from the EDU domain and testing is performed in the MED domain, the average accuracy drops to 0.71. It is important to note that values on the main diagonal of the table, which represent tests within the same domain, do not present a significant advantage over values outside the diagonal. This suggests that the augmentation and diversification of training data did not have a significant impact on classifier performance.

## 5 Conclusion

We present a generalization approach for domain adaptation in text classification in an out-of-distribution setting. It is based on DIET and XLNET and has been evaluated in detecting urgent posts. Our approach performed better than SOTA when applied to the Stanford Mooc dataset. This approach can be used to offer personalized and agile support for students, regardless of the specificities of each domain. Our investigation continues to expand the test settings. Also, considering that our current approach considered only global language models, we intend to insert local knowledge of the domain in order to further improve the performance of the proposal.

## References

1. Bakharia, A.: Towards cross-domain MOOC forum post classification. In: Proceedings of the Third ACM Conference on Learning@ Scale, pp. 253–256 (2016)
2. Candido, A.L.M., Maia, J.E.B.: Detecting urgent instructor intervention need in learning forums with a domain adaptation. In: Intelligent Systems Design and Applications, pp. 502–512. Springer Nature (2023)
3. Capuano, N.: Transfer learning techniques for cross-domain analysis of posts in massive educational forums. In: Intelligent Systems and Learning Data Analytics in Online Education, pp. 133–152. Elsevier (2021)
4. Chen, T., Guo, Y., Hao, S., Hong, R.: Semi-supervised domain adaptation for major depressive disorder detection. IEEE Trans. Multimed. 1–13 (2023)
5. Chen, Z., Liu, B., Hsu, M., Castellanos, M., Ghosh, R.: Identifying intention posts in discussion forums. In: Proceedings of NAACL-HLT, pp. 1041–1050 (2013)
6. Devika, R., Vairavasundaram, S., Mahenthar, C., Varadarajan, V., Kotecha, K.: A deep learning model based on BERT and sentence transformer for semantic keyphrase extraction on big social data. IEEE Access **9**, 165252–165261 (2021)
7. Fernando, B., Habrard, A., Sebban, M., Tuytelaars, T.: Unsupervised visual domain adaptation using subspace alignment. In: Proceedings of the IEEE International Conference on Computer Vision, pp. 2960–2967 (2013)
8. Herráez, M.A., González, P.A., Ramzan, N.: On adapting the diet architecture and the rasa conversational toolkit for the sentiment analysis task. IEEE Access, pp. 107477–107487 (2022)
9. Liu, X., et al.: Deep unsupervised domain adaptation: A review of recent advances and perspectives. APSIPA Trans. Signal Inform. Process. **11**(1) (2022)
10. Sivaramakrishnan, N., Subramaniyaswamy, V., Viloria, A., Vijayakumar, V., Senthilselvan, N.: A deep learning-based hybrid model for recommendation generation and ranking. Neural Comput. Appl. **33**, 10719–10736 (2021)

11. Song, H.J., Park, S.B.: Identifying intention posts in discussion forums using multi-instance learning and multiple sources transfer learning. Soft Comput: Methodo. Appl. **22** 8107–8118 (2018)
12. Zhu, Y., Wu, X., Qiang, J., Yuan, Y., Li, Y.: Representation learning via an integrated autoencoder for unsupervised domain adaptation (2023)

# Performance Analysis of ChatGPT in Erroneous Python Code Correction and Evaluation

Isha Ganguli(✉), Satyam Atreja, and Sanya Bhasin

Bennett University, Greater Noida, Uttar Pradesh, India
{isha.ganguli,E21CSEU0567,E21CSEU0558}@bennett.edu.in

**Abstract.** Since its inception, ChatGPT has emerged as a potent tool in various Natural Language Processing (NLP) tasks in several domains including health care, education, Business, management, etc. In particular, the education sector is experiencing significant utilization of ChatGPT. The automated correction of code plays a crucial role in the educational process within eLearning platforms and is highly valued by programmers engaged in programming activities. In this study, we conducted an analysis of the performance of ChatGPT in the automated correction and evaluation of small Python code snippets that contain syntax errors. The performance of ChatGPT in correcting and evaluating single line code has been observed to be exceptional, with minimal errors. Nevertheless, as the number of lines of code and errors increases, the performance experiences a decline. However, it is noteworthy that the system still achieves a performance rate exceeding 90% across all potential scenarios. It has been observed that ChatGPT demonstrates enhanced response quality through reinforcement learning when provided with multiple prompts, in contrast to its initial performance at the beginning of prompt execution. Furthermore, it has been observed that while ChatGPT is capable of providing corrections for specific code segments, it is not always reliable in accurately evaluating that corrected code to produce the appropriate result.

**Keywords:** ChatGPT · Program Auto-correction · Program Auto-evaluation

## 1 Introduction

The rapid progression of artificial intelligence (AI) and natural language processing (NLP) has led to the development of increasingly complex and flexible language models. The term "generative artificial intelligence" (AI) encompasses a class of AI models that exhibit the capacity to produce original data by utilizing patterns and structures obtained from existing data. Generative AI models rely extensively on deep learning techniques and neural networks to analyze, understand, and generate content that demonstrates a significant level of resemblance


A. Bajaj et al. (Eds.): HIS 2023, LNNS 1224, pp. 52–61, 2025.
https://doi.org/10.1007/978-3-031-78925-0_6

to outputs created by human beings. ChatGPT, an artificial intelligence model developed by OpenAI, has emerged as a prominent tool with broad applicability in various domains among the range of options currently available. The origins and development of ChatGPT can be attributed to its foundation as a language model that utilizes the Generative Pre-trained Transformer (GPT) architecture. This architecture is specifically designed to facilitate a wide range of natural language processing tasks, including text generation, language understanding, translation, data analysis etc. ChatGPT was trained on a large corpus of text data with 6.7 billion parameters and fine-tuned on specific tasks like generating conversational responses, which allows it to generate human-like responses to user queries. Since its initial release, researchers have been conducting investigations into the efficacy of this potent generative model across various domains [6,6,7].

The automatic correction of code is a fundamental component of the educational process in eLearning platforms for programming [8]. Errors can arise due to a multitude of factors, such as inadequate comprehension of the syntax of the coding language and impromptu typographical mistakes, among others. Moreover, acquiring hands-on guidance from instructors to address mistakes made by learners during the process of coding on an online platform presents a challenge. Consequently, the automated rectification of erroneous code blocks assumes a crucial role in aiding learners and programmers to effectively debug minor code segments during their educational and professional journey. Python has gained significant popularity among various programming languages, including C, C++, Java, and others [9]. This surge in popularity can be attributed to Python's notable advantages such as its swift execution, efficient development process, availability of a large number of libraries, and user-friendly interface. In addition, Python is utilized in various large-scale real-time applications, including web development frameworks such as Flask, Django, and Pyramid. It is also employed in game development applications like Battlefield 2, as well as popular audio-video applications such as Spotify, Netflix, and YouTube. A comprehensive computing application comprises numerous modules of code, with each module comprising smaller code blocks. It is evident that developers often make minor errors when writing small code blocks. The correction of small modules in Python code not only aids novice learners but also serves to rectify issues within larger, intricate Python modules, thereby providing assistance to experienced programmers.

In our study, we assess the efficacy of ChatGPT in the automatic correction and evaluation of small Python code errors, while considering the extensive utility of Python and the widespread usage of ChatGPT. The various types of syntax errors that may arise during Python coding have been examined, and an assessment has been conducted on the capability of ChatGPT to correct and evaluate these erroneous code blocks. A few major contributions of our works are as follows:

- Detailed analysis of the performance of ChatGPT in correcting the erroneous Python code (Single, Double, and Multiple line codes) with a single to multiple number of errors incorporated in it.
- Detailed analysis of the ability of ChatGPT in code evaluation.

This paper has been further divided into the following sections. Section 2 comprises the Related study. The Methodology and the Results and Discussions are presented in Sects. 3 and 4 respectively. Finally, the Conclusion is given in the Sect. 5.

## 2 Related Study

ChatGPT works significantly as a breakthrough in the field of NLP. This model is specifically designed for conversational interactions, making it well-suited for chat-based applications. Since its release, researchers have been investigating the ability of this powerful generative model in several domains [6,6,7]. Ranging from health care [3,10] to global warming [2], education sector [5] to solving programming bugs [4] ChatGPT has its huge applicability. Liu et al. [6] have provided a comprehensive survey of ChatGPT-related research encompassing trend analysis, word cloud representation, and distribution analysis across various application domains. The results indicate a noteworthy and growing enthusiasm in research pertaining to ChatGPT. This research primarily focuses on the direct application of natural language processing. Additionally, the findings suggest that ChatGPT exhibits substantial promise in diverse fields such as education, history, mathematics, medicine, and physics. The objective of this study is to provide a comprehensive understanding of the capabilities of ChatGPT, as well as its potential implications and ethical concerns. Additionally, this study aims to provide guidance for future advancements in this particular field of research. [11] This study investigates the diverse methods by which ChatGPT has been transforming scientific research, encompassing data processing, hypothesis generation, collaboration, and public outreach. Moreover, this study investigates the potential obstacles and ethical considerations associated with the utilization of ChatGPT in research, while emphasizing the significance of achieving a harmonious equilibrium between AI-facilitated advancement and human proficiency. Wang et. al. [13] provide a comprehensive examination of the operational principles, security and privacy risks, current advancements, and forthcoming obstacles related to the AIGC paradigm. This encompasses an exploration of the underlying technologies, the overall architecture of AIGC, as well as the various operational modes and significant attributes. Subsequently, an examination is conducted on the classification of security and privacy risks associated with Artificial Intelligence Generated Content (AIGC), with a specific focus on elucidating the ethical and societal ramifications of GPT and AIGC technologies. [7] This paper examines the historical background and technological aspects of GPT, encompassing its generative pre-trained transformer model, its capacity to execute diverse language-based tasks, and the utilization of this technology in ChatGPT to operate as an advanced chatbot.

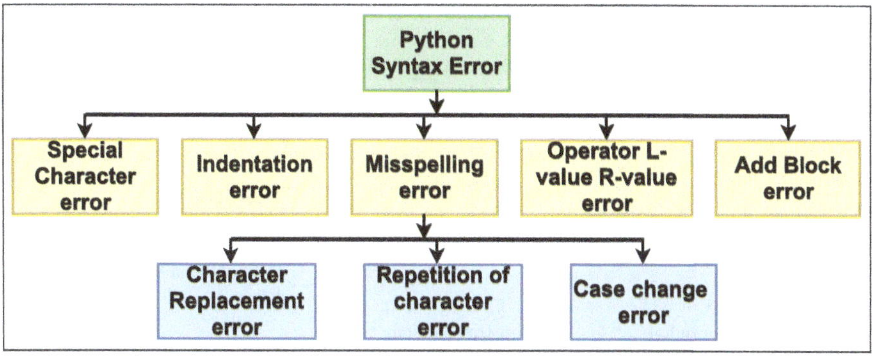


**Fig. 1.** Types of Syntax errors in Python coding language.

The discipline of computer programming necessitates an in-depth knowledge of programming languages, algorithms, and data structures due to its intricate nature. Developers often encounter difficulties in staying abreast of the most recent technologies and optimal methodologies, as well as in identifying and resolving technical issues. The paper [1] unvails that ChatGPT is a powerful and versatile tool for computer programming that can support developers and users in a wide range of tasks. Sobania et. al in their paper [12] presented an analysis of the Automatic Bug Fixing Performance of ChatGPT. In this study, we have tried to analyze the ability of ChatGPT to correct and evaluate erroneous small Python codes.

## 3 Methodology

This section presents a brief introduction to the tasks of automated program correction and program auto-evaluation. This discussion delves into various manifestations of syntax errors in Python programming through an analysis of the intrinsic characteristics of Python code. The syntax errors encountered in Python programming can be classified into the components depicted in Fig. 1.

This section presents a brief introduction to the tasks of automated program correction and program auto-evaluation. This discussion delves into various manifestations of syntax errors in Python programming through an analysis of the intrinsic characteristics of Python code. The syntax errors encountered in Python programming can be classified into the components depicted in Fig. 1.

The occurrence of the **Special character error** is rooted in the observation that, on a standard keyboard, there is a higher probability of mistakenly substituting one particular special character with another special character located on the same key, compared to the other keys on the keyboard. **Misspelling error** occurs when the words in a line of code are incorrectly typed, resulting in the generation of an incorrect word. The **Character replacement error** arises when a programmer inadvertently substitutes a character in a word of a line of

**Table 1.** Example of erroneous codes

| SL No. | Original Code | Erroneous Code | Type of Syntax error |
|---|---|---|---|
| 1 | p = 10<br>q = 20<br>if ( p!= q ) :<br>    print ( " Hi there" ) | p = 10<br>q = 20<br>if ( p != q ) ;<br>    print ( " Hi there" ) | Special Character error<br>(: ->;) |
| 2 | x = 5<br>y = 2<br>for x <10 :<br>    if ( x != y ) :<br>        y = y-x<br>        x = x+1 | x = 5<br>y = 2<br>fog x <10 :<br>    iff ( x != y ) :<br>        y = y-x<br>        x = x+1 | Misspelling errors:<br>1. Character Replacement Error (for ->fog)<br>2. Repetition of character Error (if ->iff) |
| 3 | import random<br>mylist = []<br>for i in range(0,100):<br>    x = random.randint(1,10)<br>    mylist.append(x)<br>print(mylist) | import random<br>mylist = []<br>for i in range(0,100):<br>x = random.randint(1,10)<br>    mylist.append(x)<br>print(mylist) | Indentation Error.<br>( Indentation in the forth line is erroneously missing<br>) |
| 4 | x = 0o12<br>y = 0o22<br>x = x + y<br>x-= 1<br>print(x) | x = 0o12<br>y = 0o22<br>x = x + y<br>-= 1<br>print(x) | Operator L-value R-value Error. Here L-value is missing in third line. |
| 5 | list1= [2.4,9,'yes']<br>list2 = ['pics',1.0]<br>if len( list1 ) != len( list2 ) :<br>    list3 = list1 + list2<br>    print ( list3 ) | list1= [2.4,9,'yes']<br>list2 = ['pics',1.0]<br>if len( list1 ) != len( list2 )<br>{<br>    list3 = list1 + list2<br>    print ( list3 )<br>} | Add Block Error.<br>( '{}' is wrongly placed to identify the block ) |

code with its adjacent characters, leading to a heightened likelihood of replacing a letter with the letters located at a distance of one unit on the keyboard. The occurrence of character repetition in a string often leads to typographical errors namely **Repetition of character error**, wherein a specific character is repeated two or more times. **Case change error** occurs when Often, by mistake, a lower case or upper case character in the code line is replaced by the alternate case of the same character. The **L-value R-value error** takes place when either one of the operator's L-values or R-values is omitted from the code line, causing syntax error in a Python code. The occurrence of an **Indentation error** is a result of mistakenly adding or deleting appropriate indentation. **Add Block Error** occurs due to the coder's lack of basic knowledge in Python language syntax and they mistakenly give curly braces (i.e. "{" and "}") for specific code blocks. Table 1 presents details of the different types of syntax errors in small Python codes. The correction and evaluation of small code blocks are of utmost importance in the advancement of extensive Python programming, primarily because of the progressive compilation procedure. In this work, we have analyzed the ability of ChatGPT to correct and evaluate the erroneous small Python program.

# 4 Result and Discussion

In this work, A series of erroneous Python code snippets were presented in order to evaluate ChatGPT's capacity to accurately correct and evaluate erroneous codes. To accomplish the task, we have classified the small Python codes into three primary categories based on their code length: Single Line code, Double Line code, and Multiple Line code. The maximum allowable length for the multiple-line code that has been considered is twelve. Furthermore, the analysis includes the consideration of single, double, and multiple numbers of errors for each category of Python codes.

## 4.1 Performance in Program Correction

Here, Table 2 illustrates the performance accuracy of ChatGPT in correcting code of varying lengths and with different numbers of errors. In the context of correcting a singular error, ChatGPT demonstrated commendable performance by successfully correcting code snippets containing a solitary error within a single line. It achieved an impressive correction rate of approximately 99.2% when initiated from the initial prompt. In situations where code snippets exhibited multiple errors, ChatGPT also exhibited robust correction capabilities. The code, which exhibited a significant number of errors or repeated errors, demonstrated a repair rate of approximately 98.21%. During the execution of scripts in ChatGPT, it was observed that 99.2% of code snippets containing a single error in a single line were successfully rectified, but only when the initial prompt was provided. Additionally, 98.21% of double and multiple error codes were also fixed. When initiated from the initial prompt, ChatGPT was able to correct approximately 99.15% of code snippets containing a single error in a double line. Moreover, it achieved a correction rate of approximately 98.79% for code snippets with double or multiple errors, emphasizing its consistent performance in different experimental runs. In testing programs in ChatGPT, we found that 92.5% of wrong code snippets with double and triple faults were also corrected, and over 95% of incorrect code snippets with a single error on multiple lines were fixed while only passing through the first prompt. It is noteworthy to mention that during the implementation of prompts on ChatGPT, we observed the utilization of Reinforcement Learning. This was evident as the system generated significantly improved responses after executing a substantial number of prompts, in contrast to the initial stages of prompt execution. The system consistently enhanced its responses and customized them based on our desired preferences. These observations provide crucial insights for developers and users who seek to leverage ChatGPT's functionalities for enhancing and rectifying code, as well as understanding its performance across various code-related issues.

There are some significant observations that we have made while analyzing the performance of ChatGPT for this particular task. Following are the case-wise enlisted observations:

- **1. Single Line Code**
  - *Handling Float Values:* During the analysis of single-line code, it was observed that ChatGPT encountered challenges when attempting to compare floating-point values. Significantly, instances arose where the system presented inaccurate comparisons, such as claiming that 34 was less than 34.56, thereby highlighting a deficiency in its handling of floating-point values.
  - *Correction for Bracket:* Correcting open brackets with '(' showed a specific shortcoming in ChatGPT. It consistently added '(' when given code fragments with '9' in place of an open bracket. Furthermore, it failed to replace "9" with the proper bracket and instead regarded it as an integer.
  - *Hexadecimal, binary, and octal codes handling:* ChatGPT demonstrated a notable proficiency in effectively managing binary, hexadecimal, and octal codes. Furthermore, it consistently corrected complex figures with accuracy.
  - *Taking Advice from Prompts:* An intriguing observation was made after multiple ChatGPT code assessments. With more prompts, the tool seemed to fix Python code snippets better. This suggests that ChatGPT learns as prompt data accumulates, as accuracy increased.
- **2. Double Line Code**
  - *Identical Correction Rates:* Surprisingly, the efficacy of rectifying double line code errors exhibited a high degree of consistency comparable to that observed in addressing single line code issues.
  - *Effective Manipulation:* Compared to single-line code snippets, ChatGPT handled double-line code problems effectively and frequently demonstrated a quicker rectification procedure.
  - *Correction for Brackets (Double Line):* When given with a "9' rather than an open bracket in double line code snippets, ChatGPT displayed the same bracket correction error as it did in single line code.
- **3. Multiple Line Code**
  - *Formatting for Multiple Lines of Code and Indentation:* There have been cases where indentation mistakes prevented the code output from running as intended.
  - *Managing Loops:* There were times when the tool struggled with loop handling. It had trouble comprehending unusual characters and occasionally produced indentation issues while dealing with loops, showing that it had trouble understanding loop structures.
  - *Spelling mistakes:* Spelling mistakes involving built-in functions were particularly difficult for ChatGPT to detect and fix. For instance, it had trouble changing "prinnntn(x)" to "print(x)".
  - *Differences in Expected Results:* For each query, ChatGPT gave expected output along with the generated code, but it was observed that this expected output did not always correspond to the results that were actually shown when the code was run via an online compiler. These differences made it unclear whether the output predictions made by the ChatGPT were consistent.

**Table 2.** Performance of ChatGPT in Erroneous code correction

| Code Type | Error Type | Total Number of Codes Executed | Codes Corrected by ChatGPT | Accuracy (%) |
|---|---|---|---|---|
| **Single Line Code** | **Single Error** | 4000 | 3968 | 99.2 |
| | **Multiple Error (Double/Random)** | 7000 | 6875 | 98.21 |
| **Double Line Code** | **Single Error** | 6000 | 5949 | 99.15 |
| | **Multiple Error (Double/Random)** | 12000 | 11855 | 99.79 |
| **Multiple Line Code** | **Single Error** | 8000 | 7600 | 95 |
| | **Multiple Error (Double/Random)** | 22000 | 20350 | 92.5 |

**Table 3.** Performance of ChatGPT in Erroneous code evaluation

| Code Type | Error Type | Accuracy of ChatGPT For Evaluating the Corrected code (%) |
|---|---|---|
| **Single Line Code** | **Single Error** | 100 |
| | **Multiple Error (Double/Random)** | 100 |
| **Double Line Code** | **Single Error** | 100 |
| | **Multiple Error (Double/Random)** | 99 |
| **Multiple Line Code** | **Single Error** | 100 |
| | **Multiple Error (Double/Random)** | 92 |

**Table 4. Performance of ChatGPT on example Python code snippets**

| OriginalCode | CorrectCode | CorrectedCode (ChatGPT) | ExpectedOutput | ChatGPT'S Output |
|---|---|---|---|---|
| Prnt 00112 <40,95 | print(00112<40.95) | print(00112<40.95) | FALSE | TRUE |
| B= and print 9 b ) | B=true and false print(b) | b=true and false print(9b) | FALSE | FALSE |
| print 9 90.9+37i) −965+57;) ) | print((90.9+ 37j) - (965 + 57j)) | print((9+ 90.9j) - (965 + 57j)) | (−874.1-20j) | (−956+33.900000000000006j) |

## 4.2 Performance in Program Evaluation

Here, Table 3 reveals interesting patterns in the reliability of getting the right output from ChatGPT-corrected code. Here, We have taken the codes corrected by ChatGPT and tried to find out the evaluation results of those corrected codes given by ChatGPT. The tool continuously achieved a stunning 100% accuracy in output rectification for single and double line faults. When evaluating the accuracy of output correction in instances with several errors across multiple lines, the accuracy percentage noticeably lowers to 92%. This disparity underscores a significant differentiation: while ChatGPT boasts a 100% rate of correcting code, it does not necessarily guarantee 100% accuracy in its evaluation output. This suggests the possibility of instances in which the output generated by ChatGPT deviates from the expected outcome. The discrepancy arises from ChatGPT's utilization of the corrected Python code snippet to execute a desired output within its backend environment. While this approach yields valuable insights, it is crucial to acknowledge the potential presence of model training errors that can result in disparities between the anticipated and observed outcomes. Hence, it is recommended that users should use caution and consider potential variations in the accuracy of output when employing ChatGPT for code correction and evaluation. Table 4 presents examples of erroneous Python codes and how ChatGPT is performing in erroneous code input correction and calculating the output.

In addition to ChatGPT's program correction and evaluation capabilities, we have also examined the processing and correction speed of ChatGPT for this specific task. The tool exhibited impressive efficiency in its correction capabilities, demonstrating an average correction time of approximately 2.5 s per snippet. It is worth mentioning that ChatGPT demonstrated the ability to swiftly rectify code segments containing one or two lines of errors, accomplishing this task within a time frame of less than 2 s. However, when confronted with code segments that contained numerous lines of errors, the repair process required a duration of 3 to 4 s. The prompt asserts that ChatGPT's prompt response to corrections renders it a valuable resource for software engineers seeking real-time code enhancement recommendations, thereby enhancing their productivity and efficacy.

## 5 Conclusion

In conclusion, this study has conducted a comprehensive analysis of ChatGPT's ability to correct and evaluate small erroneous Python code snippets with syntax errors. This analysis showed that ChatGPT had a noteworthy degree of accuracy, correctly identifying, correcting, and evaluating Python code. These findings underscore the proficiency of ChatGPT in the domain of Programming. Hence, it can be posited that our research holds potential benefits for the realms of education and professional development, catering to both neophyte learners and seasoned programmers, enabling them to acquire knowledge and effectively utilize ChatGPT.

## References

1. Chatgpt in computer programming: Som Biswas. Role of chatgpt in computer programming. Mesopotamian J. Comput. Sci. **2023**, 8–16 (2023)
2. Biswas, S.S.: Potential use of chat GPT in global warming. Ann. Biomed. Eng. **51**, 1126–1127 (2023)
3. Estomba, C., et al.: Exploring the potential of chat-GPT as a supportive tool for sialendoscopy clinical decision making and patient information support. In: European Archives of Oto-Rhino-Laryngology, pp. 1–6, 07 (2023)
4. Haque, M.A., Li, S.: The potential use of chatGPT for debugging and bug fixing. EAI Endorsed Transac. AI Robot. **2**, e4, 05 (2023)
5. Kalla, D., Smith, N.: Study and analysis of chat GPT and its impact on different fields of study. **8**, 03 (2023)
6. Liu, Y., et al.: Summary of chatgpt-related research and perspective towards the future of large language models. In: Meta-Radiology pp. 100017 (2023)
7. Lund, B.D., Wang, T.: Chatting about chatGPT: how may AI and GPT impact academia and libraries? Library Hi Tech News **40**(3), 26–29 (2023)
8. Malysheva, Y., Kelleher, C.: Assisting teaching assistants with automatic code corrections. In: CHI Conference on Human Factors in Computing Systems, pp. 1–18 (2022)
9. Nagpal, A., Gabrani, G.: Python for data analytics, scientific and technical applications. In: 2019 Amity International Conference on Artificial Intelligence (AICAI), pp. 140–145. IEEE (2019)

10. Parray, A.A., Inam, Z.M., Ramonfaur, D., Haider, S.S., Mistry, S.K., Pandya, A.K.: Chatgpt and global public health: Applications, challenges, ethical considerations and mitigation strategies. Global Trans. **5**, 50–54 (2023)
11. Ray, P.P.: ChatGPT: a comprehensive review on background, applications, key challenges, bias, ethics, limitations and future scope. Internet of Things and Cyber-Physical Systems (2023)
12. Sobania, D., Briesch, M., Hanna, C., Petke, J.: An analysis of the automatic bug fixing performance of chatGPT (01 2023)
13. Wang, Y., Pan, Y., Yan, M., Su, Z., Luan, T.H.: A survey on chatgpt: AI-generated contents, challenges, and solutions. arXiv preprint arXiv:2305.18339, 2023

# Potato Disease Classification Using Diverse Feature Extraction Methods and Machine Learning Models

C. Shanmukha Srinivas Sai[1], Tejus Paturu[1], B. Surendiran[1], and J. Dhakshayani[2](✉)

[1] Department of Computer Science and Engineering, National Institute of Technology Puducherry, Karaikal, India
Surendiran@nitpy.ac.in

[2] Indian Institute of Information Technology Kottayam, Kerala, India
Dhakshajey63@gmail.com

**Abstract.** Agriculture's significance in sustaining human life has never been more pronounced, with technological integration and digitalization becoming integral to enhancing productivity and ensuring food security. One critical aspect of safeguarding agricultural yields is the early detection and also the classification of various crop diseases. In this study, we hear by present a comprehensive methodology for the detection and classification of diseases affecting potato plants. Leveraging the widely accessible and reputable Plant Village Dataset, encompassing diverse potato disease images, our approach unfolds in three key stages: (1) comparison of various feature extraction methods including pixel feature extraction, LBP feature extraction, and HOG feature extraction, (2) classification of the images employing a range of machine learning models including Multinomial Naive Bayes, Random Forest, Logistic Regression, SVM model, Decision Tree, K neighbors classification, and XGB classifier, and (3) evaluation of the F1 score for each model on each disease class, with results visually depicted through graphs.

Our experimental findings reveal that the XGB model has achieved the highest accuracy of 95.3% for Pixel feature extraction, Random Forest outperforms the rest for LBP feature extraction with an accuracy of 80.3%, and HOG feature extraction excels with the SVM classifier with an accuracy of 93.3%. These results guide in choosing the most appropriate techniques for extracting features for machine learning models, maximizing the F1 score. Importantly, the trained models are envisioned to be deployed by farmers for early disease detection and classification, serving as a preventive measure against emerging diseases in potato crops. This research signifies a vital step towards sustainable agriculture, where technology and machine learning empower farmers to protect their yields and ensure food security.

**Keywords:** HOG feature extraction · LBP feature extraction · Potato leaf disease · Pixel feature extraction · Random Forest · Support vector machine · XGB classifier


A. Bajaj et al. (Eds.): HIS 2023, LNNS 1224, pp. 62–75, 2025.
https://doi.org/10.1007/978-3-031-78925-0_7

# 1 Introduction

As per the National Horticulture Board of India [1], the potato production landscape in the country has undergone a remarkable transformation, culminating in an impressive yield of 59.74 million metric tons. The Potato (Solanum tuberosum), a global agricultural cornerstone, plays a significant role in enhancing food security and furthermore underpinning agricultural economies across the world. However, the persistent specter of diseases casts a looming shadow over potato cultivation, resulting in substantial and often a devastating yield loss around the world. In the quest for sustainable potato production, the crux lies in the early and precise detection of diseases that threaten the production of this crop.

In this ever-evolving agricultural landscape, where technology, data, and machine learning are instrumental, the imperative use of timely disease detection takes centre stage here. Early identification and classification of these diseases are not just agricultural challenges; they are vital components for sustainable crop production.

Potato plants (Solanum tuberosum) commonly encounter two prevalent fungal diseases: Early Blight, linked to Alternaria solani, and Late Blight, which is the result of the Phytophthora infestans fungus [2]. Early Blight symptoms manifest as small, dark, dry papery flecks that eventually mature into brown-black spots. On the other hand, Late Blight is a severe ailment affecting not only potatoes but also tomatoes, and occasionally eggplants, all belonging to the potato family. Its symptoms include small, pale to dark green irregular spots on the leaves. These two diseases pose significant threats to potato crops.

This research endeavors to bridge the gap between technology and agriculture by offering a comprehensive methodology for the for the timely identification and the accurate categorization of ailments affecting potato crops. By leveraging the Plant Village Dataset, from Kaggle, replete with a diverse array of potato disease images. Our approach unfolds through three pivotal stages: (1) comparison of various feature extraction methods, (2) classification of the images employing a range of machine learning models, and (3) evaluation of the F1 score for each model on each disease class, with results visually depicted.

The subsequent sections of this paper provide details of the past works related to potato disease classification, methodology, empirical findings, and discussions, culminating in the identification of the most effective machine learning model paired with each feature extraction technique for accurate potato leaf disease classification.

# 2 Related Works

In the domain of plant disease detection and image analysis, researchers have explored various techniques to achieve high accuracy rates in identification and categorization of diseases in plants.

Aditi Singh et al. [3] proposed the K-means technique for image segmentation, employed gray-level co-occurrence matrix for feature extraction, and the utilization of multi-class support vector machine (SVM) technique for classification purposes. Their methodology yielded an impressive accuracy rate of 95.99%.

M.Islam et al. [4] proposed a method on integrating image processing in conjunction with machine learning for the identification of diseases in leaf images, this method leverages support vector machine techniques to demonstrate disease classification cross a dataset of over 300 images, achieving an accuracy of 95%.

Shima Ramesh et al. [5] proposed the utilization of Random Forest for distinguishing between healthy and diseased leaves within the dataset. The paper encompasses on collectively training the Random Forest model on both diseased and healthy leaf samples, here in extracting features of an image the paper states the use of Histogram of an Oriented Gradient (HOG).

Sholihati R et al. [6] proposed a method for classifying the four types of diseases affecting potato plants, based on their leaf conditions by employing deep learning techniques using the convolutional neural network architecture model VGG16 and VGG19. This experiment has demonstrated an average accuracy of 91%, suggesting the viability of also using the deep neural network approach as well.

Chaojun Hou et al. [7] suggested an approach to accurately separate the potato leaf from images, using a graph-cut algorithm. They identified the main part of the leaf and the background by examining their colors. Next, feature extraction was conducted through the utilization of the Local Binary Pattern (LBP) technique. Four distinct classifiers were employed for evaluation, namely k-Nearest Neighbor (k-NN), Support Vector Machine (SVM), Artificial Neural Network (ANN), and Random Forest (RF). The experimentation encompassed a dataset comprising 2840 images, encompassing both healthy and diseased potato leaves, the SVM classifier performed the best, the overall accuracy achieved was 97.4%, surpassing the performance of k-Nearest Neighbor (k-NN), Artificial Neural Network (ANN), and Random Forest (RF).

M. A. Iqbal et al. [8] The research paper proposed an automated system based on image processing and machine learning for the identification and classification of potato leaf diseases. In this study, image segmentation was performed on a dataset containing 450 images of both healthy and diseased potato leaves. Seven different classifier algorithms were employed for the recognition and classification of these leaves. Remarkably, the Random Forest classifier yielded an accuracy rate of 97%.

Shruthi. U et al. [9] This paper outlines the various phases of general plant diseases detection system and includes a comparative analysis on machine learning classification methods for plant disease detection. The study revealed that Convolutional Neural Network (CNN) achieved both high accuracy and the ability to detect a broader spectrum of diseases affecting multiple crops.

Existing methods for the detection of plant diseases have exhibited notable achievements in accuracy, but they grapple with limitations, including a lack of adaptability to diverse datasets and environmental conditions. To address these challenges, my proposed approach involves employing a range of feature extraction methods, such as pixel feature extraction, Histogram of Oriented Gradients (HOG), and Local Binary Patterns (LBP), and then assessing different machine learning models using accuracy, precision, and F1 score measures. This comprehensive exploration helps to identify the most effective feature-model pairs, offering a more adaptable and robust solution for plant disease detection across various scenarios.

# 3 Materials and Methods

In this study, our aim is to identify the most effective machine learning classifier model by employing three distinct feature extraction methods: HOG, pixel feature extraction, and IBP feature extraction. We calculate the F1 scores for these methods and visualize the results through graphical representations, providing insights into the performance of each classifier. Figure 1 illustrates the framework of the proposed idea.

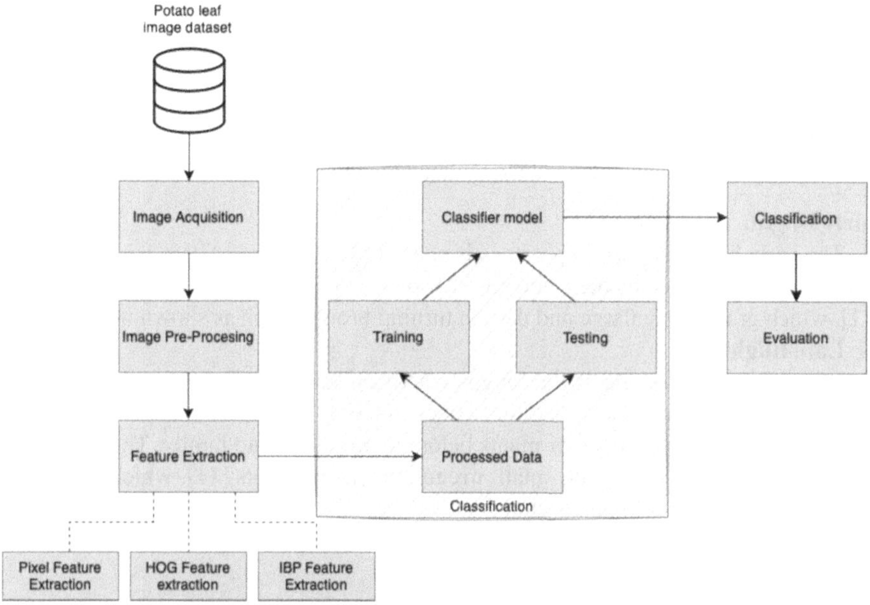


**Fig. 1.** Framework of the proposed methodology

## 3.1 Dataset Description

In our research the dataset employed for studying potato leaf diseases was sourced from the Plant Village dataset on the Kaggle website [10], a publicly accessible repository widely used for precision agricultural applications. The image dataset comprises distinct categories, encompassing two specific disease types labelled as Early Blight and Late Blight, along with a category representing healthy potato leaf images. Notably, the dataset comprises a total of 2780 images, Among the 2780 images in the dataset, there are 1020 images corresponding to Early Blight, 1000 images corresponding to Late Blight diseased leaves, and 760 images representing healthy leaves. All images in the dataset have a uniform dimension of 256 × 256 pixels. Further details regarding each disease, as well as sample images (Fig. 2), are provided below.

**Fig. 2.** Sample images from the dataset (a): healthy leaf (b): leaf afflicted by late blight (c): leaf afflicted by early blight

**Early Blight.**

The early blight disease in potatoes is caused by the fungus *Alternaria solani*. This disease is characterized by the emergence of spots which begin as small, dark, dry, flecks [11], which gradually enlarge and darken turning brown-black as shown in Fig. 2b.

**Late Blight.**

*Phytophthora infestans*, is the fungus which causes Late blight disease in potatoes and is a serious disease. This condition affects not only potatoes but also tomatoes and, on occasion, eggplants and other plants belonging to the potato family. The appearance of leaf spots usually starts with small, irregularly shaped spots [11], which can vary in color from pale to dark green as shown in Fig. 2c.

## 4 Proposed Methodology

The proposed methodology encompasses several key stages including image pre-processing, feature extraction, and classification, as visualized in Fig. 1.

### 4.1 Image Acquisition

The initial acquisition of the images is through the plant village dataset hosted on the Kaggle website [10], a reliable open-source resource database widely utilized for most precision agricultural tasks.

### 4.2 Image Pre-processing

This phase is of utmost importance within the framework. During this stage, we ensure that all images adhere to a standard size of 256 × 256, Additionally, during this phase, we determine the total number of images and inspect the dataset for any potential irregularities or discrepancies. Figure 3 Illustrates the overall count of the images in the dataset.

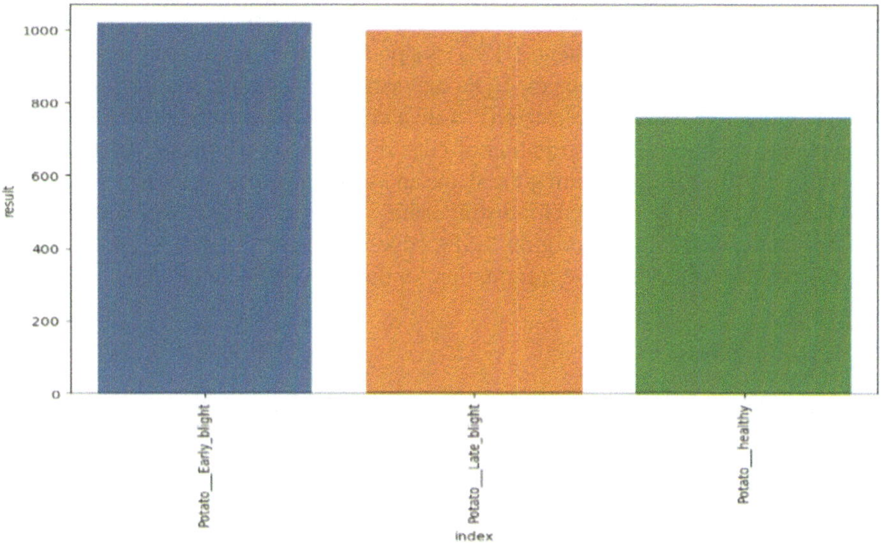


**Fig. 3.** Class wise distribution of Potato leaf dataset

### 4.3 Feature Extraction

It is a pivotal step in machine learning-based image classification, given that it transforms the raw image data into a set of meaningful features that can be used to represent and distinguish different classes. In this work, three feature extraction methods are employed to extract essential characteristics from the potato leaf images.

**Pixel Feature Extraction:**
Pixel feature extraction involves direct utilization of the pixel intensity values from the pre-processed potato leaf images. Each image is treated as a 2D array of pixels, subsequently, these pixel values are used to construct a feature vector that encapsulates the image's characteristics. This feature vector comprises of the flattened pixel intensity values, capturing the grayscale information of the entire image. While this method is relatively simple and computationally efficient, it may lack the ability to capture more intricate patterns and structural information present in these potato leaves [12].

**Local Binary Pattern (LBP) Feature Extraction:**
Local Binary Pattern (LBP) is a widely used texture descriptor that characterizes the local patterns and variations in an image. For each pixel in the image, LBP examines its neighborhood and then compares the intensity values with the center pixel. The result of this is a binary string that encodes information about the pixel's texture in that region. By pooling these binary strings over the entire image, we obtain a histogram-based feature representation that captures the texture patterns specific to each disease class. LBP is particularly effective in detecting subtle variations in leaf texture caused by diseases, and enhancing the discriminative power of the features for this disease [13].

**Histogram of Oriented Gradients (HOG) Feature Extraction:**
A Histogram of Oriented Gradients (HOG) is a powerful technique used to extract shape and edge information from images. HOG operates by computing gradient orientations and magnitudes in local image regions. It then constructs a histogram of the gradient orientations, capturing the distribution of edge directions in the image. HOG features are highly effective in representing the shape and structural characteristics of the potato leaves, making them well-suited for distinguishing different disease categories based on leaf morphology. By quantifying the leaf's edge and shape properties, HOG features provide valuable discriminative information for disease classification [14].

### 4.4 Processed Data

The results from the above feature extraction methods are first collected, which would later be subjected to a range of machine learning classification models.

### 4.5 Classification

In the classification process, the data is split into training and testing data. Subsequently, the test data will be fed into the trained classifier, which will undertake the task of categorizing the images into distinct classes, including Late Blight, Early Blight, and Healthy, we employ various range of classifiers, including Multinomial Naive Bayes [15], Logistic Regression [16], Random Forest [17], SVM model [18], Decision Tree [19], K neighbors classification [20], and XGB classifier [21].

### 4.6 Data Split

The dataset for Potato leaf disease classification has been separated into two divisions:

**Training Data:**
In the training phase the data would be obtained from the processed data, and about 80% of the data with random indexing will be considered to train the classifier model.

**Testing Data**
During the testing phase, data will be extracted from the pre-processed dataset. Approximately 20% of this data, chosen randomly through indexing, has been designated for assessing the performance of the classifier model.

**Naive Bayes (NB)**
The probabilistic classifier operates according to the principles of the Bayes classifier. In this particular variation, it starts with the assumption that the prior probabilities of the patterns are already known, and it proceeds to assign posterior probabilities to the class labels. These posterior probabilities are determined by calculating the maximum likelihood values of the data associated with a specific class label. This is achieved by multiplying the conditional probability of each feature using Bayes' theorem.

**k-Nearest Neighbors (k-NN)**
The k-NN classifier represents a prevalent supervised machine learning algorithm widely employed in the field of pattern recognition. It operates on the principle of the nearest neighbor rule, and is used in the classification process in machine learning tasks. The classifier is trained with the objective of categorizing to a test pattern by assessing the resemblance between the test pattern and each training pattern. The k-NN classifier assigns membership value to a class outcome by conducting a majority vote among its k-nearest neighbors. In essence, it allocates the object to the class label that is most frequently represented among these nearest neighbors.

**Decision Tree (DT)**
Decision trees represent a class of supervised learning algorithm applicable to both classification and regression tasks. Their methodology involves partitioning the data into smaller subsets according to diverse attributes, thus forming a tree-like structure. Typically, the Gini index or entropy serves as widely employed metrics to ascertain the optimal attribute for data partitioning. Various iterations of decision trees exist, among them ID3, C4.5, and CART, which enjoy extensive utilization within data mining and machine learning domains.

**Support Vector Machine (SVM)**
The SVM classifier employs supervised learning techniques to discern dissimilarities among data points within high-dimensional space, effectively distinguishing between different classes. It achieves this by employing a separating hyperplane. Thanks to SVM's kernel tricks, it can perform nonlinear classification. Additionally, SVM has the ability to transform features using linear, polynomial, and radial basis functions, this process can enhance the discriminability of features within a high-dimensional feature space.

**Random Forest (RF)**
This learning method constitutes an ensemble of randomized decision tree classifiers. At the training time, multiple decision trees are constructed, and the classification of the test dataset's class labels is determined through a voting mechanism involving each of these classification trees. The final classifier's result is based on this collective decision. To assemble an uncorrelated forest of trees capable of making more accurate performance predictions than a single tree, this algorithm utilizes a combination of bagging and feature randomness in the construction of each individual tree.

**Logistic Regression (LR)**
Logistic Regression a fundamental statistical method commonly employed in machine learning and statistics. While it is often used to predict binary outcomes, it can also be extended to handle multiclass classification. In contrast to the linear regression model, which forecasts continuous outcomes, the logistic regression models the probability of categorical outcomes or events. To capture this relationship, the logistic function, or sigmoid function, is employed. Logistic regression transforms its output into a probability score ranging from 0 to 1, rendering it suitable for both binary and multiclass classification tasks. Typically, a threshold, such as 0.5, is applied to make binary predictions.

Logistic regression is esteemed for its simplicity, interpretability, and computational efficiency.

**XGBoost (XGB)**

XGBoost, short for Extreme Gradient Boosting, is a highly effective machine learning classifier and known for its exceptional predictive accuracy and computational efficiency. Operating within the realm of ensemble learning, XGBoost combines the predictions from multiple decision trees to deliver remarkably precise results. What distinguishes XGBoost is its unique boosting technique, where each new decision tree is constructed to rectify the errors made by its predecessors. This process optimizes a cost function by skilfully assigning weights to individual trees and combining their outputs to produce the final predictions.

### 4.7 Evaluation

From the classifier model, we will generate the evaluation metrics, including precision, accuracy, F1-score, and support will be generated, and graphs are plotted.

## 5 Results and Discussion

In this section, we assess the performance of the aforementioned classification techniques including Naive Bayes (NB), k-Nearest Neighbors (k-NN), Decision Trees (DT), Support Vector Machines (SVM), Random Forest (RF), Logistic Regression (LR), and XGBoost (XGB) on the Plant Village dataset, The described classifiers are implemented using Python 3.8, operating on a Windows 10 system equipped with an AMD Ryzen 73750H CPU running at 2.3 GHz and 16 GB of RAM. In the evaluation of the classifiers, we utilized three distinct feature extraction techniques: Pixel analysis, Local Binary Pattern (LBP), and Histogram of Oriented Gradients (HOG).

Figure 4 presents the F1-score comparison of these classifiers for each class, along with their macro average and weighted average when using Pixel feature extraction. Figure 5 showcases the accuracy-versus-model comparison after implementing the Pixel feature extraction method. Here, the XGBoost (XGB) model performs exceptionally well with an accuracy of 95.3%. Figures 6, 7 and 8 display the F1-score comparisons with LBP and HOG feature extraction, respectively. These figures provide insights into how the choice of feature extraction method can impact the performance of the classifiers across different classes. Figures 9 and 10 illustrate the accuracy-versus-model comparison following the implementation of the LBP and HOG feature extraction techniques, respectively. In Fig. 9, the Random Forest (RF) model leads with an accuracy of 80.5%, while in Fig. 10, the Support Vector Machine (SVM) model emerges as the top performer with an accuracy of 93.3%.

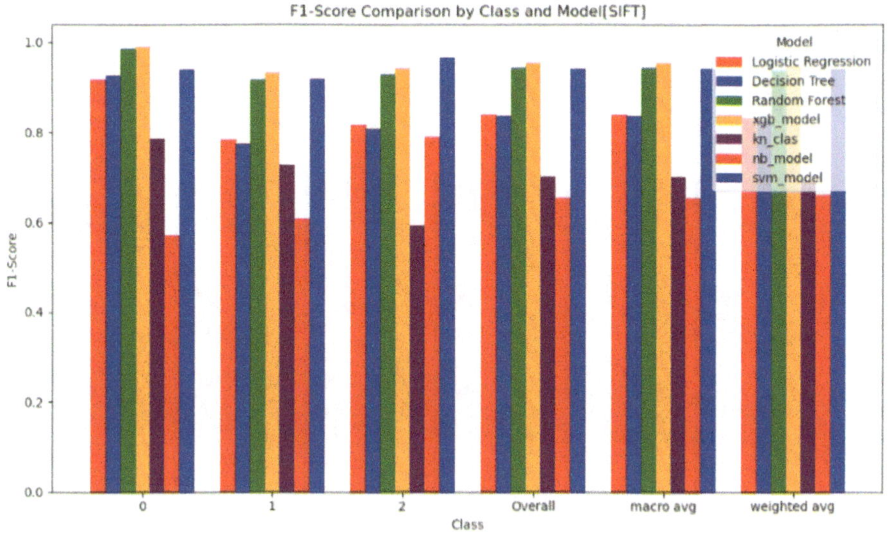


**Fig. 4.** F-1 score comparison using Pixel feature extraction method.

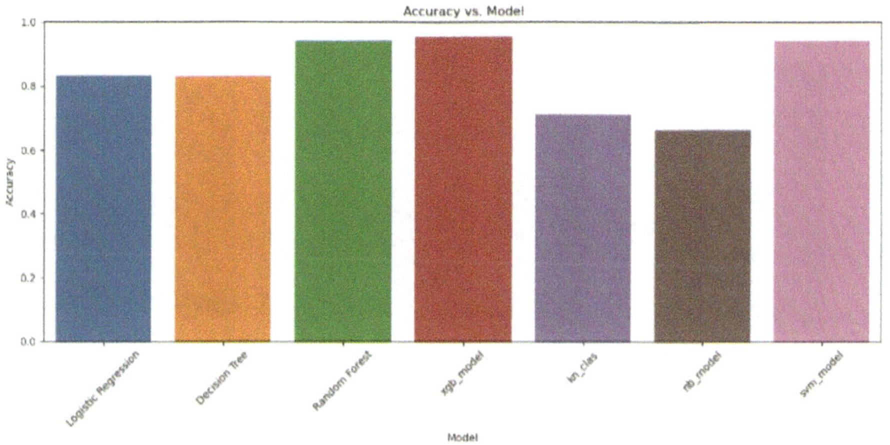


**Fig. 5.** Comparing accuracy across different models with pixel feature extraction.

These results collectively offer valuable insights into various classification techniques and feature extraction methods for potato plant disease classification. The choice of the optimal combination of these methods may depend on specific priorities, such as accuracy, computational efficiency, or interpretability, and can guide future research and applications in the fields (Table 1).

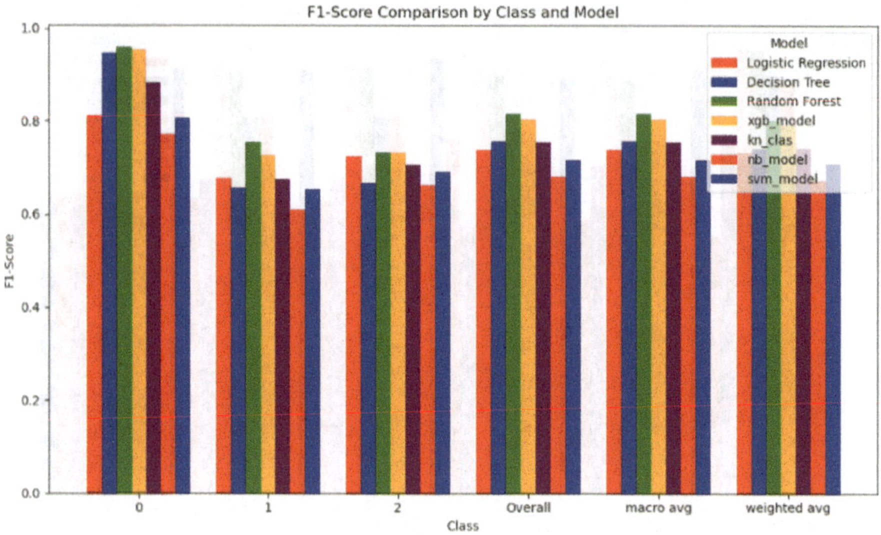


**Fig. 6.** F-1 score comparison using Local Binary Pattern feature extraction method.

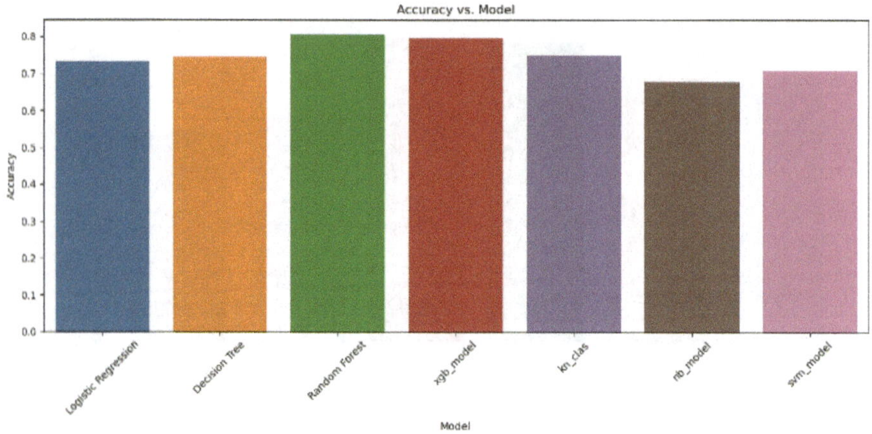


**Fig. 7.** Comparing Accuracy across different models using Local Binary Pattern feature extraction method.

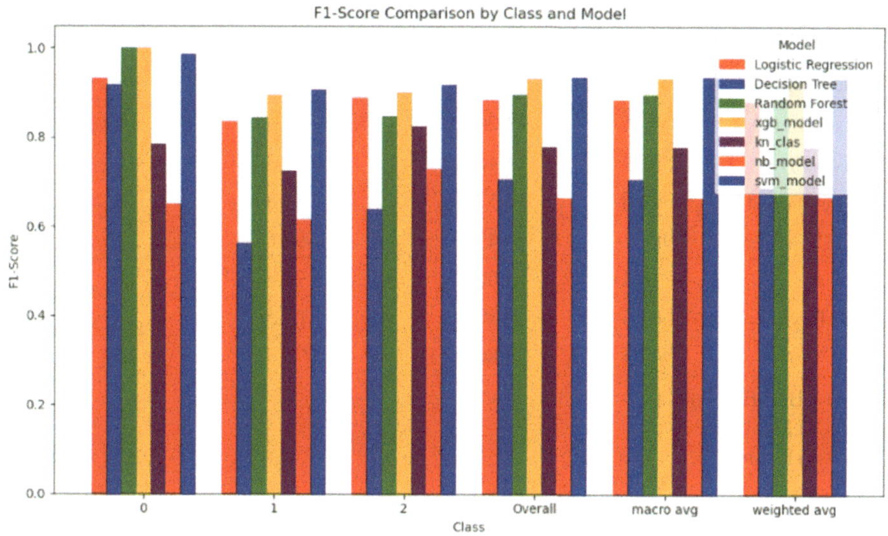


**Fig. 8.** F-1 score comparison across different models by employing the feature extraction method know as Histogram of oriented Gradient (HOG).

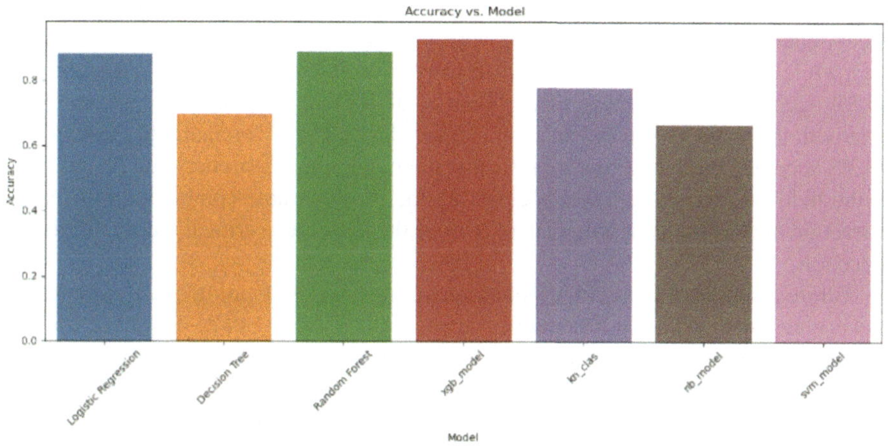


**Fig. 9.** Comparing Accuracy across different models by employing the feature extraction method know as Histogram of oriented Gradient (HOG).

**Table 1.** Accuracy (%) comparison of various ML models and feature extraction methods

| Model/method | Pixel feature | Local Binary Pattern | Histogram of oriented gradient |
|---|---|---|---|
| Logistic Regression | 83.45 | 73.20 | 88.30 |
| Decision Tree | 83.09 | 74.46 | 69.78 |
| Random Forest | 94.06 | 80.57 | 88.84 |
| XGB Classifier | 95.32 | 79.49 | 92.62 |
| KNN | 71.04 | 75.00 | 77.87 |
| Naïve Bayes | 66.36 | 67.80 | 66.54 |
| SVM | 94.24 | 70.86 | 93.34 |

Class 0 – Bird Eye Spot
Class 1 – Gray Light
Class 2 – Healthy

## 6 Conclusion

In conclusion, our experimental findings have provided valuable insights into the selection of optimal feature extraction methods and machine learning models for potato plant disease classification. The results demonstrate that the XGBoost (XGB) model exhibits exceptional performance when paired with Pixel feature extraction with 95.3% accuracy, while Random Forest emerges as the top performer for LBP feature extraction with 80.5% accuracy. Moreover, our experiments indicate that HOG feature extraction coupled with the Support Vector Machine (SVM) classifier yields remarkable results with 93.3% accuracy these findings are instrumental in guiding researchers and practitioners in the field of agriculture towards choosing the most suitable combination of feature extraction methods and classifiers to maximize the F1 score, a critical metric for disease detection.

Future research in the field of potato plant disease classification could explore several avenues such as, convolutional neural networks (CNNs) can be employed for both feature extraction and classification and also integration of remote sensing data, such as hyperspectral imaging or drones, to complement visual data potato plant varieties would be crucial for practical implementation. These avenues of future work hold great potential for advancing the field and delivering practical solutions for early disease detection and classification in potato plants and beyond.

## References

1. India production of POTATO. https://agriexchange.apeda.gov.in/India%20Production/India_Productions.aspx?cat=Vegetables&hscode=1083. Last accessed 11 Sep 2023
2. Fry, W.E.: The canon of potato science: 10. Late blight and early blight. Potato Res. **50**, 243–245 (2007). https://doi.org/10.1007/s11540-008-9046-9
3. Singh, A., Kaur, H.: Potato plant leaves disease detection and classification using machine learning methodologies. IOP Conf. Ser. Mater. Sci. Eng. **1022**, 012121 (2021). https://doi.org/10.1088/1757-899X/1022/1/012121

4. Islam, M., Dinh, A., Wahid, K., Bhowmik, P.: Detection of potato diseases using image segmentation and multiclass support vector machine. In: 2017 IEEE 30th Canadian Conference on Electrical and Computer Engineering (CCECE), pp. 1–4. IEEE, Windsor, ON (2017). https://doi.org/10.1109/CCECE.2017.7946594
5. Ramesh, S., Hebbar, R., Niveditha, M., Pooja, R., Bhat, N.P., Shashank, N., Vinod, P.V.: Plant disease detection using machine learning. In: 2018 International Conference on Design Innovations for 3Cs Compute Communicate Control (ICDI3C), pp. 41–45. IEEE, Bangalore (2018). https://doi.org/10.1109/ICDI3C.2018.00017
6. Sholihati, R.A., Sulistijono, I.A., Risnumawan, A., Kusumawati, E.: Potato leaf disease classification using deep learning approach. In: 2020 International Electronics Symposium (IES), pp. 392–397 (2020). https://doi.org/10.1109/IES50839.2020.9231784
7. Hou, C., Zhuang, J., Tang, Y., He, Y., Miao, A., Huang, H., Luo, S.: Recognition of early blight and late blight diseases on potato leaves based on graph cut segmentation. J. Agric. Food Res. **5**, 100154 (2021). https://doi.org/10.1016/j.jafr.2021.100154
8. Iqbal, Md.A., Talukder, K.H.: Detection of potato disease using image segmentation and machine learning. In: 2020 International Conference on Wireless Communications Signal Processing and Networking (WiSPNET), pp. 43–47 (2020). https://doi.org/10.1109/WiSPNET48689.2020.9198563
9. Shruthi, U., Nagaveni, V., Raghavendra, B.K.: A review on machine learning classification techniques for plant disease detection. In: 2019 5th International Conference on Advanced Computing & Communication Systems (ICACCS), pp. 281–284 (2019). https://doi.org/10.1109/ICACCS.2019.8728415
10. Potato Disease Leaf Dataset(PLD). https://www.kaggle.com/datasets/rizwan123456789/potato-disease-leaf-datasetpld. Last accessed 11 Sep 2023
11. Stearns, S.: Early blight and late blight of potato | Integrated pest management. https://ipm.cahnr.uconn.edu/early-blight-and-late-blight-of-potato/. Last accessed 10 Sep 2023
12. Due Trier, Ø., Jain, A.K., Taxt, T.: Feature extraction methods for character recognition—a survey. Pattern Recogn. **29**, 641–662 (1996). https://doi.org/10.1016/0031-3203(95)00118-2
13. Haralick, R.M., Shanmugam, K., Dinstein, I.: Textural features for image classification. IEEE Trans. Syst. Man Cybern. **SMC-3**, 610–621 (1973). https://doi.org/10.1109/TSMC.1973.4309314
14. Dalal, N., Triggs, B.: Histograms of oriented gradients for human detection. In: 2005 IEEE Computer Society Conference on Computer Vision and Pattern Recognition (CVPR'05), pp. 886–893. IEEE, San Diego, CA, USA (2005). https://doi.org/10.1109/CVPR.2005.177
15. Rish, I.: An empirical study of the Naïve Bayes Classifier. In: IJCAI 2001 Workshop on Empirical Methods in Artificial Intelligence, vol. 3 (2001)
16. LaValley, M.P.: Logistic regression. Circulation. **117**, 2395–2399 (2008). https://doi.org/10.1161/CIRCULATIONAHA.106.682658
17. Ho, T.K.: The random subspace method for constructing decision forests. IEEE Trans. Pattern Anal. Mach. Intell. **20**, 832–844 (1998). https://doi.org/10.1109/34.709601
18. Cortes, C., Vapnik, V.: Support-vector networks. Mach. Learn. **20**, 273–297 (1995). https://doi.org/10.1007/BF00994018
19. Induction of decision trees | SpringerLink. https://link.springer.com/article/10.1007/BF00116251. Last accessed 11 Sep 2023
20. Cover, T., Hart, P.: Nearest neighbor pattern classification. IEEE Trans. Inf. Theory. **13**, 21–27 (1967). https://doi.org/10.1109/TIT.1967.1053964
21. Chen, T., Guestrin, C.: XGBoost: a scalable tree boosting system. In: Proceedings of the 22nd ACM SIGKDD International Conference on Knowledge Discovery and Data Mining, pp. 785–794. Association for Computing Machinery, New York, NY, USA (2016). https://doi.org/10.1145/2939672.2939785

# GAN-Deepfakes Detection Using ELA and Deep Learning

Fatma Ben Aissa(✉), Nesrine Chaibi, Mahmoud Mejdoub, and Mourad Zaied

Research Team in Intelligent Machines (RTIM), Gabes, Tunisia
fatmabenaissa1@gmail.com, mourad.zaied@ieee.org

**Abstract.** The Generative Adversarial Network (GAN) has made significant advances in the field of deep learning, compared to other generative models. GANs have the capability to generate images that look extremely realistic, which can be exploited for deceptive purposes, such as doctoring images or videos to spread false information on social media. As GANs continue to advance, the similarity between generated and real images becomes increasingly challenging to discern. Consequently, the need to distinguish between fake images generated by GANs and real ones becomes paramount. In this paper, we suggested a new approach for the detection of fake images generated by GANs. We introduced ELA-based deep learning techniques to detect real and fake images. The images from the dataset are preprocessed using Error level analysis (ELA) to check the image compression ratio and then passed to a pre-trained convolutional neural network model (ResNet50). Based on the transfer learning principle, the obtained features will be classified using three classifiers (Softmax, SVM, and Random Forest). Two datasets of fake images generated by GANs were used to evaluate the suggested method (Real and Fake Face detection dataset and CIFAKE dataset). On both datasets, excellent results are reached. The developed approach has shown exceptional performance in the detection of fake images generated by GANs.

## 1 Introduction

The Generative Adversarial Network (GAN) [1] has made substantial strides in the domain of deep learning when compared to other generative models. As of now, GAN has been referenced in over 17,000 instances on Google Scholar. So far, it has been applied to many image subfields [2,3] including super-resolution, segmentation, steganography, translation from image to image, text-to-image synthesis, classification, and others. Generative models are used to produce samples that closely match the patterns seen in the training data. By using larger training datasets and enhancing the model's architecture, generative models can produce samples that are almost indistinguishable from real data. Researchers have noted that improving the design of neural network structures or using more precise methods for measuring differences between images can make training Generative Adversarial Networks (GANs) more reliable. As a result of these


A. Bajaj et al. (Eds.): HIS 2023, LNNS 1224, pp. 76–85, 2025.
https://doi.org/10.1007/978-3-031-78925-0_8

advancements, GANs can now create high-quality and diverse images. However, this progress also brings forth security concerns. GANs have the capability to generate images that look extremely realistic [4,5], which can be exploited for deceptive purposes, such as doctoring images or videos to spread false information on social media. As GANs continue to advance, the similarity between generated and real images becomes increasingly challenging to discern. Consequently, the need to differentiate between fake images generated by GANs and real ones becomes paramount. Recently, GAN-DeepFakes have become more popular in computer vision. Several methods for detecting fake images have been developed. Numerous deep learning-based strategies have been introduced to tackle this issue and identify fake images. The primary objective of this study was to create an effective technique for detecting fake images generated by GANs. The paper is organized as follows. Section 2 gives a number of methods applied to the detection of GAN-deepfakes. In Sect. 3, we elaborate on our proposed methodology, which relies on Error Level Analysis in conjunction with ResNet50. The different results are given in Sect. 4. In the end, we conclude the paper.

## 2 Related Works

In recent years, a number of researchers have introduced deep learning (DL) based methods for the detection of deepfake content. For instance, in their work [6], the authors presented an innovative technique aimed at discerning between authentic and manipulated videos. Their approach employed a YOLO face detector to extract facial regions from video frames. Subsequently, these extracted faces were passed through the InceptionResNetV2 convolutional neural network (CNN) model to extract visual features. These graphical features were then fed into an XGBoost classifier to distinguish between original and manipulated videos. Impressively, the proposed model attained an astounding accuracy rate of 90.73% when evaluated on the merged CelebDF-FaceForencics++ (c23) dataset.

Khalil et al. [7] proposed the iCaps-Dfake technique for identifying deepfake images and videos. This method integrated two key techniques: Local Binary Patterns (LBP) and a CNN-based modified High-Resolution Network (HRNet), along with the application of capsule neural networks (CapsNets). The authors trained their model on the DeepFakes Detection Challenge-Preview (DFDC-P) dataset and conducted testing on the DFDC-Preview and CelebDF datasets.

In the research conducted by Wang and Dantcheva [8], they investigated 3D convolutional neural network-based deep learning techniques (3DCNN), including 3D ResNet, 3D ResNeXt, and I3D, for detecting manipulated videos. Their experiments employed the Faceforensics++ dataset and focused on three distinct manipulation techniques, reporting the true classification rate (TCR) for all experiments. The study underscored the need to address various deepfake manipulation techniques in future research endeavors.

Moreover, the authors in [9] introduced a hybrid approach for the effective detection of deepfake images. This approach, referred to as pairwise learning, leveraged generative adversarial networks (GANs) for producing fake images.

Subsequently, pairwise learning was employed to capture discriminative information between fake and real images, utilizing the common fake feature network (CFFN).

Looking ahead, the research community intends to develop new deepfake detection approaches that emphasize appearance, motion, and pixel-level-based generated noise to combat manipulation algorithms effectively.

## 3 Proposed Approach

In this section, we describe our fake image detection pipeline, which is used to differentiate between real and fake images generated by GAN. We will start by applying ELA in both real and fake datasets, building a CNN model for the purpose of feature extraction and classification. Following this, the authenticated model will be employed in experiments aimed at assessing its performance. This will involve an analysis of the features extracted by the CNN, specifically from the fully connected layer of ResNet50. In the analysis, three well-established traditional classifiers, namely Support Vector Machine (SVM), Random Forest (RF), and Softmax, will be utilized for each set of features. We illustrate our model end-to-end pipeline in Fig. 1

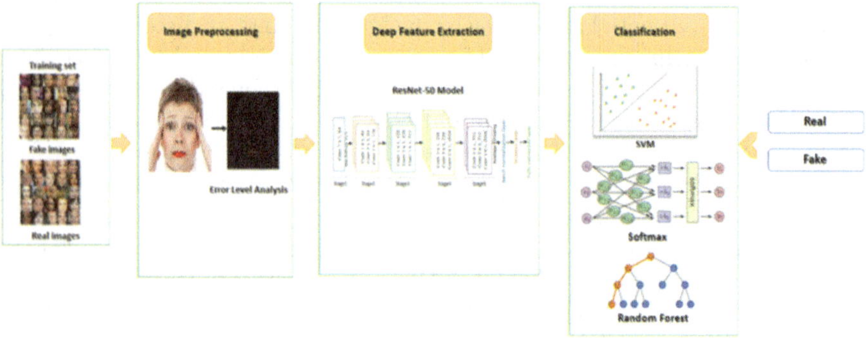


**Fig. 1.** Overview of the proposed approach.

- Data Collection: The initial stage involves gathering a dataset containing both real and fake images.
- Preprocessing: The collected real and fake images undergo preprocessing using Error Level Analysis (ELA) to highlight potential image differences or anomalies.
- Feature Extraction: Utilizing a pre-trained CNN model, specifically ResNet50 is employed to extract deep features from the preprocessed images.

- Classification: The extracted deep features are subsequently used as input into a classification process. Three different classifiers are utilized: SoftMax, Support Vector Machine (SVM), and Random Forest (RF). These classifiers aim to differentiate between real and fake images based on the learned features.
- Evaluation: The performance of the deepfake detection approach is assessed using various evaluation metrics. The results are analyzed to gauge the efficiency and effectiveness of the approach. Additionally, a comparison is made with the findings of recent studies in the field.

This structured approach encompasses data preprocessing, feature extraction, and classification, ultimately enabling the identification of deepfake images within the dataset. The evaluation metrics provide a means to quantify the performance and reliability of the detection method.

### 3.1 Image Preprocessing Using Error Level Analysis

Error Level Analysis (ELA) is a valuable tool designed to unveil manipulated regions within JPEG images [10]. This technique proves particularly useful in detecting alterations made to compressed (JPEG) images by highlighting the noise distribution that emerges after an image is recompressed at a specific compression rate. ELA was conceived by Neal Krawetz, who observed how errors propagated during the saving of JPEG images. When a portion of an image is cut out and pasted into another image, ELA often identifies this pasted section as having a more significant error, rendering it brighter than the surrounding image. While there are various implementations of the ELA algorithm, they all share a common set of steps in their execution.

In this study, the dataset originally consisted of RGB images of varying sizes. To facilitate the computational process and ensure consistency, both the real and fake images were resized to a uniform size of $200 \times 200$ pixels. Subsequently, Error Level Analysis (ELA) was employed to assess the compression ratio of both the real and fake images. ELA is particularly effective when dealing with lossy image formats, such as JPEG, which was the format used in the dataset. To apply ELA, the following steps were carried out: 1. Preprocessing: The images underwent preprocessing to prepare them for ELA. 2. Compression: Sample images were compressed at an 85% compression level. 3. Difference Calculation: The next step involved calculating the pixel-wise difference between the preprocessed images and the resaved images, both of which were in the PNG format. This process allowed for the generation of ELA images that could highlight potential variations or discrepancies in the compression ratios of the real and fake images. ELA serves as a valuable tool for identifying subtle differences between original and manipulated images, particularly in lossy image formats, which are commonly used for deepfake generation.

### 3.2 Deep Feature Extraction Using ResNet50

The architecture of the proposed pre-trained CNN model, ResNet50 [11], is structured into five Convolutional blocks (Conv blocks), pooling layers, and a Fully Connected (FC) layer. These components serve specific purposes in the model's operation:

- Convolutional and Pooling Layers: These layers are primarily responsible for feature extraction. Convolutional layers utilize local connections to detect local features within the input images, while pooling layers are used to merge similar local features into more generalized features.
- Fully Connected Layer (FC): The FC layer is employed for the final image classification stage, where it computes the output for each input image, ultimately assigning it to one of the predefined classes.

In this study, a transfer learning approach was adopted, specifically using pre-trained ResNet50. In transfer learning, the model takes advantage of the knowledge gained from training on a large dataset (in this case, ResNet50's original training dataset) and adapts it to a different task. The last three layers of the pre-trained model were fine-tuned based on the classes of the input image dataset, which is essential for domain-specific tasks like deepfake detection. Furthermore, the FC layers of ResNet50 was replaced by other classifiers, such as Support Vector Machine (SVM) or Random Forest (RF), as illustrated in Fig. 2. This transfer learning and fine-tuning approach allows the model to leverage the powerful features learned from a broader dataset while customizing the final layers to suit the specific classification requirements of the deepfake detection task. After the completion of the data collection and image pre-processing stage, the dataset is partitioned into three distinct sets: the training set, the validation set, and the testing set. The training set (a labeled dataset) trains the CNN model on a particular task, such as feature extraction, where the CNN model will generate feature vectors from the fully connected layer. After that, the feature vectors are entered into three different classifiers. The validation set serves as an impartial

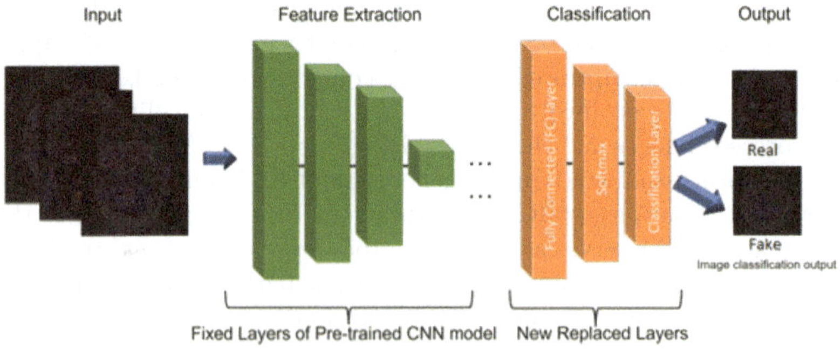


**Fig. 2.** Transfer Learning mechanism.

evaluation during the model tuning process, assessing how well the model fits the training dataset. The test set was used to evaluate the ELA-ResNet50-Softmax, ELA-ResNet50-SVM, and ELA-ResNet50-RF model approaches.

We apply a pre-trained CNN called ResNet50 using Tensorflow and Keras applications to the images instead of training a CNN from scratch, which requires a huge dataset. We selected the ResNet50 model because it is arguably the most groundbreaking work in the computer vision community in the last 5 years. ResNet enables the training of increasingly deeper networks with hundreds of layers while maintaining good performance. It won the 2015 ImageNet Large-Scale Visual Recognition Challenge (ILSVRC) and it outperformed all prior competitors and won the challenge by reducing the top-five error to 3.57%.

### 3.3 Image Classification

The CNN's fully connected (FC) layers can be replaced by other classifiers. In this study, we will assess the CNN with regard to Softmax, SVM, and RF classifiers. In the realm of deep learning, three prominent classifiers emerge as valuable tools for enhancing model performance. The Softmax classifier excels at multi-class classification by computing probability distributions over classes, making it a popular choice for the final layer in Convolutional Neural Networks (CNNs). On the other hand, the Support Vector Machine (SVM) classifier harnesses the power of hyperplane-based separation in high-dimensional feature spaces, replacing CNN's fully connected layers for efficient classification. Lastly, the Random Forest classifier, an ensemble learning method, integrates with CNN features to capture intricate relationships and optimize classification outcomes. The choice among these classifiers hinges on specific task requirements and dataset characteristics, enabling researchers to tailor their approach for optimal results.

## 4 Results and Discussion

In this section, we will present the experimental results of our study and conduct a comparative analysis with the state-of-the-art. Furthermore, We will provide a detailed description of the used dataset and the training configurations.

### 4.1 Used Dataset

To assess the robustness and performance of our approach, we conducted experiments using three diverse datasets: "Real and Fake Face Detection Dataset," and "CIFAKE," each with its own characteristics:

- Real and Fake Face Detection Dataset: This publicly available deepfake dataset was created by the Computational Intelligence Photography Lab at Yonsei University [12]. It comprises a total of 2,041 images categorized into two classes: real and fake images. The dataset includes 960 fake images and 1,081 real images. Some sample images from both categories can be seen in Fig. 3.

- CIFAKE: This dataset, introduced in [13], contains 60,000 synthetically generated images and 60,000 real images, which were collected from CIFAR-10. The dataset also consists of two classes: real and fake images.

**Fig. 3.** Examples of Real and Fake Face detection dataset.

**Fig. 4.** Some examples from CIFAKE dataset.

### 4.2 Experiments and Results

We evaluated the classification performance of the proposed model using Softmax, SVM, and RF classifiers with the Real and Fake Face Detection dataset and the CIFAKE dataset. The experiment aimed to determine the most accurate approach for the pre-trained model ELA-ResNet50. Initially, we applied transfer learning on ResNet50 using Softmax in the classifier layer. Subsequently, we tested the proposed approaches (ResNet50-Softmax, ResNet50-SVM, and ResNet50-RF). The results demonstrated that the model with the Softmax classifier outperformed SVM and RF in all performance measures. Tables 1 and 2 display the accuracy, specificity, and sensitivity of each classifier on both datasets (Fig. 4).

**Table 1.** Performance of the three classifiers in the proposed model with Real and Fake Face detection dataset

| Classifier used with ELA-Resnet50 | Accuracy | Specificity | Sensitivity |
|---|---|---|---|
| Softmax | 92% | 91% | 87% |
| SVM | 90% | 91% | 87% |
| RF | 85.7% | 88% | 89% |

**Table 2.** Performance of the three classifiers in the proposed model with CIFAKE dataset

| Classifier used with ELA-Resnet50 | Accuracy | Specificity | Sensitivity |
|---|---|---|---|
| Softmax | 93.83% | 95.69% | 92.87% |
| SVM | 91.40% | 90.42% | 92.32% |
| RF | 88.2% | 88.5% | 88% |

From Tables 1 and 2, it is evident that the proposed ELA-ResNet50 model has proven to be effective, achieving a favorable classification rate of 93.83% for the CIFAKE dataset and a classification rate of 92% for the Real and Fake Face Detection dataset. Additionally, the results indicate that the model's performance remains consistent across the three classifiers. Specifically, the Softmax classifier achieves the highest accuracy, followed by SVM as the second-best classifier, and RF as the third. This demonstrates that the proposed model's performance is robust across different datasets.

## 5 Comparison with State of the Art

In our comparison with other state-of-the-art methods (as mentioned in Table 3), it is evident that our proposed ELA-ResNet50 method outperforms the rest, particularly in terms of validation accuracy, when applied to the first dataset. Our proposed system not only delivers superior performance compared to existing methods but also utilizes lightweight and efficient models for the detection of deep fakes from images.

**Table 3.** Comparison of proposed method with existing techniques

| Ref | Year | Technique | Dataset | Acc |
|---|---|---|---|---|
| [14] | 2020 | Alex Net KNN | Real and Fake Face detection dataset | 55.8% |
| | | QIEA-FS | | 57.9% |
| | | QIEA-FS | | 58.3% |
| [15] | 2021 | ResNet-152 | | 76.79% |
| [16] | 2021 | Alex Net+KNN | | 86.8% |
| | | Alex Net+SVM | | 88.1% |
| | | Shuffle Net+KNN | | 88.2% |
| | | Shuffle Net+SVM | | 87.9% |
| **Proposed Method** | 2023 | **ELA+ResNet50+Softmax** | | **92%** |
| | | **ELA+ResNet50+SVM** | | **90%** |
| | | **ELA+ResNet50+RF** | | **85.70%** |

## 6 Conclusions and Future Work

In this research, we suggested a new approach for the detection of fake images generated by GANs. We introduced ELA-based deep learning techniques to detect real and fake images. The images from the dataset are preprocessed using ELA to check the image compression ratio and then passed to a pre-trained CNN model (ResNet50). Based on the transfer learning principle, the obtained features will be classified using three classifiers (Softmax, SVM, RF). Two datasets of fake images generated by GANs were used to evaluate the suggested method (Real and Fake Face detection dataset and CIFAKE dataset). On both datasets, excellent results are reached. Our future works will mostly focus on the integration of multiple modalities, such as audio and text, in addition to images and videos, to improve the accuracy of Deepfake detection.

## References

1. Goodfellow, I., et al.: Generative adversarial nets. In: Ghahramani, Z., Welling, M., Cortes, C., Lawrence, N., Weinberger, K.Q. (eds.) Advances in Neural Information Processing Systems, vol. 27. Curran Associates, Inc. (2014)
2. Chaibi, N., Eladel, A., Zaied, M.: Sr-Net: a super-resolution image based on DWT and DCNN. In: International Conference on Hybrid Intelligent Systems, pp. 291–301. Springer (2022)
3. Chaibi, N., Zaied, M.: A multi-level wavelet decomposition network for image super resolution. In: International Conference on Intelligent Systems Design and Applications, pp. 438–444. Springer (2022)
4. Aissa, F.B., Mejdoub, M., Zaied, M.: A survey on generative adversarial networks and their variants methods. In: Twelfth International Conference on Machine Vision (ICMV 2019), vol. 11433, pp. 1006–1012. SPIE (2020)
5. Aissa, F.B., Hamdi, M., Zaied, M., Mejdoub, M.: An overview of GAN-DeepFakes detection: proposal, improvement, and evaluation. Multimedia Tools Appl. 1–23 (2023)

6. Ismail, A., Elpeltagy, M., Zaki, M.S., Eldahshan, K.: A new deep learning-based methodology for video DeepFake detection using XGBoost. Sensors **21**(16), 5413 (2021)
7. Khalil, S.S., Youssef, S.M., Saleh, S.N.: iCaps-Dfake: an integrated capsule-based model for DeepFake image and video detection. Future Internet **13**(4), 93 (2021)
8. Wang, Y., Dantcheva, A.: A video is worth more than 1000 lies. comparing 3DCNN approaches for detecting DeepFakes. In: 2020 15Th IEEE International Conference on Automatic Face and Gesture Recognition (FG 2020), pp. 515–519. IEEE (2020)
9. Hsu, C.-C., Zhuang, Y.-X., Lee, C.-Y.: Deep fake image detection based on pairwise learning. Appl. Sci. **10**(1), 370 (2020)
10. Raković, D.N.: Error level analysis (ELA). Tehnika **78**(4), 445–451 (2023)
11. He, K., Zhang, X., Ren, S., Sun, J.: Deep residual learning for image recognition. In: Proceedings of the IEEE Conference on Computer Vision and Pattern Recognition, pp. 770–778 (2016)
12. Real and fake face detection (2019)
13. Bird, J.J., Lotfi, A.: CIFAKE: Image classification and explainable identification of AI-generated synthetic images. arXiv preprint arXiv:2303.14126 (2023)
14. Mittal, H., Saraswat, M., Bansal, J.C., Nagar, A.: Fake-face image classification using improved quantum-inspired evolutionary-based feature selection method. In: 2020 IEEE Symposium Series on Computational Intelligence (SSCI), pp. 989–995. IEEE (2020)
15. Chandani, K., Arora, M.: Automatic facial forgery detection using deep neural networks. In: Advances in Interdisciplinary Engineering: Select Proceedings of FLAME 2020, pp. 205–214. Springer (2021)
16. Rafique, R., Gantassi, R., Amin, R., Frnda, J., Mustapha, A., Alshehri, A.H.: Deep fake detection and classification using error-level analysis and deep learning. Sci. Rep. **13**(1), 7422 (2023)

# Revolutionizing Higher Education Institute Query System by Linking Custom Knowledge Base with Large Language Models

Mohammed Varaliya(✉), Mahendra Kanojia, and Subhashish Nabajja

Sheth L.U.J and Sir M.V. College, Mumbai, Maharashtra, India
mohammedvaraliya2661392@gmail.com

**Abstract.** Query systems in higher education institutions are in need of improvement to enhance communication, response times, and access to information. To address this pressing need, this research explores a novel approach to integrating large language models (LLMs) with custom knowledge bases specifically for higher education institution. We collected data from Sheth L. U. J. College of Arts and Sir M. V. College of Science and Commerce in Mumbai, India, in both structured and unstructured formats, following all higher education guidelines. Using the Llama-Index model for text embedding and OpenAI GPT-3.5 for response generation, our system achieved a remarkable average response time of 5 seconds and an average relevancy score of 0.93 across various query categories. These results unequivocally demonstrate the potential of LLM-powered systems to streamline interactions and provide exceptionally useful, relevant data in higher education settings. We strongly advocate for the implementation of these systems to significantly enhance the user experience and communication within the higher education landscape, even while acknowledging their potential cost implications. Looking ahead, our research roadmap includes the expansion of knowledge sources and the integration of open-source LLM models to further augment the effectiveness and coverage of these systems.

**Keywords:** Large language models · Chatbots · Generative pretrained transformer · Natural language processing · Question answering

## 1 Introduction

Artificial intelligence (AI) and chatbots have made significant progress in the field of natural language processing (NLP) in recent decades. Chatbots are a simple and common form of intelligent human-computer interaction, and are a typical example of an AI system Bansal and Khan [1]. Large language models (LLMs) have shown potential in many applications, creating new opportunities to solve complex problems that go beyond traditional language-based communication Tinn et al. [2]. LLMs have been used in a variety of industries, including education Kasneci et al. [3], financial news analysis Pavlyshenko [4], and public health Jo et al. [5]. In today's dynamic educational environment, effective communication and information sharing are essential for higher


A. Bajaj et al. (Eds.): HIS 2023, LNNS 1224, pp. 86–94, 2025.
https://doi.org/10.1007/978-3-031-78925-0_9

education institutions. This need arises from the wide range of people seeking information, including staff, current students, and prospective students. Enquirers often seek quick and accurate answers to their questions, which can range from the location and facilities of the institution to academics and course offerings, administration and leadership, scholarships, extracurricular activities, sports, academic support and services, career and internship opportunities, and general campus policies.

International students face unique challenges in navigating this complex information landscape. To get information and help, they would need to make appointments, adhere to the institution's hours of operation, pay for phone calls or meetings, and adjust to time zone differences. This process can be time-consuming and costly. This is where a chatbot can be very helpful. A chatbot can support real-time interactions, allowing anyone to ask questions whenever it is most convenient, regardless of their location.

LLM-based chatbots offer many benefits that higher education institutions can take advantage of. They can simplify communication processes by reducing response times and improving the relevance of the information provided. This research aims to build a responsive LLM chatbot for higher education institutions that can effectively respond to a wide range of queries. The main research goals were to increase response relevance and achieve fast response times. Through this research, we hope to improve communication channels within educational institutions and make important information easily accessible to everyone involved.

## 2 Literature Review

The capabilities and implications of Large Language Models (LLMs) have been extensively investigated in recent studies in the rapidly evolving field of Natural Language Processing (NLP). Using methods to increase their openness and security, Carlini et al. [6] investigated the effects and security issues associated with training data extraction from LLMs. Xu et al. [7] looked at methods to increase the efficiency and adaptability of LLMs in a range of NLP tasks. Tinn et al. [2] concentrated on improving strategies for the conversion of LLMs to biomedical writing, including BERT and GPT-3. Language model chatbot security issues were addressed by Deng et al. [8], who also developed a method called "JAILBREAKER" to address these dangers. With a focus on ethical and data privacy issues, Jo et al. [5] investigated the application of conversational AI using LLMs in public health intervention. Kasneci et al. [3] highlighted the advantages of personalized learning while acknowledging privacy concerns and the possibility of shallow learning. They also looked at the potential applications and challenges of ChatGPT in education. "Iinternchat," a framework that combines natural language conversation with computer vision for simplifying vision-centric tasks, was introduced by Liu et al. [9]. Chang et al. [10] discussed various methods of assessment for assessing the effectiveness and moral implications of LLMs. Dao et al. [11] examined the performance of three well-known LLMs, "Openai chatgpt," "microsoft bing chat," and "google bard," on tasks requiring natural language processing, providing details on how their performance differs. Huang et al. [12] concentrated on raising the accuracy and knowledge of multi-span question-answering models. The requirement for models that can generalize from small samples was emphasized by Brown et al. [13]. The rise of transformer-based models in

NLP and difficulties in knowledge graph matching were covered by Hertling et al. [14]. Hendy et al. [15] reviewed machine translation research with a focus on transformer-based models like GPT, and Su et al. [16] investigated optimizing text embeddings for various language comprehension applications. Chen et al. [17] made positional interpolation possible by examining limitations in the context window of big language models, enhancing long-distance connection recognition. Additionally, Heryandi [18] created a chatbot for academic record monitoring in higher education institutions, and Yang and Evans [19] investigated the advantages and disadvantages of using AI chatbots in higher education. Bill and Eriksson [20] demonstrate that reinforcement learning from human feedback (RLHF) improves the responses of a large language model (LLM) to mental health-related prompts, raising ethical considerations and potential benefits of AI in psychotherapy. These studies, taken together, improve our understanding of LLMs and their various applications.

## 3 Dataset Description

The dataset used in the proposed work is generated using the source documents gathered from Sheth L. U. J. College of Arts and Sir M. V. College of Science and Commerce, Mumbai, India, in compliance with the higher education institution's authority. Information regarding the higher education used in this study is collected from all possible sources. The collected data exists in both structured and unstructured formats.

The further process of converting this data into a Vector database is presented in Fig. 1. The source documents are first split into manageable text chunks, which are small units of text created from the source document. These chunks are then passed on to the embedding model, which creates a numerical representation of a text chunk. The numerical representations are stored in a vector database, also known as the semantic index, and are used to perform semantic searches based on the user query.

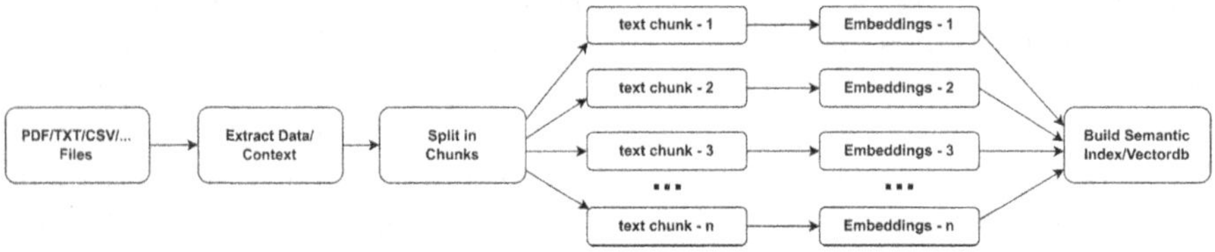


**Fig. 1.** Dataset generation flowgraph

## 4 Research Methodology

We employed two models in our research: the "llama-index" for embedding the text dataset and converting it into a vector database and the OpenAI-GPT-3.5 model "text-davinci-003" [13, 21] for generating responses. In the following section, we provide a detailed description of both models.

### 4.1 Llama Index

The Llama Index is a contextual indexing system that extends the capabilities of existing language models. It is trained on a large text dataset to build and maintain an index of passages, which enriches language models with additional context for making predictions Zirnstein [22]. To achieve this, the Llama-Index divides the input document into smaller units called indices, enabling queries to access a broader context. For Vector Store Index, embeddings of the indices are obtained by querying a Large Language Model (LLM). The prompt format may vary depending on the use case. The default Q&A prompt Zirnstein [22] is as follows:

```
"Contextual data"
"{context_str}"
"With the context information and no prior knowledge, "
{query_str}"
"respond to the query: {query_str}"".
```

The response is then synthesized using the top-k indices with the highest cosine similarity scores (see Fig. 2), following the "Create and Refine" [23] principle.

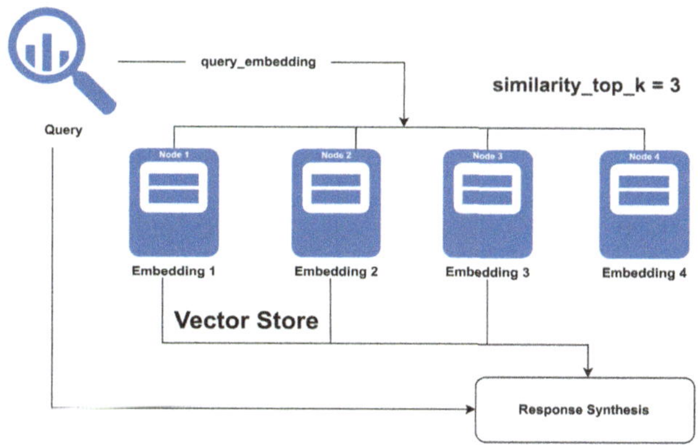


**Fig. 2.** Functionality of querying a Vector Store Index

It is possible to modify the response based on new context by passing the output of the first query to the subsequent query with the next index as context. Here is an example: Zirnstein [22].

```
"The following is the original query:"\query_str)"
"We have provided an existing answer:{existing answer}\n"
"We have the opportunity to refine the existing answer"
"(only if needed) with some more context below."
"{context_msg}"
"Given the new context, refine the original answer to better"
" answer the question. "
"If the context isn't useful return the original answer."
```

This method ensures that all potentially significant indices are considered when generating the overall response [23].

### 4.2 OpenAI GPT-3 Model "text-davinci-003"

The OpenAI GPT-3 model "text-davinci-003" is a large language model trained on an extensive dataset of text and code. It is a transformer-based model with 175 billion parameters, making it one of the most substantial and powerful language models ever created. GPT-3 is versatile and can be employed for various tasks, including text generation, translation, and question answering [33]. Transformers are renowned for their ability to capture long-range dependencies in sequential data OpenAI GPT-3: Understanding the Architecture, 2022 [24].

The GPT-3 family encompasses various models with different numbers of trainable parameters. Table 1 provides an overview of the sizes, architectures, and learning hyper-parameters of the GPT-3 models.

**Table 1.** Sizes, architectures, and learning hyper-parameters of the GPT-3 models, source: https://www.theaidream.com/post/openai-gpt-3-understanding-the-architecture

| Model name | $n_{params}$ | $n_{layers}$ | $d_{model}$ | $n_{heads}$ | $d_{heads}$ | Batch size | Learning rate |
|---|---|---|---|---|---|---|---|
| GPT-3 small | 125 | 12 | 768 | 12 | 64 | 0.5 M | $6.0 \times 10^{-4}$ |
| GPT-3 medium | 350 | 24 | 1024 | 16 | 64 | 0.5 M | $3.0 \times 10^{-4}$ |
| GPT-3 large | 760 | 24 | 1536 | 16 | 96 | 0.5 M | $2.5 \times 10^{-4}$ |
| GPT-3 XL | 1.3B | 24 | 2048 | 24 | 128 | 1 M | $2.0 \times 10^{-4}$ |
| GPT-3 2.7B | 2.7B | 32 | 2560 | 32 | 80 | 1 M | $1.6 \times 10^{-4}$ |
| GPT-3 6.7B | 6.7B | 32 | 4096 | 32 | 128 | 2 M | $1.2 \times 10^{-4}$ |
| GPT-3 13B | 13.0B | 40 | 5140 | 40 | 128 | 2 M | $1.0 \times 10^{-4}$ |
| GPT-3175B or "GPT-3" | 175.0B | 96 | 12,288 | 96 | 128 | 3.2 M | $0.6 \times 10^{-4}$ |

The OpenAI GPT-3 family of models is built on the same transformer-based architecture as the GPT-2 model, including the updated initialization, pre-normalization, and reverse tokenization, with the exception that it uses alternating dense and sparse attention patterns OpenAI GPT-3: Understanding the Architecture, (2022) [24]. The largest GPT-3175B or "GPT-3" version has a 3.2 M batch size, 96 attention layers, and 175 B Parameters. Unlike language models such as BERT Srivastava et al. [25], which use the encoder to create embeddings from raw text that can be used in other machine learning applications, the GPT family uses the decoder half, so they take in embeddings and produce text Zirnstein [22].

## 5 Proposed Models

The proposed model for revolutionizing higher education institute query systems by linking custom knowledge bases with large language models consists of two main components: a semantic index and a large language model. The semantic index is a structured repository of data designed to speed up and simplify data indexing and retrieval. It is created by dividing the information documents of the higher education institute into manageable chunks and representing each chunk as an embedding vector. Embedding vectors are numerical representations of text that capture the fundamental idea or significance of the text.

The large language model used in the proposed model is OpenAI GPT-3. GPT-3 is a powerful language model that can be used for a variety of tasks, including text generation, translation, and question answering [33].

The answer retrieval process, as shown in Fig. 3, operates as follows:

1. When an enquirer asks a question, it is run through a query embedding process using the "llama_index" model class GPTSimpleVectorIndex Zirnstein [22]. This process converts the query text into numerical vectors representations that the computer can understand.
2. The computer then looks up the query in the knowledge base/vectordb.
3. The knowledge base/vectordb returns a list of ranked results.
4. The computer sends the processed query, along with the ranked results, to the GPT-3 model (LLM Generative AI).
5. GPT-3 generates a response to the enquirer's query based on the information provided in the ranked results.
6. The computer sends the response to the enquirer.

The goal of the proposed model is to provide enquirers with the most relevant and informative answers to their questions possible.

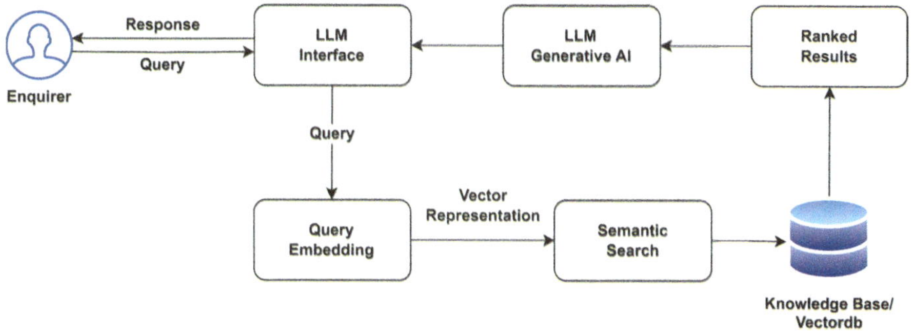


**Fig. 3.** Answer retrieval process

## 6 Results & Discussion

The proposed models for revolutionizing higher education institute query systems by linking custom knowledge bases with large language models underwent evaluation by submitting 300 queries to the model, covering various categories such as location, facilities offered, academics, course offerings, extracurricular activities, and academic support. The average response time was 5 seconds, and the average relevancy of the responses was 0.93.

As illustrated in Fig. 4, the success rate of the model for 50 queries in each category was important to note. This demonstrates the effectiveness of the model in providing relevant and informative answers to a diverse range of queries related to higher education institutions.

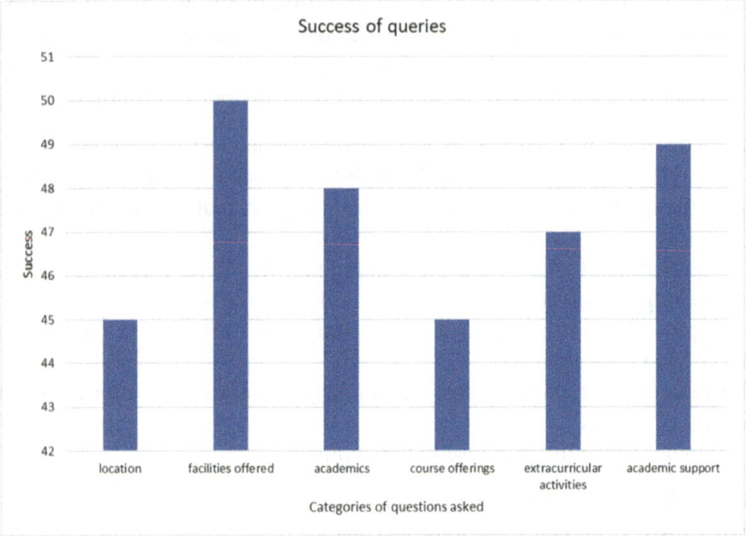


**Fig. 4.** Success of queries

Moreover, the model demonstrated the ability to respond appropriately to queries that were not relevant to the knowledge base created for the specific higher education institution under study. In such cases, the model replied with a response such as: "`It is not possible to answer this question without prior knowledge.`" This capability is crucial as it helps the system avoid generating incorrect or misleading responses, ensuring the accuracy of the information provided.

The OpenAI GPT-3 model text-davinci-003 [13, 21] was implemented using the online setup. The cost of 1k tokens, which make up the GPT-3 "text-davinci-003" model, is $0.0200 [21].

## 7 Conclusion and Recommendations

Our study presents a novel method of improving query systems in higher education institutions by linking custom knowledge bases with large language models (LLMs). The implemented system showed an impressive average response time of 5 seconds and an average relevancy score of 0.93 across various query categories. It used the Llama-Index Zirnstein [22] model for text embedding and OpenAI GPT-3.5 [13, 21] for response generation. This achievement highlights how LLM-powered chatbots have the potential to completely transform the way individuals communicate and access information in educational settings.

Going forward, we suggest expanding the system's coverage and effectiveness by investigating additional knowledge sources and integrating open-source LLM models. Furthermore, storing knowledge datasets on servers and combining data from multiple sources could improve the chatbot's functionality and increase its applicability. Our research provides a strong foundation for future advancements in responsive LLM chatbots, which will quickly and effectively provide users with precise and relevant data in a variety of industries, including healthcare and customer service in along with education.

## 8 Future Scope

Our proposed model can be expanded by hosting knowledge datasets on servers to accommodate additional context in the vector databases. The dataset used for this study was limited to a single higher educational institution. By incorporating information from various sources, we can increase the chatbot's reach by covering several educational institutions. Additionally, using more open-source LLM models can increase productivity, lower overall costs, and shorten turnaround times.

## References


1. Bansal, H., Khan, R.A.: A review paper on human computer interaction. Int. J. Adv. Res. Comput. Sci. Softw. Eng. **8**(4), 53 (2018). https://doi.org/10.23956/ijarcsse.v8i4.630
2. Tinn, R., Cheng, H., Gu, Y., Usuyama, N., Liu, X., Naumann, T., et al.: Fine-tuning large neural language models for biomedical natural language processing. Patterns. **4**(4), 100729 (2023)
3. Kasneci, E., Seßler, K., Küchemann, S., Bannert, M., Dementieva, D., Fischer, F., et al.: ChatGPT for good? On opportunities and challenges of large language models for education. Learn. Individ. Differ. **103**, 102274 (2023)
4. Pavlyshenko, B.M.: Financial news analytics using fine-tuned Llama 2 GPT model. arXiv preprint arXiv:2308.13032 (2023)
5. Jo, E., Epstein, D.A., Jung, H., Kim, Y.H.: Understanding the benefits and challenges of deploying conversational AI leveraging large language models for public health intervention. In: Proceedings of the 2023 CHI Conference on Human Factors in Computing Systems, pp. 1–16 (2023)
6. Carlini, N., Tramer, F., Wallace, E., Jagielski, M., Herbert-Voss, A., Lee, K., et al.: Extracting training data from large language models. In: 30th USENIX Security Symposium (USENIX Security 21), pp. 2633–2650 (2021)
7. Xu, R., Luo, F., Zhang, Z., Tan, C., Chang, B., Huang, S., Huang, F.: Raise a child in large language model: towards effective and generalizable fine-tuning. arXiv pre-print arXiv:2109.05687 (2021)
8. Deng, G., Liu, Y., Li, Y., Wang, K., Zhang, Y., Li, Z., et al.: Jailbreaker: automated jailbreak across multiple large language model chatbots. arXiv preprint arXiv:2307.08715 (2023)
9. Liu, Z., He, Y., Wang, W., Wang, W., Wang, Y., Chen, S., et al.: Intern-chat: solving vision-centric tasks by interacting with chatbots beyond language. arXiv pre-print arXiv:2305.05662 (2023)
10. Chang, Y., Wang, X., Wang, J., Wu, Y., Zhu, K., Chen, H., et al.: A survey on evaluation of large language models. arXiv preprint arXiv:2307.03109 (2023)

11. Dao, X.Q.: Performance comparison of large language models on VNHSGE English dataset: OpenAI ChatGPT, Microsoft Bing Chat, and Google Bard. arXiv preprint arXiv:2307.02288 (2023)
12. Huang, Z., Zhou, J., Xiao, G., Cheng, G.: Enhancing in-context learning with answer feedback for multi-span question answering. arXiv preprint arXiv:2306.04508 (2023)
13. Brown, T.B., Mann, B., Ryder, N., Subbiah, M., Kaplan, J., Dhariwal, P., et al.: Language models are few-shot learners. arXiv (Cornell University). https://arxiv.org/pdf/2005.14165.pdf (2020)
14. Hertling, S., Portisch, J., Paulheim, H.: KERMIT—a transformer-based approach for knowledge graph matching. arXiv (Cornell University). https://doi.org/10.48550/arxiv.2204.13931 (2022)
15. Hendy, A., Abdelrehim, M., Sharaf, A., Raunak, V., Gabr, M.A., Matsushita, H., Kim, Y.J., Afify, M., Awadalla, H.H. How good are GPT models at machine translation? A comprehensive evaluation. arXiv (Cornell University). https://doi.org/10.48550/arxiv.2302.09210 (2023)
16. Su, H., Kasai, J., Wang, Y., Hu, Y., Ostendorf, M., Yih, W.T., et al.: One embedder, any task: instruction-finetuned text embeddings. arXiv preprint arXiv:2212.09741 (2022)
17. Chen, S., Wong, S., Chen, L., Tian, Y.: Extending context window of large language models via positional interpolation. arXiv preprint arXiv:2306.15595 (2023)
18. Heryandi, A.: Developing Chatbot for academic record monitoring in higher education institution. IOP Conf. Ser. **879**(1), 012049 (2020). https://doi.org/10.1088/1757-899x/879/1/012049
19. Yang, S., Evans, C.: Opportunities and challenges in using AI chat-bots in higher education. In: Proceedings of the 2019 3rd International Conference on Education and E-Learning, pp. 79–83 (2019)
20. Bill, D., Eriksson, T.: Fine-tuning a LLM using reinforcement learning from human feedback for a therapy Chatbot application (2023)
21. https://openai.com/
22. Zirnstein, B.: Extended context for InstructGPT with LlamaIndex
23. Liu, J.: LlamaIndex documentation. https://gpt-index.readthedocs.io/en/latest/ (2022)
24. OpenAI GPT-3: understanding the architecture. The AI dream. https://www.theaidream.com/post/openai-gpt-3-understanding-the-architecture (2022)
25. Srivastava, H., Varshney, V., Kumari, S., Srivastava, S.: A novel hierarchical BERT architecture for sarcasm detection. In: Proceedings of the Second Workshop on Figurative Language Processing, pp. 93–97 (2020)
26. Eloundou, T., Manning, S., Mishkin, P., Rock, D.: GPTs are GPTs: an early look at the labor market impact potential of large language models. arXiv preprint arXiv:2303.10130 (2023)
27. Anand, Y., Nussbaum, Z., Duderstadt, B., Schmidt, B., Mulyar, A.: GPT4All: training an assistant-style Chatbot with large scale data distillation from gpt-3.5-turbo. GitHub (2023)
28. Wilianto, D., Girsang, A.S.: Automatic short answer grading on high school's e-learning using semantic similarity methods. TEM J. **12**(1), 297–302 (2023)
29. Pavlyshenko, B.M.: Analysis of disinformation and fake news detection using fine-tuned large language model. arXiv preprint arXiv:2309.04704 (2023)
30. https://gpt-index.readthedocs.io/en/stable/
31. Yadav, R., Mohammed, V., Kanojia, M.: AIML based automated college Chatbot system-MVLUBot. Manag. Guru J. Manag. Res. **XI**(3) (2023). ISSN 2319-2429
32. De, M., Almeida, F., Xexéo, G.: Word embeddings: a survey. arXiv (Cornell University). https://arxiv.org/pdf/1901.09069.pdf (2019)
33. Contributeurs aux projets Wikimedia: GPT-3. fr.wikipedia.org. https://fr.wikipedia.org/wiki/GPT-3 (2023)

# Towards a Deep Learning Post-traumatic Stress Disorder Dialogue System Based on Transformers

Nourchene Ouerhani[1](✉), Ahmed Maalel[1,2], Dalibor P. Drljača[3,4], and Henda Ben Ghezala[1]

[1] National School of Computer Sciences, RIADI Laboratory, University of Manouba, 2010 Manouba, Tunisia
norchene.ouerhani@ensi-uma.tn,
{ahmed.maalel,henda.benghezala}@ensi.rnu.tn
[2] Higher Institute of Applied Sciences and Technology, University of Sousse, 4003 Sousse, Tunisia
[3] Pan European University APEIRON Banja Luka, Banja Luka, Bosnia and Herzegovina
[4] University Clinical Centre of the Republic of Srpska, Banja Luka, Bosnia and Herzegovina

**Abstract.** Symptoms of post-traumatic stress disorder (PTSD) frequently occur following life-threatening emergencies such as strokes, heart attacks, and stays in the intensive care unit (ICU). In response, we are introducing "Heal-Mind", an innovative mobile application currently in active development. This application aims to harness the potential of generative artificial intelligence (AI) to provide personalized support for individuals grappling with post-emergency PTSD.

At the core of Heal-Mind lies a meticulously trained Seq2Seq Transformer model with 44.7 million parameters, which has been trained on a publicly available dataset. This generative model is specifically designed to offer empathetic and coherent responses tailored to users' unique experiences, fostering a sense of understanding and support.

While still in the developmental phase, Heal-Mind shows promising potential in addressing the distinct challenges posed by post-emergency PTSD. Users of the application will have access to an evolving range of services aimed at aiding their recovery. Notably, Heal-Mind is being developed with a strong focus on user experience, incorporating post-processing algorithms to identify contradictions, enhance coherency, and reduce repetitiveness in its responses.

As our work continues to progress, we anticipate that Heal-Mind will emerge as a mental vital tool, providing accessible and compassionate assistance to individuals navigating the complexities of post-emergency PTSD.

**Keywords:** Generative artificial intelligence · Natural language processing · Transformers · Post-Traumatic Stress Disorder · Sentiment Analysis · emergency


A. Bajaj et al. (Eds.): HIS 2023, LNNS 1224, pp. 95–104, 2025.
https://doi.org/10.1007/978-3-031-78925-0_10

## 1 Introduction

According to the National center for PTSD [11], In any given year, about 5% of U.S. Adults, or 5 out of every 100, are affected by PTSD. In 2020, an estimated 13 million Americans had experienced PTSD.

In fact, the concept of utilizing technology to provide mental health support is not new, with numerous studies highlighting its potential in increasing access to care and alleviating barriers to treatment [4]. Traditional mental health services often face limitations, such as geographical barriers, stigma, and cost, which can hinder individuals from seeking and receiving timely support [13]. Generative artificial intelligence (AI) models, particularly those employing deep learning architectures, have emerged as promising tools to bridge these gaps. These models have demonstrated success in a range of applications, including natural language understanding and generation, which are integral components of dialogue systems for mental health support [8,9].

This article introduces "Heal-Mind," a mobile application that represents an ongoing effort to harness the potential of generative AI in the context of post-emergency PTSD support. Heal-Mind's core is a Seq2Seq Transformer model designed to provide tailored, empathetic responses that resonate with users' experiences. The application aspires to provide accessible and compassionate assistance to those grappling with post-emergency trauma.

In the subsequent sections, we delve into the theoretical underpinnings of using AI dialogue systems for mental health support, outline the development of Heal-Mind and its unique generative model, discuss the potential benefits and challenges, and present preliminary insights from our ongoing work. By shedding light on the evolving landscape of AI-powered mental health interventions, this article aims to contribute to the growing discourse surrounding innovative solutions for addressing post-emergency PTSD.

## 2 Related Works

Dialogue systems, often referred to as chatbots or virtual assistants, have emerged as a transformative force in the healthcare sector, showcasing their efficacy in emergencies, healthcare guidance, and mental health support.

Dialogue systems can be divided into two main groups: those leveraging artificial intelligence and generative processes, and those relying on symbolic methods or hard coding. The pioneering application of virtual dialogue systems within mental health counseling dates back to the introduction of the ELIZA [27] program in 1964. ELIZA employed symbolic techniques to systematically process user inputs and generate responses. Contemporary innovations [8,12,14,16,17,19,21,25,26,28] employ machine learning for dialogue generation, while retaining symbolic methods for therapeutic recommendations. Transitioning to stress management, the subsequent phase of our research investigates the profound impact of virtual assistants and Natural Language Processing (NLP)[17,18,20]. Research has explored sentiment analysis models using Bidirectional Encoder Representations from Transformers (BERT)[6] to gauge the effects of external factors like emergency on people's emotional states. For example, messenger chatbots have been developed to detect and alleviate emotional stress, while others assess user depression through question-driven methodologies, contributing to mental health support. The Serena model[3], cited as an inspiration for our approach, showcases the potential of generative dialogue systems in simulating counseling sessions. Serena employs advanced NLP techniques to guide and simulate therapeutic conversations, offering users a virtual platform for discussing their concerns and receiving guidance. By adopting a generative approach, we aim to build upon Serena's foundation and tailor it to stress management scenarios.

## 3 Heal-Mind Methodology and Model

Heal-Mind is comprised of a combination of natural language processing (NLP) models and rule-based mechanisms, working in a sequential manner to derive the most appropriate responses based on input messages. The initial phase involves a prominent Seq2Seq Transformer model [7], which generates a set of potential responses. Subsequently, a selection process is conducted in the second stage, involving various specialized Transformer-based models and heuristic rules, to pinpoint the ultimate response from the pool of candidates.

Figure 1 depicts the overall architecture of Heal-Mind, consisting of three main modules: the Preprocessing module, the Generative AI model, and the Postprocessing module.

### 3.1 Preprocessing

Preprocessing involved text tokenization [10], lemmatization [1,22], lowercasing [5], and anonymization [2] to ensure user privacy and data security.

### 3.2 Generative Model Architecture:

At the heart of Heal-Mind lies a generative model established on the foundation of the Generative Transformer[23,24] trained on the IMDb sentiment dataset [15]. It is a 44.7 million parameters Transformer architecture with 12 encoder layers, 12 attention heads per layer and 768 units of hidden size.

The Transformer model, known for its sequence-to-sequence learning capabilities, is adapted to capture the nuances of therapeutic conversations. The model's self-attention mechanism is optimized to contextualize each response within the context of preceding dialogue turns, fostering coherence and relevancy.

### 3.3 Model Training and Fine-Tuning

During this process, the model learns to generate empathetic and supportive responses that resonate with users' experiences. Fine-tuning involves minimizing a loss function that considers relevance, coherence, and context. The resulting

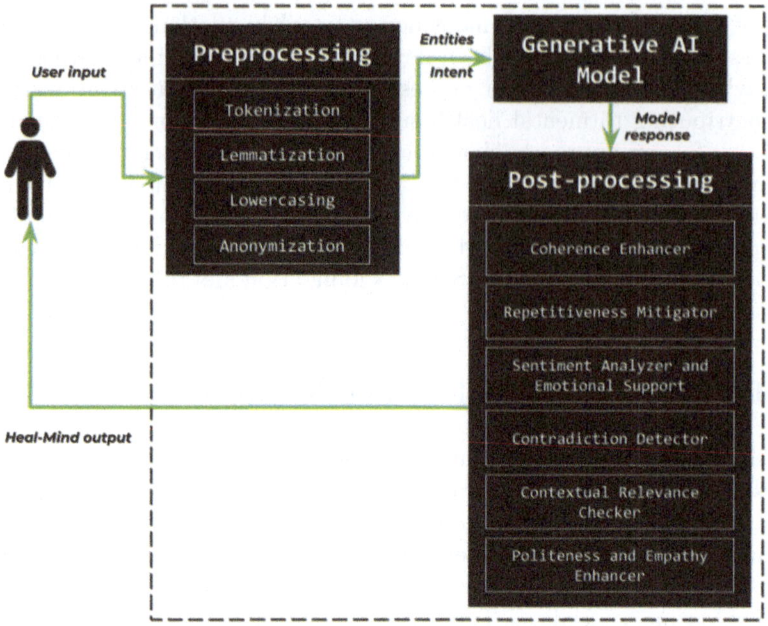


**Fig. 1.** Overall architecture of Heal-Mind.

model is thus tailored to address the unique challenges posed by post-emergency PTSD.

Fine-tuning involves adapting the pre-trained model to the specific task of providing post-emergency PTSD support. The fine-tuning process consists of several key components:

- **Loss Function:** We used the cross-entropy loss because for dialogue generation tasks like Heal-Mind, it is the common loss function, that quantifies the dissimilarity between the predicted probability distribution over tokens and the true distribution.
- **Optimizer:** Adam optimizer is chosen to adjust the model's parameters during training based on the computed loss. It adjusts the weights of the model's layers using gradients obtained through back-propagation.
- **Learning Rate Schedule:** to determine how the learning rate changes during training.
- **Gradient Clipping:** It is applied to prevent exploding gradients. This technique limits the size of the gradients during back-propagation, ensuring stable updates to the model's parameters.

Once training is complete, the fine-tuned model is ready for deployment as part of the Heal-Mind application, offering tailored and empathetic support to individuals dealing with post-emergency PTSD.

### 3.4 Post-processing Algorithms

After receiving the generated response from our AI model, Heal-Mind incorporates the following post-processing algorithms to enhance the quality of its responses:

**Coherence Enhancer.** The coherence enhancer algorithm focuses on improving the logical flow and coherence of the dialogue. It addresses any issues related to abrupt topic shifts, disjointed responses, or instances where the generated text lacks smooth transition. This algorithm analyzes the conversation history, identifies potential areas of incoherence, and employs strategies such as rephrasing, reordering, or incorporating bridging phrases to create more cohesive and contextually connected responses.

**Repetitiveness Mitigator.** Repetitive responses can negatively impact user engagement and experience. That's why we developed and applied repetitiveness mitigator algorithm to monitor the dialogue history to identify instances where the model generates redundant or overly similar responses. It employs techniques like response diversity promotion, where the model is encouraged to explore different ways of expressing similar ideas, thereby providing a more engaging and varied conversation.

**Sentiment Analyzer and Emotional Support Module.** In the context of PTSD support, understanding and addressing user emotions are paramount. A sentiment analyzer, based on the work we did in [17], is employed to assess the user's emotional state based on their messages. The emotional support module then tailors the AI's responses to provide appropriate emotional support, empathy, and validation. For instance, if a user expresses distress or sadness, the algorithm ensures that the AI response is adapted, empathetic and sensitive to their emotional needs.

**Contradiction Detector.** To enhance the credibility and reliability of our AI-generated responses, a contradiction detector algorithm is used. It scans the conversation history to identify any contradictions or inconsistencies between the user's statements and the AI's previous responses. If a contradiction is detected, the algorithm aims to resolve it by clarifying the context or seeking additional information, thus ensuring that the AI's interactions remain coherent and accurate.

**Contextual Relevance Checker.** The contextual relevance checker algorithm assesses the generated response's alignment with the ongoing conversation context. It evaluates whether the AI's output adequately addresses the user's current concerns, questions, or emotional state. If the generated response appears tangential or disconnected, the algorithm triggers corrective measures to ensure that the AI maintains relevance to the ongoing dialogue.

**Politeness and Empathy Enhancer.** Building a sense of empathy and rapport is essential in mental health support interactions. Our politeness and empathy enhancer algorithm evaluates the tone and phrasing of the AI-generated responses. It employs strategies to ensure that the AI communicates in a respectful, and supportive manner, fostering a comfortable and therapeutic conversation environment.

These post-processing algorithms work synergistically to refine and optimize the AI-generated responses, ensuring that Heal-Mind provides meaningful, contextually appropriate, and emotionally supportive interactions to users seeking PTSD support.

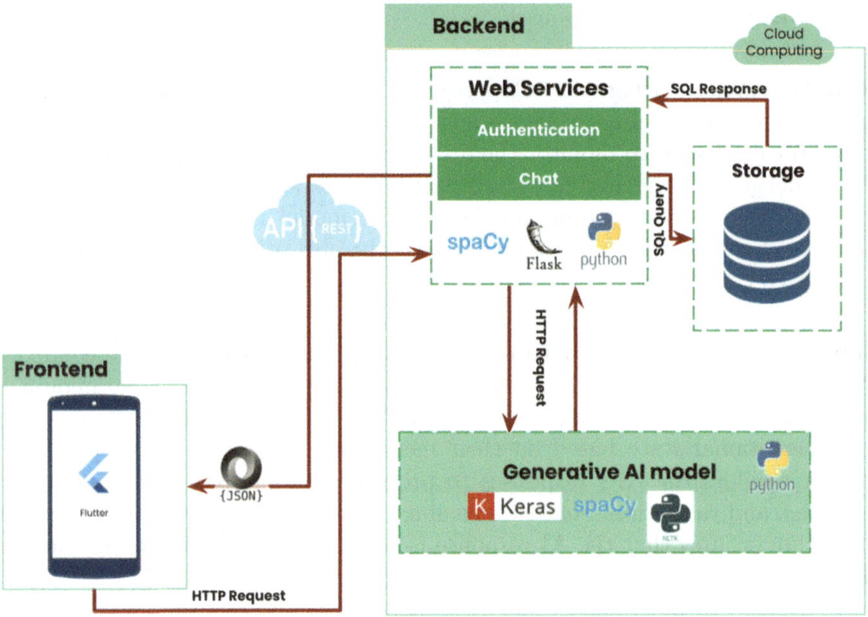


**Fig. 2.** Heal-Mind data workflow.

**Table 1.** Results

| Model | Accuracy | F1-score |
|---|---|---|
| LSTM | 62.4% | 77.75% |
| Bert | 90.62% | 80.23% |
| RoBERTa | 92.3% | 89.31% |
| GPT-2 | 91.5% | 90.7% |
| Our Model | 94.3% | 93.78% |

### 3.5 Evaluation and Implementation

We evaluated our model's results on the basis of F1-score and accuracy. After training and testing our model we got 94.3% of accuracy and 93.78 % of F1-score on a validation set (Table 1).

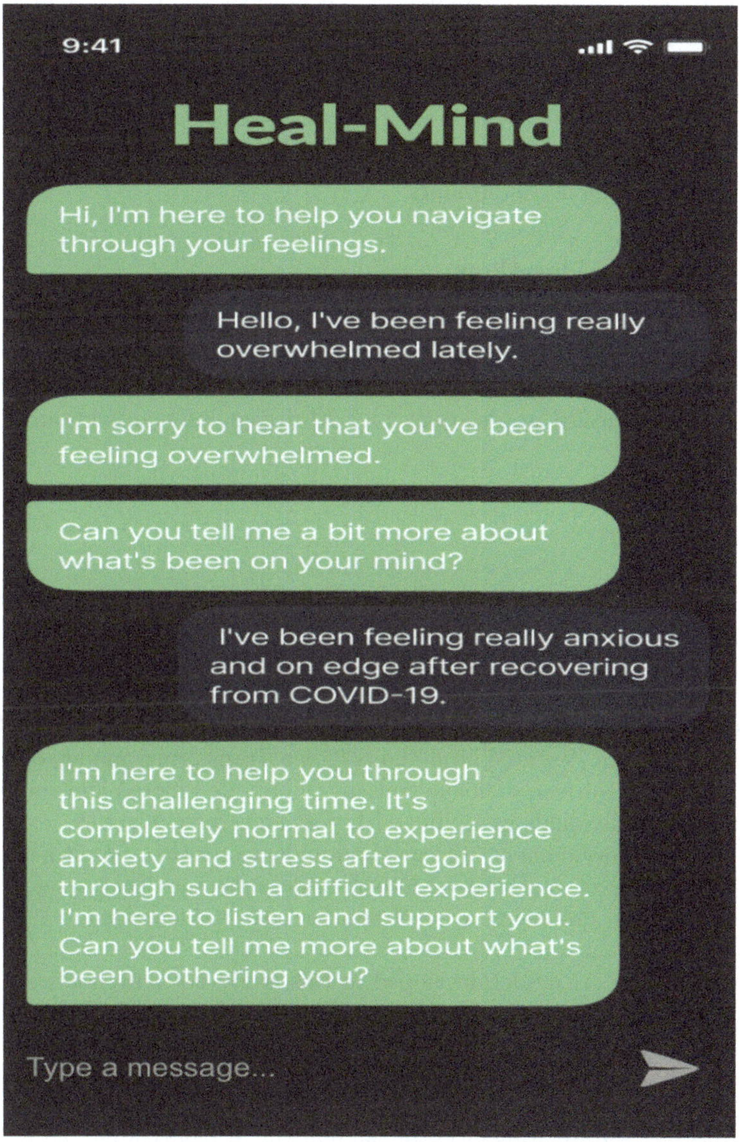


**Fig. 3.** Conversation example with Heal-Mind.

Heal-Mind is a mobile application and currently in the developmental phase, with ongoing efforts to ensure its effectiveness, user-friendliness, and ethical considerations. As part of the development process, the deployment and user accessibility aspects are carefully designed to provide a controlled and informative experience to users seeking post-emergency PTSD support.

Figure 2 illustrates the Data Flow of our mobile application. This visual representation provides an overview of how data is processed, transferred, and utilized within the application's framework. The architecture showcases the intricate interplay of various components, highlighting the seamless flow of information and interactions that contribute to the functionality and user experience of the application.

Figure 3 shows an example of a conversation with Heal-Mind showing a good understanding of the user's prompts.

### 3.6 Ethical Considerations

Heal-Mind places paramount importance on ethical considerations, adhering to strict privacy and data protection measures. User interactions are anonymized and encrypted, with no personally identifiable information stored. Ethical guidelines for mental health applications and AI-driven interventions are followed rigorously.

## 4 Conclusion

The profound impact of the emergency on global societies has underscored the imperative for innovative approaches to address the burgeoning mental health challenges arising in its wake. The intersection of artificial intelligence and mental health care presents a promising avenue for scalable, accessible, and empathetic solutions. As demonstrated in this article, the development of Heal-Mind, a mobile application centered around a Seq2Seq Transformer model, exemplifies the potential of AI-driven dialogue systems to provide tailored and compassionate support to individuals grappling with post-emergency trauma.

The ongoing development of Heal-Mind underscores the commitment to harnessing the power of generative AI to provide assistance that resonates with users' experiences, promising a novel approach to post-emergency PTSD support.

The generative AI model developed in this study not only holds significant potential for addressing post-emergency PTSD but also presents an adaptable solution applicable to a broader spectrum of emergency cases. While our focus has been on harnessing the power of AI to provide tailored and empathetic support for individuals grappling with the aftermath of the emergency, the underlying principles and functionalities of the model can be extrapolated to various emergency contexts.

However, the journey toward integrating AI into mental health care is not without challenges. Ethical considerations, data privacy, and the delicate balance between AI-driven and human-led interventions necessitate careful navigation. Despite these hurdles, the preliminary insights gleaned from our ongoing work with Heal-Mind are encouraging, suggesting the potential benefits of AI-powered dialogue systems in complementing traditional therapeutic methods. As the field continues to evolve, it is essential to foster interdisciplinary collaboration to ensure the responsible and effective implementation of AI in mental health care.

## References


1. Balakrishnan, V., Lloyd-Yemoh, E.: Stemming and lemmatization: A comparison of retrieval performances (2014)
2. Bayardo, R.J., Agrawal, R.: Data privacy through optimal k-anonymization. In: 21st International Conference on Data Engineering (ICDE'05), pp. 217–228. IEEE (2005)
3. Brocki, L., Dyer, G.C., Gładka, A., Chung, N.C.: Deep learning mental health dialogue system (2023)
4. Bults, M., van Leersum, C., Olthuis, T., Bekhuis, R., Ouden, M.: Mobile health apps for the control and self-management of diabetes type 2: A qualitative study on users' acceptability and acceptance (preprint) (2022). https://doi.org/10.2196/preprints.41076
5. Camacho-Collados, J., Pilehvar, M.T.: On the role of text preprocessing in neural network architectures: An evaluation study on text categorization and sentiment analysis. arXiv preprint arXiv:1707.01780 (2017)
6. Devlin, J., Chang, M., Lee, K., Toutanova, K.: BERT: pre-training of deep bidirectional transformers for language understanding. CoRR abs/1810.04805 (2018). http://arxiv.org/abs/1810.04805
7. Egonmwan, E., Chali, Y.: Transformer and seq2seq model for paraphrase generation. In: Proceedings of the 3rd Workshop on Neural Generation and Translation, pp. 249–255 (2019)
8. Fitzpatrick, K.K., Darcy, A., Vierhile, M.: Delivering cognitive behavior therapy to young adults with symptoms of depression and anxiety using a fully automated conversational agent (Woebot): a randomized controlled trial. JMIR Mental Health **4**, e19 (2017). https://doi.org/10.2196/mental.7785
9. Garcia-Vidal, C., Sanjuan, G., Puerta-Alcalde, P., Moreno-García, E., Soriano, A.: Artificial intelligence to support clinical decision-making processes. EBioMedicine **46** (2019). https://doi.org/10.1016/j.ebiom.2019.07.019
10. Grefenstette, G.: Tokenization. In: Syntactic wordclass tagging, pp. 117–133. Springer (1999)
11. Hamblen, J., Barnett, E.: PTSD: National center for PTSD. Behavioral Medicine, pp. 366–367 (2018)
12. Hanai, T., Ghassemi, M., Glass, J.: Detecting depression with audio/text sequence modeling of interviews, pp. 1716–1720 (2018). https://doi.org/10.21437/Interspeech.2018-2522
13. Knaevelsrud, C., Maercker, A.: Internet-based treatment for PTSD reduces distress and facilitates the development of a strong therapeutic alliance: a randomized controlled clinical trial. BMC Psychiatry **7**, 13 (2007). https://doi.org/10.1186/1471-244X-7-13

14. Lee, C.W., Wang, Y.S., Hsu, T.Y., Chen, K.Y., Lee, H.Y., Lee, L.S.: Scalable sentiment for sequence-to-sequence chatbot response with performance analysis. In: 2018 IEEE International Conference on Acoustics, Speech and Signal Processing (ICASSP), pp. 6164–6168 (2018). https://doi.org/10.1109/ICASSP.2018.8461377
15. Maas, A.L., Daly, R.E., Pham, P.T., Huang, D., Ng, A.Y., Potts, C.: Learning word vectors for sentiment analysis. In: Proceedings of the 49th Annual Meeting of the Association for Computational Linguistics: Human Language Technologies, pp. 142–150. Association for Computational Linguistics, Portland, Oregon, USA (2011). http://www.aclweb.org/anthology/P11-1015
16. Malik, T., Ambrose, A.J., Sinha, C.: Evaluating user feedback for an artificial intelligence-enabled, cognitive behavioral therapy-based mental health app (Wysa): qualitative thematic analysis. JMIR Hum. Factors **9**(2), e35668 (2022). https://doi.org/10.2196/35668
17. Ouerhani, N., Maalel, A., Ben Ghezala, H.: SMAD: smart assistant during and after a medical emergency case based on deep learning sentiment analysis. Cluster Comput. **25** (2022). https://doi.org/10.1007/s10586-022-03601-7
18. Ouerhani, N., Maalel, A., Ben Ghézala, H.: Towards a French virtual assistant for COVID-19 case psychological assistance based on NLP, pp. 199–207 (2022)
19. Ouerhani, N., Maalel, A., Ben Ghézela, H.: SPeCECA: a smart pervasive chatbot for emergency case assistance based on cloud computing. Clust. Comput. **23**, 2471–2482 (2020)
20. Ouerhani, N., Maalel, A., Ghezala, H.B.: Towards a smart pervasive conversational agent for COVID-19 psychological assistance based on NLP, pp. 59–63 (2022)
21. Ouerhani, N., Maalel, A., Ghézela, H.B.: Towards a chatbot based smart pervasive healthcare medical emergency cases, pp. 149–153. Digital Health in Focus of Predictive, Preventive and Personalised Medicine pp (2020)
22. Plisson, J., Lavrac, N., Mladenic, D., et al.: A rule based approach to word lemmatization **3**, 83–86 (2004)
23. Radford, A., Wu, J., Child, R., Luan, D., Amodei, D., Sutskever, I.: Language models are unsupervised multitask learners (2019). https://api.semanticscholar.org/CorpusID:160025533
24. Vaswani, A., et al.: Attention is all you need (2023)
25. Wang, J., Yu, L.C., Lai, K., Zhang, X.: Dimensional sentiment analysis using a regional CNN-LSTM model, pp. 225–230 (2016). https://doi.org/10.18653/v1/P16-2037
26. Wang, Y., Huang, M., Zhu, X., Zhao, L.: Attention-based LSTM for aspect-level sentiment classification, pp. 606–615 (2016). https://doi.org/10.18653/v1/D16-1058
27. Weizenbaum, J.: Eliza—a computer program for the study of natural language communication between man and machine. Commun. ACM **9**(1), 36–45 (1966). https://doi.org/10.1145/365153.365168
28. Yenter, A., Verma, A.: Deep CNN-LSTM with combined kernels from multiple branches for IMDB review sentiment analysis. In: 2017 IEEE 8th Annual Ubiquitous Computing, Electronics and Mobile Communication Conference (UEMCON), pp. 540–546 (2017). https://doi.org/10.1109/UEMCON.2017.8249013

# Human Pose Estimation for Action Recognition in Sports Video Using GNN

Nozha Jlidi(✉), Olfa Jemai, and Tahani Bouchrika

Research Team in Intelligent Machines (RTIM), National Engineering School of Gabes (ENIG) University of Gabes, 6029 Gabes, Tunisia
nozha.aljlidi@isimg.tn, olfa.jemai@isimg.tn

**Abstract.** Human action recognition in sports is an important research area that contributes to the well-being and performance analysis of individuals. During the preceding years, the recognition of human movements and gestures in sports has become essential for evaluating, guiding, and improving athletic performance. It aids in identifying injuries and potential physical issues in the human body. In this paper a novel approach is presented based on two main phases; first phase which is features extraction and representation in which human pose estimation were detected by an algorithm called MediaPipe, once we obtained these poses, we contributed by representing these poses into graphs where nodes represent the coordinates of the poses and edges are the links between them. In the second steps we used Graph Neural Networks to classify these activities. The developed approach was tested on Olympics sports dataset and has shown exceptional performance in recognizing human activities, surpassing the current standards in human action recognition.

## 1 Introduction

Recognition of human actions in sports videos is a significant area of research and development [1,2]. It involves identifying and understanding the various movements and gestures performed by athletes during sporting events. This recognition process aims to analyze and interpret the actions, enabling applications such as performance evaluation, activity monitoring, injury prevention, and tactical analysis. Recognizing human activities in videos is divided of two phases. First phase which is, features extraction and representation: this step play a key role in action recognition tasks. In which meaningful features are extracted from data (videos, frames, sensors...) and representing it in a format that is suitable for analysis and classification. In literature, numerous techniques exist for extracting and representing features each with its own strengths and limitations such as Histogram-based features, such as Histogram of Oriented Gradients (HOG), efficiently capture local image or video motion information by binning gradient orientations into histograms. This makes them suitable for representing spatial details in static images or videos [3]. Furthermore, optical flow-based features involve computing dense motion vectors between consecutive frames,


A. Bajaj et al. (Eds.): HIS 2023, LNNS 1224, pp. 105–114, 2025.
https://doi.org/10.1007/978-3-031-78925-0_11

provides a means of representing the temporal aspects of the action. Examples of optical flow-based techniques include Motion History Images (MHI) [4] and dense Trajectories. Moreover, 3D Convolutional Neural Networks (3D CNNs) which are a class of deep learning models that extend traditional 2D CNNs to include temporal information by processing video data as spatiotemporal volumes. They automatically capture spatial and temporal features, demonstrating iremarkable efficacy in tasks related to recognizing actions [5]. Lastly, pose-based features focus on extracting human pose information from images or videos [6,7]. Pose estimation algorithms, such as OpenPose [8] PoseNet [9,10] and MediaPipe [11] can detect key body joints and their spatial relationships. The derived pose data can serve as features for the identification of actions.

In the first step of the proposed approach, we used the MediaPipe algorithm to detect human skeleton from videos. Once these skeletons (poses) where detected we extracted (x,y) coordinates of each articulations to represent them as nodes and edges. The features extraction and representation phase will be provided in Sect. 3.1. On the other hand, in recent years, activity classification phases have increasingly involved the use of deep learning models [12,13]. These models have demonstrated exceptional performance in tasks related to action recognition, and we can mention some notable existing models as example Convolutional Neural Networks (CNNs) are the favored option for visual-based action recognition, as they extract features from images through different CNN layers. Recurrent Neural Networks (RNNs), especially variants like Long Short-Term Memory (LSTM), excel in modeling temporal dependencies in sequential data, making them well-suited for capturing the dynamics and context of actions over time. Additionally, pre-trained models and transfer learning, which involve models trained on large-scale image or video datasets like ImageNet, have become common practice. Fine-tuning these models on action recognition datasets with limited labeled data can significantly enhance performance and eliminate the necessity of starting the training process anew. Furthermore, Graph Neural Networks (GNNs) have recently emerged as a promising approach for activity classification. Since GNNs are specifically made to analyze data arranged in a graph, action recognition tasks can benefit from their use in modeling the relationships between entities. For the second phase which is activity classification, we used the GNN for classifying the activity played by the human. Furthermore, the graph in GNN is constructed to represent the relationships between different components involved in an action, such as body joints or objects. Each node in the graph corresponds to a component, and the edges represent the connections between them. As we mentioned previously, nodes will represent the articulations of the human skeletons in a given frame and edges are links between nodes (frames). To sum up the contributions of these articles can be presented as follows:

- We developed an approach based on human poses, we extracted skeleton coordinates from videos and represented them as graphs where nodes (frame) are the key points of each skeleton and edges presents the links between these nodes.

- The constructed graph was classified using the GNN concept to predict the action played by the human.
- The developed approach was tested on the Olympic Sports dataset and showed comparable performance to state-of-the-art action recognition.

The rest of the paper is structured as follows: Sect. 2 offers an extensive overview of the relevant research conducted in the domain of action recognition. In Sect. 3, we present the methodology used in our study, feature extraction and representation techniques, and the classification model employed. Section 4 showcases the experimental results and analysis. We assess the effectiveness of our proposed approach using the Olympics Sports dataset and compare it with existing methods, showcasing its effectiveness in accurately classifying activities. Finally, in Sect. 6, we will wrap up the paper by summarizing the primary contributions of the developed approach.

## 2 Related Works

Action recognition refers to the physical movements performed by humans. It includes a varied range of activities, such as walking, running, jumping, dancing, and various sports related movements. Human actions are observable behaviors that involve the coordinated activation of muscles and body parts to achieve a specific goal. In the context of activity recognition, the goal is to analyze and classify these human actions. In this section, we will present an overview of the relevant literature in the field of activity recognition. We will discuss the different approaches and methodologies employed by previous researchers. Shiwei Zhang, Changxin Gao, Jing Zhang, Feifei Chen and Nong Sang [14] proposed a method called Maximum Margin Part (MMP) and a recursive framework for selecting the most discriminative action parts in videos for human action recognition. They showed that choosing a discriminative subset of parts based on their correlation is an effective approach. Additionally, they used a recursive part elimination (RPE) scheme to refine candidate parts and address candidate clutter. However, because their method uses the original videos directly and may be lacking in semantic information because of the low-level descriptors that were collected, it could have limitations.

Heng Wang and Cordelia Schmid [15] presented an approach to enhance dense trajectories by accurately estimating camera motion. They removed background trajectories and applied a robust estimated holography to wrap optical flow. The developed approach was tested on four challenging action datasets (i.e., Hollywood2 [16], HMDB51 [17], Olympic Sports [18] and UCF50 [19]) which significantly exceeds the current state of the art. Motion keypoint Trajectory (MKT) and Trajectory-Based Covariance (TBC) descriptor for human action recognition were designed by Yun Yi and anli Wang [20]. The MKT approach monitors motion keypoints across various spatial scales and the TBC descriptor based on the covariance matrix representation of trajectory captures the linear relationships between the derivations of dense optical flow. Ghazal Shamsipour, Jamshid Shanbehzadeh, and Abdolhossein Sarrafzadeh [21] utilized three randomly selected frames instead of employing all the frames and Convolutional Neural Network to extract relevant characteristics and recognize the video objects. They obtained respectively 93.1% and 87.5% on Olympic Sports and UCF Sports Action datasets. Ke Xu et al. [22] presented two-stream dictionary in their architecture. First, they used Interest Patches (IP) detectors which is based on pixel values, optical flow and gradients, and other types of detectors to extract human contours. Then they calculated the similarities between sequences. In [10] introduced the PTLHAR approach for human activity recognition (HAR). The suggested approaches uses joints of the human body, and PoseNet algorithm [9] is employed to extract the human skeleton. The coordinates of each joint are derived from the skeleton. In the activity classification phase, a transfer learning model is utilized to classify and recognize the activities. The performance of the developed system was evaluated on two datasets, KTH [23] and RGBD-HudAct [24].

## 3 Proposed Approach

The approach we propose, as outlined in Fig. 1 is divided into two phases: features extraction and representation and activity classification. In this section, we will describe and elaborate on the methodology of the developed approach. We will provide explanation of the process, highlighting the key components and techniques utilized in our approach. The details will encompass the feature extraction and representation, and activity classification.

### 3.1 Features Extraction and Representation

Human pose estimation involves the detection and estimation of the positions of different body parts in an image or video. These body parts include ankles, elbows, head, hips, knees, shoulders, and wrists. There are various methods available for extracting human pose, including PosNet [9] and MediaPipe [11]. In this section, we will focus on the first phase, which is data features extraction and representation. To accomplish this phase, we extracted the human pose for each frame of the video. For this purpose, we employed the MediaPipe framework. Human pose estimation has wide-ranging applications in computer vision such as human-computer interaction, action recognition, and surveillance. In our study, we utilized the MediaPipe framework to extract human poses from videos. MediaPipe [11] is an open-source framework that proposes developers a comprehensive platform for making real-time multi-modal machine learning pipelines. By leveraging the MediaPipe algorithm [11], we were able to predict the positions of 33 pose landmarks, such as the nose, right index finger, and left knee, as well as the background segmentation mask, on the entire body from RGB video frames (Fig. 2).

After acquiring the skeletal data predicted by the MediaPipe algorithm, our next step involved extracting the (x, y) coordinates of every articulation point (key point) within each frame of the input video. This process results in a matrix

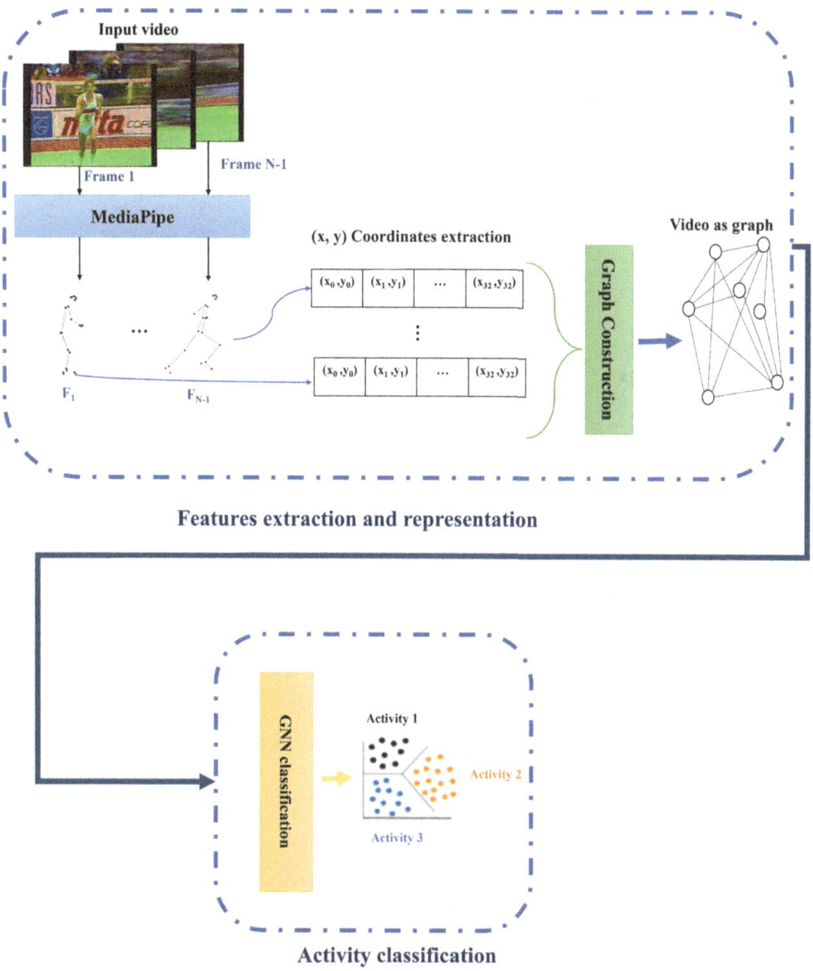


**Fig. 1.** Overview of the proposed approach.

consisting of 66 distinct characteristics. These obtained features were then meticulously preserved in a CSV file. In this file, rows corresponded to the number of frames in the video, and columns encompassed the extracted features, the associated class label, and a unique video identifier (a total of 68 columns). After the feature extraction stage, we proceeded to the representation step, wherein we visualized the extracted features as a graph. Within this graph, each node encapsulated the features extracted from a specific video frame, while the edges symbolized connections between these frames. Notably, we linked each frame to its respective video and its corresponding class label.

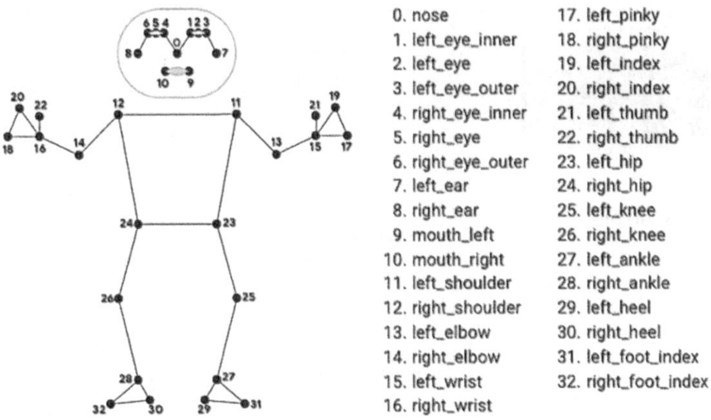


**Fig. 2.** Poses predicted by MediaPipe [19].

### 3.2 Activity Classification

Activity classification involves the identification of specific activities from a predefined set of classes, such as walking, running, cycling, or hand clapping. In recent years, deep learning has emerged as a prominent field for activity classification. As mentioned earlier, we employed Graph Neural Networks (GNNs) to classify the skeleton coordinates extracted from MediaPipe [11].

## 4 Results and Discussion

In this section, We will exhibit the experimental results of our study and conduct a comparative analysis with the state-of-the-art. Furthermore We will provide a detailed description of the used dataset, and the training configurations.

### 4.1 Used Dataset

To assess the efficiency of our suggested method, we used the Olympic Sports action dataset. The Olympic Sports Dataset contains a variety of video sequences featuring athletes practicing 16 different sports as shown in Fig. 3.

Each class has videos that are taken from YouTube, and they have been realistically annotated with the help of Amazon Mechanical Turk. The shots within the videos show changing camera movements and distance from the object, as well as occlusions.

### 4.2 Training Configurations

In this experimental section, we utilized Google Colab for our research within the computational framework. The primary aim was to highlight the significance of our proposed methodology by assessing its performance, specifically

**Fig. 3.** Samples of Olympics Sports dataset.

its ability to accurately classify various human activities. The experiments were conducted on the Olympic Sports dataset [18], a robust benchmark for activity recognition. Testing spanned 120 epochs, optimized through rigorous trial and error. We employed the Adam optimization algorithm, a well-established choice in deep learning. The results, depicted in Fig. 4, showcase the effectiveness of our approach through loss and accuracy curves. These results serve as compelling evidence of the potential of our system in advancing the field of activity recognition.

**Fig. 4.** Loss and accuracy vs epochs.

## 5 Comparison with State of the Art

The core of our approach lies in two fundamental components. Firstly, we meticulously extract and represent human pose coordinates using the MediaPipe algorithm. This step is crucial as the accuracy and quality of pose representation significantly influence the overall performance of action recognition. Secondly, we applied Graph Neural Networks (GNN) for action classification, a method designed to handle data structured as graphs, such as our pose representations. The remarkable outcome of our evaluation is the achieved accuracy 98.23% on the Olympic Sports Dataset. This level of accuracy underscores the effectiveness of our approach in precisely classifying human actions in video sequences as presented in Table 1.

**Table 1.** Descriptive of Olympic Sports dataset

| Authors | Accuracy on Olympic sports |
|---|---|
| Shiwei et al. [14] | 94.6% |
| Heng Wang and Cordelia Schmid [15] | 91.1% |
| Yun Yi and anli Wang [20] | 93.2% |
| Ghazal, Jamshid and Abdolhossein [21] | 93.1% |
| Ke Xu et al. [22] | 88.81% |
| **Proposed approach** | **98.23%** |

## 6 Conclusions

In summary, the rapidly advancing field of human action recognition in sports has profound implications for analyzing athletic performance and sports-related healthcare. Our approach to Human Action Recognition in sports videos, utilizing human poses extracted by the MediaPipe framework and Graph Neural Networks (GNNs) for activity classification, demonstrated a substantial performance improvement in extensive experiments on the Olympics Sports dataset compared to current state-of-the-art approaches.

**Acknowledgement.** The authors would like to acknowledge the financial support of this work by grants from General Direction of Scientific Research (DGRST), Tunisia, under the ARUB.

## References

1. Wu, F., et al.: A survey on video action recognition in sports: Datasets, methods and applications. CoRR abs/2206.01038 (2022)
2. Wensel, J., Ullah, H., Munir, A.: ViT-ReT: vision and recurrent transformer neural networks for human activity recognition in videos. IEEE Access **11**, 72227–72249 (2023)

3. Patel, C.I., Labana, D., Pandya, S., Modi, K., Ghayvat, H., Awais, M.: Histogram of oriented gradient-based fusion of features for human action recognition in action video sequences. Sensors **20**(24), 7299 (2020)
4. Santhosh Kumar, S., John, M.: Human activity recognition using optical flow based feature set. In: IEEE International Carnahan Conference on Security Technology, ICCST 2016, Orlando, FL, USA, October 24-27, 2016, pp. 1–5. IEEE (2016)
5. Ji, S., Wei, X., Yang, M., Kai, Yu.: 3D convolutional neural networks for human action recognition. IEEE Trans. Pattern Anal. Mach. Intell. **35**(1), 221–231 (2013)
6. Snoun, A., Jlidi, N., Bouchrika, T., Jemai, O., Zaied, M.: Towards a deep human activity recognition approach based on video to image transformation with skeleton data. Multimedia Tools Appl. **80**(19), 29675–29698 (2021). https://doi.org/10.1007/s11042-021-11188-1
7. Snoun, A., Bouchrika, T., Jemai, O.: View-invariant 3D skeleton-based human activity recognition based on transformer and spatio-temporal features. In: De Marsico, M., di Baja, G.S., Fred, A.L.N. (eds.) Proceedings of the 11th International Conference on Pattern Recognition Applications and Methods, ICPRAM 2022, Online Streaming, February 3-5, 2022, pp. 706–715. SCITEPRESS (2022)
8. Cao, Z., Hidalgo, G., Simon, T., Wei, S.-E., Sheikh, Y.: OpenPose: realtime multi-person 2D pose estimation using part affinity fields. CoRR, abs/1812.08008 (2018)
9. Papandreou, G., Zhu, T., Chen, L.-C., Gidaris, S., Tompson, J., Murphy, K.: PersonLab: person pose estimation and instance segmentation with a bottom-up, part-based, geometric embedding model. In: Ferrari, V., Hebert, M., Sminchisescu, C., Weiss, Y. (eds.) Computer Vision - ECCV 2018 - 15th European Conference, Munich, Germany, September 8-14, 2018, Proceedings, Part XIV, vol. 11218, pp. 282–299. Springer (2018)
10. Jlidi, N., Snoun, A., Bouchrika, T., Jemai, O., Zaied, M.: PTLHAR: PoseNet and transfer learning for human activities recognition based on body articulations. In: Osten, W., Nikolaev, D.P. (eds.) Twelfth International Conference on Machine Vision, ICMV 2019, Amsterdam, The Netherlands, 16-18 November 2019, vol. 11433, p. 114330Q. SPIE (2019)
11. Lugaresi, C., et al.: MediaPipe: a framework for building perception pipelines. CoRR, abs/1906.08172 (2019)
12. Jarray, R., Snoun, A., Bouchrika, T., Jemai, O.: Deep human action recognition system for assistance of Alzheimer's patients. In: Abraham, A., et al. (eds.) Hybrid Intelligent Systems - 20th International Conference on Hybrid Intelligent Systems (HIS 2020), Virtual Event, India, December 14-16, 2020, vol. 1375 of Advances in Intelligent Systems and Computing, pp. 484–493. Springer (2020)
13. Snoun, A., Bouchrika, T., Jemai, O.: Deep-learning-based human activity recognition for Alzheimer's patients' daily life activities assistance. Neural Comput. Appl. **35**(2), 1777–1802 (2023)
14. Zhang, S., Gao, C., Zhang, J., Chen, F., Sang, N.: Discriminative part selection for human action recognition. IEEE Trans. Multim. **20**(4), 769–780 (2018)
15. Wang, H., Schmid, C.: Action recognition with improved trajectories. In: IEEE International Conference on Computer Vision, ICCV 2013, Sydney, Australia, December 1-8, 2013, pp. 3551–3558. IEEE Computer Society (2013)
16. Marszalek, M., Laptev, I., Schmid, C.: Actions in context. In: 2009 IEEE Computer Society Conference on Computer Vision and Pattern Recognition (CVPR 2009), 20-25 June 2009, Miami, Florida, USA, pp. 2929–2936. IEEE Computer Society (2009)

17. Arandjelovic, R., Zisserman, A.: Three things everyone should know to improve object retrieval. In: 2012 IEEE Conference on Computer Vision and Pattern Recognition, Providence, RI, USA, June 16-21, 2012, pp. 2911–2918. IEEE Computer Society (2012)
18. Niebles, J.C., Chen, C.-W., Fei-Fei, L.: Modeling temporal structure of decomposable motion segments for activity classification. In: Daniilidis, K., Maragos, P., Paragios, N. (eds.) ECCV 2010. LNCS, vol. 6312, pp. 392–405. Springer, Heidelberg (2010). https://doi.org/10.1007/978-3-642-15552-9_29
19. Reddy, K., Shah, M.: Recognizing 50 human action categories of web videos. Mach. Vis. Appl. **24**, 07 (2013)
20. Yi, Y., Wang, H.: Motion keypoint trajectory and covariance descriptor for human action recognition. Vis. Comput. **34**(3), 391–403 (2018)
21. Shamsipour, G., Shanbehzadeh, J., Sarrafzadeh, H.: Human action recognition by conceptual features (2017)
22. Ke, X., Jiang, X., Sun, T.: Two-stream dictionary learning architecture for action recognition. IEEE Trans. Circuits Syst. Video Technol. **27**(3), 567–576 (2017)
23. Schüldt, C., Laptev, I., Caputo, B.: Recognizing human actions: a local SVM approach. In: 17th International Conference on Pattern Recognition, ICPR 2004, Cambridge, UK, August 23-26, 2004, pp. 32–36. IEEE Computer Society (2004)
24. Ni, B., Wang, G., Moulin, P.: RGBD-HuDaAct: a color-depth video database for human daily activity recognition. In: IEEE International Conference on Computer Vision Workshops, ICCV 2011 Workshops, Barcelona, Spain, November 6-13, 2011, pp. 1147–1153. IEEE Computer Society (2011)

# An Agricultural Domain Based Question Answering System Using Natural Language Processing and Deep Learning Methodologies

Krithikha Sanju Saravanan(✉) and B. L. Velammal

Department of Computer Science and Engineering, College of Engineering, Guindy, Anna University, Chennai, India
Krithikhasanju3008@gmail.com

**Abstract.** Question Answering System (QAS) is one of the information extraction techniques that is used to retrieve the answer from the given text paragraph for the given question. In the proposed QAS system for agricultural domain (AQAPS) involves both Question Generation and Answer Extraction from the domain based input document. First, a Knowledge Graph (KG) is generated from the given input text document using relationship extractor. From the generated KG, questions are generated using encoder-decoder neural network and answers are extracted using Recurrent Neural Network (RNN). Appropriate question-answer pairs are generated using answer selection methodology. The proposed AQAPS shows accuracy of 87.3% and while comparing with existing systems, the AQAPS performs better.

**Keywords:** Question generation · Answer extraction · Knowledge graph · Neural network · CNN · RNN

## 1 Introduction

QAS is one of the important fields of information retrieval and Natural Language Processing (NLP) technique that can automatically answer the questions from the input text. Data in the World Wide Web (WWW) is increasing daily because of the enhancement of social media. Increasing text data makes the users impossible to read the entire contents. As part of information retrieval process, the agricultural document is processed with generating the knowledge graphs which is one of the effective ways of representing the data. Usually, most of the systems focus on extracting the answer for the given query from structured or flat files. For successive information extraction, focus on both question and answer extraction is necessary. Question generation involves issues such as what the question type is and how to successfully stick with the context of the QAS. The proposed system receives the agricultural domain based document and process it to generate KG using relationship extraction. Queries are generated for the given entities from the KG using Neural Networks and answers are extracted from the KG using Answer Generation Network. The most appropriate answer is mapped with the query using Question-pair generation. The proposed system aims at generating question answer pairs for the given agricultural domain based document. The contributions of the research work are as follows:


A. Bajaj et al. (Eds.): HIS 2023, LNNS 1224, pp. 115–126, 2025.
https://doi.org/10.1007/978-3-031-78925-0_12

- Generating Knowledge Graph for input text document.
- Generating agricultural-based questions from the given features of KG.
- Extracting question-answer pairs using answer selection methodology.

By addressing the issues that lacks in understanding the context from the current systems involving in creating the above-mentioned contributions, the research work constructs an effective question-answering system which involves question generation and answer generation system with domain based KG as intermediate a knowledge representation. Effectiveness of knowledge representation method increases the efficiency of question generation and answer generation phases. Incorporating advanced learning techniques in phases of the AQAPS boosts the efficiency. The proposed AQAPS framework is implemented and results are obtained. The results are compared with current systems and it is observed that the system outperforms the other compared systems.

Section 2 gives the survey on the contemporary works related to the proposed AQAPS research work. Section 3 explains the proposed AQAPS work in detail. Section 4 gives the experimental setup and result analysis of the research work and Sect. 5 concludes the research work.

## 2 Literature Survey

Question answering framework involves the information representation and language generation models. So, the literature is divided into two subsections namely, Knowledge Graph Generation and Question Answering System. KG generation involves the review of works on constructing a KG from the sources and Question Answering System involves the review of works on knowledge representation based question answering system.

### 2.1 Knowledge Graph Construction

[1] proposed a name entity recognition method to tag horticulture corpus. Conditional Random Field is used for tagging the seven fine-grained types which includes Organism, Trait, Method/equipment, Chemical, Gene, Environment and Miscellaneous. [2] proposed coreference resolution for 10 different scientific disciplines from Science, Technology and Medicine. KG population is done for each domain to study the impact of co-reference resolution on it. Finally, an evaluation on the four KG populating strategies is carried out. [3] proposed a software tool called FoodKG to give efficient search results on food, energy and water domains. Specialized Tensor model is used to predict new relations in the graph embedding. AGROVOC is used to produce the graph and compare the semantic relations.

[4] proposed a method for constructing knowledge graph based on vector variance algorithm. The Bi-GRU remote supervised relationship extraction model is used to obtain classification relationships. Food domain knowledge is used to validate the final proposed knowledge graph. [5] proposed an approach to generate semantic based synthetic knowledge graph construction. Graph Differential Dependencies (GDD) is employed to develop the KG generation algorithm. The relationship generation between graph patterns is done by multi-stage process. [6] proposed an automatic framework to construct

knowledge graph of Power Terminal Equipment (PTE). Relevant algorithms for constructing knowledge graph of PTE in the ubiquitous power Internet of Things UPIoT environment is established.

[7] proposed a semantic data model called Research Contribution Model from three Scientific domains Medicine, Computer Science and Agriculture. The concept of Knowledge Graph Cells is introduced to learn Open Research Knowledge Graph (ORKG). [8] proposed a pipeline based on semantic web and NLP methods to generate knowledge graphs in agriculture domain. The inclusion of domain ontologies in the graph proved the potential of the model.

### 2.2 Question Answering System (QAS)

[9] developed a chat room and Chat bot to ease the task of decision making by farmers on farming. The model is trained using Artificial Intelligent Markup Language (AIML) which predicts intent based on the existing example. Chat bot uses cloud platform thus reducing the need for computational resources. [10] proposed a Question Answering (QA) system built upon NLP and semantic web technologies for agriculture domain. The knowledge base for the QA system is built using ontology. The primary task of the system is to analyse the user queries given in natural language and provides precise answers.

[11] proposed a framework using NLP tools to develop a Question Answering System for a restricted domain. The primary objective of the model is to extract precise answers from large dataset for the user queries. The framework for the system is divided into 4 modules such as, question processing module, document processing module, paragraph extraction module and answer extraction module. [12] built an entity extraction algorithm and question answering system with the help of food knowledge graph. The motto of this work is to help people analyse the potential problems about food safety. SPARQL language is used to represent the query in knowledge graph and the results are stored in gStore database. Bhagat et al. [13] proposed to build an automated FAQ retrieval system based on Interactive Voice Response (IVR) in the domain of sexual and reproductive health and rights (SRHR). The main goal of the voice bot is to provide precise answers from existing corpus of questions and answers.

[14] designed a question answering system based on knowledge graph to provide accurate answers to student's academic questions. The teaching application mode is built based on the Successive Approximation Model (SAM model). The knowledge graph and question-answering technology are integrated to build the system. [15] proposed a hybrid approach based named entity recognition for agriculture documents. Two unsupervised approaches called LDA and AGROVOC is used to identify the named entities without using annotated corpus.

## 3 Materials and Methods

The proposed AQAPS system receives the agricultural text document as the input and generates the Question-Answer Pairs from the input texts. A relationship extraction algorithm is proposed for extracting the agricultural relationships between the entities in sentences that uses Bidirectional Encoder Representations from Transformers (BERT) model which is customized for the agricultural domain. An Agricultural Domain-based Knowledge Graph (ADKG) is constructed using the extracted relationships. The entities in the knowledge graph are fed to an unique Question Generation Network for generating the agricultural domain based questions which involves encoder-decoder network. For the generated query, the answers are extracted from the knowledge graph using the novel answer extraction network. For one query, there could be many answers, the most appropriate answer is selected using the proposed Question-Answer Pair Generator by a pair of Convolutional Neural Network (CNN) and binary classifier which is similar to the answer selection phase in a QAS. The appropriate question-answer pairs are stored in the database for the later use. The overall system design in shown in Fig. 1.

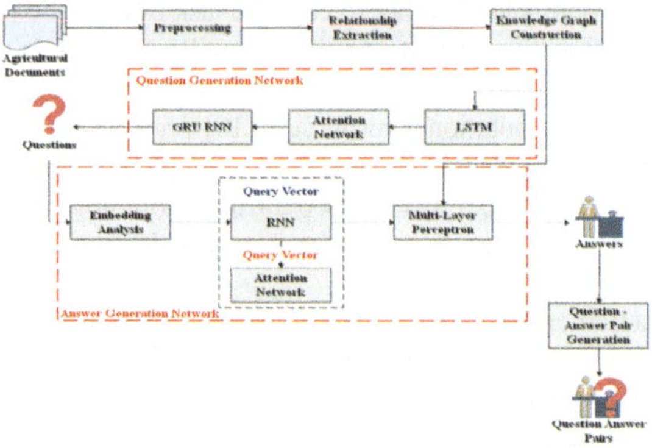


**Fig. 1.** Overall diagram of the proposed work.

### 3.1 Preprocessing

Preprocessing in the proposed AQAPS system receives the agricultural document as the input. Entity annotation is performed on every sentence in the given document. The annotated sentences in the document have syntactical and semantic features of the words in the corresponding sentences. The entity annotations identify the end-to-end linking of entities and entity disambiguation. The agricultural document is also segmented based on the sentences and the sentences are classified using binary classifier, random forest. Random forest algorithm classifies the sentences that it belongs to agricultural domain or not. The classification in this stage acts as a filter mechanism to restrict the other domain sentences inside the system which is not the scope of the proposed system. The algorithm for preprocessing is shown in Algorithm.

```
Algorithm 1 Preprocessing
Input: Agriculture Document, D
Output: Annotated Statements Sannotation, Agricultural State-
ments Sagri
Sagri = ϕ
Sannotation = ϕ
```

```
S[] = Tokenization(D)
for s in S[] do
if Random Forest(s) == 1 then
Sannotation.append(Linguistic Model(s))
Sagri.append(s)
end if
end for
return Sannotation, Sagri
```

### 3.2 Relationship Extraction

From the classified sentences with their annotations, relationships available in the sentence are extracted using the Relationship Extractor. A dictionary for the agricultural domain is manually created by adding the contextual words that belong to the domain. The dictionary consists of the various agricultural relationships such as "cultivated", "disease", "vitamins", "deficiency" and so on. The dictionary of agricultural domain at this phase has the direct impact in the relationship extraction for constructing ADKG. To extract new types of relationships for KG, corresponding words and keywords need to be added to this dictionary. For extracting the agricultural relationships, the sentences are tokenized into words and the tokens are given as input. The words that belong to dictionary are masked and the masked words directly appear at the end of the relationship extractor. The sentence with tokenized and masked words, are passed through the pre-trained BERT model for BERT Masked Language Modelling (BERT MLM). The BERT MLM extracts the relationships and masked words from the sentences which are used for KG construction in the upcoming functions. The output of the relationship extractor contains the entities needed for constructing ADKG which includes all masked words and other agricultural words in the token list.

### 3.3 Knowledge Graph Construction

An agricultural domain based knowledge graph is generated in this phase of the proposed AQAPS work. ADKG comprised of nodes, edges and labels. Nodes represents the agricultural entities, presence of edges between two nodes represents the existence of relationship between the nodes and labels in the ADKG defines the verbal relationships between the nodes, similar to the edge values in a graph. From relationship extractor, the words are classified into nodes and labels using Named Entity Recognition (NER) tag and syntactic relationship from the sentence annotations with the classified words, the ADKG is constructed. In the resulting KG, any relationship in a node is represented using three levels of features that are node (entity), edge and label.

**Algorithm 2 Knowledge Graph Generation.**

```
Input: Entities from Relationship Extractor (ϵ)
Output: Knowledge Graph K Graph
Global K Graph
V ectorent[] = filter(Word2vec(sentences)[word] word ∈ Entities
ϵ)
```

```
for V ∈ V ectorent do
V.type= Random Forest(V )
end for
\\Words (Entities) grouping is done based on Word Ordering in given
Document
for WGroup ∈ EntityGroups do
K Graph = K Graph U Graph(WGroup)
end for
return K Graph
```

### 3.4 Question Generation Network

Question Generation Network is based on the Encoder-Decoder technique which can generate questions from the triplet features of a node in the ADKG. The encoder consists of attention layers which converts the triplet into a continuous value in which label is also included so, the question generated would not deviate from the context of the node. The attention layer calculates a vector which decides context of the questions to be generated. The vector V (t) at time step t is calculated using weighted average function on hidden vector h(i). Mathematically the vector V (t) is represented using Eq. 1.

$$\mathrm{V(t)} = \sum_{i=0}^{N} \mathrm{w\ (t-1, j)h(i)} \tag{1}$$

The weight vector w(t, i) is calculated using Eq. 2.

$$\mathrm{w(t,i)} = \frac{\exp\{z[S(t), h(i), V(t-1)]\}}{\sum_{i=1}^{H} \exp\{z[S(t), h(i), V(t-1)]\}} \tag{2}$$

Where, S(t) is the sentence considered at the time t and z() is a non-linear function which gives the importance of current hidden state h(i). Further normalization is performed using softmax function. For the compound question generation and for generating questions that involves a greater number of node pairs, each pair of nodes is considered as single vector. The combination of vectors for generating complex questions is taken care by decoder phase.

The decoder recognizes the encoded vector and generates the question using Gated Recurrent Unit (GRU). GRU in turn generates the vector for generating the question as the result. A list of vocabulary in which contain nearly 20K words used as a dictionary of English language is utilized for generating the question using GRU. Since the dictionary size is huge, the low-frequency words in the knowledge graph suffers an issue of occurrence in the sentences. It has been proven that the copying mechanism is an effective way to replicate the low-frequency words in sequence to sequence learning problems. For effectiveness of the neural network, the neural gate function is implemented using the Eq. 3.

$$\gamma(\mathrm{t}) = \sigma\left[\mathrm{W}\gamma(\mathrm{s(t), V\ (t), s(0)}, \zeta(\mathrm{t}-1)) + \beta\mathrm{qg}\right] \tag{3}$$

Where σ is the sigmoid function, ζ(t − 1) is the loss function at time t − 1 and βqg is the hyper parameter of the model. The words to be generated for the question have words from KG using the copying mechanism. The decoder generates the word either using the English vocabulary set via standard softmax or from the masked words in the Relationship Extraction phase via the copying mechanism.

### 3.5 Answer Generation Network

To generate answer for the query, Answer generation network is used. Answer generation network performs embedding analysis for the query by vectorization in the Bidirectional Recurrent Neural Network (RNN). Bi-directional RNN produces the query vector which comprises of values similar to the entities in KG and thus it makes the Out of Dictionary (OoD) words to map with KG entities. Bidirectional RNN is also trained as like Question Generation Phase. The query vector is processed with attention layer and Multi-Layer Perceptron (MLP), that is mapped with the nodes in KG using the combination of linear layers where the entities of KG are already vectorized. MLP results in selecting multiple answers for the given query. For a single query, multiple answers are mapped.

### 3.6 Question-Answer Pair Generation

The query is generated in the Question Generation Phase and corresponding answers for the queries are extracted in Answer Generation Network. Such that, one query has n number of answers. The most relevant answers should be selected and the Question-Answer Pairs should be generated for the agricultural domain. Aiming that, Query Corpus Network and Query Relationship Network are used in this phase to select the most relevant answers.

**Query Corpus Network.** Query Corpus Network consists of Siamese Neural Networks which takes the query with answers as one input and the sentences from the corpus as another input to generate the feature vectors for the neural network. The structure of the constructed neural network is shown in Fig. 2. The sentences from corpus are extracted using QAS which takes question as input and extracts the relevant answers from the corpus using syntactic level features. The score reflects the vector convolution between the query, answer and the corpus sentences. The sentences from the corpus are selected using the co-occurrences of words between the query and the corpus sentences.

**Query Relationship Network.** Query Relationship Network consists of CNN which takes query and extracted labels in Answer Generation Phase as the input. The structure of the proposed CNN is shown in Table 1. A feature vector is generated using Query Relationship Network in which relationship between the query and label is estimated. The feature vectors from the query corpus and query relationship vectors are unified and processed with binary classifiers. In this research work, Support Vector Machine (SVM) as classifier is used. For each Question-Answer Pair, SVM decides the whether the answer is correct or not. Then, the correct Question-Answer Pairs are stored in the database for the future references. The information's are structured and extracted successfully from the flat files.

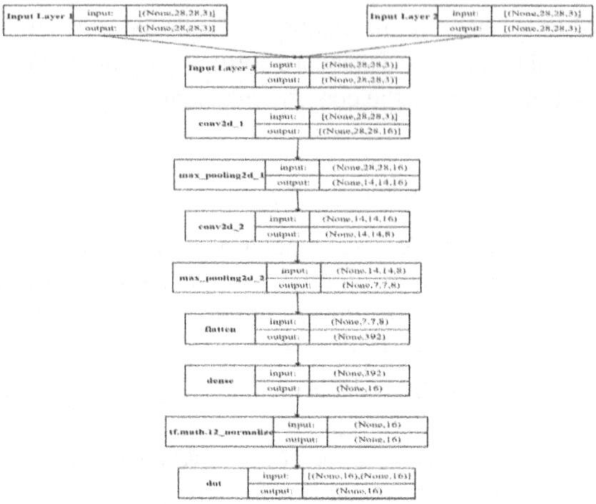

**Fig. 2.** Constructed Query Corpus Neural Network.

**Table 1.** Architecture of Query Relationship Network.

| Layer | Size | Stride | Activation function |
|---|---|---|---|
| Input Layer | 156 × 156 × 1 | – | – |
| 1D Convolution Layer | 154 × 154 × 32 | 3 | ReLU |
| 2D Pooling Layer | 76 × 76 × 32 | 2 | – |
| 2D Convolution Layer | 74 × 74 × 64 | 3 | ReLU |
| 2D Pooling Layer | 36 × 36 × 64 | 2 | – |
| Fully Connected Layer | 5 | – | Softmax |

## 4 Result and Analysis

### 4.1 Dataset

Since there is no domain specific complete dataset, Various texts are collected from the newspaper articles and other agricultural blogs from internet sources.

### 4.2 Numerical Results

The overall system as the AQAPS is experimented with the dataset and results are observed. The sample input and the question answer pairs for that input is shown in Table 2. The BM25 [16] is the score of a query generated from the table with plain text. WordCnt and CNN [16] are the feature based approach that is used to identify the query score for questions from a web table. MT based PP [17] is a phrase based translation model and encoding algorithm to compare different phrase based translation models. The results are taken with the metrics of Mean Average Precision (MAP) and Accuracy. The performance of the proposed system is compared with the existing systems and

**Table 2.** Sample input and the question answer pairs for that input generated by the proposed AQAPS framework.

| | | | |
|---|---|---|---|
| Sample Paragraph | Paddy, also called rice paddy, small, level, flooded field used to cultivate rice in southern and eastern Asia. Wet-rice cultivation is the most prevalent method of farming in the Far East, where it utilizes a small fraction of the total land yet feeds the majority of the rural population. Rice was domesticated as early as 3500 BC, and by about 2000 years ago it was grown in almost all of the present-day cultivation areas, predominantly deltas, floodplains and coastal plains, and some terraced valley slopes | The average world consumption of orange is 12 kg per person and year, although this figure varies depending on the countries. The developed countries have an average of 28 kg/person and year, whereas in the developing countries the consumption is 6 kg/person and year. The highest consumption figures are in the United States, reaching 60 kg, whereas in Western Europe it is 40 kg. These figures contrast with the consumption in the East European Countries, 5 kg per person | Apple varieties, of which there are thousands, fall into three broad classes: (1) cider varieties; (2) cooking varieties; and (3) dessert varieties, which differ widely but tend to emphasize colour, size, aroma, smoothness, and perhaps crispness and tang. Many varieties are relatively high in sugar, only mildly acidic, and very low in tannin. Apples provide vitamins A and C, are high in carbohydrates, and are an excellent source of dietary fibre |
| Generated Question Answer pairs | [{'answer: rice paddy', 'question': 'What is another name for paddy?"}, {'answer': 'Wet-rice', 'question': 'What is the most prevalent method of farming in the Far East?"}, {'answer': '3500 BC', 'question': 'When was rice domesticated?"}] | [{'answer': '12 kg', 'question': 'What is the average world consumption of orange per person and year?'}, {'answer': '6 kg, question': 'What is the average consumption of oranges in developing countries?"}, {"answer': '60 kg', 'question': "What is the average orange consumption in the United States?'},{'answer': '5 kg per person', 'question': 'What is the average consumption of orange in the East European countries?"}] | [{'answer': 'three', 'question': 'How many broad classes do apple varieties fall into?"}, {'answer': 'tannin', 'question': 'What is low in many varieties of apples?"}, {'answer': 'carbohydrates', 'question': 'What are apples high in?'}] |

the proposed system shows the encouraging results. The results of MAP and Accuracy comparisons are shown in Table 3.

**Table 3.** Comparing the MAP and accuracy performance measures of the proposed AQAPS framework with existing systems

| Method | MAP | Accuracy |
|---|---|---|
| BM25 [16] | 0.429 | 0.294 |
| WordCnt [16] | 0.318 | 0.19 |
| MT based PP [17] | 0.327 | 0.213 |
| CNN [16] | 0.359 | 0.238 |
| Proposed AQAPS | 0.869 | 0.873 |

The Question Generation in the proposed system generates the question for the given knowledge graph entities. Since it involves natural language generation, Bilingual Evaluation Understudy (BLEU) shows the exact efficiency of the system effectively. The results are obtained from the proposed system by conducting various experiments and the results are compared with the existing systems. The Seq2Seq [18] is a basic encoder decoder learning system that uses LSTM for question generation. The GCN [19]is a generative model that uses adversarial nets to determine the sample distribution. The Seq2Seq para [20] model uses the encoder decoder paradigm to generate the paraphrase for given sentence. The dual model [21] uses dual-NMT mechanism that generate the bilingual sentence pairs to train NMT system. The comparison with respect to BLUE score is shown in Table 4 and the proposed system shows an encouraging result.

**Table 4.** Comparing BLUE score of the proposed AQAPS framework with existing systems.

| Method | BLEU |
|---|---|
| Seq2Seq [18] | 8.87 |
| Dual [21] | 9.31 |
| Seq2SeqPara [20] | 9.16 |
| GCN [19] | 9.22 |
| Proposed AQAPS | 9.71 |

Answer Extraction system in the proposed system extracts the answer for the given query from the Knowledge graph using the triplets. CDSSM [22] is a convolutional latent semantic modelling technique that captures the important contextual information. ABCNN [23] is an attention integrated CNN model that is used to generate sentence pair modelling. The performance of the Answer Extraction System is shown in Table 5 with the comparison of existing systems. The proposed AQAPS framework shows an encouraging result.

**Table 5.** Comparing the MAP and precision performance measures of the proposed AQAPS framework with existing systems.

| Model | MAP | Precision |
|---|---|---|
| WordCnt [16] | 0.395 | 0.179 |
| CDSSN [22] | 0.442 | 0.228 |
| ABCNN [23] | 0.469 | 0.263 |
| Dual [21] | 0.484 | 0.275 |
| Seq2SeqPara [20] | 0.47 | 0.26 |
| GCN [19] | 0.468 | 0.257 |
| Proposed AQAPS | 0.869 | 0.88 |

## 5 Conclusion

The proposed system generates knowledge graph from the agricultural text and questions-answer pairs are generated using Question Generation phase and Answer Extraction Phase. For the given entities of the knowledge graph, the query is generated which can be complex questions. The generated queries are fed into answer extraction phase that extracts the answer from the generated ADKG. The question-answer pairs are generated using the answer selector networks. Various experiments are conducted to assess the performance of the system and compared with the existing systems where encouraging results are obtained. The work can be extended to agricultural domain based Interactive Voice Response (IVR) system to give automated answers for the given speech queries.

## References

1. Liu, Z., Luo, M., Yang, H., Liu, X.: Named entity recognition for the horticultural domain. J. Phys. Conf. Ser. **1631**(1), 012016 (2020)
2. Brack, A., Muller, D.U., Hoppe, A., Ewerth, R.: Coreference resolution in research papers from multiple domains. arXiv preprint arXiv:2101.00884 (2021)
3. Gharibi, M., Zachariah, A., Rao, P.: Foodkg: a tool to enrich knowledge graphs using machine learning techniques. Front. Big Data. **3**, 12 (2020)
4. Yu, H., Li, H., Mao, D., Cai, Q.: A domain knowledge graph construction method based on Wikipedia. J. Inf. Sci. **47**(6), 016555152093251 (2020)
5. Feng, Z., Mayer, W., He, K., Kwashie, S., Stumptner, M., Grossmann, G., Peng, R., Huang, W.: A schema-driven synthetic knowledge graph generation approach with extended graph differential dependencies. IEEE Access. **9**, 5609–5639 (2020)
6. Su, Z., Hao, M., Zhang, Q., Chai, B., Zhao, T.: Automatic knowledge graph construction based on relational data of power terminal equipment. In: 2020 5th International Conference on Computer and Communication Systems (ICCCS), pp. 761–765. IEEE, New York (2020)
7. Vogt, L., D'Souza, J., Stocker, M., Auer, S.: Toward representing research contributions in scholarly knowledge graphs using knowledge graph cells. In: Proceedings of the ACM/IEEE Joint Conference on Digital Libraries in 2020, pp. 107–116 (2020)

8. Abad-Navarro, F., Bernabe-Diaz, J.A., Garcia-Castro, A., Fernandez-Breis, J.T.: Semantic publication of agricultural scientific literature using property graphs. Appl. Sci. **10**(3), 861 (2020)
9. Ekanayake, J., Saputhanthri, L.: E-AGRO: intelligent Chat-Bot. IoT and artificial intelligence to enhance farming industry. AGRIS Online Pap. Econ. Inf. **12**, 15–21 (2020)
10. Devi, M., Dua, M.: Adans: an agriculture domain question answering system using ontologies. In: 2017 International Conference on Computing, Communication and Automation (ICCCA), pp. 122–127. IEEE, New York (2017)
11. Biswas, P., Sharan, A., Malik, N.: A framework for restricted domain question answering system. In: 2014 International Conference on Issues and Challenges in Intelligent Computing Techniques (ICICT), pp. 613–620. IEEE, New York (2014)
12. Qin, L., Hao, Z., Yang, L.: Question answering system based on food spot-check knowledge graph. In: Proceedings of 2020 the 6th International Conference on Computing and Data Engineering, pp. 168–172 (2020)
13. Bhagat, P., Prajapati, S.K., Seth, A.: Initial lessons from building an IVR-based automated question-answering system. In: Proceedings of the 2020 International Conference on Information and Communication Technologies and Development, pp. 1–5 (2020)
14. Yang, Z., Wang, Y., Gan, J., Li, H., Lei, N.: Design and research of intelligent question-answering (Q&A) system based on high school course knowledge graph. Mobile Netw. Appl. **26**, 1884–1890 (2021)
15. Gangadharan, V., Gupta, D.: Recognizing named entities in agriculture documents using LDA based topic modelling techniques. Procedia Comput. Sci. **171**, 1337–1345 (2020)
16. Sun, Y., Yan, Z., Tang, D., Duan, N., Qin, B.: Content-based table retrieval for web queries. Neurocomputing. **349**, 183–189 (2019)
17. Koehn, P., Och, F.J., Marcu, D.: Statistical phrase-based translation. University of Southern California Marina Del Rey Information Sciences Inst, Tech. Rep. (2003)
18. Du, X., Shao, J., Cardie, C.: Learning to ask: neural question generation for reading comprehension. arXiv preprint arXiv:1705.00106 (2017)
19. Goodfellow, I., Pouget-Abadie, J., Mirza, M., Xu, B., Warde-Farley, D., Ozair, S., Courville, A., Bengio, Y.: Generative adversarial nets. In: Advances in Neural Information Processing Systems, vol. 27. Springer, Berlin (2014)
20. Egonmwan, E., Chali, Y.: Transformer and seq2seq model for paraphrase generation. In: Proceedings of the 3rd Workshop on Neural Generation and Translation, pp. 249–255. Association for Computational Linguistics (2019)
21. He, D., Xia, Y., Qin, T., Wang, L., Yu, N., Liu, T.-Y., Ma, W.-Y.: Dual learning for machine translation. In: Advances in Neural Information Processing Systems, vol. 29. Springer, Berlin (2016)
22. Shen, Y., He, X., Gao, J., Deng, L., Mesnil, G.: A latent semantic model with convolutional-pooling structure for information retrieval. In: Proceedings of the 23rd ACM International Conference on Information and Knowledge Management, pp. 101–110 (2014)
23. Yin, W., Schutze, H., Xiang, B., Zhou, B.: Abcnn: attention-based convolutional neural network for modeling sentence pairs. Trans. Assoc. Comput. Linguist. **4**, 259–272 (2016)

# Enhancing Operational Efficiency and Decision-Making Through NLP Analysis of Machine Data

Welinton Trentin[1](✉) and Rafael Stubs Parpinelli[2]

[1] WEG Electric Machines, Jaragua do Sul, Brazil
wtrentin@weg.net
[2] Graduate Program in Applied Computing, Santa Catarina State University (UDESC), Joinville, Brazil
rafael.parpinelli@udesc.br

**Abstract.** This paper investigates the utilization of Natural Language Processing (NLP) techniques for analyzing and extracting insights from textual data concerning machines within a particular company. By leveraging NLP, the primary goal is to enhance operational efficiency and decision-making processes by understanding machine status, behavior, and patterns comprehensively. Adopting this tool is anticipated to have a lasting impact, empowering the company to improve its products and make informed decisions across diverse domains for sustained success.

## 1 Introduction

Over the years, artificial intelligence has gained significant popularity and importance in various aspects of our lives. With the increasing importance of data, many companies are utilizing this tool to gain insights into what is happening and how to respond to different situations. Machine Learning and Natural Language Processing, both subfields of artificial intelligence, work with data to provide valuable tools and knowledge [1].

Machine Learning and Deep Learning techniques have brought about a revolution in various business sectors. Moreover, these learning techniques have made significant positive contributions to numerous technologies, one of them being Natural Language Processing. NLP is an interdisciplinary field that combines elements of Artificial Intelligence (AI), linguistics, and computer science. It equips machines with the ability to understand, interpret, and generate human language, enabling effective communication between machines and humans. This advancement in Natural Language Processing has bridged the communication gap between machines and their human counterparts, unlocking new possibilities and applications in various domains [2–4].

One of the crucial phases in the new product development cycle is to analyze and learn from past mistakes, evaluating both internal and field failures of current products. This analysis involves studying an extensive list of failures and


A. Bajaj et al. (Eds.): HIS 2023, LNNS 1224, pp. 127–136, 2025.
https://doi.org/10.1007/978-3-031-78925-0_13

defects organized into 26 categories, along with detailed descriptions. Despite the difficulties, valuable data can be found in the open-text section, although converting it into useful information is a time-consuming process. For example, a previous procedure took three months of full-time work to cover a single product line over four years. This paper focuses on the practical application of Natural Language Processing (NLP) techniques to examine and comprehend past events and incidents related to a specific company's machinery. The primary goal is to harness NLP to extract insights regarding the machines' status, behaviors, and recurring patterns, ultimately enhancing operational efficiency and informing more informed decision-making processes.

This paper is structured as follows: In Sect. 2, we present an in-depth overview of machine learning and NLP. Section 3 provides a comprehensive review of three notable studies that have effectively employed Natural Language Processing (NLP) in practical applications. Section 4 delves into the intricacies of our approach, offering insights into its inner workings and research methodology. Section 5 presents the experiments conducted, the corresponding results, and a detailed analysis of our findings. Finally, Sect. 6 concludes the paper, summarizing our key insights and offering recommendations for future research endeavors.

## 2 Machine Learning and NLP

Machine Learning and Natural Language Processing (NLP) represent pivotal subfields of artificial intelligence, commanding increasing attention and application across various sectors. Machine learning, fundamentally, empowers computers to address challenges for which explicit programming would be impractical or unfeasible. This is especially pertinent in real-world scenarios where it's arduous to foresee all potential input types and devise algorithms to solve complex problems effectively. On the other hand, NLP is a multidisciplinary fusion of linguistics and computer science, dedicated to equipping computers with the capability to comprehend and interpret human language, both spoken and written. To achieve this, NLP encompasses a range of intricate stages, including morphological analysis, syntactic analysis, semantic analysis, discourse analysis, and pragmatic analysis, each contributing to the machine's proficiency in understanding and processing natural language [1,2].

## 3 Related Works

This paper delves into the practical applications of Natural Language Processing (NLP) techniques in diverse domains. It encompasses studies conducted in industrial settings, software engineering, and workplace safety. The first study [5] focuses on utilizing NLP for defect detection in railway signaling manufacturers' requirements documents. By comparing NLP techniques with traditional manual techniques, the study provides insights to enhance NLP capabilities and modify practices for more effective utilization. The second study [6] addresses the challenge of linking artifacts in software engineering, evaluating NLP techniques' performance in identifying equivalent requirements. Principles for reliable evaluation are proposed, and a case study in the defense and aerospace domain demonstrates their application. The third study [7] investigates workplace safety in the construction sector, employing text mining and NLP techniques to classify accident causes. The study highlights the effectiveness of the methods employed and suggests future improvements. Collectively, these studies contribute to the understanding and advancement of NLP techniques in various domains, showcasing their potential for enhancing operational efficiency, decision-making processes, and safety measures.

## 4 Proposed Approach

The workflow of the proposed approach is represented in Fig. 1.

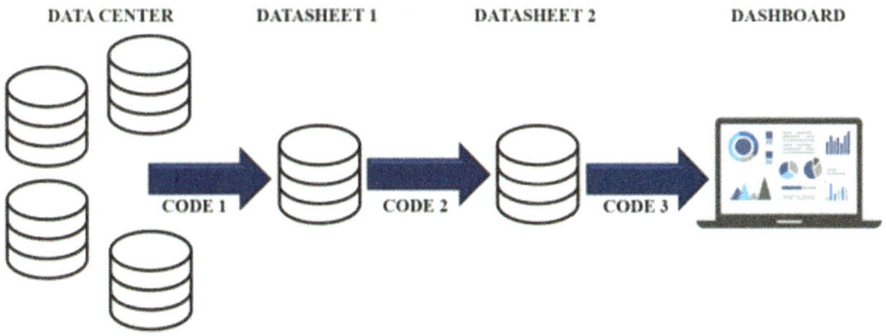


**Fig. 1.** Data Workflow.

Code 1 retrieves data from multiple databases and consolidates it into a single datasheet, as shown in Fig. 1. The collected data includes structured information as well as unstructured text data from both internal and external situations. The data encompasses notes about events within and outside the company, along with details such as the location, machine characteristics, connected areas, and note values. The focus of this work is on handling and analyzing the unstructured text data within the notes.

The natural language processing is used in code 2 and its pipeline is represented in Fig. 2.

- First Clean

In this part of the code, several operations have been performed on the text. Blank spaces, punctuation marks, and stop words have been removed. Stop words are commonly used words in a language that are often considered insignificant for analysis purposes. Additionally, in this specific case, the authors have decided to add some extra words to the list of stop words, considering them as not important in the given situation.

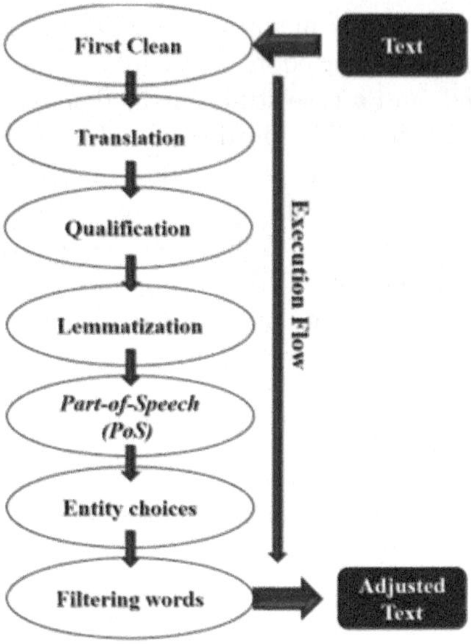


**Fig. 2.** Code 2 pipeline.

- Translation

Most of the texts in the data are in Portuguese, but in some cases, the sentence is entirely in English, or there may be a mix of Portuguese and English within a single sentence. To handle the translation, the code utilizes the Transformers library with the MarianMT model, specifically using the pre-trained model 'Helsinki-NLP/opus-mt-tc-big-en-pt' [8].

- Qualification

In most cases, the sentences contain words that serve a connecting function, but these words are not crucial for understanding the meaning of the sentence when considering answers related to what is happening. To address this, the code utilizes the scikit-learn library to qualify or evaluate the importance of words within the sentence.

- Lemmatization

In natural language processing, tokenization and lemmatization are important processes carried out at the morphological stage. Tokenization involves breaking a text into individual words or tokens. Lemmatization, on the other hand, is the process of reducing each word to its base or dictionary form, known as the lemma, by removing affixes. This helps in standardizing the words and reducing variations, making it easier for machines to process and analyze the text accurately. Lemmatization ensures that words with similar meanings are represented by their common base form, improving the effectiveness of language processing tasks [9,10]. Even in this part, it was necessary to use stop words, because of the translation step.

- Part-of-Speech

Part-of-speech (POS) tagging is a fundamental process in natural language processing (NLP) that involves assigning a specific part-of-speech label to each word in a given text. The parts of speech can include nouns, verbs, adjectives, adverbs, pronouns, prepositions, conjunctions, and more. POS tagging helps in understanding the syntactic structure of a sentence, enabling further analysis and interpretation of the text [11].

- Entity choices

After performing part-of-speech tagging, it is necessary to select which labeled parts of speech will be helpful in obtaining insights from the text.

- Filtering words

In this final step, repeated words will be removed from the sentences, and the words will be adjusted to align with the Portuguese language if necessary.

The dashboard was developed in code 3 and Fig. 3 shows the workflow.

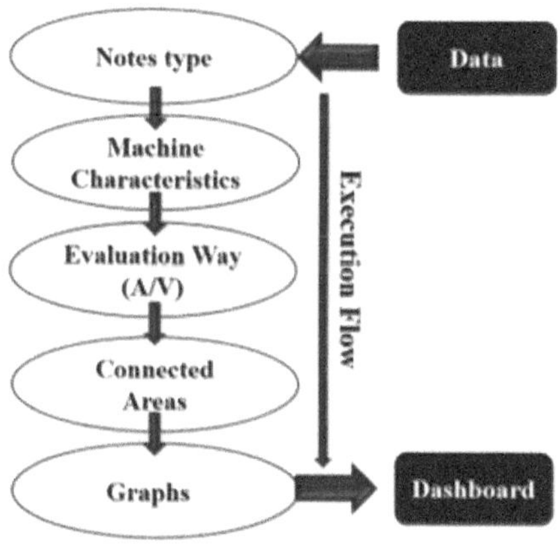


**Fig. 3.** Code 3 workflow.

Code 3 retrieves information from datasheet 2 and transforms the unstructured data (text) into structured data. The resulting structured data is then displayed in a dashboard, providing a comprehensive view of all the information from datasheet 2 (Fig. 1).

- Notes type

The user has the option to select whether they want to analyze internal or external notes.

- Machine Characteristics

The user has the option to choose the type of machine and additional characteristics such as frame, poles, month, and year. These characteristics will be automatically selected based on the chosen machine type. The user also has the flexibility to remove any unwanted item from the selection. It is possible to select multiple machines with their corresponding characteristics for analysis.

- Evaluation Way (A/V)

The user can choose to view the graphs based on the number of notes (A) or the value (cost) of notes (V).

- Connected Areas

Similar to frames and poles, connected areas will be displayed on the screen when the user selects the type of machine(s).

- Graphs

With the selected information, the system will generate the corresponding graphs.

## 5 Experiments and Results

Within the context of our experimental setup, Python emerges as the predominant programming language for the development of sophisticated solutions in the domain of Natural Language Processing (NLP). Our toolkit includes essential libraries such as scikit-learn (sklearn), Transformers, NLTK, and spaCy, which collectively form a robust framework for text analysis and manipulation. These libraries facilitate a wide range of NLP tasks, encompassing sentiment analysis, named entity recognition, and language modeling. To conduct our experiments effectively, we harnessed a computational platform boasting 32GB of RAM and 16 processing cores, operating on the Windows platform. This computational environment empowered us to explore the capabilities and insights offered by this potent amalgamation of NLP tools and technologies.

The dataset used in this study is generated through Code 1, employing the VBA programming language. This code is designed to extract data from various modules within the SAP software, which serves as the repository for the company's critical information. The complexity of the task lies in deciphering the relevance of the data and its specific locations within the software. This dataset comprises an extensive collection of more than 50 thousand data points, reflecting the intricate nature of the data extraction process and its potential significance for our research and analysis. Due amount data and many steps in code 2, this code was processed in parallel.

The developed software facilitates a comprehensive understanding of the operational status of machines both internally and externally, across various domains. By leveraging artificial intelligence (natural language processing) for the analysis of machine data, this tool offers access to valuable, data-driven insights that empower informed decision-making processes.

The dataset under examination spans a substantial time frame, encompassing data from 2018 through May 2023. Users are granted the flexibility to choose between two distinct Word Cloud visualizations, each based on specific criteria: Amount or Cost. As illustrated in Fig. 4, the left panel displays the Word Cloud generated based on the "Amount" criterion, while the right panel showcases the Word Cloud visualization derived from the "Cost" criterion. It is worth noting that the words depicted within these visualizations are purely illustrative in nature, owing to the stringent compliance regulations governing sensitive company data.

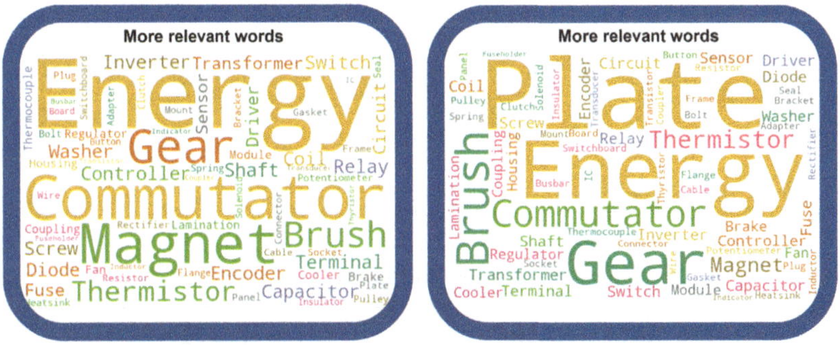


**Fig. 4.** Word Cloud - By Amount of Notes and by Cost.

By analyzing Fig. 4, we can identify the main issues affecting the Machine. Notably, there is a significant association between some words and the machine's performance problems. In software, there are values charts (near Word Cloud) that provide a numerical representation of the information contained in the Word Cloud, and with this is possible to measure how these words are affecting the company even in the number of notes and cost.

In order to enhance the visual representation of the data, the user has the option to remove certain words from the word cloud and values chart (Fig. 4). This allows for a cleaner and more focused display of relevant information. By eliminating unimportant words, the user can gain better insights and a clearer understanding of the data being analyzed.

The user has the flexibility to adjust the number of words displayed in the word cloud, ranging from 5 to 100. Additionally, in the values chart, it is possible to search for specific words even if they do not appear in the word cloud.

The platform provides the ability to identify the countries and continents where failures are occurring, based on either the criteria of Amount or Cost (Fig. 5).

Furthermore, it enables users to identify the specific sections that are linked to the notes and analyze the cumulative cost and cumulative number of notes over the years (Fig. 6).

The user has the option to customize their analysis by removing unwanted areas from the dataset. This allows them to focus on specific areas of interest and analyze only the relevant information. By removing unwanted sections, the user can streamline their analysis and obtain insights that are specifically tailored to their needs and objectives (Fig. 6).

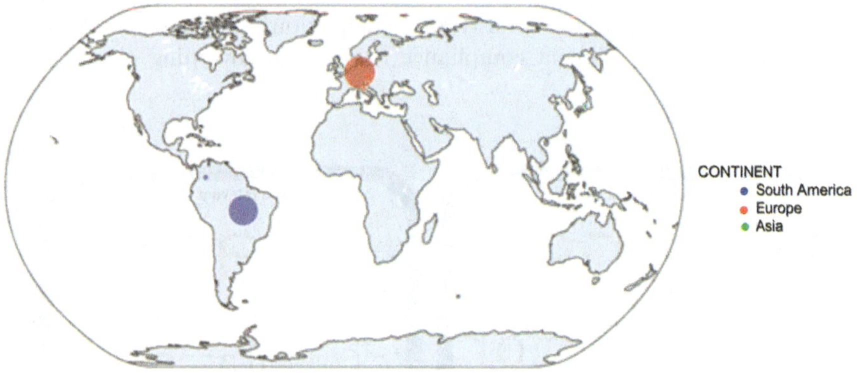


**Fig. 5.** Notes in World - By Continent.

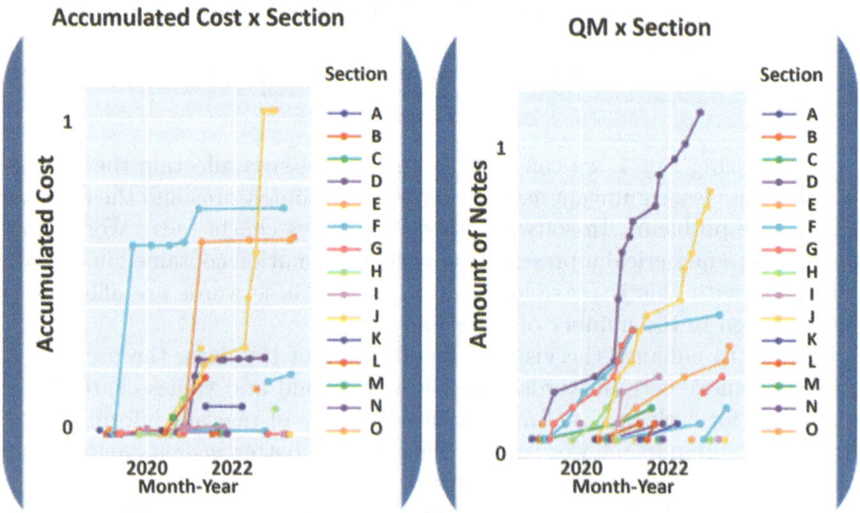


**Fig. 6.** Accumulated Cost and Amount of Notes by Section.

**Fig. 7.** Notes in World - By Section.

It is possible to identify the countries and continents where failures are occurring in relation to specific sections, based on either the criteria of Amount or Cost (Fig. 7).

After selecting machine characteristics, period, and connected areas and analyzing the dashboard information, users can download a filtered datasheet containing specific details. This datasheet provides valuable insights, such as the number of notes related to particular situations, allowing users to delve deeper into the identified problems. This process facilitates a better understanding of the situation and empowers users to gain meaningful insights into ongoing events and occurrences, aiding in problem-solving and decision-making.

## 6 Conclusion and Future Works

In conclusion, the software discussed in this study is a powerful tool for gaining detailed insights into machine operations, covering various aspects within diverse domains. Leveraging artificial intelligence, especially natural language processing (NLP), has greatly advanced quality management, particularly in analyzing generator and motor performance data.

Moreover, the software's flexibility and customization options empower users to tailor their analyses to specific goals. Whether adjusting Word Cloud settings, conducting targeted searches in the values chart, or applying data filters to focus on specific areas, this adaptability is invaluable.

Lastly, the software's ability to extract well-organized datasheets with precise details empowers users to uncover meaningful insights into identified issues, supporting data-driven decision-making and effective problem-solving. The collaborative effort between authors and requester has resulted in a tool that significantly enhances operational efficiency and decision-making, effectively transforming unstructured data into actionable information, and facilitating comprehensive analysis across multiple product lines over several years.

As future work, we point to the integration of n-grams in word cloud analysis. Incorporating n-grams, which are contiguous sequences of n-words, into word cloud analysis can offer a more nuanced understanding of the textual data. By connecting words in meaningful combinations, n-grams can reveal additional patterns, relationships, and context that may be overlooked in traditional word clouds.

**Acknowledgments.** This work received financial support from the Coordination for the Improvement of Higher Education Personnel - CAPES - Brazil (PROAP/AUXPE - CONSOLIDA 3/4), from the FAPESC agency, and from the CNPq (The National Council for Scientific and Technological Development).

## References

1. Nagarhalli, T.P., Vaze, V., Rana, N.K.: Impact of machine learning in natural language processing: a review. In: 2021 Third International Conference on Intelligent Communication Technologies and Virtual Mobile Networks (ICICV), pp. 1529–1534. IEEE (2021)
2. Brownlee, J.: Deep learning for natural language processing. Mach. Learn. Mystery Vermont Australia **322** (2017)
3. Clark, A., Fox, C., Lappin, S.: The handbook of computational linguistics and natural language processing. John Wiley Sons **118** (2012)
4. Jurafsky, D.: Speech language processing. Pearson Education India (2000)
5. Rosadini, B., et al.: Using NLP to detect requirements defects: an industrial experience in the railway domain. In: Requirements Engineering: Foundation for Software Quality: 23rd International Working Conference, REFSQ 2017, Essen, Germany, February 27–March 2, 2017, Proceedings, pp. 344–360. Springer (2017)
6. Falessi, D., Cantone, G., Canfora, G.: Empirical principles and an industrial case study in retrieving equivalent requirements via natural language processing techniques. IEEE Trans. Softw. Eng. **39**(1), 18–44 (2011). IEEE
7. Zhang, F., Fleyeh, H., Wang, X., Lu, M.: Construction site accident analysis using text mining and natural language processing techniques. Autom. Constr. **99**, 238–248 (2019). Elsevier
8. Tiedemann, J.: The Tatoeba Translation Challenge–Realistic Data Sets for Low Resource and Multilingual MT. arXiv preprint arXiv:2010.06354 (2020)
9. Balakrishnan, V., Lloyd-Yemoh, E.: Stemming and lemmatization: A comparison of retrieval performances. Year of Publication (2014)
10. Webster, J.J., Kit, C.: Tokenization as the initial phase in NLP. In: COLING 1992 Volume 4: The 14th International Conference on Computational Linguistics (1992)
11. Martinez, A. R.: Part-of-speech tagging. Wiley Interdisciplinary Reviews: Computational Statistics **4**(1), 107–113 (2012). Wiley Online Library

# Email Spam Classification: A Machine Learning Approach

Atta Rahman[1](✉), Linah Saraireh[2], Mustafa Youldash[3], Wafa Hantom[1], Dania Alkhualifi[1], Majed Nabil[1], Ashraf Saadeldeen[4,5], Maqsood Mahmud[6], Asiya Abdus Salam[6], Mohammed Salih Ahmed[3], and Mohammed Gollapalli[7]

[1] Department of Computer Science, College of Computer Science and Information Technology (CCSIT), Imam Abdulrahman Bin Faisal University (IAU), Dammam 31441, Saudi Arabia
aaurrahman@iau.edu.sa

[2] Department of Management Information System, College of Business Administration, Imam Abdulrahman Bin Faisal University (IAU), Dammam 31441, Saudi Arabia

[3] Department of Computer Engineering, College of Computer Science and Information Technology, Imam Abdulrahman Bin Faisal University (IAU), Dammam 31441, Saudi Arabia

[4] Department of Management Information Systems, College of Business, Al-Baha University, Al-Baha, Saudi Arabia

[5] Faculty of Computer Science and Information Technology, Omdurman Islamic University, Omdurman, Sudan

[6] School of Computing, Ulster University, Belfast, Northern Ireland, UK

[7] Department of Information Technology & Engineering, Sydney Metropolitan Institute of Technology, Sydney, NSW 2000, Australia

**Abstract.** Email spamming is one of the most common scams that are occurring nowadays. In this method, an email, seemingly sent from a trusted address or sender, can be used for various purposes, including advertisement, commercial and non-commercial intents. Furthermore, it may contain content harmful to the recipient, a tactic known as phishing. Despite several email spam blocking mechanisms employed by the email service providers, it is still a major issue to handle. In this paper, several machine learning algorithms have been investigated over a realistic dataset including (but not limited to) Naïve Bayes (NB), Support Vector Machine (SVM), and Decision Tree (DT) classifiers. Among them Naïve Bayes classifier exhibited the highest accuracy at 99.8% followed by AdaBoost at 96.7%. Analyses showed scheme was promising in terms of accuracy and it outperformed various state-of-the-art approaches in the literature upon comparison.

**Keywords:** Email Spam Classification · Machine Learning · Naïve bayes (NB) · Support Vector machine (SVM) · Decision tree (DT)

## 1 Introduction

We are currently in the era of information and communication technologies (ICT). Every field of human life from personal to business to enterprises is greatly influenced by these technologies. With several benefits, there are numerous issues and threats imposed in


A. Bajaj et al. (Eds.): HIS 2023, LNNS 1224, pp. 137–146, 2025.
https://doi.org/10.1007/978-3-031-78925-0_14

terms of cybersecurity. Email spamming is one of the prevalent scams today. Scammers are becoming increasingly adept at camouflaging their emails to appear as if they are from trusted sources, some scammers generate numerous emails and send them to their targets, this tactic often results in people losing important information from their email inboxes, which can also become cluttered with spam, such emails waste network bandwidth and, more importantly, the user's time in deleting them. Today, there are improved methods to address this issue, such as using Artificial Intelligence and filters to detect spam emails, either moving them to the spam folder or deleting them. However, there is still a need to investigate further classification methods to improve the existing state of the art in terms of accuracy. In this regard, in the current research, we have utilized public datasets available on the Kaggle, and several machine learning-based classifiers have been investigated in the experiment. For instance, the main classifiers are Naïve Bayes, Support Vector Machine, and Decision Tree. Subsequently, the outcomes of all significant performing classifiers are compared.

The rest of the paper is sectioned as follows: Section 2 contains literature review on email spam detection and classification using machine learning. Section 3 provides materials and methods; Section 4 presents results and discussion while Sect. 5 concludes the paper.

## 2 Literature Review

This section provides a comprehensive review of the current state of the art in email spam classification using various approaches in general and machine learning in specific. According to the study in [1], email is one of the primary communication methods. However, it also introduces threats such as the increasing prevalence of spam emails. While email filters can delete these unwanted emails, this method has limitations due to the adaptability of email tools. Therefore, there is a need for further refinement of existing classifiers. The study suggests using the Support Vector Machine (SVM) algorithm and emphasizes the importance of finding optimal parameters to enhance result accuracy. SVM is a supervised classifier that can build complex relationships between classes. The dataset is divided into 70% training and 30% testing. The experiment revealed that the proposed SVM classifier achieved an accuracy of 94.06%, outperforming all other classifiers tested on the same dataset. Another study in [2] highlights that the prevalence of spam emails is currently increasing, leading to users wasting time identifying and managing them in their inboxes. The First Naïve Bayes and the second is a support vector machine. The Naïve Bayes classifier, a commonly used supervised machine learning method, offers rapid classification. Vector machines, being supervised classifiers, have a unique ability to establish complex relationships among classes. The test was conducted on emails from the Enron corpus and the dataset consisted of 6000 emails. It was divided into 3000 emails for spam and the other 3000 emails for ham. The tests have been run numerous times. Out of the 200 test emails, 100 were used to test ham, and the remaining 100 were used as a dataset for spam. The result from the experiment is that for 6000 training emails, the winner was the support vector machine which obtained 95.5% accuracy for naïve bayes got 94.5% accuracy.

In a similar study by [3], the authors highlighted the increasing issue of spam emails. To address this, they proposed a technique using deep learning, specifically the long short-term memory (LSTM) algorithm, which is a variant of the recurrent neural network (RNN). Before employing LSTM, which can abstract features, it was necessary to convert words into semantic word vectors using the word to vector (word2vec) approach, a renowned natural language processing (NLP) method. Subsequently, the results were compared with other classifiers, such as SVM and Naïve Bayes, on two datasets: a Twitter dataset and an SMS spam collection dataset. The experimental results revealed that the LSTM classifier achieved the highest scores with a precision of 95.54%, recall of 98.37%, accuracy of 95.09%, and an F1-score of 96.84%. In another study by [4], it is highlighted that email remains a crucial communication tool. However, the prevalence of unwanted spam emails poses challenges, wasting users' time and reducing network efficiency. This inefficiency can negatively impact legitimate emails due to storage constraints. While machine learning offers a solution, its effectiveness can be compromised with limited datasets, leading to classification issues. To address this, the study introduces a method based on the document labeling concept to classify emails as spam or ham. There are two primary approaches to email classification: knowledge engineering and machine learning. Knowledge engineering classifies emails based on predefined protocols, while machine learning analyzes data patterns for classification. The latter is deemed more effective in this study. Three algorithms, namely Decision Tree (DT), Naïve Bayes (NB), and Random Forest (RF), were employed for classification. These classifiers were trained on labeled instances and tested on unlabeled ones. The experiments were conducted using the PyCharm IDE with three datasets: Enron, Ling-Spam, and PU. Data was split with 70% for training and 30% for testing. Among the classifiers, Random Forest achieved the highest accuracy, scoring 92.97%. According to [5], emails are among the most powerful means of communication, serving both informal and formal purposes. While English is the mainstream language, there are still other native languages in various regions, such as Urdu. With the increased use of Urdu as a medium for communication, there has been a rise in spam emails, posing a challenge. Numerous studies have aimed to detect these spam emails using existing machine learning algorithms, including Naïve Bayes, SVM, CNN, and LSTM classifiers. To test these classifiers, a dataset [1] consisting of 5000 ham and spam emails was used. This dataset was divided into two portions: 80% for training and 20% for testing. In comparison, LSTM achieved the highest accuracy of 98.4%, followed by Naïve Bayes at 98%, SVM at 97.5%, and CNN at 96.2%. In the study presented by [6], various machine-learning methods are explored for detecting spam emails. While these methods excel at classifying emails as unsolicited or solicited, they often struggle with handling high-dimensional data and tend to have a lower detection rate. By integrating the logistic regression model with an artificial bee colony algorithm, significant improvements in accuracy were observed. Tests were conducted using multiple datasets, including CSDMC2010 and TurkishEmail. Results indicated that the combination of the artificial bee colony with logistic regression outperformed other methods in terms of accuracy. While many machine learning methods are efficient and straightforward to implement, they can be computationally intensive, especially with high-dimensional data. The Artificial Bee Colony (ABC) algorithm comprises three types of artificial bees: onlookers, scout bees, and employed bees. The primary objective of this algorithm is to identify the food source with the highest nectar content. The highest accuracy was obtained by SVM with 98.75% for Turkish email datasets overall followed by the Naïve bayes algorithm that performed in the range of 93.71% to 96.63% for the diverse datasets. The objective of the study presented [7] is to evaluate the effectiveness of twelve commonly used feature selection (FS) methods on a spam email dataset. Several FS methods, including those involving the radial basis function (RBF) kernel, have been tested on the spam dataset to determine the most effective approach. The results of the study indicate that filter methods such as CFS, GR, Chi-squared, ReliefF, SU, IG, and oneR enabled the classifiers to achieve the highest increase in classification prediction accuracy due to the removal of unnecessary attributes. Furthermore, the results also demonstrate that using feature selection can improve the accuracy of support vector machine classifiers when applied correctly. This paper [8] outlines the successful implementation of an ontology-based email classification system. The motivation was to introduce a new aspect of email classification on the semantic web, as this approach is generic in nature and can be applied to any system. The challenge was to interface two independent systems and create a prototype that classifies emails through ontologies. The results were based on the default settings, and further experiments could be conducted using different algorithms and feature selection techniques [9]. The paper also proposes an adaptive ontology as a spam filter, which can be customized and scaled to suit the user's needs. The purpose of this research reported in [10] is to provide a comparative technique for phishing email detection that combines Natural Language Processing (TF-IDF, Word2Vec, and BERT) with Machine Learning (Random Forest, Decision Tree, Logistic Regression, Gradient Boosting Trees, and Naive Bayes). The test was carried out on two datasets, one balanced and one unbalanced, both of which contained emails from the well-known Enron corpus and the Nazario phishing corpus. Their highest-performing algorithm for the TF-IDF balanced dataset was Logistic Regression with an accuracy of 92.9%. For the Word2Vec balanced dataset, the highest result was 98.9% with Random Forest. For the TF-IDF imbalanced dataset, the highest accuracy was 89.9% with Logistic Regression. For the Word2Vec imbalanced dataset, the highest result was 92.4% with Logistic Regression. The research in [11] aims to employ a distributional representation, specifically TFIDF+SVD, to numerically represent phishing emails. The study involved a comparison of traditional machine learning approaches, including Random Forest, AdaBoost, Naive Bayes, Decision Tree, and SVM. Among these, SVM achieved the highest accuracy at 99.4%. Likewise, the research conducted in [12] aims to describe a method for detecting spam emails. This method utilizes machine learning algorithms enhanced with bio-inspired techniques. Using a dataset of 50,000 emails, various machine learning models, including Naive Bayes, Support Vector Machine, Random Forest, Decision Tree, and Multi-Layer Perceptron, were evaluated. To optimize classifier performance, Particle Swarm Optimization and the Genetic Algorithm were employed. Naïve Bayes model accurately identified 98% of the emails. The study in [13] applied two machine learning classifiers for spam email detection: Support Vector Classifier and Random Forest Classifier. The data they used came from the UC Berkeley spam database. The Random Forest Classifier had the highest accuracy (91.36%). Similarly, in the report presented in [14] the data was first preprocessed by removing distinguishing symbols, URLs, and HTML elements. They selected the SVM classifier and achieved an accuracy of 98%. Moreover, in the study by [15], natural language processing (NLP) was employed to enhance model accuracy. The research utilized two open-source datasets: the UCI Spambase dataset with 5569 emails and the Kaggle Spam filter dataset comprising 5728 emails. The datasets were preprocessed by removing stop words and punctuation, followed by tokenization. Among the classifiers tested, which included KNN, NB, BiLSTM, and Bert Base Cased (BBC) which achieved the highest accuracy at 98.67% and an impressive F1 score of 98.66%. The study detailed in [15] aims to assess the efficacy of twelve prevalent feature selection (FS) methods on a dataset of spam emails. Various FS techniques, including those incorporating the RBF kernel, were evaluated on this dataset to identify the most potent method. The findings revealed that filter methods such as CFS, GR, Chi-squared, ReliefF, SU, IG, and oneR enhanced the classifiers' predictive accuracy by eliminating superfluous attributes. Additionally, the study underscores that the judicious application of feature selection can bolster the precision of support vector machine classifiers. Another study in [16] presented Arabic twitter spam detection approach using MARBERT model. The proposed approach was able to detect sentiment as well as the spam tweets using machine learning algorithms on the Arabic dataset. The dataset belonged to STC a telecom company in Saudi Arabia. Based on the literature review, it can be concluded that email spam detection is among the hottest areas of research in cybersecurity with NB, DT, SVM are among the most frequently used algorithms. Still there is room for improvement in terms of accuracy and investigation of various machine learning algorithms over the diverse datasets.

## 3 Materials and Methods

This section presents the algorithms utilized in the current research for email spam classification. That are Naïve Bayes, Support Vector Machine (SVM) and Decision Trees.

### 3.1 Naïve Bayes

Naive Bayes is a probabilistic machine learning algorithm primarily used for classification tasks. It operates based on the Bayes theorem and adopts the 'naive' assumption that all features of a data point are independent of each other. Simple to implement, it's suitable for both binary and multi-class classification. The algorithm calculates the probability of a data point belonging to a specific class based on its features and uses this probability for classification. Due to its speed and accuracy, Naive Bayes is employed in various applications, such as text classification and spam filtering [17–20]. Naïve Bayes algorithm is based on the conditional probabilistic model. The Naïve Bayes algorithm's working principle is to pick the most likely hypothesis, which is also known as the maximum a posteriori (MAP) probability and can be expressed as given in Eq. 1:

$$y = \arg\max yP(y)\prod P(P(x_i)) \tag{1}$$

### 3.2 Support Vector Machines (SVMs)

The SVM is a type of supervised machine learning algorithm used for both classification (discrete) and regression (continuous) tasks. They excel in applications like image classification, text categorization, and time series forecasting. They are fast and dependable classification algorithms. It is supervised machine learning that shows to give excellent and accurate performance results. SVMs operate on the principle of maximizing the margin between two classes, enhancing the model's generalization. The algorithm maps data into a space where it can define a hyperplane that best separates the classes. This hyperplane is then used to predict the class label of unseen data points [21–25]. It is equally good with binary as well as non-binary classifications. Moreover, SVM works better when the dimensional spaces get higher as the separation of the margin is more apparent between the classes, while it is not a good choice for large datasets. Equation 2 illustrates the optimal hyperplane in SVM, where w is the weight vector, x represents the input vector, and b is the bias, which affects the classifier so that it won't always go through the origin.

$$w \bullet x + b = 0 \tag{2}$$

### 3.3 Decision Tree Algorithm

Decision tree algorithm also known as J48 is a popular data mining technique used for creating predictive models [26–32]. These algorithms utilize a tree-like structure where each node represents a decision based on the data, and each branch signifies the possible outcomes of that decision. The algorithm determines the best way to split the data into subsets and create branches accordingly. Once the tree is constructed, it can generate predictions based on new data inputs. Mathematically it is written as,

$$Entropy = -\sum_{i=1}^{n} p_i \log(p_i) \tag{3}$$

In the current study, we selected these algorithms based on several criteria. Primarily, we believe they are well-suited for the dataset being investigated, and its characteristics are compatible with these methods. Additionally, many other studies have successfully implemented these algorithms using data mining techniques on similar datasets. Since our dataset is supervised, we opted for classification methods over unsupervised approaches.

## 4 Results and Discussion

We utilized nine algorithms in our study, of which three are highlighted here: Naive Bayes, Decision Tree (DT), and Support Vector Machines (SVM). Among the three classifiers of Naive Bayes, we employed the Multinomial classifier, achieving a balanced accuracy of 92%. In contrast, the related work reported an accuracy of 94.76%, approximately 95%. A challenge we encountered was the issue of imbalanced classes, which can lead to an imbalanced model. To address this, we implemented the cross-validation technique with ten folds, resulting in a better accuracy. Unlike the related work that used regular accuracy, we opted for a different metric to better represent our model's performance, especially when dealing with imbalanced classes. We suspect that the related work might have overfitted their model predictions and did not address this potential issue. Unbalanced classes pose a challenge to both the Decision Tree (DT) and Support Vector Machines (SVM) algorithms. To address this, we applied a balanced accuracy score as an evaluation metric for our model's performance. Our results showed a balanced accuracy of 85% for the DT method and 80% for the SVM algorithm. In comparison, the related work reported accuracy results of 64% for DT and 46.6% for SVM. A significant factor contributing to the discrepancy in accuracy is the related work's inadequate preprocessing. They overlooked several crucial components in data preparation, leading to their models' reduced accuracy. In the experiments with various algorithms, we obtained the following results after fine-tuning and optimizing the hyperparameters, all of which are detailed in Table 1.

- Naïve Bayes: Achieved an accuracy of 99.8% without any modifications or tuning with the lowest computational complexity inherently.
- Decision Tree: Using the cross-validation technique to split the dataset, we obtained an accuracy of 91%.
- Random Forest: With the default depth and using cross-validation for testing, we achieved an accuracy of 92%.
- SVM: After extensive tuning and adjusting the Regularization parameter multiple times, we achieved an accuracy of 92%
- Bagging Classifier: After fine-tuning and experimenting with different numbers of base estimators in the ensemble settings, we achieved an accuracy of 93%.
- AdaBoost: We achieved an accuracy of 96.7% after experimenting with different numbers of base estimators in the ensemble, learning rates, and algorithms.
- AdaBoost with Decision Tree (AWDT): We achieved an accuracy of 91% after experimenting with different numbers of base estimators in the ensemble, learning rates, and algorithms.
- Gradient Boosting: Without any alterations, we obtained an accuracy of 92%. This was achieved after dividing the data using cross-validation.
- Extremely Randomized Tree: We attained an accuracy of 92% after testing with different numbers of base estimators in the ensemble and adjusting the class weight.

**Table 1.** Result Summary

| Class | Accuracy |
|---|---|
| Naïve Bayes | 99.8% |
| Decision Tree | 91% |
| Random Forest | 92% |

*(continued)*

**Table 1.** (*continued*)

| Class | Accuracy |
|---|---|
| SVM | 92% |
| Bagging Classifier | 93% |
| AdaBoost | 96.7% |
| AWDT | 91% |
| Gradient Boosting | 92% |
| Extremely Randomized Tree | 92% |

## 5 Conclusion

We conclude that email spamming is a real problem that must be dealt with because it wastes the user's time in determining whether an email is legitimate or not. It also consumes network bandwidth and poses potential cyber threat to the receiver. Several studies have been conducted in literature; however, it still requires more research and studies to be investigated. In this paper, after conducting adequate literature reviews of email spamming, we figured out the research gap, commonly used algorithms, and techniques. Consequently, we investigated various machine learning algorithms on the diverse datasets collected from open sources. Upon experimental evaluation, it was found that Naïve Bayes classifier achieved the highest accuracy of 99.8% followed by AdaBoost at 96.7% while other classifiers also performed above 91% in the experiment, outperforming other classifiers in the literature. Moreover, due to its naïve nature, Naïve Bayes exhibits little computation complexity in contrast to other algorithms. All the schemes are fine-tuned by optimizing the hyperparameters and are promising in terms of accuracy. In future, more diverse, and amalgamated datasets can be utilized, together with feature selection. Also, various deep, transfer and hybrid learning as well as real time approaches can be investigated [33–43]. That will help further fine tune the results.

## References

1. Olatunji, S.: Improved Email Spam Detection Model Based on Support Vector Machines. Springer, Dammam (2017)
2. Ma, T.M., Yamamori, K., Thida, A.: A comparative approach to naive Bayes classifier and support vector machine for email spam. In: Global Conference on Consumer Electronics (GCCE), Miyazaki (2020)
3. Jain, G., Sharma, M., Agarwal, B.: Optimizing Semantic LSTM for Spam Detection. CrossMark, Buona (2018)
4. Gaurav, D., Tiwari, S.M., Gandhi, N., Abraham, A., Goyal, A.: Machine Intelligence-Based Algorithms for Spam Filtering on Document Labeling. Springer Nature, Punjab (2019)
5. Siddique, Z.B., Khan, M.A., Din, I.U., Almogren, A., Mohiuddin, I., Nazir, S.: Machine Learning-Based Detection of Spam Emails. Hindawi, Haripur (2021)
6. Dedeturk, B.K., Akay, B.: Spam Filtering Using a Logistic Regression Model Trained by an Artificial Bee Colony Algorithm. ScienceDirect, Kayseri (2020)

7. Youn, S., McLeod, D.: Spam email classification using an adaptive ontology. J. Softw., Los Angeles.
8. Hantom, W.H., Rahman, A.: Arabic spam tweets classification: a comprehensive machine learning approach. AI **5**(3), 1049–1065 (2024). https://doi.org/10.3390/ai5030052
9. Sharma, P., Bhardwaj, U.: Machine Learning Based Spam E-Mail Detection. INASS, Rohtak (2017)
10. Bountakas, P., Koutroumpouchos, K., Xenakis, C.: A Comparison of Natural Language Processing and Machine Learning Methods for Phishing Email Detection. ACM, Vienna (2021)
11. Ravi, H.N.B.V., Kp, S.: A machine learning approach towards phishing email detection CEN-Security@IWSPA 2018. In: ACM Conference on Data and Application Security and Privacy (CODASPY 2018), Arizona (2018)
12. Jacob, S.M., Gibson, S., Issac, B., Zhang, L.: Detecting Spam Email with Machine Learning Optimized with Bio-Inspired Metaheuristic Algorithms. IEEE Access, Middlesbrough (2020)
13. T. E. O and E. S. P: A model to detect spam email using support vector classifier and random Forest classifier. Int. J. Comput. Sci. Math. Theory, Nigeria. (2020)
14. Reddy, A., Uma Maheswari. M., Viswanathan, A., Vikram, G.: Using support vector machine for classification and feature extraction of spam in email. Int. J. Innov. Eng., Secunderabad. (2022)
15. AbdulNabi, I. and Yaseen, Q.: Spam email detection using deep learning techniques. The 2nd International Workshop on Data-Driven Security, Jordan (2021)
16. Alotaibi, A., Rahman, A., Alhaza, R., Alkhalifa, W., Alhajjaj, N., Alharthi, A., et al.: Spam and sentiment detection in Arabic tweets using MarBert model. **9**(6), 1574–1582 (2023)
17. Gollapalli, M., Rahman, A., Osama, A., Alfaify, A., Yassin, M., Alabdullah, A.: Data mining and visualization to understand employee attrition and work performance. In: 3$^{rd}$ International Conference on Computing and Information Technology (ICCIT), Tabuk, Saudi Arabia, pp. 149–154 (2023). https://doi.org/10.1109/ICCIT58132.2023.10273889
18. Musleh, D.A., Olatunji, S.O., Almajed, A.A., Alghamdi, A.S., Alamoudi, B.K., Almousa, F.S., Aleid, R.A., Alamoudi, S.K., Jan, F., et al.: Ensemble learning based sustainable approach to carbonate reservoirs permeability prediction. Sustain. For. **15**, 14403 (2023)
19. Ahmed, M.I.B., Zaghdoud, R.A., et al.: Ensemble machine learning based identification of adult epilepsy. Math. Model. Eng. Probs. **10**(1), 84–92 (2023)
20. Alghamdi, A.S., Rahman, A.: Data mining approach to predict success of secondary school students: a Saudi Arabian case study. Educ. Sci. **13**, 293 (2023)
21. Ahmed, M.I.B., Alabdulkarem, H., Alomair, F., Aldossary, D., Alahmari, M., Alhumaidan, M., Alrassan, S., Rahman, A., Youldash, M., Zaman, G.: A deep-learning approach to driver drowsiness detection. Safety. **9**, 65 (2023). https://doi.org/10.3390/safety9030065
22. Qamar, R.A., Sarfraz, M., Rahman, A., Ghauri, S.A.: Multi-criterion multi-UAV task allocation under dynamic conditions. J. King Saud Univ.-Comput. Inf. Sci. **35**(9), 101734 (2023)
23. Aljameel, S.S., Rahman, A.: Enhancing multi-user detection in multicarrier 5G and beyond: a space-time spreading approach with parallel interference cancellation. Math. Model. Eng. Probs. **10**(4), 1207–1215 (2023)
24. Ahmed, M.S., Rahman, A., AlGhamdi, F., AlDakheel, S., Hakami, H., AlJumah, A., AlIbrahim, Z., Youldash, M., Alam Khan, M.A., Basheer Ahmed, M.I.: Joint diagnosis of pneumonia, COVID-19, and tuberculosis from chest X-ray images: a deep learning approach. Diagnostics. **13**, 2562 (2023). https://doi.org/10.3390/diagnostics13152562
25. Ahmed, M.I.B., Alotaibi, R.B., Al-Qahtani, R.A., Al-Qahtani, R.S., Al-Hetela, S.S., Al-Matar, K.A., Al-Saqer, N.K., et al.: Deep learning approach to recyclable products classification: towards sustainable waste management. Sustain. For. **15**, 11138 (2023)

26. Farooqui, M., Rahman, A., Alorefan, R., Alqusser, M., Alzaid, L., Alnajim, S., et al.: Food classification using deep learning: presenting a new food segmentation dataset. Math. Model. Eng. Probs. **10**(3), 1017–1024 (2023)
27. Ibrahim, N.M., Gabr, D.G., Rahman, A., Musleh, D., AlKhulaifi, D., AlKharraa, M.: Transfer learning approach to seed taxonomy: a wild plant case study. Big Data. Cogn. Comput. **7**, 128 (2023). https://doi.org/10.3390/bdcc7030128
28. Gollapalli, M., Rahman, A., Alkharraa, M., Saraireh, L., AlKhulaifi, D., Salam, A.A., Krishnasamy, G., et al.: SUNFIT: a machine learning-based sustainable university field training framework for higher education. Sustain. For. **15**, 8057 (2023)
29. Alqarni, A., Rahman, A.: Arabic tweets-based sentiment analysis to investigate the impact of COVID-19 in KSA: a deep learning approach. Big Data Cogn. Comput. **7**, 16 (2023)
30. Sajid, N.A., Ahmad, M., Rahman, A., Zaman, G., Ahmed, M.S., Ibrahim, N., et al.: A novel metadata based multi-label document classification technique. Comput. Syst. Sci. Eng. **46**(2), 2195–2214 (2023)
31. Mahmud, M., Rahman, A., Lee, M., Choi, J.Y.: Evolutionary-based image encryption using RNA codons truth table. Opt. Laser Technol. **121**, 105818 (2020)
32. Atta-ur-Rahman, Dash, S., Luhach, A.K., et al.: A neuro-fuzzy approach for user behaviour classification and prediction. J. Cloud Comput. **8**, 17 (2019)
33. Qureshi, M.M., Yunus, F.B., Li, J., Ur-Rahman, A., Mahmood, T., Ali, Y.A.A.: Future prospects and challenges of on-demand mobility management solutions. IEEE Access. **11**, 114864–114879 (2023). https://doi.org/10.1109/ACCESS.2023.3324297
34. Qureshi, M.A., Asif, M., Anwar, S., Shaukat, U., et al.: Aspect level songs rating based upon reviews in English. Comput. Mater. Continua. **74**(2), 2589–2605 (2023)
35. Rahman, A.: Memetic computing based numerical solution to Troesch problem. J. Intell. Fuzzy Syst. **36**(6), 1–10 (2019)
36. Rahman, A.: Optimum information embedding in digital watermarking. J. Intell. Fuzzy Syst. **36**(6), 553–564 (2019)
37. Rahman, A., Alhaidari, F.: An electronic data interchange framework for educational institutes. ICIC Express Lett. **13**(9), 831–840 (2019)
38. Rahman, A.: Efficient decision based spectrum mobility scheme for cognitive radio based V2V communication system. J. Commun. **13**(9), 498–504 (2018)
39. Jan, F., Rahman, A., Busaleh, R., Alwarthan, H., Aljaser, S., Al-Towailib, S., Alshammari, S., Alhindi, K.R., Almogbil, A., Bubshait, D.A., et al.: Assessing acetabular index angle in infants: a deep learning-based novel approach. J. Imaging. **9**, 242 (2023)
40. Abbas, S., Raza, S.A., Khan, M.A., Rahman, A., Sultan, K., Mosavi, A.: Automated file labeling for heterogeneous files organization using machine learning. Comput. Mater. Continua. **74**(2), 3263–3278 (2023)
41. Zaman, G., Mahdin, H., Hussain, K., Atta-Ur-Rahman, J.A., Mostafa, S.A.: An ontological framework for information extraction from diverse scientific sources. IEEE Access. **9**, 42111–42124 (2021). https://doi.org/10.1109/ACCESS.2021.3063181
42. Rahman, A., Alhaidari, F.: Querying RDF data. J. Theor. Appl. Inf. Technol. **26**(22), 7599–7614 (2018)
43. Farooq, M.S., Abbas, S., Rahman, A., et al.: A fused machine learning approach for intrusion detection system. Comput. Mater. Continua. **74**(2), 2607–2623 (2023)

# Models for Short-Term Electricity Price Forecasting

Lídio Mauro Lima de Campos[1,2](✉)

[1] Universidade Federal do Pará, Belém, Brazil
[2] Instituto de Ciências Exatas e Naturais - Programa de Pós Graduacão em Ciência da Computacão - PPGCC, Belém do Pará, Pará, Brazil
lidio@ufpa.br
https://www.ufpa.br/,https://www.computacao.ufpa.br/

**Abstract.** Currently, due to the competitive energy markets, more and more energy producers, concessionaires, and those who sell energy are exposed to risks caused by the volatility of electricity prices. Due to dependence on renewable sources, weather, and global financial instability, accurate electricity price forecasting is increasingly important. The main objective of this work is to evaluate the performance of forecasting methods: Deep neural networks and LSTM neural networks, using a time series of electricity prices in five regions of Brazil. The results for Deep feedforward networks show an excellent ability to predict peaks and satisfactory accuracy according to the Root Mean Squared Error of up to 9.8 R$/Mwh. Preliminary studies show that LSTM networks can predict the PLD value with a percentage Root Mean Squared Error of up to 0.13 R$/Mwh, lower than that obtained by deep feedforward networks and other methods. In 66.66% of the cases, the LSTM network presents more minor prediction errors than the Deep Feedforward network.

**Keywords:** Deep Neural Network · Times Series Prediction · wind speed prediction

## 1 Introduction

Electricity price forecasting is an important issue for all market participants to decide on the most suitable bidding strategies and establish bilateral contracts that maximize their profits and minimize their risks. Currently, some forecasting methods have been proposed in the literature [1,2].

Electric energy is vital for the industrial infrastructures of nations, companies in the electricity sector are increasingly looking to monitor and control it, to provide improvements in the management and planning of the sector. Specifically in Brazil, privatizations and deregulation changed the sector's scenario, making it evolve through the institution of a market structure, aimed at free competition in the purchase and sale of electricity. Thus, generation, transmission, and distribution activities began to be carried out independently and autonomously.

Federal University of Pará.


A. Bajaj et al. (Eds.): HIS 2023, LNNS 1224, pp. 147–156, 2025.
https://doi.org/10.1007/978-3-031-78925-0_15

The Fig. 1 shows the trajectory of each tariff component in the period from 2010 to 2019.

Although the average price of electricity in several countries has fallen, there has been an increase in volatility in energy prices on the spot market in several countries, such as the USA (California), Australia, and Brazil. The energy market has physical characteristics that make the occurrence of sudden and temporary peaks (price spikes) in the price to be greater in any other commodity markets [3].

So, developing price forecast models in the short-term market to reduce possible associated risks is a major challenge that the electricity sector is facing. The prediction of future energy price values and their peaks is of great importance for the decision-making process and the elaboration of energy commercialization strategies.

Currently, different methods have been applied in order to minimize costs and seek greater system efficiency, in this direction some research has been developed. As a result, machine learning techniques have begun to be increasingly used, due to the increased accessibility of data, in the modeling and analysis of energy systems.

Therefore, uncertainty modeling in electricity systems is increasingly necessary. In particular, ARMA, ARIMA, ARMAX, and ARIMAX are methods most used classics in time series. Among the existing machine learning methods, the deep learning method can be considered one of the most successful ones, especially in image and text mining areas. However, its application to time series in electric power systems is less reported. This work presents a study on power market price forecasting by deep learning is proposed. Two of the most successful deep learning frameworks, the LSTM (Long short-term memory) neural network and Deep Feedforward are utilized.

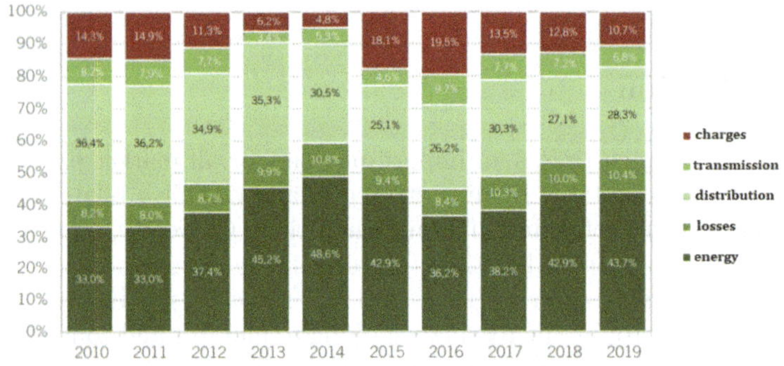


**Fig. 1.** Evolution of tariff components in the period from 2010 to 2019 (percentage share), Font:[7]

This paper is organized as follows. Section 2 dicusses related work, Sect. 3 presents the background of the study, Sect. 4 presents Material and Methods. Lastly, simulation results and conclusions are presented in Sects. 5 and 6, respectively.

## 2 Related Work

The author [4] studied different machine learning methods that can be used to predict electricity prices and articulated the importance of adopting price forecasting to regulate chaining. The supply forecast is quite static in the short term. However, renewable energy sources: solar and wind, are volatile in the short term, but highly predictable, as they are highly correlated with the climate. As can be shown by the correlation between wind production and spot prices in Denmark [5].

The objective of [6]'s work was to reduce electricity consumption for consumers with a focus on the residential sector in periods of greater demand. For this, the author proposed a methodology to statistically analyze the energy demand data and for the end user to better organize their consumption. Different methods were tested, the Auto-Regressive Integrated Moving Average model was selected because it allowed to forecast production under load demand with good accuracy.

[8] used a hybrid neural model (HIRA model) for short-term electricity price forecasting, which was also applied and tested in two electricity markets: German and Hungarian. In Hungary, thermal plants have 90% of the installed capacity and represent a stable source of electricity, but generally involve a large installed capacity per unit. Hungary uses a bid-based market, its work calculates electricity price forecast in three levels: Hourly forecast error, Peak load forecast error, and Base load forecast error.

## 3 Brazilian Energy Market

Brazil uses a cost-based market rather than a bid-based market and adopts a restricted pool model with centralized dispatch and lower cost organized by the National System Operator (ONS) [9]. This scheme was adopted due to the peculiarities of the country, which has an installed capacity of 121 GW where 65.96% corresponds to hydroelectric generation. The water system is composed of several reservoirs subject to multiannual regulation located on the same river with different owners.

The difference between the amount of contracted energy and that effectively consumed or produced by the agents is accounted for in the short-term market based on the spot price called PLD (settlement price for differences). The PLD is calculated weekly, based on the marginal operating cost of the system obtained in an optimization process for dispatching generators, at three load levels ("Medium", "Heavy", "Light"). The PLD was established by the National

Electric Energy Agency (ANEEL) [10] and is evaluated for each submarket associated with the country's regions: North, Northeast, Midwest/Southeast, and South.

# 4 Material and Methods

This section presents the methodology used for Electricity price forecasting in the five Brazilian regions (north, south, midwest, northeast, and southeast). Throughout this section, the main performance indices that are used to quantify how satisfactory the results of the model's predictions will be discussed. In addition, the used architectures of deep neural networks are presented.

## 4.1 Machine Learnings Models

**Neural Networks.** In this section, we describe the Neural Network architectures that were used in the experiments, which were deep feedforward and LSTM.

**Deep FeedForward NN.** This model, characterized by a three-layer or more network of simple processing units connected by acyclic links (Fig. 2(a)), is commonly employed in systems biology studies, including evolutionary studies. Information flows through the ANN in discrete time.

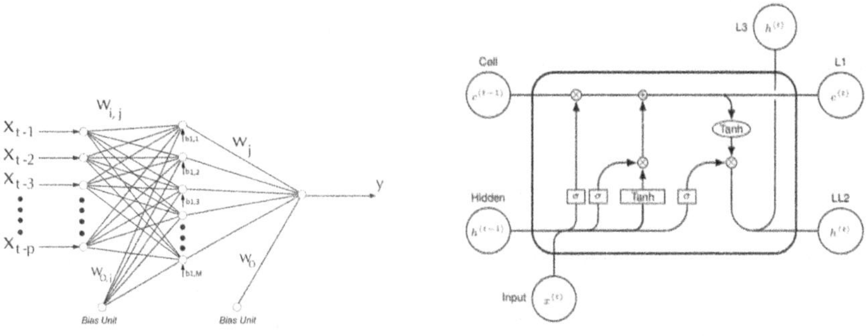


(a) Deep Feed Forward NN (b) LSTM NN

**Fig. 2.** Wind Speed observed and predicted with Deep Feed Forward and LSTM networks.

**Long Short-Term Memory-LSTM.** The Recurrent neural network dynamics can be formulated by deterministic transitions from previous to current hidden states. The deterministic state transition is described in Eq. 1 The architecture of the LSTM neural network is shown in Fig. 2(b). A typical LSTM network is composed of memory cells. The cell state and the hidden state are transferred to the next cell. A gate is analogous to a layer or a chain of matrix operations carrying discrete weights. LSTMs are projected to prevent the long-term dependency problem of using gates to regulate the memorizing process.

$$RNN : h_t^{l-1}, h_{t-1}^l \rightarrow h_t^l \tag{1}$$

### 4.2 Dataset Description

The dataset used in this research contains data on electricity prices from 2002 to 2009, presented weekly, for each sub-market: North, Northeast, South, and Midwest/Southeast, provided by the National Electric Energy Agency [10], in addition to data on the explanatory variables taken from the website of the National Operator of the Brazilian System [9].

**Data Preparation.** Many problems are involved in analyzing rare patterns of occurrences. In the dataset used, some patterns occur more frequently than others. In addition, most of the time the price remains at low values, below R$ 100.00 and the price of energy rarely reaches values above R$ 300.00. However, neural networks are sensitive to unbalanced datasets, as they cause difficulties in the learning process and can deteriorate model performance. Therefore, it was necessary to perform data balancing during the data preparation process.

### 4.3 Performance Indices

In time series forecasting problems, an important task is to quantify the quality of the prediction obtained. This allows, for example, comparing different algorithms and different model structures using performance indices. The following are some of the performance indices used in the prediction of the time series of energy consumption: Root mean squared error (RMSE), Mean absolute error (MAE), and Mean absolute percentage error (MAPE). These quantities are calculated by:

$$MAE = \frac{1}{N} \Sigma_{i=1}^{N} |p_i true - p_i forecast| \tag{2}$$

$$RMSE = \sqrt{\frac{1}{N} \Sigma_{i=1}^{N} \left(p_i true - p_i forecast\right)^2} \tag{3}$$

$$MAPE = \frac{100\%}{N} \sum_{I=1}^{N} \left| \frac{p_i true - p_i forecast}{p_i true} \right| \tag{4}$$

where: N is the number of samples, $p_i true$ is the actual price and $p_i forecast$ is the forecasted price.

# 5 Simulation Results

This section presents the results of predicting the price of electricity in the five Brazilian regions, after balancing the data.

## 5.1 Deep FeedForward Neural Network (DFFNN)

Table 1 presents the best results with balanced data for the South region. Table 2 illustrates the average values of prediction errors (RMSE, MAE, and MAPE) obtained by the DFF neural networks for the North, South, Southeast, and Northeast regions.

The values for the South region reveal a mean square error (RMSE) of 8.20 R$/ Mwh, MAE and MAPE of 5.80 R$/Mwh, and 0.19 R$/Mwh, respectively. For the North region, the mean square error (RMSE) was 5.02 R$/Mwh, and the MAE and MAPE were 3.55 R$/Mwh and 0.12 R$/Mwh, respectively. Therefore, the errors (RMSE, MAE, and MAPE) obtained for the southern region were higher than those obtained for the northern region.

Data for the Southeast region reveal a mean squared error (RMSE) of 9.8 R$/Mwh, MAE and MAPE obtained were 6.93 R$/Mwh and 0.14 R$/Mwh, respectively. For the Northeast region, the mean squared error (RMSE) was 5.8 R$/Mwh, and MAE and MAPE obtained were 4.1 R$/Mwh and 0.17 R$/Mwh, respectively. Only the MAPE value obtained for the Southeast region was lower than that obtained for the Northeast region, the other RMSE and MAE values were higher. Figure 3(a) shows the short-term price observed and predicted by the DDFNN 36 weeks ahead for the Northeast region.

**Table 1.** Best results with balanced data for the South region, where HLN 1 and HLN 2 are number of hidden layer neurons 1 and 2, HLAF 1 and HLAF 2 are the Hidden layer 1 and 2 activation function, OLAF is Output layer activation function.

| PLD | HLN 1 | HLN 2 | HLAF 1 | HLAF 2 | OLAF | RMSE | MAE | MAPE |
|---|---|---|---|---|---|---|---|---|
| Light | 25 | 12 | sigmoid | sigmoid | linear | 7,52294 | 5,31952 | 0,17856 |
| Average | 13 | 12 | sigmoid | sigmoid | linear | 8,52302 | 6,02668 | 0,198 |
| Heavye | 29 | 27 | sigmoid | sigmoid | linear | 8,573619 | 6,066428 | 0,19821 |

## 5.2 LSTM Neural Network LSTM NN

In this section, we present the results obtained by the LSTM neural networks. Recurrent neural networks (RNNs) have gained wide use in sequential data modeling. The Long Short-Term Memory (LSTM) network is a type of RNN in deep learning because very large architectures can be successfully trained. The work [11] shows the robustness of LSTM networks in handling unbalanced data. The

**Table 2.** Energy price error measures obtained with the DFFNN 36-weeks ahead.

| Average PLD | South | | | North | | |
|---|---|---|---|---|---|---|
| | RMSE | MAE | MAPE | RMSE | MAE | MAPE |
| 36 steps forward | 8,2073 | 5,8034 | 0,1982 | 5,0203 | 3,5513 | 0,1288 |
| Average PLD | Northeast | | | Southeast | | |
| | RMSE | MAE | MAPE | RMSE | MAE | MAPE |
| 36 steps forward | 5,80 | 4,1070 | 0,1704 | 9,8764 | 6,9303 | 0,1460 |

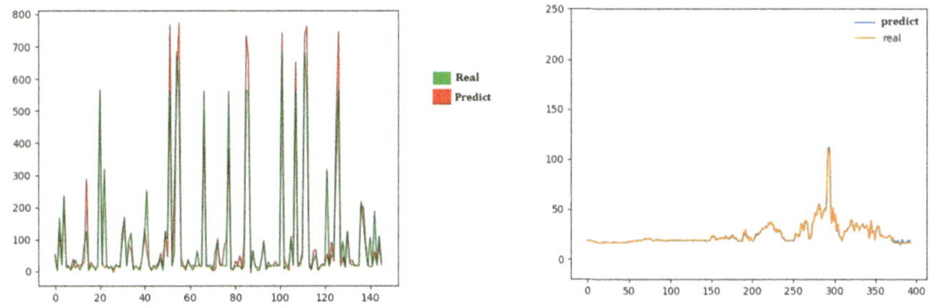


(a) Deep Feed Forward NN (b) LSTM NN

**Fig. 3.** Energy price observed and predicted with Deep Feed Forward and LSTM networks for Northeast and Southeast regions respectively.

empirical studies conducted and reported in [12] show that algorithms based on deep learning, such as LSTM, outperform traditional algorithms such as the ARIMA model.

Table 3 presents the best results with balanced data for the Southeast region. Table 4 presents the prediction results for the Southeast and Northeast regions, for 12, 24, and 26 steps ahead. It is possible to conclude that the values of MAE and RMSE errors were lower than those obtained by DFFNN, except for the MAPE value. However the MAPE values can be interpreted as follows, $< 10\%$ very good, 10%–20% good, 20%–50% acceptable and $> 50\%$ not good.

Table 5 presents the prediction results for the North and South regions, for 12, 24, and 26 steps ahead. It is possible to conclude that the values of MAE and RMSE errors were lower than those obtained by DFFNN, except for the MAPE value.

The results using the HIRA model [13] in the Hungarian market, three days in advance for day d, in the 3rd interval for the base load energy price (it is the hourly average price for all 24 h) and in the peak (is the delivery of the same amount of energy in the period from 8:00 am to 8:00 pm) indicate a percentage error of around 0.56% to 2.96%. Our results for all regions revealed a percentage error of around 0.09% to 0.7333%, lower than the values of the HIRA model [13].

**Table 3.** Best results with balanced data for the Southeast region, where HLN 1 and HLN 2 are number of hidden layer neurons 1 and 2, HLAF 1 and HLAF 2 are the Hidden layer 1 and 2 activation function, OLAF is Output layer activation function.

| PLD | HLN 1 | HLN 2 | HLAF 1 | HLAF 2 | OLAF | RMSE | MAE | MAPE |
|---|---|---|---|---|---|---|---|---|
| Light | 20 | 23 | H. tangent | H. tangent | Relu | 0,24 | 0,06 | 0,29 |
| Average | 31 | 26 | H. tangent | H. tangent | Relu | 0,34 | 0,11 | 0,44 |
| Heavy | 29 | 15 | H. tangent | H. tangent | Relu | 0,2762 | 0,0763 | 0,3906 |

**Table 4.** Results for LSTM networks.

| Heavy PLD | Southeast——-Northeast | | | | | |
|---|---|---|---|---|---|---|
| | MAE | MAPE | RMSE | MAE | MAPE | RMSE |
| 12 steps forward | 0,0483 | 0,1208 | 0,6167 | 0,0167 | 0,0063 | 0,14 |
| 24 steps forward | 0,1333 | 0,7333 | 0,3600 | 0,2083 | 0,5800 | 0,2966 |
| 36 steps forward | 0,0800 | 0,3733 | 0,2833 | 0,0456 | 0,1900 | 0,2133 |

### 5.3 Comparison Between DFFNN and LSTM

In addition, the significances of the results obtained by DFFNN and LSTM NN, models, by means of $t$ tests at the 95% confidence level (i.e., $p$-value below 0.05). We statistically compared the algorithm performances by three criteria. The primary (CC1), second (CC2), and third (CC3) criteria were the average RMSE, MAPE, and MAE. After, we compared pairs of models (DFFNN and LSTM NN). When there were no significant statistical differences between the CC1, CC2, and CC3 values of two models on a given simulation, they considered that both models performed equally well. In this case, both models are awarded 1 point. On the other hand, if two models obtain significantly different CC1 or CC2 or CC3 scores, the better performing algorithm is awarded 2 points and the other zero points. Consequently, we reject the null hypothesis $H_0$ and accept the alternative hypothesis $H_1$. In deciding whether two performances differ, we test the significance of the difference between $u_1$ and $u_2$ ($p < 0.05$). As it is a parametric test, the t-test requires that the data have a normal distribution. Before running the test, we check normality using the Shapiro-Wilk test for each pair of error metrics obtained by the DFFNN and LSTM NN.

**Table 5.** Results for LSTM networks.

| Average PLD | North———South | | | | | |
|---|---|---|---|---|---|---|
| | MAE | MAPE | RMSE | MAE | MAPE | RMSE |
| 12 steps forward | 0,0536 | 0,0949 | 0,2233 | 0,1633 | 0,1373 | 0,2533 |
| 24 steps forward | 0,0470 | 0,1333 | 0,2133 | 0,0560 | 0,3233 | 0,2400 |
| 36 steps forward | 0,0760 | 0,4801 | 0,2766 | 0,1166 | 0,7333 | 0,3366 |

The overall performance of each model was then calculated by summing all points achieved in the pairwise comparisons of all simulations. The results are shown in Table 6.

**Table 6.** Statistical comparison between DFFNN and LSTM NN models

| | DFFNN | LSTM NN |
|---|---|---|
| RMSE | 0 | 8 |
| MAE | 0 | 8 |
| MAPE | 8 | 0 |
| Overall Performance | 8 | 16 |

## 6 Conclusions

Linear and non-linear computational intelligence and identification techniques were applied to obtain models to estimate the price of electrical energy in five Brazilian regions. For the modeling, two representations were used, direct and recurrent neural networks. This research proposes an approach for short-term electricity price prediction applied to the Brazilian market. The results of the DFFNN and LSTM models are compared, they present a sufficiently good level of accuracy in comparison with other methods, and the balancing of the data was essential for the good results obtained. It is common to find in the literature that LSTM networks are more robust than deep feedforward networks, a hypothesis that was confirmed in our results. Thus, the use of LSTM networks expanded the possibilities of solving the problem using recurrent neural networks.

Furthermore, we compared the average percentage error obtained by our system with the HIRA method [13], which was also applied and tested in two electricity markets: German and Hungarian. It can be concluded from the presented results that the LSTM network can predict the PLD value with an error of up to 0.7333%, which is lower than the value generated by multilayer perceptron networks and the HIRA model [13].

**Acknowledgment.** We would like to thank the Federal University of Pará for having supported this study.

## References

1. Samuel, O.W., Asogbon, G.M., Sangaiah, A.K., Li, G.: Chapter 5 - computational intelligence enabling the development of efficient clinical decision support systems: case study of heart failure. In: Sangaiah, A.K., Sheng, M., Zhang, Z. (eds.). Computational Intelligence for Multimedia Big Data on the Cloud with Engineering Applications, Academic Press, pp. 123–133 (2018). https://doi.org/10.1016/B978-0-12-813314-9.00005-0

2. Sharma, M., Gupta, S., Sharma, P., Gupta, D.: Bio-inspired algorithms for diagnosis of breast cancer. Int. J. Innovative Comput. Appl. **10**(3-4), 164–174 (2019)
3. Mercados de Eletricidade, M.R.: Produção, Transmissão, Comercialização, Consumo e Derivativos de Energia. Rio de Janeiro: Synergia (2021)
4. Mohammadi, S., Hezamsadeh, M.R.: Econometric modeling of intraday electricity market price with inadequate historical data (2022)
5. Hu, W., Chen, Z., Bak-Jensen, B.: The relationship between electricity price and wind power generation in Danish electricity markets. In: Asia-Pacific Power and Energy Engineering Conference. Chengdu, China vol. 2010, pp. 1–4 (2010). https://doi.org/10.1109/APPEEC.2010.5448739
6. Panagiotidis, P., Effraimis, A., Xydis, G.A.: An R-based forecasting approach for efficient demand response strategies in autonomous micro-grids. Energy Environ. **30**(1), 63–80 (2019). https://doi.org/10.1177/0958305X18787259
7. Intituto acende Brasil. https://acendebrasil.com.br/. Accessed 7 Nov 2023
8. Cerjan, M., Petricić, A., Delimar, M.: HIRA model for short- term electricity price forecasting. Energies **12**, 3 (2019). https://doi.org/10.3390/en12030568
9. Operador Nacional do Sistema Eletrico. http://www.ons.org.br/
10. Agencia Nacional de Energia Eletrica. http://www.aneel.gov.br/
11. Akkaradamrongrat, S., Kachamas, P., Sinthu- pinyo, S.: Text generation for imbalanced text classification. In: 2019 16th International Joint Conference on Computer Science and Software Engineering (JCSSE). 181–186 (2019). https://doi.org/10.1109/JCSSE.2019.8864181
12. Siami-Namini, S., Tavakoli, N., Namin, A.S.: A comparison of ARIMA and LSTM in forecasting time series. In: 2018 17th IEEE International Conference on Machine Learning and Applications (ICMLA), pp. 1394–1401 (2018). https://doi.org/10.1109/ICMLA.2018.00227
13. Cerjan, M., Petričić, A., Delimar, M.: HIRA model for short-term electricity price forecasting. Energies **12**(3), 568 (2019). https://doi.org/10.3390/en12030568

# Predicting Health and Financial Stability Based on Goal Setting Status Using Classification Techniques

Dhruvi Khanna and Prafulla Bafna(✉)

Symbiosis Institute of Computer Studies and Research, Symbiosis International Deemed University, Pune, India
{drk2022047,prafulla.bafna}@sicsr.ac.in

**Abstract.** Goal setting is a fundamental and transformative process that plays a pivotal role in personal and professional development. It serves a multitude of purposes, each of which contributes to individual growth, organizational success, and the realization of dreams and aspirations.

Goal setting directly impacts health and financial stability. Several parameters like measurable progress and clarity of focus are associated with goal setting and in turn Health and Financial stability.

We predict health and financial status using 15 parameters related to goal setting. The research is useful because it equips human beings with different parameters' and measures that are necessary to have satisfied health and financial situations.

SVM and KNN classifiers were implemented on the dataset. The training-to-testing dataset ratio is 80: 20. Both of the classifiers are implemented on varied-size datasets and the precision of the algorithms is compared. SVm proved to be better as it gives consistently better accuracy of prediction for the dataset varied from 50 to 250 with a minimum accuracy of 0.7 and maximum accuracy of 0.9.

**Keywords:** Prediction · classifier · accuracy · health and finance

## 1 Introduction

Goal setting is a fundamental and transformative process that plays a pivotal role in personal and professional development. It serves a multitude of purposes, each of which contributes to individual growth, organizational success, and the realization of dreams and aspirations. In this comprehensive exploration, we delve deep into the multifaceted purposes of goal setting, highlighting its significance in a wide range of contexts [1]. Many classification algorithms have been used in business scenarios. Due to their relevant results presented, K-Nearest Neighbor (KNN), Support Vector Machine (SVM) [3]. Confusion matrix is used to evaluate classifiers. Value of Precision and sensitivity analysis should be more than 0.5 to prove technique's betterment [5].

At its core, goal setting provides clarity and focus. When individuals or organizations set specific goals, they define their desired outcomes with precision. This clarity


A. Bajaj et al. (Eds.): HIS 2023, LNNS 1224, pp. 157–165, 2025.
https://doi.org/10.1007/978-3-031-78925-0_16

helps in crystallizing thoughts and ideas, reducing ambiguity, and establishing a clear direction for action. Without clear goals, people may drift aimlessly, expending energy and resources without purpose [6].

Imagine a ship navigating the vast ocean. Without a destination or course plotted on a map, it would drift aimlessly, subject to the whims of wind and currents. Similarly, without goals, individuals and organizations risk drifting through life, subject to external forces and lacking a sense of purpose.

Goals serve as powerful motivators. They ignite the spark of inspiration, providing a reason to get out of bed each day and work tirelessly towards something meaningful. When individuals have well-defined goals, they are more likely to feel motivated, driven, and enthusiastic about their endeavours [7].

Psychologically, having a goal triggers the release of dopamine, a neurotransmitter associated with pleasure and reward. The pursuit of goals, therefore, creates a positive feedback loop, reinforcing the desire to continue working towards them.

Another crucial purpose of goal setting is the ability to measure progress. Goals are typically framed in specific, measurable terms, making it easier to track how far one has come and how much further they need to go. This measurement not only provides a sense of accomplishment but also helps individuals and organizations make informed decisions about their strategies and efforts [8].

Imagine running a marathon without mile markers or a finish line. The uncertainty of progress can be disheartening, leading to confusion and potential burnout. In contrast, a well-defined goal with measurable indicators allows runners to track their progress and adjust their pace accordingly, leading to a more successful and satisfying race.

In a world filled with endless tasks and distractions, setting goals helps prioritize efforts. It enables individuals and organizations to determine what is most important and where resources should be allocated. Without goals, there's a risk of spreading oneself too thin, chasing numerous endeavors without making significant progress in any one area [9].

Picture a juggler attempting to keep multiple balls in the air simultaneously. Without prioritization, they may struggle to manage all the objects, and eventually, some may fall. Setting goals is akin to deciding which balls are the most important to keep in the air while temporarily setting others aside [10].

Goals create a sense of accountability. When individuals or teams establish objectives, they are more likely to hold themselves responsible for their actions and decisions. The commitment to achieving specific outcomes provides a framework for evaluating performance and taking ownership of results.

Consider a student setting a goal to achieve an "A" grade in a challenging course. This goal not only motivates the student to study diligently but also holds them accountable for attending classes, completing assignments, and seeking help when needed. Without the goal, the student might be less inclined to exert the necessary effort [11].

## 2 Literature Review

Supervised learning methods are described in the context of both regression and classification, beginning with the simplest case of linear models, then presenting more complex modeling with trees, neural networks, and support vector machines, and concluding with some methods only for classification [4].

Effective resource allocation is a critical aspect of goal setting. With defined objectives, individuals and organizations can allocate their most valuable resources—time, money, and manpower—more efficiently. Goals serve as a compass, guiding decisions on where to invest resources for the greatest impact.

Imagine a startup company with limited capital. Setting the goal to develop a minimum viable product (MVP) before seeking investors enables the team to focus their financial resources on product development rather than frivolous expenditures. This strategic allocation increases the likelihood of achieving the goal and securing funding [13].

Setting and pursuing goals fosters personal and professional growth. It encourages individuals to stretch their abilities, acquire new skills, and overcome challenges. The pursuit of goals often pushes people out of their comfort zones, leading to personal transformation and increased self-confidence.

Consider an aspiring writer setting a goal to complete a novel. Along the journey, they may face writer's block, rejection letters, and self-doubt. However, each setback presents an opportunity for growth. Over time, the writer hones their craft, gains resilience, and becomes a more accomplished author, thanks to the pursuit of their goal [14].

In professional settings, goal setting ensures team alignment. When all team members share a common understanding of the objectives and priorities, it reduces confusion and enhances collaboration. Team members can coordinate their efforts more effectively, working towards a shared vision.

Imagine a sports team with a goal to win the championship. Each player understands their role within the team, whether it's scoring goals, defending, or assisting others. The collective alignment towards the championship goal enhances teamwork and maximizes the team's chances of success [15].

Goals provide a framework for decision making. When individuals or organizations face choices, they can evaluate which option aligns best with their goals. This alignment helps in making informed decisions that contribute to goal attainment.

Think of a business considering whether to expand into a new market. If the goal is to increase market share, the decision-making process becomes clearer. The company assesses whether entering the new market aligns with this objective and supports its overall growth strategy [16].

Goals are often instrumental in realizing a long-term vision. They break down complex, aspirational objectives into manageable, actionable steps. These incremental goals serve as milestones on the path to achieving a broader vision or mission.

Imagine an environmental organization with a vision of a world with clean oceans. Their long-term goal might be to eliminate plastic pollution. To achieve this, they set intermediate goals, such as organizing beach clean-ups, advocating for policy changes, and promoting plastic alternatives. These smaller goals contribute to the realization of their grand vision [17].

Achieving goals can bring a profound sense of satisfaction and well-being. It provides a tangible sense of accomplishment, boosting self-esteem and overall happiness. Goal achievement is a reminder of one's capabilities and resilience, fostering a positive outlook on life.

Think of an individual who sets a fitness goal to run a marathon. Crossing the finish line represents a significant achievement, instilling a sense of pride and well-being. It serves as a testament to their dedication and perseverance, which can extend beyond running and positively impact other areas of life [18].

Regularly setting and reviewing goals can foster a culture of continuous improvement. It encourages individuals and organizations to reflect on their performance, identify areas for enhancement, and adapt strategies accordingly. This commitment to ongoing growth and development leads to sustained success.

Consider a tech company striving for innovation. By setting goals for research and development, they continually push the boundaries of technology. Regularly reviewing progress and adjusting strategies based on feedback leads to the creation of cutting-edge products and services.

## 3 Results and Discussion

A dataset was collected consisting 250 records and 15 attributes and such as long-term vision financial stability and so on. SVM and KNN classifiers were implemented on the dataset. The training-to-testing dataset ratio is 80: 20. Both of the classifiers are implemented on varied size datasets and the precision of the algorithms is compared. Figure 1 shows the research methodology.

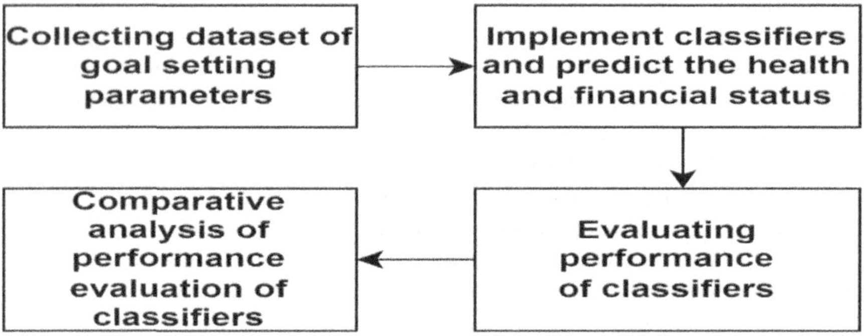


**Fig. 1.** Research methodology

```
> library(readxl)
> classification <- read_excel("H:/2023-2024/classification.xlsx")
> View(classification)
> names(classification)
 [1] "1. Clarity and Focus"            "2. Motivation"
 [3] "3. Measurable Progress"          "4. Time Management:"
 [5] "5. Increased Productivity:"      "6. Better Decision-Making:"
 [7] "7. Personal Growth:"             "8. Enhanced Self-Esteem"
 [9] "9. Resilience:"                  "10. Improved Relationships:"
[11] "11. Financial Stability:"        "12. Health and Well-Being:"
[13] "13. Long-Term Vision"            "14. Sense of Achievement:"
[15] "15. Happiness and Fulfilment"
> |
```

**Fig. 2.** Dataset attribute description

Figure 2 shows total 15 attributes present in the dataset. Attributes and its description is given below

1. **Clarity and Focus:** Setting clear and specific goals helps you identify what you want to achieve. It provides a sense of direction and focus, making it easier to prioritize tasks and make decisions aligned with your objectives.
2. **Motivation:** Goals provide motivation and a sense of purpose. When you have something to work towards, you're more likely to stay committed and persevere through challenges.
3. **Measurable Progress:** Goals allow you to track your progress and measure your success. This sense of achievement can boost your confidence and drive to continue working towards your objectives.
4. **Time Management:** Goal setting encourages effective time management. When you have set deadlines and priorities, you're less likely to waste time on less important activities.
5. **Increased Productivity:** Setting goals helps you become more productive as you become more focused on tasks that align with your goals, reducing distractions and procrastination.
6. **Better Decision-Making:** Goals provide a framework for decision-making. You can evaluate choices based on whether they bring you closer to or further from your goals.
7. **Personal Growth:** Pursuing goals often involves learning and self-improvement. This can lead to personal growth, increased skills, and a sense of accomplishment.
8. **Enhanced Self-Esteem:** Achieving your goals can boost your self-esteem and self-confidence. It reinforces the belief that you can set and accomplish meaningful objectives.

9. **Resilience:** Goals can help you develop resilience and adaptability. When facing setbacks or failures, having a clear goal can keep you motivated to overcome obstacles.
10. **Improved Relationships:** Goal setting can also be applied to relationships, whether in personal or professional contexts. Setting shared goals with others can strengthen bonds and improve collaboration.
11. **Financial Stability:** Financial goals, such as saving for retirement or paying off debt, can lead to greater financial stability and security.
12. **Health and Well-Being:** Setting health and wellness goals can lead to a healthier lifestyle, improved physical fitness, and reduced stress.
13. **Long-Term Vision:** Goals can help you create a long-term vision for your life. They allow you to think beyond the present and plan for the future.
14. **Sense of Achievement:** Achieving your goals provides a sense of accomplishment and satisfaction, which can contribute to a more fulfilling life.
15. **Happiness and Fulfilment:** Pursuing meaningful goals that align with your values can lead to a sense of happiness and fulfillment.

Figures 3 and 4 shows the execution of KNN and SVM classifier on the sample dataset respectively. In Fig. 3, X axis shows better decision-making value and the Y axis shows personal growth. Sillhouut's width constant for KNN is 0.46 for 41 records.

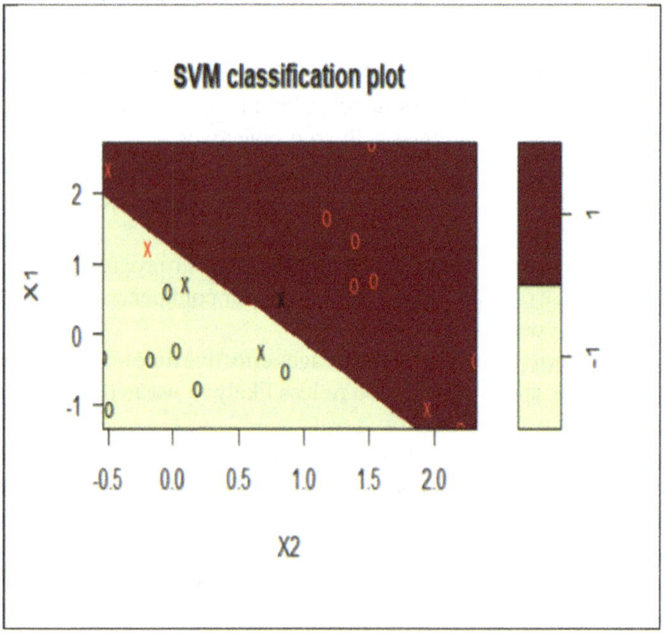


**Fig. 3.** KNN implementation on sample dataset

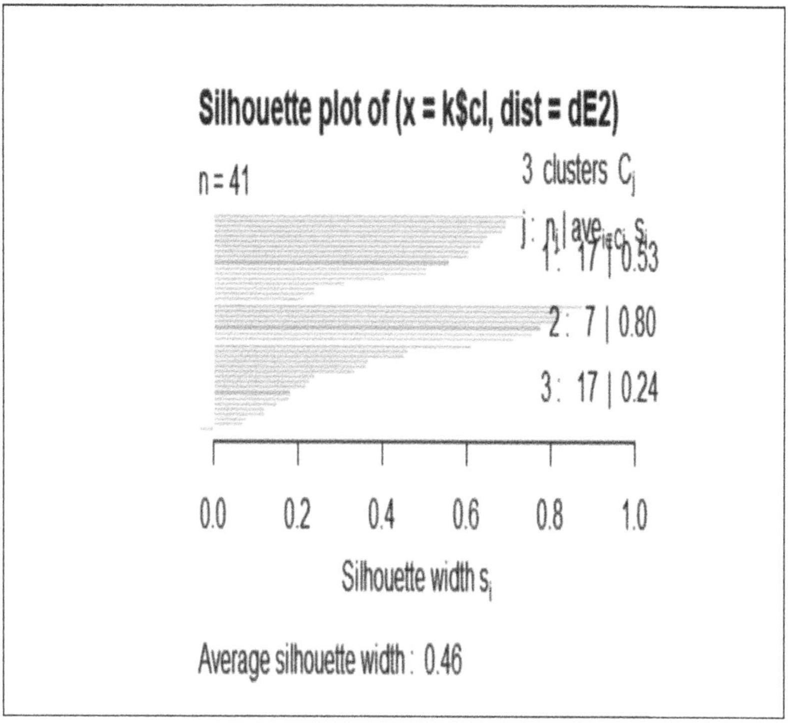


**Fig. 4.** SVM classifier implemeantaion on 41 records

Figure 5 shows the Distance matrix depicting relation between goal-setting parameters.

It means that the relation between all goal-setting parameters is established. All the parameters are independent of each other, hence we could execute SVM on this dataset.

Table 1 shows comparative analysis depicting accuracy of prediction. Its easily observable that SVM show better accuracy that is better sensitivity and precision for varied dataset consistently. It ranges for 0.7 to 0.9.

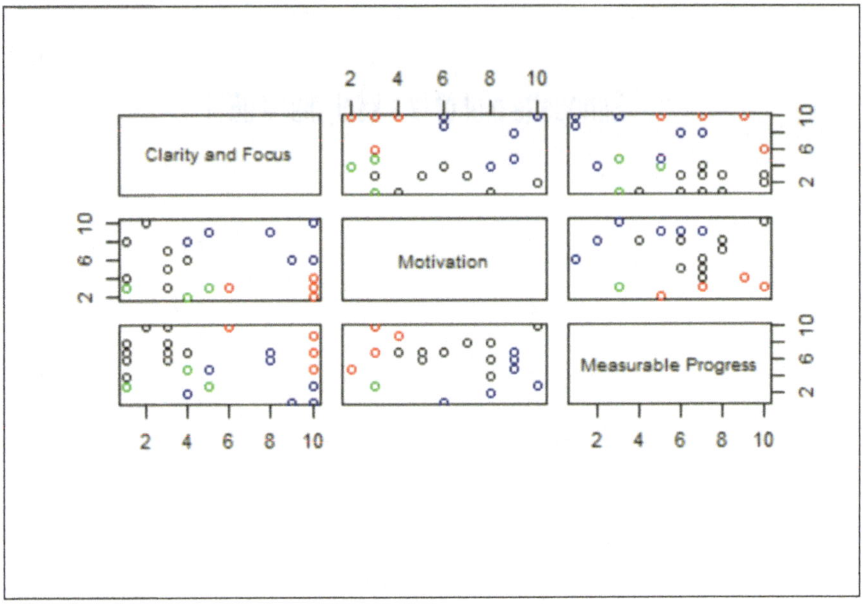


**Fig. 5.** Distance matrix showing relation between goal setting parameters

**Table 1.** Performance evaluation of implemented algorithms.

| Dataset size | KNN | | SVM | |
|---|---|---|---|---|
| | Sensitivity | Precision | Sensitivity | Precision |
| 50 | 0.8 | 0.8 | 0.9 | 0.9 |
| 150 | 0.7 | 0.7 | 0.8 | 0.8 |
| 250 | 0.6 | 0.6 | 0.7 | 0.7 |

## 4 Conclusions

Goal setting has a direct impact on your health and financial stability. Certain metrics like measurable progress and clear direction are relevant to goal setting and, therefore, these are relevant to financial health and stability. We predict health and financial status using 15 parameters related to goal setting.

Research is useful because it provides people with various parameters and measures needed to have a satisfactory health and financial situation. SVM and KNN classifiers were implemented on the dataset. The ratio between training and testing dataset is 80:20. Both classifiers are implemented on datasets of different sizes and the accuracy of the algorithms is compared. SVM is considered better as it gives better prediction accuracy for data set varying from 50 to 250 with minimum accuracy of 0.7 and maximum accuracy of 0.9.

## References


1. Gupta, S.K., Khang, A., Somani, P., Dixit, C.K., Pathak, A.: Data mining processes and decision-making models in the personnel management system. In: Designing Workforce Management Systems for Industry 4.0, pp. 85–104. CRC Press (2023)
2. Tsui, K.L., Chen, V., Jiang, W., Yang, F., Kan, C.: Data mining methods and applications. In: Springer Handbook of Engineering Statistics, pp. 797–816. Springer London, London (2023)
3. Hernández Pérez, S.N., Pérez Reynoso, F.D., Gutiérrez, C.A.G., Cosío León, M.D.L.Á., Ortega Palacios, R.: EOG signal classification with wavelet and supervised learning algorithms DT. Sensors. **23**(9), 4553 (2023)
4. Mariani, N.W.R., Sudjana, I.M.: Classification of tourism business certification institution auditors based on the decision tree algorithm C4. 5: case study of LSUP XYZ. Jurnal Mantik. **7**(1), 112–123 (2023)
5. Locke, E.A., Latham, G.P.: The development of goal setting theory: a half century retrospective. Motiv. Sci. **5**(2), 93 (2019)
6. Grant, A.M.: An integrated model of goal-focused coaching: an evidence-based framework for teaching and practice. In: Coaching Researched: a Coaching Psychology Reader, pp. 115–139 (2020)
7. Ajzen, I., Kruglanski, A.W.: Reasoned action in the service of goal pursuit. Psychol. Rev. **126**(5), 774 (2019)
8. Chaplot, D.S., Gandhi, D.P., Gupta, A., Salakhutdinov, R.R.: Object goal navigation using goal-oriented semantic exploration. Adv. Neural Inf. Proces. Syst. **33**, 4247–4258 (2020)
9. Herrera, V.: Reconciling global aspirations and local realities: challenges facing the sustainable development goals for water and sanitation. World Dev. **118**, 106–117 (2019)
10. Xu, Y., Zeng, G.: Corporate social performance aspiration and its effects. Asia Pac. J. Manag. **38**, 1181–1207 (2021)
11. Ansong, D., Eisensmith, S.R., Okumu, M., Chowa, G.A.: The importance of self-efficacy and educational aspirations for academic achievement in resource-limited countries: evidence from Ghana. J. Adolesc. **70**, 13–23 (2019)
12. Al-Bahrani, M.A., Allawati, S.M., Abu Shindi, Y.A., Bakkar, B.S.: Career aspiration and related contextual variables. Int. J. Adolesc. Youth. **25**(1), 703–711 (2020)
13. Zhang, P.: Target interactions and target aspiration level adaptation: how do government leaders tackle the "environment-economy" nexus? Public Adm. Rev. **81**(2), 220–230 (2021)
14. Genicot, G., Ray, D.: Aspirations and economic behavior. Annu. Rev. Econ. **12**, 715–746 (2020)
15. Janke, S., Dickhäuser, O.: A neglected tenet of achievement goal theory: associations between life aspirations and achievement goal orientations. Personal. Individ. Differ. **142**, 90–99 (2019)
16. Abdeldayem, M.M., Aldulaimi, S.H.: Trends and opportunities of artificial intelligence in human resource management: aspirations for public sector in Bahrain. Int. J. Sci. Technol. Res. **9**(1), 3867–3871 (2020)
17. Goel, S., Karri, R.: Entrepreneurial aspirations and poverty reduction: the role of institutional context. Entrep. Reg. Dev. **32**(1–2), 91–111 (2020)
18. Martela, F., Bradshaw, E.L., Ryan, R.M.: Expanding the map of intrinsic and extrinsic aspirations using network analysis and multidimensional scaling: examining four new aspirations. Front. Psychol. **10**, 2174 (2019)

# Multimodal Learning Applications on Digital Marketing: A Review

Inês César[1], Ivo Pereira[1,2,3(✉)], Fátima Rodrigues[1], Vera Miguéis[4], Susana Nicola[1], and Ana Madureira[1]

[1] ISRC - Interdisciplinary Studies Research Center, ISEP, 4249-015 Porto, Portugal
{icdoa,mfc,sca,amd}@isep.ipp.pt, ivopereira@ufp.edu.pt
[2] Universidade Fernando Pessoa, 4249-004 Porto, Portugal
[3] E-goi, 4450-190 Matosinhos, Portugal
[4] Faculdade de Engenharia da Universidade do Porto, 4200-465 Porto, Portugal
vera.migueis@fe.up.pt

**Abstract.** The effectiveness of digital marketing relies on the seamless integration of intelligent technology, enabling encounters that closely resemble those experienced with physical vendors in the real world. Thus, the importance of scalable artificial intelligence (AI) systems guided by a multimodal approach cannot be overstated, as they can be used to gain a deeper understanding of user preferences and engagement behaviors. The investigation conducted concerning multimodal learning in this review uncovers a variety of benefits and limitations on the available data, presenting consistency in finding the relationship between modalities. The results suggest multimodality as a topic with a noticeable dearth of research, yet a promising path to reduce uncertainty and develop innovative perspectives on decision-making for Digital Marketing improvement tasks. The complexity inherent in data processes like analysis, processing, and granular modulation requires a lot of effort for researchers to build accurate multimodal representations while trying to suppress imprecision in these new elements. Therefore, our approach aims to explore how theoretical foundations are successfully applied to learning operational procedures, considering real-life case comprehension, the technical challenges of the learning process, and the importance given to each feature. Even so, comparing the restrictions found in the state-of-the-art made possible the reformulation of limitations to this particular type of technology and encouraged the search for more guidelines on the entire process.

## 1 Introduction

Digital Marketing structures its path to becoming successful by maximizing and solidifying the interactions with customers through the combination of intelligent technologies and tools in E-commerce, gaining the capacity of a real vendor [1–5]. This emerging trend is driving product owners to deliver effective and scalable artificial intelligence (AI) solutions, building an environment where user


A. Bajaj et al. (Eds.): HIS 2023, LNNS 1224, pp. 166–176, 2025.
https://doi.org/10.1007/978-3-031-78925-0_17

preferences, behaviors, and emotions shape service provision [3,6–9]. By 2025, Gartner, Inc. [1,2] expects multimodal strategies as a valuable approach to AI applications, creating interfaces specialized for the conversational tasks of virtual assistants (VAs). The new future of implementing multimodal perspectives as a standard feature in these interactive tools [2] assembles thriving assumptions on having more than 50% of implementations be specific to the industry domain, providing guidance for 30% of knowledge-workers [1]. The importance of receptivity towards human interactions conducts an extensive analysis to improve Digital Marketing strategies, considering different types of sensors to capture verbal and non-verbal user data and different signals from each subjective modality expression [9].

According to Liang et al. [10], the multimodal learning process can be defined as an exploit of data sample characteristics, regarding the heterogeneity composition and the logical interconnections among them to create new elemental representations. As so, the acquisition of copious amounts of the symbiotic interplay of many sensory modalities in human interactions points out a complex yet innovative journey to explore multimodality in Digital Marketing strategies [11]. By maximizing the use of this kind of methodology, Digital Marketing applications enrich valuable products considering the analysis of customers' attention, pattern behaviors and preferences, compiled in a multimodal treasured admonition for future campaigns and services [10,11]. As opposed to the traditional unimodal system assumed, multimodality benefits the aggregation of a great wealth of information concerning the straightforward representation of information combining modalities [12], warranting the presence of relevance and the absence of ambiguity in the proposed conclusions by leveraging their interdependencies [22]. Due to that, keeping logical criteria to uphold an equitable contribution of each fragmented sample for a multimodal representation portrays a practical maneuver to prevent compromising or disregarding pertinent input during data manipulation [13,15].

With the ultimate goal of assembling cohesive knowledge, new novel methodologies are developed with the special purpose of addressing common challenges and difficulties faced during raw data processing [15,18]. On account of that, the process of knowledge acquirement faces limitations at an early stage, considering basic input composition and granularity as crucial issues to the dimensional analysis and management of data [22]. Hence, this article aims to provide a comprehensive review of multimodal learning applications in Digital Marketing, owing great detail to the inherent subjectivity of the actual use case expressed in modalities. The methodology built assembles relevant studies to an integrity evaluation, verifying proper use of multimodal theoretical and practical techniques, relating each proposed solution and their findings for the identification of limitations and hereafter tasks.

This article is structured as follows: in Sect. 2, a summary of the applications is provided, focusing on their specifications regarding multimodal implementation, a discussion on the related work, highlighting the similarities and differences between learning processes, and the main gaps of the literature and future directions are outlined. Finally, the conclusions are derived in Sect. 3.

## 2 Multimodal Learning Applications

Multimodal Learning represents a prolific route to achieve better trustworthiness in the definition of customer engagement range [18], by personalizing their experience through the capture of multiple formats of data instead of only one [16,19]. Digital Marketing strategies can develop real-time solutions by combining a multimodal perspective in data analysis, even with implications regarding robustness [15].

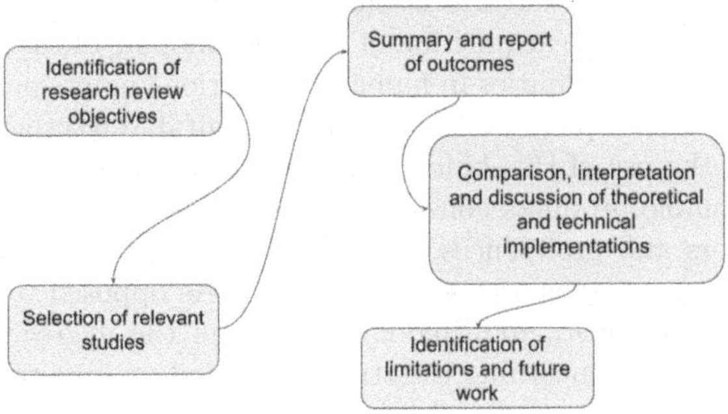


**Fig. 1.** Adopted methodology

Figure 1 describes the methodology adopted for this review elaboration, where the whole process is acquired by integrating beneficial parts of state-of-the-art systematic research methods. According to the developments made in Digital Marketing strategies with a multimodal learning perspective, this methodology combines literature and mapping approaches for the aggregation of information in order to highlight the impact of each solution to this matter and the quantity and quality of published knowledge [14]. For instance, the concerns on identifying the main objectives, described in Sect. 1, mark the first phase of this methodology and are used to select relevant studies to fulfill them. A brief summary of each is developed to contextualize the problem, followed by a comparable and interpretative discussion of the results to evaluate each contribution. In the end, the overall limitations and the future work are presented as possible guidelines for the development of further assignments.

This review explores the application of multimodal learning in real-world scenarios, highlighting advancements and challenges. It provides an overview of academic contributions, contextualizes problems, and analyzes technical qualities. The study emphasizes the importance of addressing digital issues.

### 2.1 Digital Marketing Implementations

The study by Yu et al. [15] explores multimodality in sentiment recognition using bots in customer conversations. The study aims to improve the recognition of a customer's emotional intention during interactions. The study integrates text, video, and audio modalities, including dialog, emotional recognition, and client interaction modules, to enhance the effectiveness of the assistance system. The interaction between the client and dialogue module involves a two-way exchange of responses, considering an intermediate deviation for sentiment recognition score by combining modalities to calculate valence, enhancing the credibility of the bot's replies.

Guo et al. [16] analyzed the affective association of online news samples, identifying sentiment valence based on image, text, and format similarities. They used the LD-MAN model to classify news based on subject and substance, addressing the goal of Artificial Intelligence in emotion analysis. The study proceeded to the creation and use of the RON and DMON databases to address the lack of resources for emotions and their valences. The data was evaluated using statistical computations and impartial volunteers. The multimodal fusion technique was used to predict sentiment based on pre-established dimensions, with each modality's influence on predictive performance. The procedure, databases, and corresponding programming resources are available to the scientific community.

According to Wagle et al. [17], the relationship between customer experience data and information for another customer is crucial for creating intelligent mechanisms for automatic clarification. A model is formed from a thorough examination of data, establishing a quantifiable measure of credibility. The model includes four factors: the author's publishing experience, platform credibility, information quality, and reader response cues. Scores are assigned to these factors and incorporated into a database. These scores are then assessed to determine overall credibility and justify the value gained by defining the significance of the most relevant aspects.

Khandelwal et al. [18] propose a method to tackle information overload in e-commerce campaigns by extracting attributes and incorporating them into the MXT model. The model consists of three components: MAG for processing and integrating modalities, Xception network for additional information on images, and T5 transformer for attribute-based responses. The model adjusts the level of detail in the material to create a novel depiction of garment categorization and desired attributes. To assess the model's accuracy, modifications were made to datasets, and scalability was enhanced to enable conclusions based on a broader range of factors.

Sukel et al. [19] propose using multimodality to improve product demand forecasts. They use text, images, and geographic coordinates to gather insights on products, purchasing locations, and user behavior. This approach addresses cold-start issues and reduces the need for extensive information. The Multimodal Temporal Fusion Transformer (MTFT) pipeline aims to forecast customer behavior, improve stock management, reduce waste, and maintain market competitiveness. The system uses convolutional networks to process product photos, graph-based networks to process spatial relationships between customers and retailers, and a transformer model to consolidate all data into a unified representation.

The study by Hebert et al. [20] developed a framework for integrating deep learning architectures in a multimodal learning setting for hate speech. The Multi-Modal Discussion Transformer (mDT) model combines textual and visual elements to understand emotional context. The model uses multiple datasets and images as optional elements for training. The study compares the mDT model with the BERT model to improve assessment metrics. The study also examines the impact of various factors, finding positive outcomes when they are included. The mDT model enhances understanding of hate speech context.

Sahin et al. [21] propose two techniques for the 2023 Multimodal Hate Speech Event Detection, focusing on identifying hate speech within textual content embedded within images and identifying specific targets. The first step involves extracting text, analyzing it, and encoding it. A fusion process is then used to integrate the text with the image, allowing for emotional distance between parts associated with hatred. The data is processed to identify entities and classify them as hate speech. This multimodal approach contributes to developing strategies to combat hate speech.

Guo et al. [22] propose a solution for news recommendation that integrates multimodality and reinforcement learning techniques. They focus on accurately capturing detail in news articles and propose a system that assesses data within news samples. The system is integrated with a Q network to optimize news recommendations. The study examines a neural network that evaluates text and image fusion using multimodal integration conditions. The system agent collects user behavior patterns and uses the UMAP algorithm to organize data into two clusters based on dispersion and trend. The authors conclude that incorporating multimodality enhances the efficiency of news recommendations.

Abousaleh et al. [23] propose a framework for predicting image popularity on social networks. The study uses a fusion of neural networks and convolutional networks to predict the popularity of interactions on visual evidence. The dataset includes image quality, content, user information, metadata, and publication timestamp. The methodology consists of two phases: data extraction and manipulation, and the implementation of a VSCNN deep learning model. The first phase extracts relevant segments from the dataset, creating vectors for each modality. The second phase evaluates the performance of the neural network layers, resulting in popularity-level classification for each image.

### 2.2 Related Work Discussion

After confirming that the aforementioned studies included several forms of data, it became crucial to ascertain certain factors that would facilitate a comparison of the practical implementation of multimodal approaches. Therefore, this study explored the modalities identified in Table 1, i.e. the operational and structural aspects of the linked granularities, and the level of complexity exhibited by the offered techniques. The intricacy of the solutions generated can be discerned by considering the presence of sentiment recognition, as either the main focus or an associated module.

**Table 1.** Multimodal Applications Related Work

| Ref. | Modalities | | | | | Granularity | Approach | |
|---|---|---|---|---|---|---|---|---|
| | Text | Image | Video | Audio | GPS | | Sentiment | Assoc. |
| [15] | ✓ | | ✓ | ✓ | | | ✓ | ✓ |
| [16] | ✓ | ✓ | | | | ✓ | ✓ | |
| [17] | ✓ | ✓ | | | | | ✓ | ✓ |
| [18] | ✓ | ✓ | | | | ✓ | | ✓ |
| [19] | ✓ | ✓ | | | ✓ | ✓ | | ✓ |
| [20] | ✓* | ✓ | | | | ** | ✓ | ✓ |
| [21] | ✓ | ✓ | | | | | ✓ | ✓ |
| [22] | ✓ | ✓ | | | | ** | | ✓ |
| [23] | ✓ | ✓ | | | | | | ✓ |

* Principal modality analyzed
** Granularity personalized

Table 1 provides a concise overview of the critical points of comparison among the papers under consideration for this assessment. The initial step involves establishing the definitions of the prevalent modes, while considering the modalities mentioned in the multimodality contribution by [10,11]. The predominant modalities employed in various studies are text and image, as indicated by most sources. However, [15] stands out by using additional forms of evidence, such as video and audio, while [19] takes into account geographical coordinates in their analysis. Overall, the studies included highlight how important is to use multiple modalities and enhance data types to achieve a more comprehensive representation of knowledge for machine processing.

The process of analyzing various data forms presents challenges, requiring a detailed assay of unrefined samples with granularity variations, demanding data conversion into suitable formats to ensure valid representations. As demonstrated in [16,18,19], it is imperative to specify how the linked subjectivity of granularity impacts the overall learning experience and enhances a need to understand how features are received by the model. The boundary setting in [16] by considering text granularity as sentences, composed by a maximum of 200 words, promotes both outlooks for text mining. However, both [18,19] remodeled data granularity into a new blended feature [18], categorizing products, warehouses, and temporal order moments as multimodal high dimensional representation, or a target one [19], generating a unique product-type attribute for an improved e-commerce experience. Regardless, both [20,22] present evidence of granularity issues due to the feature management, providing a non-conservative plan of data processing. Detected in [20], the inclusion of image modality may be discretionary, resulting in the creation of a redundant multimodal unit as it is equivalent to the unimodal variable of text, yielding a distinct level of detail. In a similar vein, [22] styles their input to the suggested neural network as a spare inclusion, introducing duplicate representations of each modality, taking in each singularly and fusion processed.

The decision-making process in Digital Marketing is influenced by various elements, with emotions and sentiments playing a significant role in subconscious decision-making. Understanding the complexity of idealized methods is crucial for studies like [15–17,20,21] that look into a comprehensive understanding of human multimodal judgment. All of above have identified specific modules or parts within the framework that focus on analyzing the emotional payload as one of the foundational backbones of the solution. The study [15] integrates an examination of established dialogues into the chatbot system. The authors create specific categories for identifying polarities, which are then generalized as positive, weak positive, negative, weak negative, and neutral. The aforementioned concept of category specification is also applicable in [16], wherein a dataset is organized based on the three established polarities, with the inclusion of an additional one. In the present study, an additional dataset was generated to incorporate a separate method of emotional charge discerning associated with multimodality. This was achieved by introducing the identification of one of the eight fundamental emotions to distinguish different emotional states. The study [17] examines a binary polarity that is conveyed through the existence or absence of credibility regarding the received data. This polarity is subsequently utilized to compute a final score. However, [20,21] diverge from this prevailing pattern by solely recognizing the emotion of "hate" within the modalities examined in their respective investigations. While the latter studies [22,23] do not explicitly address the identification or examination of emotions, both textual and visual analysis methodologies involve extracting elements that facilitate the semantic and contextual examination of hate speech in the introductory phase. In conjecturing approaches, multimodality infers proficient results as it can be associated with different modules of the same task inference. The analysis and detection of emotion can be the main focus of the paper [16] but it can also be granted as an associated module for the overall result improvement [15,17,20,21]. Still, multimodal learning is prevailing as a positive affiliation to Digital Marketing in studies like [18,19], launching new solutions to an unexplored advantage of the attention-based aspects problems. This assigns Multimodal Learning a rising research area to establish future contributions, bring a more sophisticated angle of completeness by mixing various e-commerce clients activity signals.

Although certain articles handle emotion labels, each of them primarily adopts a specific method that involves incorporating modules, such as emotion recognition, to enhance the outcomes achieved. Only a limited number of studies provided a comprehensive account of the implementation specifics and the learning process. Several studies [16–18,21,23] discuss the distribution of data points in detail, providing information on key aspects such as epochs and batch size. Batch size selection depends on record collection, computational resources, and model complexity. To define multimodality, evaluations of modalities' influence and effectiveness are crucial. Among the studies mentioned, only [18,21,23] specifically investigate the influence of modalities. These three studies primarily focus on the fusion of textual and visual information to achieve multimodality, indicating limited exploration of the effects of other feature types. However, [19,20,22] offer a comprehensive examination of the impact different types of data have on models, emphasizing their fundamental nature and cohesiveness.

### 2.3 Limitations and Future Directions

According to the research, multimodal learning is routed through increasing complexity, which might typically limit improved final model performances. The availability of several data kinds presents substantial issues in modeling claims, as it necessitates a careful evaluation of both the quality and quantity of knowledge included in the results. All deductions may be optimized by using a balanced approach that incorporates the individual and collective advantages of modalities. The joint record of limitations previously stated by the authors highlights the need to critically report major constraints for multimodal future applications:

- **Real use cases' impact on modalities and their granularities** - The aforementioned studies have accumulated empirical evidence suggesting that the adoption of multimodal techniques represents a novel avenue for investigation, hence enhancing the efficiency of prior endeavors aimed at impacting Digital Marketing. With the exception of two isolated instances, the majority of the suggested frameworks aim to improve the utilization of text and images, mostly through their conversion into various analytical activities. Nevertheless, the developer's interpretation plays a significant role in determining the granularity utilized to structure a scalable solution for each real-world use-case problem. This subjectivity, without concrete validation, limits the model's ability to produce an accurate representation of the information [15]. Therefore, it is imperative to prioritize the consideration of this issue when utilizing outcomes and assessment methodologies derived from contemporary models. [22] exemplifies a case where the granularity involves individual modalities as inputs, as well as their multimodal fusion with user behavior data. Generating distinct representations of the same knowledge without a comprehensive understanding of the implications is a misleading path to multimodal learning success.
- **Lack of technical knowledge on other multimodal learning challenges** - As previously referenced, theoretical and practical aspects of multimodal learning examine taxonomy, principles, and challenges that shed light on the important information to be analyzed [10,11]. However, it is evidenced that most of the articles gathered for this review rely on approaches of multimodal representation that do not fully apply coherently the foundational

requirements mentioned above. Evaluating the achieved outcomes by developing multimodal features leaning only on fusion phase underscores probable discrepancies in the developments and achieved outcome. When considering the structure of knowledge representation and quantification in model training, it is essential to address certain challenges and prior validation to ensure the lack of redundancy and misleading results.
- **Ethical and biased concerns** - The exponential nature of concerns around ethical and biased evidence for each multimodal undertaking is amplified when addressing the foundational modalities. The ongoing assessment of these models and databases, with the aim of making them completely accessible to the community, provides a comprehensive understanding of the data collection process, its utilization, and the impact it has on the performance of the model. Simultaneously, providing a reliable perspective is an effective approach to obtaining further voluntary contributions of fresh data. This serves as a means to mitigate such challenges and establish a dependable knowledge foundation on multimodal tools for enhancing customer experience customization.

## 3 Conclusions

In conclusion, the review of multimodal learning applications within the realm of Digital Marketing has unveiled a lack of scientific contributions with significant possibilities for utilizing diverse data modalities to enhance marketing tactics and enhance consumer experiences. However, the existing body of literature for this review mostly focuses on the integration of different modalities and the coverage of potential research applications, demanding more comprehensive approaches that consider the level of detail required in practical applications.

Using sentiment analysis has been increasingly prominent as a valuable instrument for comprehending customer feelings and preferences. As one of the examples of multimodal learning applications, it is filled with a wider array of obstacles towards the data core. By acknowledging a multimodal representation for emotive analysis, it tackles fundamental steps, such as data integration, the extraction of distinct modality features, the alignment between embeddings, and posterior scalability. In order to fully harness the capabilities of multimodal learning, future research should not only enhance fusion techniques but also give precedence to attaining awareness of the technical challenges that underlie this field.

The rising importance of the confluence of multiple data sources and modalities is evident as the Digital Marketing landscape undergoes continuous evolution. Attaining success in this particular domain necessitates a deliberate endeavor to reconcile the disparity between theoretical frameworks and tangible execution, hence facilitating the emergence of inventive and efficacious multimodal marketing approaches. This evolutionary development holds the potential to change the dynamics between companies and consumers by leveraging various senses and data kinds to facilitate more tailored and appealing interactions.

**Acknowledgements.** This work is funded by national funds through the FCT—Fundação para a Ciência e Tecnologia, within PHYNHANCAI project: 2022.01303.PTDC

## References

1. Casey, D., et al.: Emerging Tech Impact Radar: Conversational Artificial Intelligence. Gartner, Inc
2. Jump, A., Casey, D., Lee, A.: Emerging Technologies: Tech Innovators in Advanced Virtual Assistants. Gartner, Inc (2023)
3. Nguyen, T., Jump, A., Casey, D.: Emerging Tech Impact Radar: 2023. Gartner, Inc (2022)
4. César, I., Pereira, I., Madureira, A., Coelho, D., Rebelo, M.Â., de Oliveira, D.A.: Analysing and modeling customer success in digital marketing. In: International Conference on Innovations in Bio-Inspired Computing and Applications, pp. 404–413. Cham, Springer Nature Switzerland (2022)
5. Paulo, M., Miguéis, V.L., Pereira, I.: Leveraging email marketing: using the subject line to anticipate the open rate. Expert Syst. Appl. **207**, 117974 (2022)
6. Araújo, C., Soares, C., Pereira, I., Coelho, D., Rebelo, M.Â., Madureira, A.: A novel approach for send time prediction on email marketing. Appl. Sci. **12**(16), 8310 (2022)
7. Rebelo, M.Â., Coelho, D., Pereira, I., Fernandes, F.: A new cascade-hybrid recommender system approach for the retail market. In International Conference on Innovations in Bio-Inspired Computing and Applications, pp. 371–380. Cham: Springer International Publishing (2021)
8. Singh, A., Casey, D.: Emerging Tech: Benchmark Conversational AI Investments to Drive Revenue and Customer Acquisition. Gartner, Inc (2023)
9. Zimmermann, A., Vashisth, S.: Emerging Tech: Computer Vision. Gartner Inc, Voice Analysis and CGI Evolve Into Emotionally Intelligent Virtual Beings (2022)
10. Liang, P.P., Zadeh, A., Morency, L.P.: Foundations and recent trends in multimodal machine learning: Principles, challenges, and open questions. arXiv preprint arXiv:2209.03430 (2022)
11. Baltrušaitis, T., Ahuja, C., Morency, L.P.: Multimodal machine learning: a survey and taxonomy. IEEE Trans. Pattern Anal. Mach. Intell. **41**(2), 423–443 (2018)
12. Haouhat, A., Bellaouar, S., Nehar, A., Cherroun, H.: Modality Influence in Multimodal Machine Learning. arXiv preprint arXiv:2306.06476 (2023)
13. Rahman, W., et al.: Integrating multimodal information in large pretrained transformers. In: Proceedings of the Conference. Association for Computational Linguistics. Meeting (Vol. 2020, p. 2359). NIH Public Access (2020)
14. Haddaway, N.R., Bernes, C., Jonsson, B.-G., Hedlund, K.: The benefits of systematic mapping to evidence-based environmental management. Ambio **45**(5), 613–620 (2016). https://doi.org/10.1007/s13280-016-0773-x
15. Yu, D., Tian, J., Su, T., Tu, Z., Xu, X., Wang, Z.: Incorporating multimodal sentiments into conversational bots for service requirement elicitation. In: 2021 IEEE International Conference on Service-Oriented System Engineering (SOSE), pp. 81–90 (2021)
16. Guo, W., Zhang, Y., Cai, X., Meng, L., Yang, J., Yuan, X.: LD-MAN: layout-driven multimodal attention network for online news sentiment recognition. IEEE Trans. Multimedia **23**, 1785–1798 (2020)

17. Wagle, V., Kaur, K., Kamat, P., Patil, S., Kotecha, K.: Explainable AI for multimodal credibility analysis: case study of online beauty health (MIS)-information. IEEE Access **9**, 127985–128022 (2021)
18. Khandelwal, A., Mittal, H., Kulkarni, S.S., Gupta, D.: Large scale generative multimodal attribute extraction for e-commerce attributes. arXiv preprint arXiv:2306.00379 (2023)
19. Sukel, M., Rudinac, S., Worring, M.: Multimodal Temporal Fusion Transformers Are Good Product Demand Forecasters. arXiv preprint arXiv:2307.02578 (2023)
20. Hebert, L., Sahu, G., Sreenivas, N.K., Golab, L., Cohen, R.: Multi-Modal Discussion Transformer: Integrating Text, Images and Graph Transformers to Detect Hate Speech on Social Media. arXiv preprint arXiv:2307.09312 (2023)
21. Sahin, U., Kucukkaya, I.E., Ozcelik, O., Toraman, C.: ARC-NLP at Multimodal Hate Speech Event Detection 2023: Multimodal Methods Boosted by Ensemble Learning, Syntactical and Entity Features. arXiv preprint arXiv:2307.13829 (2023)
22. Guo, N., Fu, Z., Zhao, Q.: Multimodal news recommendation based on deep reinforcement learning. In: 2022 7th International Conference on Intelligent Computing and Signal Processing (ICSP), pp. 279–284 (2022)
23. Abousaleh, F.S., Cheng, W.H., Yu, N.H., Tsao, Y.: Multimodal deep learning framework for image popularity prediction on social media. IEEE Trans. Cogn. Dev. Syst. **13**(3), 679–692 (2020)

# Multivariate Analysis of Products Tipology Data - A Case Study

Nelson Costa[1,2(✉)], Alzira Mota[1,3], and Inês Sousa[2,4]

[1] Polytechnic of Porto - School of Engineering (ISEP), Porto, Portugal
[2] University of Minho - School of Sciences, Braga, Portugal
nfc@isep.ipp.pt
[3] ISRC -Interdisciplinary Studies Research Center, Porto, Portugal
[4] CMAT - Centre of Mathematics, Research, Braga, Portugal

**Abstract.** Small, medium, and large organizations collect vast amounts of data with the expectation of using it to generate commercial value. Machine learning is a powerful tool for extracting valuable insights from this data and serves as a pivotal sales strategy for companies to maximize profits. This paper seeks to analyze sales data and discern patterns in sales among products that exhibit similarities, such as boxes and bags. In order to achieve this goal, was used unsupervised learning methods that allow the segmentation of groups, specifically Principal Component Analysis (PCA), k-means algorithms, and hierarchical clustering. PCA was used to identify correlated variables and find hidden patterns in the data, particularly pertaining to product families with similar sales. Elbow, Silhouette, and 30 indices methods were applied to determine the optimal number of clusters. Based on these results, it was determined the optimal number of clusters. Validation methods were employed to identify the clustering algorithm exhibiting the best performance. Stability measures evaluated the consistency of the clusters, while the cophenetic coefficient aided in determining the most effective data grouping method. After validation, the clustering algorithms were implemented. The results indicated that all clustering algorithms effectively segmented the data, with particular emphasis on the performance of the k-means algorithm. This study identified product groups with similar sales patterns and key products that impact the company's global sales. Multivariate analysis provided a deeper understanding of sales dynamics, enabling the company to implement targeted marketing strategies and optimize resource allocation to boost bag and box sales in Portugal and other countries.

**Keywords:** Machine Learning · PCA · k-Means · Hierarchical Clustering · Sales

## 1 Introduction

Advances in technology and the availability of data in recent decades have made it possible to apply machine learning techniques to analyze and interpret large


A. Bajaj et al. (Eds.): HIS 2023, LNNS 1224, pp. 177–186, 2025.
https://doi.org/10.1007/978-3-031-78925-0_18

amounts of data. Machine learning can be defined as a set of algorithms and techniques that enable a system to learn from data and make predictions and decisions without being explicitly programmed to do so.

In recent years, there has been a growing awareness of the wealth of information contained within business data. This awareness has driven the search for advanced analytical methods capable of identifying complex patterns and hidden relationships in the data. One such method is Principal Component Analysis (PCA), a widely used dimensionality reduction technique in multivariate analysis [3]. PCA can identify and quantify the relative contribution of different variables within a dataset, thus revealing underlying trends [1].

The purpose of this paper is to carry out a multivariate analysis of the boxes and bags marketed by a Portuguese company, with the aim of gaining in-depth knowledge and contributing to the optimization of sales strategies [1,2]. Products sales analysis comprehensives a six-year period, from 2015 to 2021.

Furthermore, this study aims to fill a knowledge gap by highlighting the importance of multivariate analysis in product sales. While many previous studies have focused on general sales performance metrics, overlooking the nuances of different products, this approach focuses on product segmentation to identify specific patterns. Product segmentation is central to strategic decision-making, allowing companies to tailor their marketing, inventory management, and production strategies based on the distinct characteristics of each product category [8,9].

The specific objectives of this study are to identify groups of products that share similar sales characteristics, to recognize products that are influential in overall sales, and analyzing interactions among different product categories. This detailed approach will not only provide the company with useful insights into its past sales performance but will also be the basis for the improvement of its future sales strategies. Thereby, this study contributes to the understanding of the dynamics of product sales and the effective application of multivariate analysis techniques in this specific business context.

The following sections of this paper, detail the methodology, present the results, and discuss the implications of these findings for the company and the broader field of product sales analysis will be discussed.

## 2 Related Work

Principal Component Analysis (PCA) is a commonly used machine learning technique to reduce data dimensionality and identify patterns in multivariate datasets. By recognizing the main features of the data, PCA can help simplify and visualize the data, making it more easily interpretable. An example of PCA application in sales can be found in a recent study by [12], who used PCA to identify the main features of sales data of a luxury goods company. Another common machine learning technique in sales is clustering, which involves dividing a dataset into homogeneous groups based on similar characteristics. Clustering can be useful for segmenting customers based on their preferences and buying behaviors, allowing companies to personalize their marketing efforts and improve customer satisfaction. For instance, [11] used K-means clustering to segment customers of an online retail company based on their product preferences and buying behaviors. Another common clustering technique in sales is hierarchical clustering, which involves creating a tree-like structure of nested clusters. This technique can be useful for identifying subgroups within larger clusters and visualizing the relationships between clusters. In [13], the authors used hierarchical clustering to categorize customers of a large-scale retail chain based on their purchase records. The study showed that the clusters generated were highly correlated with demographic and geographic characteristics. This indicates that hierarchical clustering can be a valuable tool for customer segmentation and customized marketing in the retail sector. By combining PCA and clustering, it is possible to conduct a more comprehensive and detailed analysis of sales data, allowing companies to identify patterns and trends that may not be visible through traditional data analysis methods. For example, [10] applied PCA and K-means clustering to segment customers of an e-commerce company based on their product preferences and buying behaviours, identifying consumption patterns and providing useful insights for the company's marketing strategy. In summary, machine learning has become a prevalent tool in the sales industry, with PCA and clustering being among the most commonly used techniques for analyzing large volumes of data and identifying patterns and trends. These techniques are utilized in various types of sales data, such as retail, online, and luxury goods sales. Overall, the use of machine learning techniques in the sales field has proven essential insights to help companies improve customer satisfaction and marketing strategy.

## 3 Data

The data from this study were provided by a Portuguese company dedicated to the production and sale of boxes and bags. The data collected is over a seven-year period (2015–2021).

The data was made available in Microsoft Excel, and data preprocessing was performed using R software.

The main objective of this study is to conduct an exploratory analysis of the sales data and apply unsupervised learning techniques to segment the data. The dataset under consideration comprises a total of 17 variables and 3465 observations. Among these variables, five are categorical (Reference, Type, Code, Measure and Year), and the rest are quantitative.

The description of variables is presented below:

- **Reference**: A product's family, i.e. a specific product is characterized by a specific group;
- **Type**: A product's typology, that is, it characterizes the product as being a bag or a box;
- **Code**: Identify the product in its essence, that is, assign a unique identifier to it;

- **Measure**: Information on the dimensions of the product (mm);
- **Year**: Identifies the year in which the data was collected;
- **Qt_Country**:[1] Quantity of sales made by the company for a particular product in a given country (units);
- **Vl_Country**:[2] Sales revenue generated by the company for a particular product in a given country (euros).

The quantitative variables are associated with a specific country (Portugal, Spain, France, Germany, and England) and are also related to total sales in these countries. Therefore, for each country, we find two associated variables (quantity and sales value).

The company provided seven databases in Excel format (.xls), each referring to a year between 2015 and 2021. Initially, each instance data was processed. The product family columns were disaggregated, the sales quantity and sales value columns were disaggregated for each country, and irrelevant columns were removed from the dataset for analysis. The initial data processing is crucial for loading data into R and for extracting relevant information about a particular product (by isolating entries).

Regarding the original dataset, in addition to the disaggregation of columns and entries, the color column and the column comparing sales values with previous years were removed by the company's decision. Additionally, two new columns were generated to identify the entries in the database as either box products or bag products, and to identify the year in which the product sales values were registered. After processing the original dataset, new databases were generated and saved in .csv format.

The explanatory variables for this study are Reference, Type, Measure, Code, and Year, and the variables of interest are Qt_Country and Vl_Country.

The databases for all years were loaded into the R environment and merged into a single database using the rbind() command. The original database was then divided into two new databases (one for boxes and another for bags) based on the Type variable. The database contains 1294 observations related to boxes and 2171 observations related to bags.

## 4 Principal Component Analysis

Principal Component Analysis (PCA), is a widely used multivariate technique for reducing the dimensionality of data, transforming it into an equal or fewer number of variables known as "principal components" [2,3]. These principal components are linear combinations of the original data that capture most of the variability in the data [4]. Typically, there is a reduction in the dimensionality of the original data to two or three components, allowing for a more simplified representation of the data [5].

[1] $Qt\_Country \in \{Qt\_Pt, Qt\_Sp, Qt\_Fr, Qt\_Eng, Qt\_Ger, Qt\_Tot\}$.
[2] $Vl\_Country \in \{Vl\_Pt, Vl\_Sp, Vl\_Fr, Vl\_Eng, Vl\_Ger, Vl\_Tot\}$.

PCA is a powerful tool that helps identify hidden patterns in data. It also assists in reducing redundancy in the data, eliminating unnecessary information while retaining the most informative features [6]. Furthermore, PCA is effective in identifying correlated variables, revealing important relationships among different variables [7].

Initially, only the quantitative variables from the boxes dataset were selected (England's data was excluded due to the absence of box sales in this country), and then the data was normalized.

The variability explained in each of the components is measured by the eigenvalues, which can be extracted using the *get_eigenvalue*() function in the R software. Table 1 contains the information retained through the application of this function.

**Table 1.** Principal components of normalized PCA

| Dimension | Eigenvalue | Variance (%) | Accumulated (%) |
|---|---|---|---|
| 1 | 2.67170 | 33.39619 | 33.39619 |
| 2 | 1.97825 | 24.72807 | 58.12426 |
| 3 | 1.83066 | 22.88329 | 81.00756 |
| 4 | 1.20868 | 15.10848 | 96.11603 |
| 5 | 0.19202 | 2.40029 | 98.51632 |
| 6 | 0.07233 | 0.90421 | 99.42054 |
| 7 | 0.02376 | 0.29700 | 99.71755 |
| 8 | 0.02260 | 0.28245 | 100.00000 |

Eigenvalues greater than 1 indicate that the variance of the component is greater than what would reprcsent the variance of the original data, and it can even be used as a cutoff point to decide how many components to use. Table 1 shows that 8 principal components were created, with the first three components explaining 81.00756% of the total data variability.

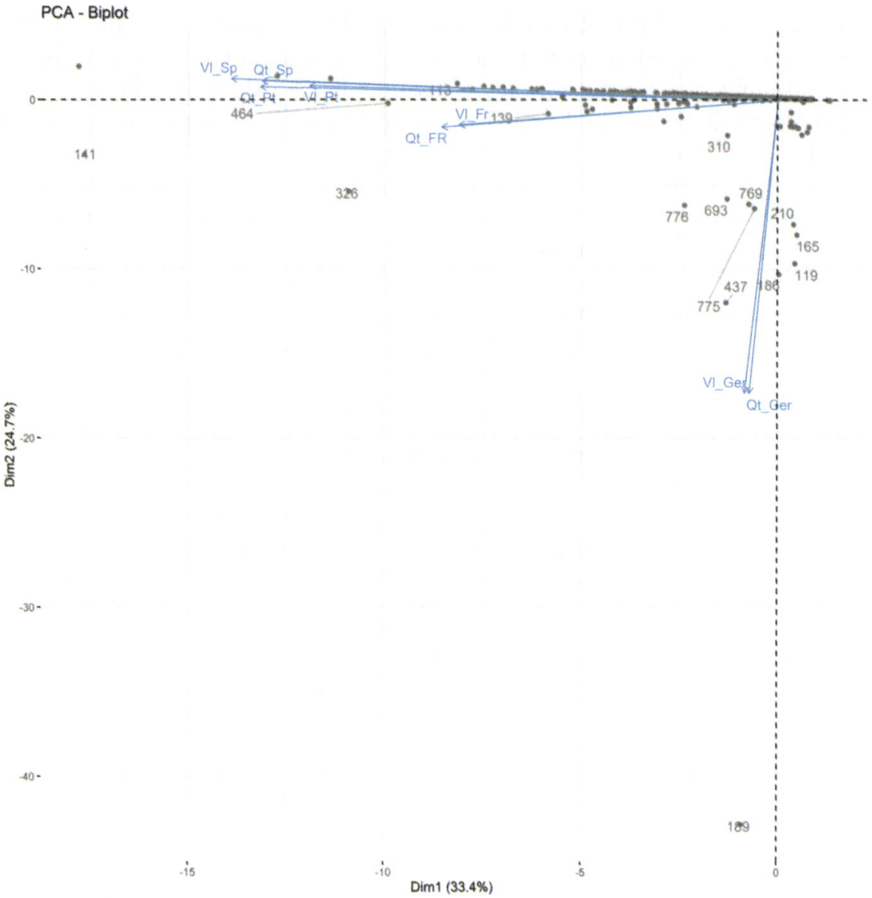


**Fig. 1.** Biplot - Graph of variables and individuals (samples)

Figure 1 depicts a biplot chart with variables and individuals (samples), allowing us to observe that the correlation between several variables is very high, such as, *Vl_Sp*, *Qt_Sp*, *Qt_Pt*, and *Vl_Pt*. These variables point in the same direction as Component 1, reinforcing its importance for this component. On the other hand, it is evident that variables *Qt_Ger* and *Vl_Ger* point in the direction of Component 2, which contributes the most.

Figure 2 contains a graphical representation with product family labels. In general, some overlap in the data representation can be observed. However, it is easy to discern certain groups of boxes with similarities in sales, particularly the products in pink, green, and orange.

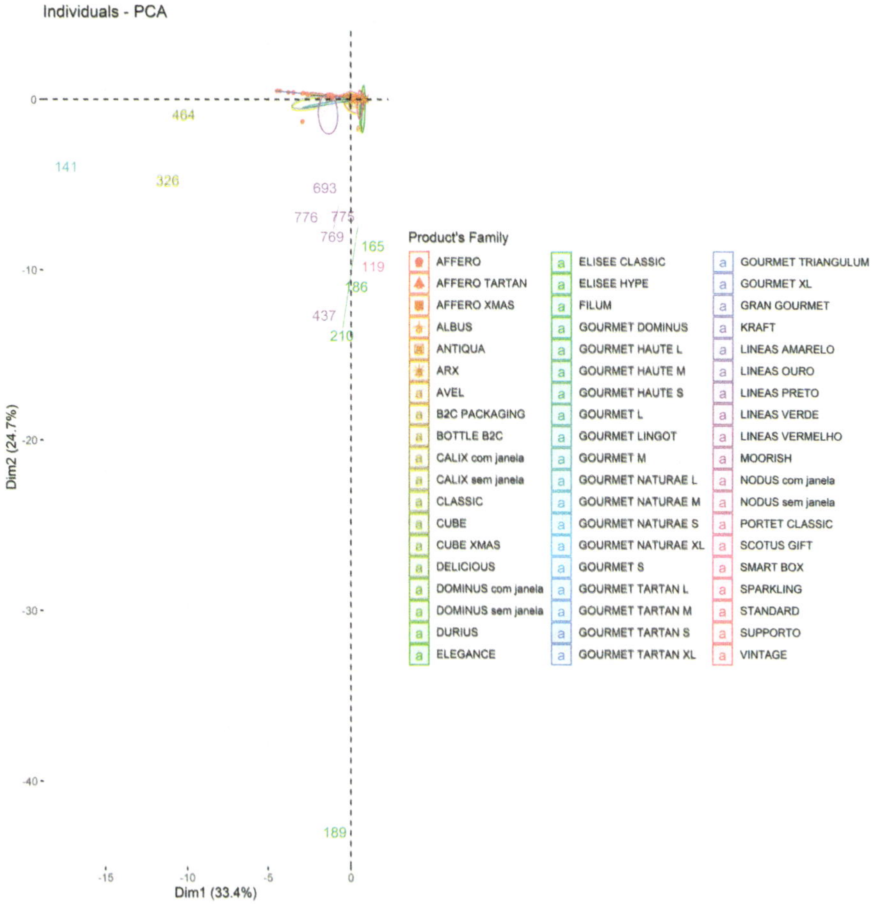


**Fig. 2.** Graphical representation with product family labels

# 5 Clustering Analysis

## 5.1 K-Means

To apply the k-Means algorithm, it is necessary to determine the optimal number of clusters. The number of clusters is subjective and depends on the method used to measure similarities and the parameters used for partitioning. To perform this validation, the following methods were used: elbow, silhouette, and 30 indices to determine the best number of clusters [14,15]. Table 2 presents the optimal number of clusters ($k$) determined using each of these methods.

**Table 2.** Optimal number of clusters

| Method | k |
|---|---|
| Elbow | 5 |
| Silhouette | 2 |
| 30 indices | 2 |

After determining the optimal number of clusters, the k-Means algorithm was implemented with 2 clusters (number of clusters determined by the Silhouette method and 30 indices to determine the best number of clusters) and 5 clusters (number of clusters determined by the Elbow method). Figure 3 presents the result of the k-Means algorithm for k = 2 and k = 5. With this representation, it is possible to observe that with only 2 clusters, there are two well-segmented groups, contrary to what is observed with 5 clusters, where some groups are identified but they exhibit more overlap among themselves.

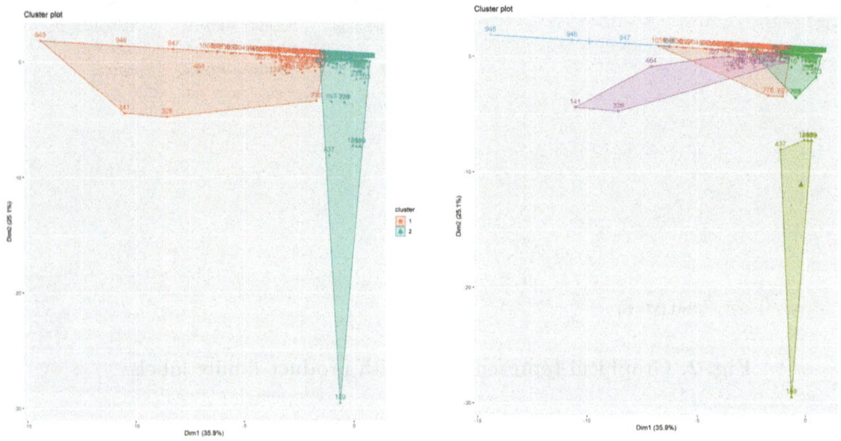


**Fig. 3.** k-Means result with 2 (left side) and 5 clusters (right side)

### 5.2 Hirarchical Clustering

Another attempted approach was to implement the hierarchical clustering algorithm with the 'ward' method, aiming to visually detect the optimal number of clusters through the dendrogram [16]. By analysing the dendrogram in Fig. 4, it can be observed that the dendrogram can be divided into three clusters. The result of implementing this algorithm with 3 clusters can be seen in Fig. 4 on the right-hand side.

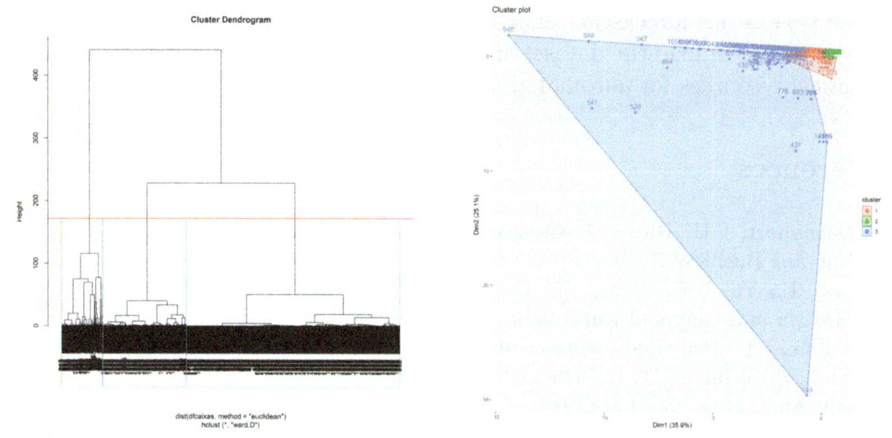


**Fig. 4.** Representation of the dendrogram (left side) and the clusters (right side) obtained with the hierarchical clustering method 'ward'.

### 5.3 Cluster Evaluation

The k-Means algorithm for k = 2 achieved a good silhouette coefficient (around 0.78), which effectively defined the clusters. The k-Means algorithm for k = 5 had a silhouette coefficient of approximately 0.75. Despite the hierarchical clustering algorithm using the 'ward' method obtaining the lowest silhouette coefficient (0.52), it satisfactorily segmented the data into three clusters.

## 6 Discussion and Conclusion

In this study, several specific objectives were achieved, providing valuable insights for sales and product management strategies. Groups of products with similar sales patterns were identified, along with key products that significantly impact the company's global sales, enabling more efficient resource allocation.

One of the main conclusions of this study was the effectiveness of the multivariate analysis approach, including the application of Principal Component Analysis (PCA). This approach reduced data redundancy, identified correlated variables, and unveiled hidden sales patterns, offering the company a deeper understanding of product sales dynamics. Furthermore, the k-Means algorithm, when applied with 2 clusters, stood out with an impressive silhouette coefficient, indicating the quality of the sales data clustering.

As a result of this analytical study, the company is now positioned to implement targeted marketing strategies to boost sales of bags and boxes in Portugal and other countries by leveraging product groups that share similar sales characteristics.

As future work, it is suggested to generate new explanatory variables for more effective group segmentation analysis, to collect data consistently over time to enable time series forecasting analysis, and to create a graphical interface that allows users to upload the dataset and provide histograms and other exploratory statistical resources for automated analysis.

## References


1. Davenport, T.H., Harris, J.: Competing on Analytics: The New Science of Winning. Harvard Business Review Press (2007)
2. Cao, L., Yu, P.S., Zhang, C.: Data-Intensive Text Processing with MapReduce. Morgan and Claypool Publishers (2013)
3. Jolliffe, I.T.: Principal Component Analysis. Wiley (2014)
4. Abdi, H., Williams, L.J.: Principal component analysis. Wiley Interdisc. Rev. Comput. Stat. **2**(4), 433–459 (2010)
5. Peng, H., Long, F., Ding, C.: Feature selection based on mutual information: criteria of max-dependency, max-relevance, and min-redundancy. IEEE Trans. Pattern Anal. Mach. Intell. **27**(8), 1226–1238 (2015)
6. Tipping, M.E., Bishop, C.M.: Probabilistic principal component analysis. J. R. Stat. Soc. Series B (Statistical Methodology) **61**(3), 611–622 (1999)
7. Gorban, A.N., Zinovyev, A.Y., Popova, T.G.: How many eigenvectors do we need to analyze diffusion maps? In: European Conference on Machine Learning Springer, pp. 548–562 (2016)
8. Kotler, P., Keller, K.L.: Marketing Management (15th ed.). Pearson (2015)
9. Wedel, M., Kamakura, W.A.: Market Segmentation: Conceptual and Methodological Foundations (2nd ed.). Springer (2012)
10. Han, Y., Li, Z., Li, X., Li, Y.: An application of PCA based K-means clustering for customer segmentation in one luxury goods company. J. Comput. Data Sci. **3**, 17–22 (2020)
11. Hwang, J., Kim, J.: A study on customer segmentation using k-means clustering and RFM analysis in online retail industry. J. Digital Convergence **18**, 463–472 (2020)
12. Kaur, A., Singh, P.: PCA based segmentation of sales data of a luxury goods company: a case study. Int. J. Comput. Sci. Mob. Comput. **10**, 250–255 (2021)
13. Liu, X., Xu, H., Sun, Y.: Customer segmentation based on hierarchical clustering and its application in retail industry. Adv. Intell. Syst. Comput. **1118**, 330–340 (2020)
14. Bholowalia, P., Kumar, A.: EBK-means: a clustering technique based on elbow method and k-means in WSN. Int. J. Comput. Appl. **105**, 9 (2014)
15. Matt, O.: 10 Tips for Choosing the Optimal Number of Clusters. https://towardsdatascience.com/10-tips-for-choosing-the-optimal-number-of-clusters-277e93d72d92. Accessed May 2022
16. Reddy, C.: Understanding the concept of Hierarchical clustering Technique. https://towardsdatascience.com/understanding-the-concept-of-hierarchical-clustering-technique-c6e8243758e. Accessed May 2022

# SMS Send Frequency Prediction: A Case Study

Vasco Azevedo[1,2], Ana Madureira[3], Ivo Pereira[2,3,4(✉)], Duarte Coelho[2], Miguel Ângelo Rebelo[2,5], and Daniel Oliveira[2]

[1] ISEP - Instituto Superior de Engenharia do Porto, 4249-015 Porto, Portugal
1202016@isep.ipp.pt
[2] E-goi, 4450-190 Matosinhos, Portugal
ivopereira@ufp.edu.pt, {dcoelho,mrebelo,dalves}@e-goi.com
[3] ISRC - Interdisciplinary Studies Research Center, ISEP, 4249-015 Porto, Portugal
amd@isep.ipp.pt
[4] Universidade Fernando Pessoa, 4249-004 Porto, Portugal
[5] i3s, Rua Alfredo Allen, 208, 4200-135 Porto, Portugal

**Abstract.** Nowadays, companies are actively endeavoring to adjust their strategies in response to the dynamic patterns of customer behavior, thereby embracing the utilization of artificial intelligence as a means to accomplish this objective. Given the increasing necessity for individualized communication to capture customer attention, the utilization of artificial intelligence emerges as a crucial factor in optimizing marketing strategies. This study primarily centers on the creation and implementation of a machine learning model aimed at forecasting the optimal day and hour for sending smart SMS marketing campaigns. Additionally, it encompasses an extensive examination and preprocessing of the dataset's data. The dataset utilized in this study originates from an authentic corporate entity, and has been employed for the purpose of training the model to generate predictions that are applicable within a real-world context.

## 1 Introduction

Digital marketing refers to the strategic utilization of various technologies to facilitate and enhance marketing efforts [3]. In contemporary times, organizations are consistently endeavoring to explore novel strategies for effectively interacting with prospective clientele, while simultaneously striving to sustain and nurture existing customer relationships. The adoption of the online realm is increasingly imperative to optimize the efficacy of conventional marketing approaches [7]. Utilizing contemporary technology and platforms inside a marketing plan can confer a competitive advantage within the industry. The potential avenues for engaging with consumers in a virtual context are extensive, encompassing many strategies such as social media advertising, email marketing, content creation, and collaborations with influencers.


A. Bajaj et al. (Eds.): HIS 2023, LNNS 1224, pp. 187–196, 2025.
https://doi.org/10.1007/978-3-031-78925-0_19

Artificial Intelligence (AI) enables firms to effectively analyze large quantities of data pertaining to client demographics and browsing patterns. This information grants them a comprehensive comprehension of their target demographic, enabling them to enhance their marketing efforts with greater accuracy. The utilization of predictive analytics proves advantageous in predicting client behavior [4,5,8], so facilitating organizations in identifying prospective individuals with a greater probability of conversion and even pinpointing specific customers who exhibit a higher inclination to engage with marketing initiatives. This projection serves as a guiding tool for marketing strategies, facilitating efficient allocation of resources and ultimately resulting in a higher return on investment (ROI). Chatbots, which are software applications infused with AI, provide real-time interactions with consumers [1]. These chatbots are commonly positioned as pop-up features located in the corner of webpages. These highly intelligent entities provide efficient and precise client care, facilitating the development of brand loyalty and enhancing consumer happiness. Marketers often encounter challenges in content production, primarily due to the substantial time commitment it requires. Nevertheless, organizations are progressively resorting to the domain of AI in order to address this issue, thereby diminishing the level of work required to attain desired outcomes [11].

This paper aims to analyze the concepts of Send Frequency Prediction and Send Time Optimization, providing an overview of the current advancements in forecasting personalized and individualized communications. The utilization of AI and Machine Learning techniques are explored to develop models that can accurately predict the optimal timing for sending SMS communications. The obtained results from the developed models will be analyzed and interpreted by examining the available data. Additionally, the potential integration of the SMS sending prediction functionality with the existing E-mail Send Frequency Prediction mechanism will be investigated, with a focus on determining the most effective channel for communication.

This paper is structured as follows: Sect. 2 defines the problem; Sect. 3 describes the related work; Sect. 4 details the proposed methodology; Sect. 5 presents the computational study; and, finally, Sect. 6 presents the conclusions.

## 2 Problem Definition

The objective of this study is to determine the most advantageous moment for the sending of SMS messages. Acknowledging the fact that subscribers display varied online behavior, such as opening their SMS messages at different times of the day (morning, afternoon, or evening), it is imperative to strategically select optimal time frames that maximize visibility and effectively engage customers. By gaining insight into the temporal preferences of individual subscribers, marketing campaigns can expect to observe a rise in the rate at which emails are opened, thereby resulting in enhanced conversion rates and ultimately bolstering the sales performance.

The implementation of targeted marketing campaigns via email and SMS has emerged as a crucial factor in driving the financial prosperity of firms. Nevertheless, the inappropriate scheduling of promotional messages may lead to their delivery to users' devices during periods of low engagement, such as when individuals are occupied with work or do not have immediate access to their phones or other technological devices. While the intrinsic quality of the SMS campaign may be commendable, a mere temporal discrepancy might lead to a decline in open rates. This underscores the crucial need of delivering targeted and well-timed messages. The challenge at hand is acknowledged, and our aim is to address it by constructing a predictive model. This model will enable us to determine the most suitable day and time for delivering marketing campaigns to individual users through smart SMS. By doing so, we anticipate achieving increased open rates and fostering stronger connections between companies and their intended audiences.

## 3 Related Work

The available information on the most effective timing for sending SMS campaigns is limited. In a related study, Araújo et al. [2] conducted research that examined the optimal timing for marketing efforts, specifically focusing on email communication as opposed to SMS. The focus of Araújo's work was the development of a predictive model aimed at identifying the optimal day and time for the implementation of email marketing campaigns. In order to accomplish this objective, the researcher utilized a range of regression methodologies, such as the Random Forest Regressor, Multiple Linear Regression, K-Neighbors Regressor, and Support Vector Regressor. In order to assess the effectiveness of the model, various important metrics were employed, including the Mean Absolute Error (MAE), coefficient of determination (R2 score), Mean Squared Error (MSE), and Root Mean Squared Error (RMSE). The K-Neighbors Regressor (KNN) exhibited the highest level of accuracy among the models that were tested, as seen by its R2 score of 0.898. This outcome highlights the predictive capabilities of the KNN model. The study employed an ensemble methodology, which involved combining the outcomes of many regression models through averaging and stacking approaches. This strategy was utilized to generate a definitive forecast, resulting in an R2 accuracy score of 0.91.

In a separate study conducted by Droomer and Bekker [6], the primary objective was to utilize machine learning techniques in order to forecast the future date of purchase for a certain retail consumer. This study is pertinent to the project since it also aims to utilize a subscriber's interaction history in order to forecast the most suitable moment for a certain event. Droomer's study emphasizes the application of machine learning (ML) techniques, including Recurrent Neural Networks (RNN), XGBoost, and Neural Networks (NN), in the context of predicting tasks. Among the methods that were examined, it was seen that the NN algorithm exhibited superior performance compared to the other algorithms in terms of absolute error, as assessed in days. Following the Neural Network algorithm, the XGBoost strategy demonstrated rather favorable results.

## 4 Proposed Methodology

The selected methodology is the Cross Industry Standard Process for Data Mining (CRISP-DM) [10], which is widely acknowledged in the field of data science, in order to facilitate the development of an effective system. The process, seen in Fig. 1, consists of six separate steps, each demanding significant attention and cannot be disregarded as it may jeopardize the final product's overall quality.

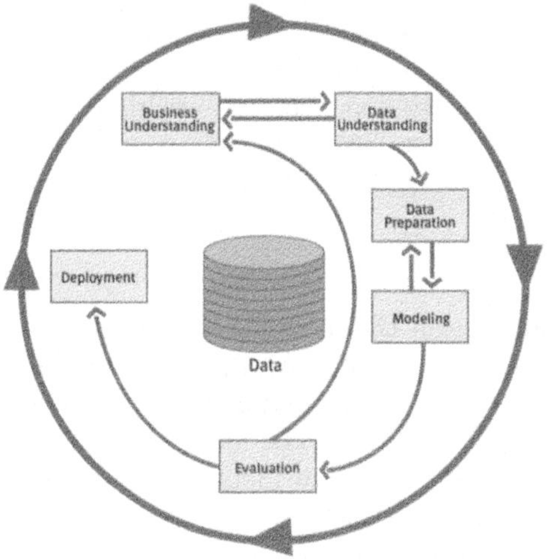


**Fig. 1.** Cross Industry Standard Process for Data Mining (CRISP-DM) methodology [10]

Delving further into the intricacies of each phase, the first phase, referred to as "Business Understanding," involves gaining a comprehensive understanding of the business context and objectives. The primary focus of this initial phase is to acquire a thorough understanding of the project's goals and objectives. A comprehensive comprehension of the business establishes a strong basis for the succeeding stages, creating favorable conditions for the success of the project.

Data understanding is the process of comprehending and gaining insights from data, focusing on a comprehensive understanding of both internal and external sources. To handle a large data repository, a query was used to extract interactions and organize them into a CSV file. The CSV file format offers practical benefits, such as reducing server strain, simplifying data manipulation, and overcoming network delays and overhead associated with remote database queries. CSV files can be maintained using various tools, offering versatility for data manipulation.

The process of preparing data for analysis involves manipulating final datasets, eliminating duplicate attributes, ensuring data reliability, and organizing it for analysis. Quality assurance is crucial, requiring data cleansing, transformation, and structural refinement before integration into the system. The method described in Fig. 2 is initiated by doing an initial scan to identify Not a Number (NaN) values, which are then removed. Data type conversions are performed to improve runtime performance and reduce resource utilization. Correlation analysis is used to examine variables' associations, removing those with no correlation. Validation occurs when the customer leaves the dataset if a threshold is not met.

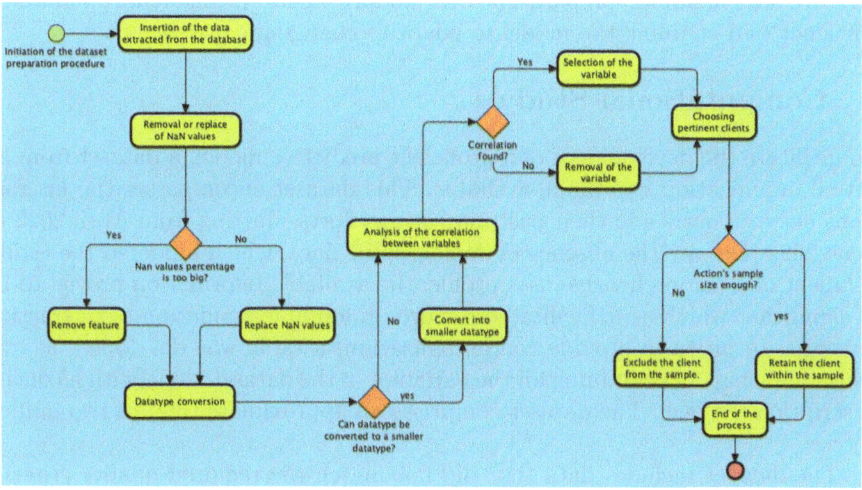


**Fig. 2.** Business Process Model and Notation (BPMN) for data preparation

Data modeling is a crucial process in information technology, involving the creation of a conceptual representation of data and its relationships within a system. It helps organizations make well-informed business decisions by identifying patterns and trends in data. The first stage involves selecting an algorithm and analyzing data characteristics, such as features and data structure. A specific algorithm is chosen and trained using preprocessed information to identify patterns and anomalies in users' interactions with SMS marketing campaigns. The primary objective is to determine the most advantageous timing for sending SMS messages. As the model evolves, the accuracy improves and the discrepancy between predicted values and actual real-world values decreases.

The evaluation phase is a crucial stage in the development of predictive models. It helps organizations determine the most suitable model for their goals and optimal performance. The model is tested to ensure its efficacy and quality, and its performance is evaluated in real-world scenarios. The model's predictive outputs are compared with unobserved values. Various metrics are introduced to assess the model's performance, including the Mean Absolute Error (MAE) and the Root Mean Squared Error (RMSE), which evaluate the precision of regression predictions.

The deployment phase involves integrating a predictive model into an organization's operations, continuously monitoring and modifying it as needed. This systematic approach helps organizations efficiently oversee projects, improve decision-making, and achieve organizational objectives. The primary objective is to enhance smart SMS delivery timing efficiency. Docker, a containerization platform, has been used to encapsulate the paradigm, ensuring consistent deployment across diverse settings. This approach provides portability and simplifies deployment, allowing for diverse workloads. Using scripting in automation can improve data handling, preparation, and model execution, enhancing operational efficiency and enabling the model to adapt to changing datasets.

## 5 Computational Study

To facilitate the development of a proficient predictive model, a dataset from an actual organization was made available. This dataset encompasses the interactions between users and their past marketing efforts, starting from April 2023 to June 2023. Despite the absence of demographic data, which hindered the establishment of a comprehensive user profile, the available information proved to be of significant worth as it facilitated a more profound examination of interaction patterns. In order to provide contextualization, a table was developed, as evidenced in Chap. 6, containing all the variables of the dataset along with a concise description for each. The dataset comprises an approximate total of two million entries.

The dataset includes data that did not match the required quality criteria, necessitating a process of cleaning and refining. The initial stage in developing an effective prediction model is ensuring the data is of the highest quality. The company's database underwent a minor transition, leading to the duplication of data pertaining to action 1 and 2. In order to address this concern, a filtering process was conducted, taking into account the action that exhibited the greatest number of interactions. The elimination of these replicated interactions was relevant, as they had the potential to impact the model's capacity to generate precise predictions.

Initially, a query was performed in order to identify and eliminate characteristics that exhibited a significant number of NaN values, as they would not contribute to the project's objectives. Following this exclusion, the sole columns that retained any significance were the operating system and the program utilized to access the hyperlink embedded within the intelligent SMS. There is a possibility that certain interactions present in the dataset may have been attributed to user "missclicks". Taking into consideration the aforementioned, it was necessary for clicks originating from the same user and within the same campaign to have a minimum time interval of 15 min in order to avoid their deletion.

The datatypes of the features that were capable of being transformed to a smaller size were indeed converted. Although seemingly insignificant, even modest adjustments hold significance when handling large volumes of data. This modification facilitates the liberation of memory during the process of training the system.

The preprocessing phase involved an exploratory analysis to identify hidden information and patterns. The experiment focused on actions linked to clicks on links, with each interaction accompanied by a timestamp. The day and hour components were extracted and transformed into sine and cosine values. This was necessary due to the system's inability to understand the cyclical nature of time and the need to visualize user clicks through a bi-dimensional scatter graph. The system's inherent limitation in understanding time resulting in the failure to recognize minute intervals was addressed.

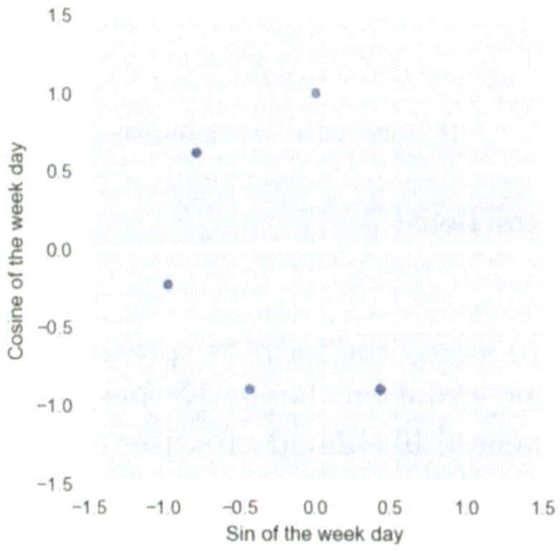


(a) Interaction during the week

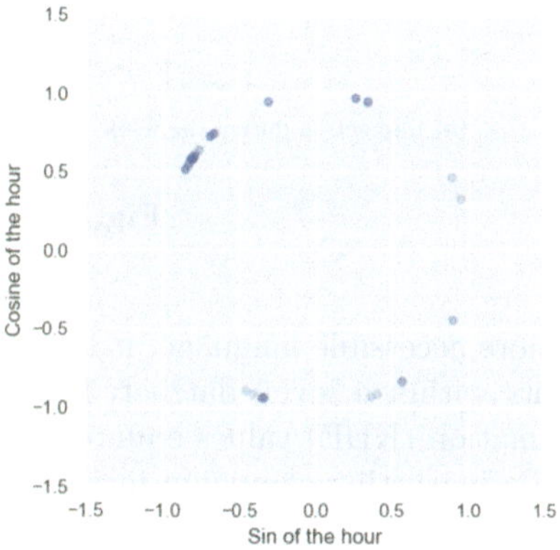


(b) Interaction during the day

**Fig. 3.** User 1 interactions

To better understanding, two users interactions are shown. The charts illustrate that User 1 (Fig. 3) exhibits a lack of activity on Tuesday and Wednesday, as indicated by the absence of scatter points on the right side of the image (specifically at sin $= 0.781$, cos $= 0.623$, and sin $= 0.974$, cos $= -0.222$). It becomes evident that there is a considerable dispersion of values. Nevertheless, an absence of data is observed between the time intervals of 15:00 and 19:00, as indicated on the left side of the graph. User 1 exhibits non-traditional behavior by actively participating in marketing initiatives throughout the late night hours, specifically between 3:00 AM and 4:00 AM.

As depicted in Fig. 4, the engagements of User 2 are extensively distributed across the entire week. The discrepancy in data availability occurs over a distinct time period, namely between 22:00 and 7:00, as opposed to User 1. This phenomenon might likely be attributed to the establishment of a well-regulated sleep routine, wherein the individual minimizes their exposure to digital devices.

The model was constructed solely based on the subscribers' interaction history, as no significant connections were found even when included demographic data. Predictions were generated with a kernel-density estimate employing Gaussian kernels. Kernel density estimation (KDE) is a statistical technique employed to estimate the probability density function of a continuous random variable [9].

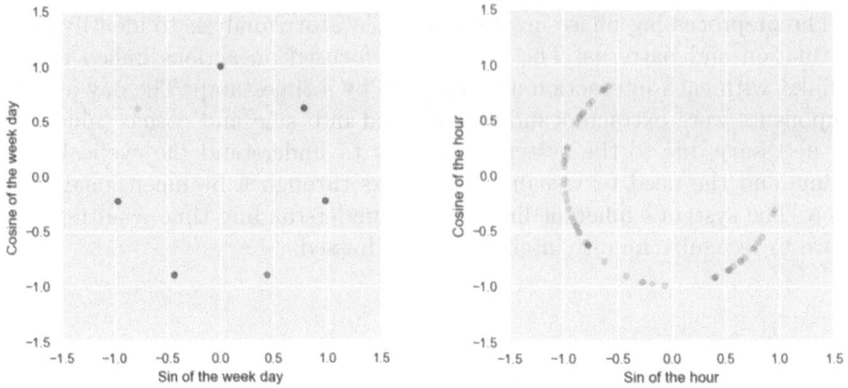


(a) Interaction during the week

(b) Interaction during the day

**Fig. 4.** User 2 interactions

In more accessible language, it is employed to assess the form or spread of data points within a given dataset. In the present situation, larger Kernel Density Estimation (KDE) values equate to an augmented likelihood of a user engaging with a marketing campaign inside the associated time period.

Initially, the data that was previously handled necessitated a series of preparatory procedures prior to its compatibility with the "stats.gaussian_kde" function. Only the interactions that involved the number 1 were retained, as they were the sole interactions deemed pertinent to this model. All further characteristics were removed, except for the user IDs and interaction timestamps. Subsequently, the data was organized into a dictionary, wherein each unique identifier was linked to its respective timestamps. Subsequently, the date and time components were extracted from each timestamp and subsequently converted into their respective sine and cosine representations. The modified data was segregated into distinct arrays and subsequently combined. The implementation of a weight modification was intended to enhance the importance of recent interactions. This is because the early interactions of a subscriber may vary from their current ones due to changes in their daily routines or habits, which can be influenced by various causes.

The argmax function from the numpy library was utilized to determine the point of maximum density inside the Gaussian distribution. Following this, the values were subsequently transformed from their sine and cosine representations back to their original format, denoting the day and hour. In order to minimize resource consumption, predictions were exclusively generated for user IDs that were supplied as arguments. This approach aimed to facilitate the functioning of the system and decrease the time required for prediction generation. To evaluate the extent of model development, an application programming interface (API) was implemented. The API includes an authentication mechanism that serves to ensure that the user possesses sufficient authorization to make predictions. In the event that all operations are executed without encountering any complications, the API will provide the user with the user ID, as well as the projected date and time, accompanied by the requisite authorizations.

The model's precision was evaluated using the most recent timestamp from a dictionary as a reference value. However, this approach has limitations, as the final timestamp may represent an anomaly in user behavior. The test may be more challenging to complete due to the cyclical nature of time, which results in a difference of just 2 h between 23:00 and 01:00. To address this, sine and cosine functions were used to represent timestamps as angles on a circular scale, allowing mathematical operations to consider the cyclic aspect of time. However, this approach may result in the recurrence of angle values throughout a certain period. To address this, the absolute angular difference between the anticipated value and the previous timestamp was computed, focusing solely on the size of the error. The inaccuracy was then transformed into units of days and hours for presentation of the findings.

The MAE value of 1.2338 suggests that, on average, the deviations between the predicted and actual optimal days for sending SMS marketing campaigns amount to around 1.23 days. Although the model exhibits a satisfactory level of accuracy in its predictions, there exists potential for further enhancement. The RMSE for the variable "days" is calculated to be 1.6416. This number is marginally higher than the MAE, but it remains quite low. Consequently, it can be inferred that the model's predictions closely align with the actual values. The MAE of 4.4617 indicates that, on average, the model's estimates for the optimal hour to send SMS campaigns deviate by around 4.46 h from the actual optimal hour. The RMSE for the variable "hours" is calculated to be 5.7496. This value is found to be higher than the MAE, suggesting the presence of variability in the predictions, including some instances of larger errors. Despite the greater metrics associated with hours, this can be attributed to the ratio of four hours per day compared to one day per week.

## 6 Conclusions

Organizations are employing AI to adjust to client behavior and enhance marketing efforts. The primary objective of this research is to develop a machine learning model that can accurately anticipate the optimal timing for smart SMS marketing campaigns. To do this, a dataset obtained from a genuine corporate organization will be utilized to make predictions that are applicable in real-world scenarios.

One of the obstacles encountered over the course of this research pertained to the dataset. The absence of significant correlations between elements posed a challenge in constructing a default user profile, despite the abundance of available information.

Considering the objective of this project was to be incorporated into a larger initiative referred to as "The next best channel," it would be highly beneficial to do a more comprehensive analysis of user engagement with the marketing campaigns. The "next optimal channel" refers to the dedicated method of delivering marketing campaigns to people, taking into consideration the most effective communication channel available, such as email, smart SMS, or web-mail. Nevertheless, the extent of user interactions with smart SMS and web-mail is rather restricted in comparison to email. In order to develop a precise predictive model, it is important to enhance the level of engagement with web-mail and smart SMS platforms. Subsequently, the matter becomes more amenable to resolution through the implementation of a switch mechanism that takes into account consumers' interactions and subsequently guides the allocation of marketing campaigns.

## References

1. Adamopoulou, E., Moussiades, L.: An overview of chatbot technology. In: IFIP International Conference on Artificial Intelligence Applications and Innovations, pp. 373–383. Springer (2020)
2. Araújo, C., Soares, C., Pereira, I., Coelho, D., Rebelo, M.Â., Madureira, A.: A novel approach for send time prediction on email marketing. Appl. Sci. **12**(16), 8310 (2022)
3. Bala, M., Verma, D.: A critical review of digital marketing. Int. J. Manag. IT Eng. **8**(10), 321–339 (2018)
4. César, I., Pereira, I., Madureira, A., Coelho, D., Rebelo, M. Â., de Oliveira, D.A.: Analysing and modeling customer success in digital marketing. In: International Conference on Innovations in Bio-Inspired Computing and Applications, pp. 404–413. Cham: Springer Nature Switzerland (2022)
5. Davenport, T., Guha, A., Grewal, D., Bressgott, T.: How artificial intelligence will change the future of marketing. J. Acad. Mark. Sci. **48**, 24–42 (2020)
6. Droomer, M., Bekker, J.: Using machine learning to predict the next purchase date for an individual retail customer. S. Afr. J. Ind. Eng. **31**(3), 69–82 (2020)
7. Parsons, A.J., Zeisser, M., Waitman, R.: Organizing for digital marketing. McKinsey Q. (4), 185 (1996)
8. Paulo, M., Miguéis, V.L., Pereira, I.: Leveraging email marketing: using the subject line to anticipate the open rate. Expert Syst. Appl. **207**, 117974 (2022)
9. Scipy gaussian kde? https://docs.scipy.org/doc/scipy/reference/generated/scipy.stats.gaussian_kde.html. Accessed 25 July 2023
10. Shearer, C.: The CRISP-DM model: the new blueprint for data mining. J. Data Warehousing **5**(4), 13–22 (2000)
11. Vlačić, B., Corbo, L., Costa e Silva, S., Dabić, M.: The evolving role of artificial intelligence in marketing: a review and research agenda. J. Bus. Res. **128**, 187–203 (2021)

# Load Balancing in Parallel Machines: A Literature Review

Guilherme Santos[1], Rita Guerreiro[1], André S. Santos[2(✉)], Anabela Tereso[3], José A. Oliveira[3], and João Bastos[4]

[1] Institute of Engineering, Polytechnic Institute of Porto - ISEP/IPP, Porto, Portugal
{1180794,1180543}@isep.ipp.pt
[2] Interdisciplinary Studies Research Center (ISRC), Institute of Engineering, Polytechnic Institute of Porto - ISEP/IPP, Porto, Portugal
abg@isep.ipp.pt
[3] ALGORITMI Research Centre/LASI, University of Minho, Guimarães, Portugal
{anabelat,zan}@dps.uminho.pt
[4] INESC TEC, Institute of Engineering, Polytechnic Institute of Porto - ISEP/IPP, Porto, Portugal
jab@isep.ipp.pt

**Abstract.** This literature review explores the critical role of load balancing in parallel machines and distributed systems, with a focus on strategies, challenges, and real-world applications. The paper comprises six main sections, covering load balancing problems, a comparative analysis of identical and unrelated parallel machines, a bibliometric analysis of existing literature, in depth paper and case study discussions, and insights into future research directions. This paper highlights the importance of load balancing in optimizing resource allocation and system stability. It emphasizes the need for adaptive and sustainable load balancing algorithms in evolving computing paradigms, making it a valuable resource for researchers, practitioners, and policymakers. It concludes by exploring future paths in this topic, namely emerging areas like edge and fog computing with swarm-like models, expanding load balancing applications and machine learning algorithms, enabling real-time load prediction and management, addressing evolving load distribution complexities.

**Keywords:** Parallel Machines · Load Balancing · Heuristics · Machine Learning · Swarm Computing

## 1 Introduction

Load balancing in parallel machines is a critical factor in optimizing performance, resource utilization, and efficiency. As modern industries and data centers rely on parallel processing for many tasks, the importance of efficiently distributing workloads among machines and collaborators becomes clear. The fundamental challenge lies in ensuring that each machine or operator works near its maximum capacity, avoiding both underutilization and overburdening. Achieving this equilibrium demands advanced strategies,


A. Bajaj et al. (Eds.): HIS 2023, LNNS 1224, pp. 197–207, 2025.
https://doi.org/10.1007/978-3-031-78925-0_20

algorithms, and frameworks, which have been the focus of extensive research in recent years. This literature review delves into the evolving landscape of workload balancing in parallel machines, offering a comprehensive examination of the state of the art, emerging trends, and the implications for various domains, from manufacturing to computer networking.

The structure of this literature review is designed to provide a comprehensive exploration of workload balancing in parallel machines. It unfolds across six key sections, each contributing distinct perspectives and insights. We commence with a foundational contextualization of 'Load balancing' and the contrast between 'Unrelated parallel machines' and 'Identical parallel machines.' Following this, a bibliometric analysis of the subject 'Load balancing in parallel machines' will be undertaken to offer a panoramic view of the field's status. Subsequently, several articles will be scrutinized in depth to glean a deeper understanding, followed by a succinct overview of other noteworthy contributions. The subsequent chapter, 'Future Paths,' explores the promising frontiers where novel developments and applications of load balancing algorithms are currently unfolding. Finally, the paper culminates in a conclusive section that encapsulates the key takeaways and implications drawn from this review.

## 2 Literature Review

### 2.1 Load Balancing

In the realm of industrial operations and computing systems, the efficient allocation of tasks across parallel machines stands as an indispensable endeavor, with far-reaching implications for productivity, performance, and resource utilization [1]. The intrinsic value of load balancing on parallel machines is paramount, with the potential to revolutionize manufacturing processes, computing clusters, and workforce management.

Load balancing is at the heart of optimizing the allocation problem—tasking the right machine with the right job. At its core, this challenge seeks to address the disparities in workloads across parallel machines [2]. In the context of modern manufacturing scenarios, it carries profound ramifications. Here, the equitable distribution of tasks across machines contributes not only to the maximization of machine utilization but also to makespan minimization (maximum completion time). Reduced makespan translates to quicker production cycles, optimized resource usage, and more cost-efficient operations [2].

Moreover, the importance of load balancing takes on a human dimension. In contexts involving labor distribution, equitable load balance is the linchpin for fair work distribution among employees on a fixed salary. The repercussions of disproportionate workloads can extend to employee morale, engagement, and overall job satisfaction [3].

### 2.2 Identical Parallel Machines Vs. Unrelated Parallel Machines

Load balancing is a critical component of efficient task allocation in parallel machine environments. The nature of these challenges, however, differs significantly depending on whether the machines are identical or unrelated. In this section, we explore the fundamental distinctions between identical parallel machines and unrelated parallel machines, emphasizing the complexities introduced by the latter.

In identical parallel machine problems, tasks are allocated to a set of machines that share common characteristics. These machines are, in essence, clones of each other, featuring identical processing capabilities, speeds, and capacities. This uniformity simplifies the task allocation process as there is no need to account for variations in machine capabilities. Identical parallel machines pose challenges, but the focus is primarily on maximizing individual machine efficiency and overall load balancing. Metrics such as makespan (minimizing the time it takes to complete all tasks), workload amplitude, and workload standard deviation (equitably distributing tasks among machines) are often used for evaluation.

Unrelated parallel machine scheduling introduces a complex layer to task allocation. In this scenario, tasks are assigned to a set of machines with diverse capabilities, rendering them distinct from one another. These machines may exhibit variations in processing speeds, capacities, or other characteristics. Consequently, the task of load balancing becomes a multifaceted challenge. The core distinction of unrelated parallel machine systems lies in the diversity of machine capabilities. Different machines can complete tasks at varying speeds, making it necessary to consider not only individual machine efficiency but also the collective workload minimization [4]. In these cases, the traditional objective of minimizing makespan must coexist with the additional objective of achieving equitable load distribution.

## 3 Bibliometric Analysis in Load Balancing

For the bibliometric analysis, the Dimensions software [5] was used, a comprehensive research information system, to collect a substantial dataset of scholarly publications related to load balancing on parallel machines utilizing the search string "Load Balancing in Parallel Machines" and a date range off 1974 to 2023. Dimensions facilitated data retrieval and organization through tailored searches. The analysis was further refined using the "bibliometrix library" in R [6], a versatile tool for bibliometric examination. This combination allowed for the systematic exploration of research trends, author contributions, and emerging areas of interest within the field.

In Fig. 1 it is possible to observe an exponential growth in the number of publications about the topic of Load Balancing in Identical and Unrelated Parallel Machines.

An explosive growth is observed in the mid to late 2000s.

In Fig. 2 the areas of study with more publications on this topic are "Information and computing sciences", followed by "Biomedical and Clinical Sciences", "Human Society" and "Commerce, Management, Tourism and Services". These areas are expected to be closely related to studies on load balancing and resource allocation optimization due to the nature and scarcity of the type of resources in the scope of these topics.

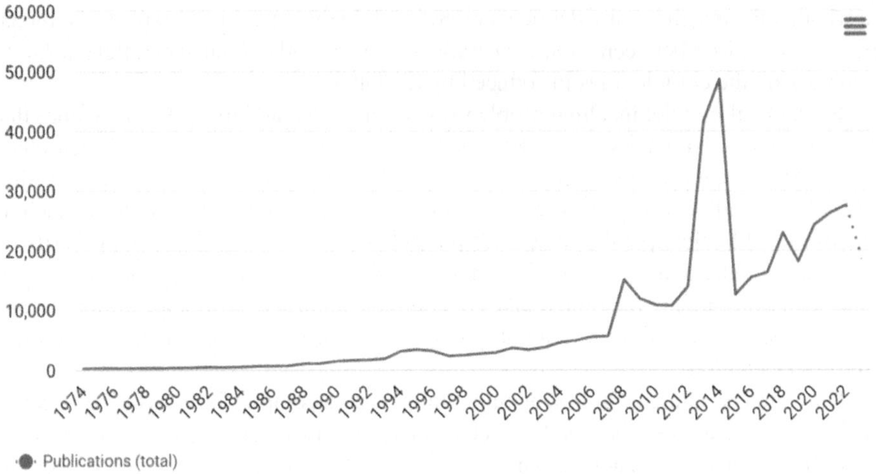


**Fig. 1.** "Load Balancing" publications per year

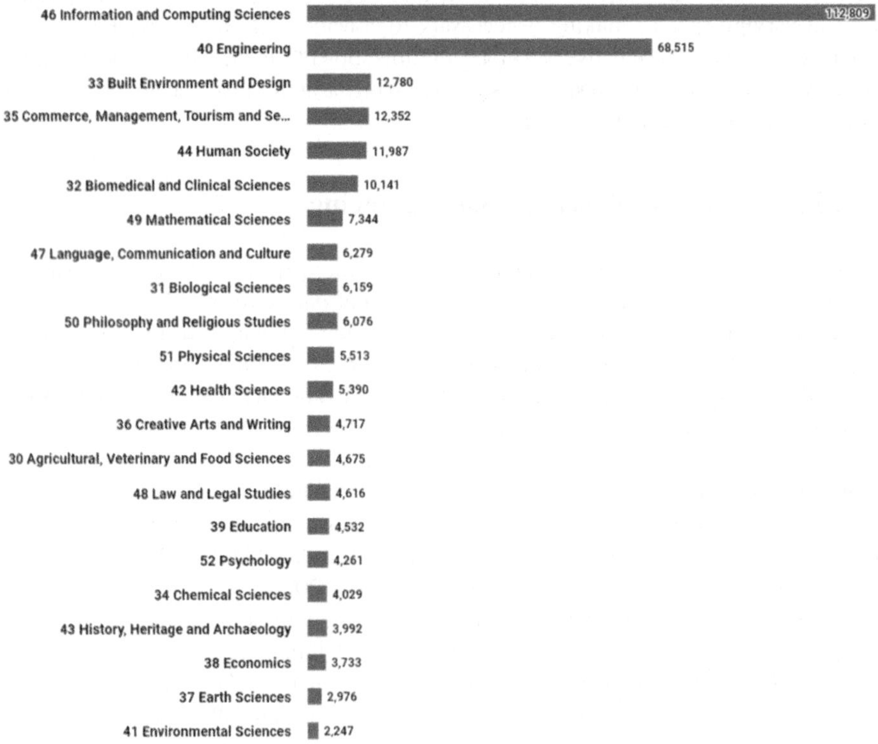


**Fig. 2.** Number of publications related to "Load balancing" per research category.

# 4 Case Studies

In this section of the literature review, a meticulous analysis of pertinent articles will be undertaken. The review will commence by conducting in-depth analysis of four key articles, delving into their methodologies, findings, and contributions. These comprehensive assessments will provide a solid foundation for understanding the nuances of the subject matter. Subsequently, an additional eight articles will be summarized, offering concise overviews of their key insights and relevance to the research theme.

## 4.1 A Modified Variable Neighborhood Search for Minimizing the Makespan on Identical Parallel Machines

In [7] a modified version of the Variable Neighborhood Search (MVNS) algorithm is applied to identical parallel machines scheduling problems with the objective of minimizing makespan (maximum completion time). Identical parallel machines scheduling is a well-known combinatorial problem, and the authors highlight the impracticality of obtaining optimal schedules for larger instances using exact algorithms due to their exponential time complexity. Heuristic algorithms offer a practical balance between time and memory requirements. Overall, the paper's findings are promising, particularly for the application of VNS to sequencing and permutation problems.

The paper presents the MVNS algorithm and tests it on 150 randomly generated instances. The results are compared with those obtained using Genetic Algorithm (GA) and Longest Processing Time (LPT) solutions. The paper concludes that the MVNS algorithm outperforms both GA and LPT in terms of solving the identical parallel machines scheduling problem with the makespan criterion.

The performance of the MVNS algorithm is highlighted by its ability to solve 96.33% of instances to the optimum. Furthermore, the study marks the first reported application of the Variable Neighborhood Search algorithm in the identical parallel machines scheduling problem with the makespan criterion.

## 4.2 Ordered Minimum Completion Time Heuristic for Unrelated Parallel-Machine Problems

This paper [8] delves into the intricate domain of scheduling problems on unrelated parallel machines, with a primary focus on the unrelated parallel machines makespan minimization problem (Rm||Cmax). Recognized as NP-hard, these problems often need heuristic approaches for practical resolution [8]. The paper introduces the Ordered Minimum Completion Time (OMCT) heuristic as a novel solution strategy, building upon the foundations of the Minimum Completion Time (MCT) heuristic. The key innovation lies in refining task allocation order and addressing the influence of allocation sequence on solution quality.

It presents the OMCT heuristic as an advanced solution, marking a significant departure from conventional approaches. OMCT prioritizes task allocation order based on standard deviation, ensuring that tasks with wider completion time variability are allocated first. Furthermore, the Suffrage index is introduced to account for tasks' likelihood of allocation to their preferred machines. This dual criterion approach, combining standard deviation and Suffrage, optimizes allocation sequencing and augments the MCT algorithm's performance.

The conclusion underscores the sensitivity of the MCT heuristic's solution quality to the random order of task allocation, prompting the need for a refined approach. Here, the OMCT heuristic is introduced as a solution to this challenge, offering a systematic allocation sequence strategy. OMCT outperforms the MCT heuristic by placing greater emphasis on tasks with greater completion time dispersion and considering the likelihood of task allocation to preferred machines.

### 4.3 A Genetic Algorithm for Unrelated Parallel Machine Scheduling Minimizing Makespan Cost and Electricity Cost under Time-of-Use (TOU) Tariffs with Job Delay Mechanism

This study [9] focuses on unrelated parallel machine scheduling under dynamic Time-of-Use (TOU) electricity pricing. The driving force behind this research is the fluctuating electricity prices in distinct time periods, which necessitates an approach that not only optimizes makespan but also minimizes the associated electricity costs. To tackle this challenge, the study leverages the Genetic Algorithm (GA) as a solution method. A pivotal contribution of this paper is the development of chromosome decoding techniques, inspired by cost considerations, aimed at transforming individuals into feasible schedules. Furthermore, the proposed schedules are refined through a job delay mechanism, allowing the strategic shifting of jobs to avoid high electricity cost periods. Preliminary results suggest a high effectiveness and efficiency of this approach in addressing the problem.

The conclusion encapsulates the paper's contributions and initial findings. It reiterates the promising outcomes of deploying GA with a job delay mechanism to improve schedules by strategically delaying certain jobs. The effectiveness of cost-based assignment rules is highlighted as a promising approach. However, the conclusion acknowledges that these preliminary findings must be confirmed through extensive statistical analysis and compared with alternative approaches, such as the Hybrid Insertion Algorithm (HIGA). Future research avenues include exploring different heuristics for initial population generation and integrating local search within the GA computation.

### 4.4 Job Sequencing for Unrelated Parallel Machines with Fuzzy Processing Time and Fuzzy Due Date

In [10] the realm of job sequencing within a fuzzy production environment, specifically addressing unrelated parallel machines. The study introduces a fuzzy scheduling model and its corresponding algorithm, tailored to the challenges posed by uncertain job processing times and due dates. It leverages the concepts of possibility and necessity measures within fuzzy theory to craft a tardiness credibility index, enabling the estimation of each job's likelihood of tardiness. A mixed integer programming model for unrelated parallel machines is constructed to minimize the average tardiness credibility of jobs. To tackle this model, a novel Hybrid Fuzzy Genetic Algorithm (HFGA) is developed. The paper concludes with a case study that effectively demonstrates the algorithm's efficacy in real-world scenarios.

The paper formulates a mixed integer programming model for fuzzy scheduling, aimed at minimizing the average tardiness credibility of jobs. Leveraging the properties of optimal fuzzy scheduling, the Hybrid Fuzzy Genetic Algorithm (HFGA) is devised. In a simulation experiment involving 15 jobs and 5 machines, the HFGA is compared to the basic Genetic Algorithm (GA). The results conclusively demonstrate the superior performance of the HFGA in addressing uncertain scheduling for unrelated parallel machines.

In Table 1 are presented other case studies analyzed.

**Table 1.** Other Case Studies

| Authors | Summary |
|---|---|
| [11] | This article introduces a problem involving scheduling independent jobs on unrelated parallel machines with sequence-dependent setup times and varying ready times. The objective is to minimize the total weighted number of tardy jobs. To address this problem, the paper presents an Iterated Local Search (ILS) approach that combines a swap shaking procedure and a Tabu Search algorithm with hybrid neighborhoods to enhance efficiency and effectiveness. Comparative experiments with heuristic rules, a conventional Tabu search algorithm, and optimal solutions indicate that the proposed ILS method shows promise as an effective solution for this problem. |
| [12] | A Joint Load Balancing Algorithm for Software Defined Networks is introduced. It addresses both data link and server load balancing, utilizing a Dynamic Server Load Balancing Algorithm to select the best server and a Hybrid Routing Algorithm that combines the Shortest Path Algorithm with the Ant Colony System to determine the optimal data path. Implementing this algorithm as a load balancing module in the OpenDayLight controller and testing with Mininet, the results demonstrate significant improvements in network performance, including higher throughput and reduced delay, compared to other algorithms. |
| [13] | This paper discusses a dynamic load balancing algorithm in distributed computing. It employs the divisible load technique to improve parallel task completion times by distributing tasks evenly across heterogeneous nodes. The algorithm introduces an information policy for nodes without neighbors to predict and manage load information. Simulation results show significant reductions in job response times. |
| [14] | In this article, scheduling non preemptable jobs on nonhomogeneous parallel machines is approached and researched. It employs a constraint-based model and a hybrid algorithm combining constraint programming (CP) and tabu search. CP validates solutions and sets constrained variables, while tabu search handles the search process. The paper introduces an efficient method called forward time slack for validating neighborhood moves. Computational experiments are performed on a standard PC. |

*(continued)*

**Table 1.** (*continued*)

| Authors | Summary |
|---|---|
| [15] | This research paper focuses on scheduling challenges in multi-chip package (MCP) assembly, where various chip modules are assembled within a single package. The assembly process involves multiple stages with diverse parallel machines and setup times. The objective is to minimize setups while maintaining on-time delivery (OTD) performance. Proposed strategies allocate machine capacity to products, virtual lots, or process stages based on production progress, leading to reduced setups with acceptable OTD performance trade-offs. In conclusion, the paper addresses the complex scheduling issues in MCP assembly lines, aiming to optimize machine capacity allocation for setup reduction without significantly impacting OTD performance, while also emphasizing the need for further improvements in OTD performance for the process step allocation (PSTEP) strategy. |
| [16] | This paper introduces a novel load balancing mechanism called "soft" load balancing, aimed at enhancing resource utilization in emerging heterogeneous overlay networks. The approach divides a user's internet protocol (IP) traffic into subflows, directing them to various access networks. Soft load balancing combines load-sharing and handover techniques. Numerical analysis yields an optimal load balancing ratio (LBR) for directing traffic to different networks in a multicell overlay environment. This optimization enhances channel transmission reliability and reduces the outage probability for user traffic. |
| [17] | This paper tackles the challenge of stochastic parallel machine scheduling in just-in-time manufacturing. It introduces a two-stage genetic algorithm with optimal computing budget allocation (OCBA) and enhanced Monte-Carlo Policy Evaluation (MCPE). OCBA is adapted for non-normally distributed processing times, and MCPE balances global and local search. The two-stage approach initially estimates performance and then uses OCBA for evaluations. Simulations confirm its efficiency and effectiveness compared to existing methods, making it a practical solution for solving parallel machine scheduling problems with stochastic processing times, especially in semiconductor manufacturing. |
| [18] | This article applies the discrete bat algorithm to flowshop scheduling problems, comparing it to Simulated Annealing, Local Search, and heuristic-based variants. The bat algorithm, originally for continuous problems, is evaluated for efficiency in discrete contexts. After implementation and parameterization, testing across 30 instances shows the hybrid Bat Algorithm with CDS (Campbell, Dudek and Smith) Algorithm outperforming Simulated Annealing, while Local Search falls short. |

## 5 Future Paths

As technology continues to evolve and the demands on parallel machines grow, researchers and practitioners are exploring several exciting avenues to address load balancing challenges. In this section, we outline some of the key future paths in this field and provide references for further exploration.

### 5.1 Fog and Edge Computing

With the increasing proliferation of edge and fog computing, load balancing is becoming critical for resource-constrained, latency-sensitive environments. Future research will need to address load balancing techniques tailored for these distributed and heterogeneous settings, ensuring efficient processing at the network edge. Relevant studies such as [19] and [20] provide insights into load balancing in edge and fog computing scenarios.

### 5.2 Machine Learning Load Balancing

The integration of machine learning techniques into load balancing algorithms is a promising avenue for future research. Machine learning models can adaptively learn from historical data and make real-time load balancing decisions. Real-time load balancing and scheduling are techniques that have already been implemented with more traditional solvers as surveyed in [21]. Reinforcement learning algorithms, for instance, can be used to create self-adjusting load balancers that optimize system performance based on changing workloads and resource availability. Researchers have already begun exploring this intersection, as seen in works like [22].

## 6 Conclusion

This literature review delves into the critical role of load balancing in parallel machines and distributed systems, emphasizing its significance in optimizing resource allocation and system stability. It underscores the necessity for adaptive and sustainable load balancing solutions in the ever-evolving landscape of computing paradigms, highlighting different strategies utilized to achieve said solutions, be it the utilization of heuristics, metaheuristics, or mathematical programming.

In conclusion, this paper emphasizes the intricate challenges posed by unrelated parallel machine problems, where careful consideration of trade-offs between collective workload minimization and load balance is essential due to their often adversarial nature. Secondly, it acknowledges the exponential growth in load balancing techniques, particularly within Information and Computing Science, as a direct result of the internet's widespread use. This is due to a demand for efficient load and work distribution among servers and computers. Lastly, the review points to promising horizons where edge and fog computing, with their swarm-like computational models, expand the applications of load balancing algorithms. Machine learning algorithms offer newfound potential in real-time load prediction and management, providing invaluable tools to address the evolving complexities of load distribution in parallel machines.

## References


1. Ouazene, Y., Nguyen, N.-Q., Yalaoui, F.: Workload balancing on identical parallel machines: theoretical and computational analysis. Appl. Sci. **11**(8), 3677 (2021). https://doi.org/10.3390/app11083677
2. Santos, A.S., Madureira, A.M., Varela, M.L.R.: An ordered approach to minimum completion time in unrelated parallel-machines for the makespan optimization. In: Sixth World Congress on Nature and Biologically Inspired Computing (NaBIC 2014), Porto (2014)
3. Inegbedion, H., Inegbedion, E., Peter, A., Harry, L.: Perception of workload balance and employee job satisfaction in work organisations. Heliyon. **6**, e03160 (2020). https://doi.org/10.1016/j.heliyon.2020.e03160
4. Serra e Santos, A., Madureira, A.M., Varela, M.L.R.: An ordered heuristic for the allocation of resources in unrelated parallel-machines. Int. J. Ind. Eng. Comput. **6**(2), 145–156 (2015)
5. DSRS: Software dimensions. Digital Science & Research Solutions Inc. Accessed: oct.13, 2023. (2023) [Online]. Available https://www.dimensions.ai/
6. Posit: Software R Studio. Posit Software (2023) Accessed: Oct13, 2023 [Online]. Available: https://posit.co/
7. Sevkli, M., Uysal, H.: A modified variable neighborhood search for minimizing the makespan on identical parallel machines. In: 2009 International Conference on Computers & Industrial Engineering, Troyes, France, pp. 108–111 (2009). https://doi.org/10.1109/ICCIE.2009.5223485
8. Santos, A.S.E., Madureira, A.M.: Ordered minimum completion time heuristic for unrelated parallel-machines problems. In: 2014 9th Iberian Conference on Information Systems and Technologies (CISTI), Barcelona, Spain, pp. 1–6 (2014). https://doi.org/10.1109/CISTI.2014.6876939
9. Kurniawan, B., Gozali, A.A., Weng, W., Fujimura, S.: A genetic algorithm for unrelated parallel machine scheduling minimizing makespan cost and electricity cost under time-of-use (TOU) tariffs with job delay mechanism. In: 2017 IEEE International Conference on Industrial Engineering and Engineering Management (IEEM), Singapore, pp. 583–587 (2017). https://doi.org/10.1109/IEEM.2017.8289958
10. Yang, H., Chen, Z., Wu, M.: Job sequencing for unrelated parallel machines with fuzzy processing time and fuzzy duedate. In: 2009 2nd International Conference on Power Electronics and Intelligent Transportation System (PEITS), Shenzhen, China, pp. 291–294 (2009). https://doi.org/10.1109/PEITS.2009.5406783
11. Chen, C.-L.: An iterated local search for unrelated parallel machines problem with unequal ready times. In: 2008 IEEE International Conference on Automation and Logistics, Qingdao, pp. 2044–2047 (2008). https://doi.org/10.1109/ICAL.2008.4636499
12. Li, J., Yang, L., Wang, J., Yang, S.: Research on SDN load balancing based on ant colony optimization algorithm. In: 2018 IEEE 4th Information Technology and Mechatronics Engineering Conference (ITOEC), Chongqing, China, pp. 979–982 (2018). https://doi.org/10.1109/ITOEC.2018.8740366
13. Lim, J.W.Y., Hoong, P.K., Yeoh, E.-T.: Neighbor's load prediction for dynamic load balancing in a distributed computational environment. In: TENCON 2012 IEEE Region 10 Conference, Cebu, Philippines, pp. 1–6 (2012). https://doi.org/10.1109/TENCON.2012.6412190
14. He, R.-J.: Parallel machine scheduling problem with time windows: a constraint programming and tabu search hybrid approach. In: 2005 International Conference on Machine Learning and Cybernetics, Guangzhou, China, vol. 5, pp. 2939–2944 (2005). https://doi.org/10.1109/ICMLC.2005.1527445
15. Lee, S.-J., Lee, T.-E.: Scheduling a multi-chip package assembly line with reentrant processes and unrelated parallel machines. In: 2008 Winter Simulation Conference, Miami, FL, USA, pp. 2286–2291 (2008). https://doi.org/10.1109/WSC.2008.4736332

16. Son, H., Lee, S., Kim, S.-C., Shin, Y.-S.: Soft load balancing over heterogeneous wireless networks. IEEE Trans. Veh. Technol. **57**(4), 2632–2638 (2008). https://doi.org/10.1109/TVT.2007.912324
17. Cao, Z., Lin, C., Zhou, M., Zhou, C., Sedraoui, K.: Two-stage genetic algorithm for scheduling stochastic unrelated parallel machines in a just-in-time manufacturing context. IEEE Trans. Autom. Sci. Eng. **20**(2), 936–949 (2023). https://doi.org/10.1109/TASE.2022.3178126
18. Sousa, B., Guerreiro, R., Santos, A.S., Bastos, J.A., Varela, L.R., Brito, M.F.: Bat algorithm for discrete optimization problems: an analysis. In: International Conference Innovation in Engineering, pp. 161–172. Springer (2022)
19. Shi, W., Cao, J., Zhang, Q., Li, Y., Xu, L.: Edge computing: vision and challenges. IEEE Internet Things J. **3**(5), 637–646 (2016). https://doi.org/10.1109/JIOT.2016.2579198
20. Yi, S., Li, C., Li, Q.: A survey of fog computing: concepts, applications, and issues, 37–42 (2015). https://doi.org/10.1145/2757384.2757397
21. Guerreiro, R., Santos, A.S., Tereso, A.: Online scheduling: a survey. In: 2023 18th Iberian Conference on Information Systems and Technologies (CISTI), pp. 1–6. IEEE (2023)
22. Anna Victoria, C.R., Oikawa, V.F., Castro, M.C., Pilla, L.L.: Adaptive load balancing based on machine learning for iterative parallel applications. In: 28th Euromicro International Conference on Parallel, Distributed and Network- Based Processing (PDP), Mar 2020, Västerås, Sweden. ff10.1109/PDP50117.2020.00021ff.ffhal-02570549f

# Sylheti to Standard Bangla Neural Machine Translation: A Deep Learning-Based Dialect Conversion Approach

Tabia Tanzin Prama(✉) and Md Musfique Anwar

Jahangirnagar University, Savar, 1342 Dhaka, Bangladesh
{prama.stu2017,manwar}@juniv.edu

**Abstract.** The Bangla language has been a focal point in the realm of natural language processing, with numerous commendable research endeavors. Despite the wealth of studies, there exists a noticeable gap in significant research dedicated to its intricately enriched dialects, primarily hindered by resource scarcity and the intricate grammatical structures inherent to these dialects. In this study, we introduce a Neural Machine Translation (NMT) system designed specifically for one of the prevalent Bangla dialects, namely the Sylheti dialect, with the objective of facilitating seamless translation into standard Bangla. To achieve this, we explore four different Seq2Seq models: Long Short Term Memory (LSTM), Gated Recurrent Unit (GRU), Bidirectional LSTM (BiLSTM), and Bidirectional GRU (BiGRU). To facilitate our study, we gathered a dataset consisting of 600 complete sentence examples and 6500 Sylheti words, each paired with their corresponding translations in standard Bangla. Through our experimentation, the BiLSTM model emerged as the most effective for Sylheti to standard Bangla Neural Machine Translation, outperforming other models. Translation quality is evaluated using the BLEU and ROUGE metrics. On the BiLSTM model, we achieved BLEU scores of 57.4, 45.8, 32.0, and 22.8 for BLEU-1, BLEU-2, BLEU-3, and BLEU-4, respectively which demonstrates the effectiveness of our proposed approach in translating the Sylheti dialect to standard Bangla.

**Keywords:** Neural Machine Translation · Standard Bangla · Sylheti Dialect · Seq2Seq Models

## 1 Introduction

Languages commonly manifest diverse regional or social variations referred to as dialects, characterized by distinctions in phonology, grammar, and vocabulary from the standard language. While the standard language is prevalent in both spoken and written forms across a nation, dialects are geographically or socio-culturally confined. In Bangladesh, a linguistically varied country, numerous dialects can be grouped into four primary clusters: North Bengal dialects (such as Dinajpur, Rajshahi, Bogra, and Pabna), Rajbanshi from Rangpur, East Bengal dialects (encompassing Dhaka, Mymensingh, Tripura, Barisal, Sylhet, Faridpur, Jessore, and Khulna), and South Bengal dialects


A. Bajaj et al. (Eds.): HIS 2023, LNNS 1224, pp. 208–217, 2025.
https://doi.org/10.1007/978-3-031-78925-0_21

(including Chittagong and Noakhali) [1]. These dialects possess intricate nuances that often present challenges for non-native speakers of standard Bangla, as they contain unique words and phrases. Consequently, individuals are increasingly relying on English and standard Bangla, resulting in a gradual decline in the usage of these enriched dialects. Despite their significance, little attention has been given to the development of conversion systems that can effectively translate between standard Bangla and the Sylheti dialects. To address this gap, our research introduces a Neural Machine Translation (NMT) system that facilitates seamless mapping between standard Bangla and regional words, providing bidirectional translation capabilities. Machine translation is essential for the processing of natural language, facilitating communication and information exchange across speakers of diverse native languages, thereby bridging linguistic gaps. Neural Machine Translation (NMT) stands out as a cutting-edge approach in machine translation, employing neural networks trained on extensive datasets.NMT models demand less memory than conventional Statistical Machine Translation (SMT) models, while also outperforming them in terms of accuracy [2].

Our primary objective is to enable effective translation between standard Bangla and the Sylheti dialects. To achieve this, we focus on four Seq2Seq models for the Standard Bengali-Shylheti Bangla language pair: Seq2Seq with Long Short Term Memory (LSTM), Gated Recurrent Unit (GRU), Bidirectional LSTM (BiLSTM), and Bidirectional GRU (BiGRU). The Seq2Seq LSTM model utilizes LSTM layers to produce hidden cells for the input sequence, utilizing these hidden cells from preceding time steps for decoding the target language. GRU functions as a recurrent unit, employing reset and update gates to manage the information flow from the previous state to the current input. Bidirectional long-short-term memory (BiLSTM) processes input in both forward and backward directions, setting it apart from a standard LSTM. Similarly, the Bidirectional GRU (BiGRU) functions with two GRUs, incorporating only the input and forget gates. This study conducts experiments with these models, assessing their performance through the ROUGE score and BLEU evaluation metrics [3]. By evaluating the effectiveness of these models in translating between standard Bangla and the Sylheti dialects, we aim to contribute to the preservation and appreciation of linguistic diversity in Bangladesh. In summary, the work's primary contributions are as follows:

- We curated a comprehensive Digital and Structured Sylheti Dialects Dataset of Bangladesh. Our dataset served as the foundation for our investigation into machine translation.
- We employed four distinct sequential models: LSTM, GRU, BiLSTM, and BiGRU.
- To assess the efficiency of our implemented models, we utilized widely recognized metrics in the generation task: BLEU and ROUGE scores. These metrics provided valuable insights into the quality and accuracy of the machine translations produced by our models.

## 2 Related Works

### 2.1 Machine Translation Systems in Different Language Pairs

Several studies have explored machine translation systems for different language pairs, with a focus on improving translation quality and performance. Siddique et al. [4] introduced a recurrent neural network (RNN) with encoder and decoder components designed for translating English to Bengali. Their approach incorporated a context vector based on knowledge to facilitate sentence mapping, demonstrating improved outcomes through linear activation in the encoder layer and tanh activation in the decoder layer. The research concluded that, in their experiments, GRU exhibited superior performance compared to LSTM. Tiwari et al. [5] assessed two NMT models for English-Hindi translation, one using LSTM and the other employing a combination of CNN and LSTM with attention mechanisms. Their evaluation revealed that the ConvS2S model performed the best, as determined by BLEU scores. In a study by Corallo et al. [6], a GRU-based RNN was employed to convert German news stories into English sentences using the WMT-2021 English-German dataset. The system demonstrated effective translation capabilities and served as a testbed for future research and applications. Studies such as [9], [10], and [11] have individually utilized different neural networks architectures like LSTM, GRU, and BiLSTM for machine translation in various language pairs, such as English to Hindi or Arabic. Mumin et al. [13] discovered that NMT surpassed phrase-based statistical machine translation in both directions for English-Bangla translation. They observed notable enhancements in BLEU scores, and further improvements were realized by incorporating subword segmentation using byte pair encoding to handle unconventional words.

### 2.2 Conversion Systems for Bangla Dialects

While machine translation research has been extensive for various language pairs, the development of conversion systems for dialects in the Bangla language has received limited attention. However, some researchers have explored conversion systems in other languages, offering valuable insights for addressing this gap. Arpita Goswami [14] conducted a study on Sylheti grammar in 2021, providing useful resources for understanding the Sylheti dialect. Hoque et al. [15] discussed the Chittagonian language, including its grammar and word pronunciation, which could be relevant to our research on Bangla dialects. Nahid [16] introduced a bidirectional transformation system for the Chittagonian dialect, employing binary search, word-to-word mapping, and morphological transformation. Similarly, Hafizur et al. [17] devised a conversion system for Chittagonian through word-to-word mapping, rule-based morphological transformation, and an auxiliary word suggestion module. In different linguistic contexts, Singh et al. [18] devised a system for transforming Punjabi dialects employing a rule-based method and a bilingual dictionary. Marimuthu et al. [19] demonstrated a technique for converting dialectal Tamil text to standard Tamil text using a Finite State Transducers approach. Gaphai et al. [20] attained accuracy through a rule-based approach with a vocabulary of 9386 words. Bakr et al. [21] suggested a hybrid approach for transforming Egyptian colloquial to Modern Standard Arabic, achieving an 88% accuracy in 2008. Chowdhury [22] explored Bangla to English machine translation incorporating POS tagging and grammar rules, integrating Tag Vectors in the conversion process. Overall, while machine translation systems for various language pairs have seen considerable advancements, there is a need for more research and attention toward developing effective conversion systems for the diverse dialects of the Bangla language.

# 3 Methodology

This section gives a detailed overview of the entire process carried out in our study, as depicted in Fig. 1.

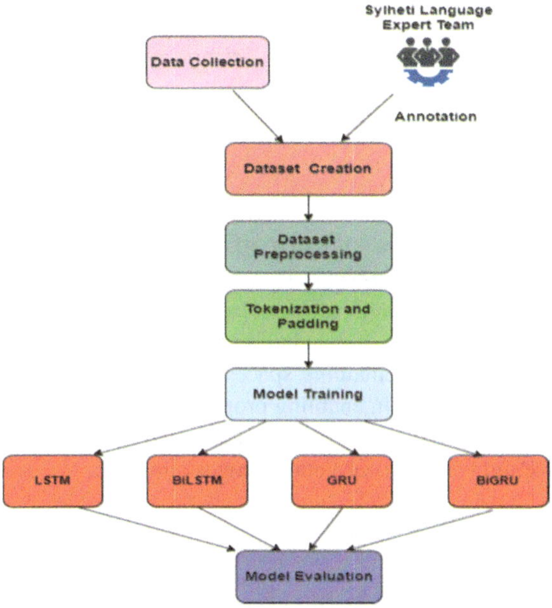


**Fig. 1.** Work Flow of Neural Machine Translation system

## 3.1 Dataset Description

Despite being rich in vocabulary, ethnic music, and folklore, the literary expressions in the Sylheti dialect are notably limited in conventional written forms, particularly in digital mediums. Consequently, we needed a bilingual dictionary dataset. Given the scarcity of digital records, we relied on Sylheti words and their Bangla equivalents extracted from well-known Bangladeshi national newspapers, social media posts, and voluntary comments. Table 1 shows examples of the dataset we employed.

Our custom combined Sylheti Bn-Standard Bn dataset comprises 600 parallel sentences in both Standard Bangla and Sylheti dialect, arranged in tab-delimited format. Given our limitations in resources and the limited dataset, we assigned 95% of the data for training and the rest 5% for testing. Within the training set, 10% of the data is reserved for validation, ensuring a robust model evaluation during the training phase.

## 3.2 Preprocessing

Tokenization is a step which divides the input paragraph into individual sentences first. Subsequently, each sentence is further broken down into individual

**Table 1.** Example of dataset sample.

| Sylheti Sentence | Standard Bangla Sentence |
|---|---|
| সিলেটের মাতও যেলা রস | সিলেটের কথাতেও যেমন রস |
| খালি হাতকরাও নি | খালি সাতকরাই নাকি |
| বুঝইনও না আর হুনইনও না | বোঝেনও না আর শুনেনও না |
| মাইনষে হুনিয়াও আরাম ফাইব | মানুষে শুনেও আরাম পাবে |
| আমার দ্যাশে আমোই স্যারা | আমার দেশে আমিই সেরা |

words or tokens, which serve as the fundamental units for subsequent processing. For instance, the sentence 'মাইনষে হুনিয়াও আরাম ফাইব'is tokenized into 'মাইনষে', হুনিয়াও','আরাম'and 'ফাইব'. These tokens are then forwarded to the following stages of the system to enable efficient language processing and translation. Due to the varying maximum lengths of standard (24) and Shylheti Bangla sentences (22), we applied padding to fill the remaining spaces with zeros. This process ensured that all sentences had the same length, enabling proper handling and processing within our neural network models. Consequently, the input data was effectively vectorized, shaped, and made ready for utilization in our neural network models.

### 3.3 Model Training

For the sequential models, training begins after applying sentence padding. The embedding layer receives parameters such as the Bengali source vocabulary, input length, and output dimension. The models include LSTM, GRU, BiLSTM, and BiGRU layers, with softmax activation for output [4]. Training involves calculating errors and improving accuracy through backpropagation, using the Categorical Cross-Entropy loss function and the Adam optimizer, a hybrid of RMSprop and Stochastic Gradient Descent [25]. Both LSTM and GRU models have a consistent architecture with 300 embedding and repeat vector layers, accompanied by two 300 neural network layers. The resulting output undergoes processing through a time-distributed layer featuring softmax activation. Likewise, in the case of BiLSTM and BiGRU models, the overall architecture remains consistent, with bidirectional neural network layers surrounding it. A batch size of 32 is configured, and a learning rate of 0.001 is maintained for the training of all models.

Figure 2 illustrates the training process of the Bidirectional LSTM (BiLSTM) model using a mixed sample of Sylheti and Standard tokenized sentences. The neural network takes these sentences as input, generates a mapped output, and feeds it into the NMT model to produce the translated sentence.

### 3.4 Performance Metrics

ROUGE serves as a metric for assessing the quality of automatically generated summaries or translations by comparing them to reference summaries. It includes ROUGE-1, ROUGE-2, and ROUGE-L, assessing precision, recall, and F1-score for n-grams. Higher

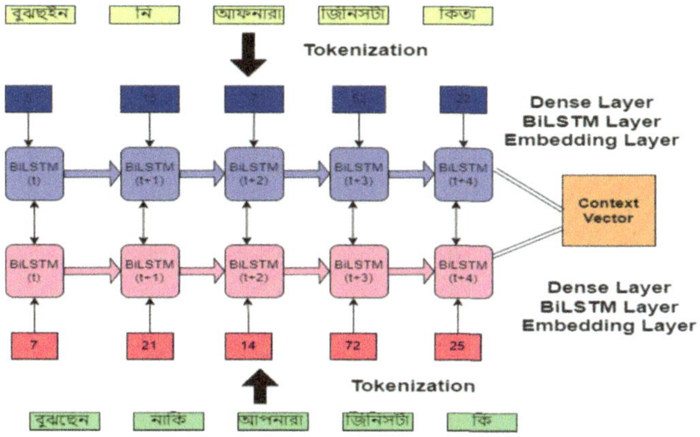


**Fig. 2.** Training Process of BiLSTM model

ROUGE-N scores indicate better alignment with reference summaries [23]. BLEU, on the other hand, rates machine translations on a scale from 0 to 1, with higher scores signify a closer alignment with human-generated reference translations, thus measuring translation quality and similarity [24].

## 4 Results and Discussion

To evaluate the performance, we analyzed training accuracy, validation loss, and validation accuracy. Additionally, we utilized ROUGE and BLEU scores to assess the translation quality. Figure 3 shows the Accuracy vs Loss graph for the BiLSTM model, while Table 2 presents the ROUGE scores. Furthermore, The BLEU-1 scores that each model obtained on the testing samples are shown in Fig. 4.

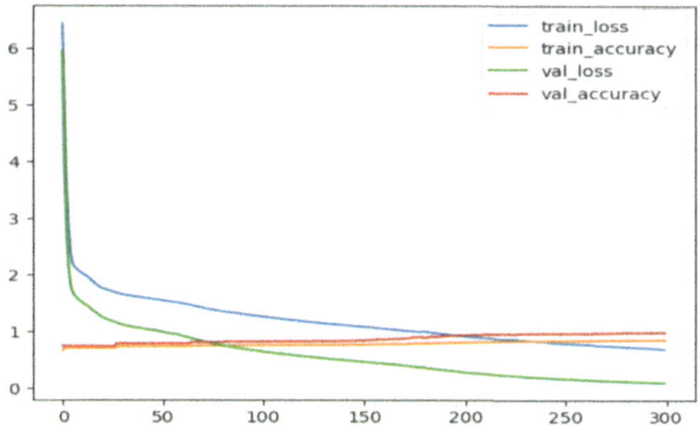


**Fig. 3.** Graph of accuracy vs loss for the BiLSTM model

The experimental results indicate significant differences in the performance of the four models. BiLSTM stands out as the fastest to achieve a commendable training accuracy of 98.5% within only 250 epochs, demonstrating efficient convergence. The validation accuracy of BiLSTM also reached a respectable 92.4%. This highlights the

**Table 2.** Rough Score of test samples

| Evaluation Metrices | | Model | | | |
|---|---|---|---|---|---|
| | | *LSTM* | *BiLSTM* | *GRU* | *BiGRU* |
| ROUGE-1 | Recall | 0.50 | **0.75** | 0.40 | 0.60 |
| | Precision | 0.67 | **0.75** | 0.57 | 0.72 |
| | F1-score | 0.57 | **0.75** | 0.47 | 0.65 |
| ROUGE-2 | Recall | 0.20 | **0.67** | 0.15 | 0.59 |
| | Precision | 0.20 | **0.67** | 0.15 | 0.59 |
| | F1-score | 0.20 | **0.67** | 0.15 | 0.59 |
| ROUGE-L | Recall | 0.50 | **0.75** | 0.40 | 0.60 |
| | Precision | 0.67 | **0.75** | 0.57 | 0.72 |
| | F1-score | 0.57 | **0.75** | 0.47 | 0.65 |

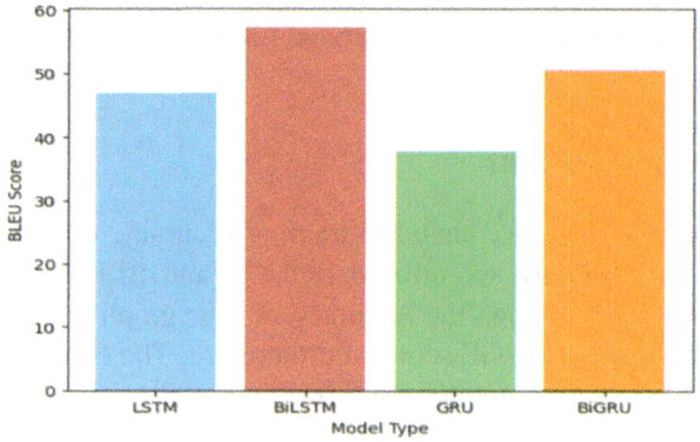


**Fig. 4.** BLUE Score of each model

superiority of BiLSTM over LSTM, GRU, and BiGRU in terms of speed and accuracy. BiGRU closely follows BiLSTM in terms of performance, achieving a slightly higher accuracy of training is 96.3% and validation is 86.6%. While it required a few more epochs to reach this level of accuracy, it still outperformed LSTM and GRU. Besides, LSTM exhibited the slowest convergence rate among the models, necessitating over 290 epochs to score a decent training accuracy of 93% and validation accuracy of 85.3%. This relatively slower convergence suggests that LSTM might require more data and tuning to attain optimal performance compared to the other models. Considering the average performance across all metrics, BiLSTM emerges as the most promising choice among LSTM, GRU, and BiGRU, striking a balance between training speed and accuracy.

From Table 2, we observe that BiLSTM, BiGRU, and LSTM are all competing closely for the best performance. However, GRU falls behind, leaving us to consider BiLSTM, BiGRU, and LSTM as potential candidates. Upon comparing LSTM and BiGRU, we notice that BiGRU outperforms LSTM in specific cases, such as recall and F1-score, where BiGRU shows around 10-6% higher accuracy. This narrows down our options to BiLSTM and BiGRU. Analyzing the ROUGE-1 (Unigram) scores, we find that BiLSTM achieves a recall, precision, and F1-score of 0.75, whereas BiGRU falls approximately 8–10% lower. A similar trend is evident for ROUGE-2 (Bigram) and ROUGE-L (longest matching sequence). Upon closer inspection of the ROUGE scores, an interesting observation emerges regarding the disparity between BiLSTM and BiGRU. Initially, there is a 15% recall difference in ROUGE-1, which diminishes to 8% for ROUGE-2. However, for precision, the difference escalates from 3% to 8%. This suggests that as the evaluation considers more consecutive words, the performance gap between BiLSTM and BiGRU widens, with BiLSTM gradually surpassing BiGRU.

In Fig. 4, it is evident that the BiLSTM model attains the highest score of 57.4% for BLEU-1 among all models, whereas the BiGRU model achieves a score of 50%. This suggests that BiLSTM exhibits superior performance when evaluating individual words or unigrams, as corroborated by both ROUGE and BLEU scores.

Table 3 shows five sample outputs from the BiLSTM model. The first output shows flawless translation, while the second and third contain errors like duplicate and incorrect words. Challenges emerge during the translation of uncommon words, highlighting the necessity for additional enhancements to the model.

**Table 3.** Sample outputs of the BiLETM model

| Sylheti Sentences | Reference Standard Bangla Sentence | Generated Sentence |
|---|---|---|
| সিলটি ভাষা হিকইন | সিলটি ভাষা শিখেন | সিলটি ভাষা শিখেন |
| এর লাগিওতো সিলেটি মানুষ লন্ডন | এর জন্যই সিলেটি মানুষ লন্ডন | এর জন্যই সিলেটি মানুষ মানুষ |
| গেছইন না কের লাগি | গেছেন না কিসের জন্য | গেছেন না কাছের জন্য |
| তুমি কারেক্টারটা বুঝিছিলে | তুমি ক্যারেক্টারটা বুঝেছিলেন | তুমি ক্যারেক্টারটা বুঝেছিলেন |
| বুঝছইন নি আফনারা জিনিসটা কিতা? | বুঝছেন নাকি আপনারা জিনিসটা কি? | বুঝছেন না আপনারা জিনিস কি? |
| জ্বী অয় | জ্বী হ্যাঁ | জ্বী হ্যাঁ |

## 5 Conclusion

This study presents a conversion system for translating between the Sylheti dialect and the standard Bangla language. The BiLSTM model demonstrated superior performance in handling consecutive words, achieving a promising BLEU score of 57.4%. However, further investigation with a larger Sylheti dataset containing rare words remains an open research topic. Translation accuracy, especially in capturing sentence meaning and maintaining grammatical errors, poses challenges that require extensive computational power and rich vocabulary resources. Future advancements using multiple dense layers, parameter tuning, and large datasets hold the potential for more robust results. The ongoing efforts involve developing a generic dialect translation system adaptable to various dialectal datasets within the Bangla language. Additionally, work is underway to create bidirectional speech-to-text and text-to-speech systems for comprehensive language conversion.

## References


1. Banglapedia contributors. Dialect. In Banglapedia. Retrieved September 15, 2021,from https://en.banglapedia.org/index.php/Dialect
2. Bahdanau, D., Cho, K., Bengio, Y.: Neural machine translation by jointly learning to align and translate. arXiv. preprint arXiv:1409.0473. (2014)
3. Sutskever, I., Vinyals, O., Le, Q.V.: Sequence to sequence learning with neural networks. In: Advances in Neural Information Processing Systems, p. 27 (2014)
4. Siddique, S., Ahmed, T., Talukder, M., Azam, R., Uddin, M., et al.: English to bangla machine translation using recurrent neural network. arXiv. preprint arXiv:2106.07225. (2021)
5. Tiwari, G., Sharma, A., Sahotra, A., Kapoor, R.: English-hindi neural machine translation-lstm seq2seq and convs2s. In: 2020 International Conference on Communication and Signal Processing (ICCSP), pp. 871–875. IEEE (2020)
6. Corallo, L., Li, G., Reagan, K., Saxena, A., Varde, A.S., Wilde, B.: A framework for german-english machine translation with gru rnn. In: EDBT/ICDT Workshops (2022)
7. Wu, Y., Schuster, M., Chen, Z., Le, Q.V., Norouzi, M., Macherey, W., Krikun, M., Cao, Y., Gao, Q., Macherey, K.: Google's neural machine translation system: bridging the gap between human and machine translation. arXiv. preprint arXiv:1609.08144. (2016)
8. Hasan, M.A., Alam, F., Chowdhury, S.A., Khan, N.: Neural machine translation for the bangla-english language pair. In: 2019 22nd International Conference on Computer and Information Technology (ICCIT), pp. 1–6. IEEE (2019)
9. Bensalah, N., Ayad, H., Adib, A., Ibn El Farouk, A.: Lstm vs. gru for arabic machine translation. In: Proceedings of the 12th International Conference on Soft Computing and Pattern Recognition (SoCPaR 2020) 12, pp. 156–165. Springer (2021)
10. Shalu, P., Meera, M.: Neural machine translation for english to hindi using gru. Available at SSRN 3851323, (2021)
11. Roy, A., Dhar, A.C., Akhand, M., Kamal, M.A.S.: Bangla-english neural machine translation with bidirectional long short-term memory and back translation. Int. J. Comput. Vis. Signal Process. **11**(1), 25–31 (2021)
12. Ranathunga, S., Lee, E.-S.A., Skenduli, M.P., Shekhar, R., Alam, M., Kaur, R.: Neural machine translation for low-resource languages: a survey. In: ACM Computing Surveys (2021)
13. Al Mumin, M.A., Seddiqui, M.H., Iqbal, M.Z., Islam, M.J.: Neural machine translation for low-resource english-bangla. J. Comput. Sci. **15**(11), 1627–1637 (2019)
14. A. Goswami: Marked geminates as evidence of sonorants in sylheti bangla: an optimality account. In: Acta Linguistica Asiatica, 11, 1, pp. 99-112, doi: https://doi.org/10.4312/ala.11.1.99-112(2021)
15. Hoque, M.A.: Chittagonian variety: dialect, language, or semi-language?. In: CRP, International Islamic University Chittagong,Bangladesh, (2015)
16. Hossain, N., Milon, H.R., Uddin Sabbir, S.N., Inan, A.: Inclusive bidirectional conversion system between Chittagonian and standard Bangla. Bulletin of Electrical Engineering and Informatics. **11**(1), 396–404 (2022). https://doi.org/10.11591/eei.v11i1.3237
17. Milon, H.R., Sabbir, S.N.U., Inan, A., N.: Hossain: a comprehensive dialect conversion approach from Chittagonian to standard Bangla. In: 2020 IEEE Region 10 Symposium (TENSYMP), Dhaka, Bangladesh, pp. 1–6 (2020)

18. Singh, A., Singh, P.: Punjabi dialects conversion system for Malwai and Doabi dialects. Indian J. Sci. Technol. **8**, 1–6 (2015)
19. Marimuthu, K., Devi, S.L.: Automatic conversion of dialectal Tamil text to standard written Tamil text using FSTs. In: Proceedings of the 2014 Joint Meeting of SIGMORPHON and SIGFSM, pp. 37–45 (2014)
20. Al-Gaphari, G.H., Al-Yadoumi, M.: A method to convert Sana'ani accent to modern standard Arabic. IJISM. **8**, 39–49 (2012)
21. Bakr, H.A.: Khaled Shaalan and Ibrahim Ziedan: a hybrid approach for converting written Egyptian colloquial dialect into diacritized Arabic. In: The 6th International Conference on Informatics and Systems. Cairo university (2008)
22. Chowdhury, M.S.A.: Developing a Bangla to English machine translation system using parts of speech tagging. J. Modern Sci. Technol. **1**, 113–119 (2013)
23. Lin, C.Y.: ROUGE: a package for automatic evaluation of summaries. In: Annual Meeting of the Association for Computational Linguistics (2004)
24. Papineni, K., Roukos, S., Ward, T., Zhu, W.-J.: Bleu: a method for automatic evaluation of machine translation. In: Proceedings of the 40th Annual Meeting of the Association for Computational Linguistics, pp. 311–318 (2002)
25. Kingma, D.P., Ba, J.: Adam: a method for stochastic optimization. arXiv. preprint arXiv:1412.6980. (2014)
26. Anwar, M.M., Zabed Anwar, M., Al-Amin Bhuiyan, M.: Syntax analysis and machine translation of Bangla sentences. IJCSNS. **9**, 317–326 (2009)
27. Ashrafi, S.S., Kabir, M.H., Anwar, M.M., Noman, A.K.M.: English to Bangla machine translation system using context-free grammars. IJCSI. **10**, 144–153 (2013)
28. Anwar, M.M.: Bangla to English machine translation using fuzzy logic. Int. J. Comput. Sci. Inf. Secur. **16**, 156–165 (2018)

# Classifying Security Change Requests in IOT-Based Systems Using LSTM Deep Learning Model

Yosra Abbes, Zaineb Sakhrawi(✉), Asma Sellami, and Nadia Bouassida

Higher Institute of Computer Science and Multimedia, MIRACL, University of Sfax, Sfax, Tunisia
{zeinab.sakhraoui,asma.sellami,nadia.bouassida}@isims.usf.tn

**Abstract.** Changes during the life cycle of IoT projects may occur, the security of an Internet of Things (IOT)-based system might be compromised due to a lack of awareness and the incorrect selection of relevant test cases. As a result, there is an increasing need for a solution that categorizes security requirements and ensures the efficacy of continuous security testing on IoT devices. Deep learning has been applied successfully in a variety of applications, including requirement classification automation and IOT systems. This study contains two major components. First, we build a dataset. Second, we used Long Short-Term Memory (LSTM) to classify the security requirements of IOT devices into multi-classes as defined by ISO 25010: confidentiality, integrity, non-repudiation, accountability, and authenticity. The model was trained and evaluated using our collected dataset. The testing findings indicated that the suggested classification model can achieve a classification performance of 79% accuracy.

**Keywords:** Deep learning · LSTM · Change Requests · IoT Devices · software testing

## 1 Introduction

As the IoT continues to grow, the number of devices connected to the Internet increases, making the security of these devices a top priority. IoT devices offer convenience and automation to our daily lives, but they also present a significant challenge in terms of security against cyber-attacks. Security vulnerabilities in IoT devices can lead to severe consequences, including breaches of personal information, financial loss, and even physical harm. One way to address these security concerns is to explore approaches to the security testing of IOT-based systems where change requests are frequent.

A security change request is a request for modifications to be made to a system to improve its security posture [1]. IoT-based devices consist of sensors and actuators, protocols and gateways, user devices, or data processing [2]. A security change request may include software updates, configuration changes, or


A. Bajaj et al. (Eds.): HIS 2023, LNNS 1224, pp. 218–225, 2025.
https://doi.org/10.1007/978-3-031-78925-0_22

other modifications to improve security. It may affect one of the security sub-characteristics (integrity, confidentiality, non-repudiation, accountability, and authenticity) as described in the ISO 25010 quality model.

Existing security change request (CR) classification approaches are based on IOT standard components using both machine learning and deep learning techniques. However, these approaches do not take into account the ISO 25010 security decomposition. The classification of security change requests (that are described in the form of security test cases) will facilitate the selection of the relevant TC. It also avoids the waste of time, cost, and effort associated with the execution of test suites that are not relevant for testing the related IOT-based component.

The goal of this study is to assist software testers in prioritizing the customer's changes requested to improve the security of IoT devices. This paper has two main contributions. The first contribution consists of collecting data related to security requirements in the context of IoT systems. Second, we propose an LSTM deep learning-based approach for classifying security CR into authenticity, integrity, accountability, confidentiality, or non-repudiation.

The remainder of this paper is organized as follows: Section 2 discusses the related work. Section 3 describes our proposed research methodology. Section 4 discusses the experimental results. Section 5 presents the threats to the study's validity. Section 6 concludes the paper and outlines some of its future works.

## 2 Related Work

This section discusses how machine learning algorithms perform in requirement classification tasks. For instance,

- Navarro-Almanza et al. [3] proposed a convolutional neural network (CNN)-based deep learning system for requirement classification. A Pre-computed word embedding using word2vec was used along with the classification model to automatically classify twelve functional and non-functional requirement (NFR) classes.
- Cody Baker et al. [4] proposed an Artificial Neural Network (ANN) and CNN to classify five NFRs: maintainability, portability, performance, security, and usability, using two well-known datasets [4]. Results showed that the CNN gives higher precision, recall, and F1 scores.
- Cody Baker et al. [5] proposed a CNN and recurrent neural network gated recurrent units (GRU) and LSTM that were compared in classifying ten NFR classes. The authors used the word2vec algorithm and preprocessing steps such as lemmatization and stopword removal to improve the performance of the classifier. In this research, the LSTM outperformed the CNN and GRU models.
- Zijad Kurtanovic et al. [6] proposed a CNN model to classify NFR into different pre-defined categories where text was converted to low dimensional vector space using the word2vec algorithm. The requirement text was represented using embedding vectors, which were fed into the CNN model. The

proposed model's effectiveness was assessed using the PROMISE dataset. To improve accuracy and reduce bias in prediction, the researchers used a k-fold cross-validation strategy. The technique used various statistical analyses to compare the proposed model results to those of others. The results demonstrated that while CNN can efficiently classify NFR into different categories, the RNN provided a better accuracy. However, none of these researches have take into account the software security aspect in the IoT devices, and when changes are frequently requested.

## 3 Overview of the Research Methodology

This section introduces the research questions and gives an overview of the research method. The main steps of our approach are illustrated in Fig. 1. The proposed approach begins with data collection, accompanied by data cleaning and data transformation. Thereafter, it continues with classification model construction and evaluation. Next, each step will be explained.

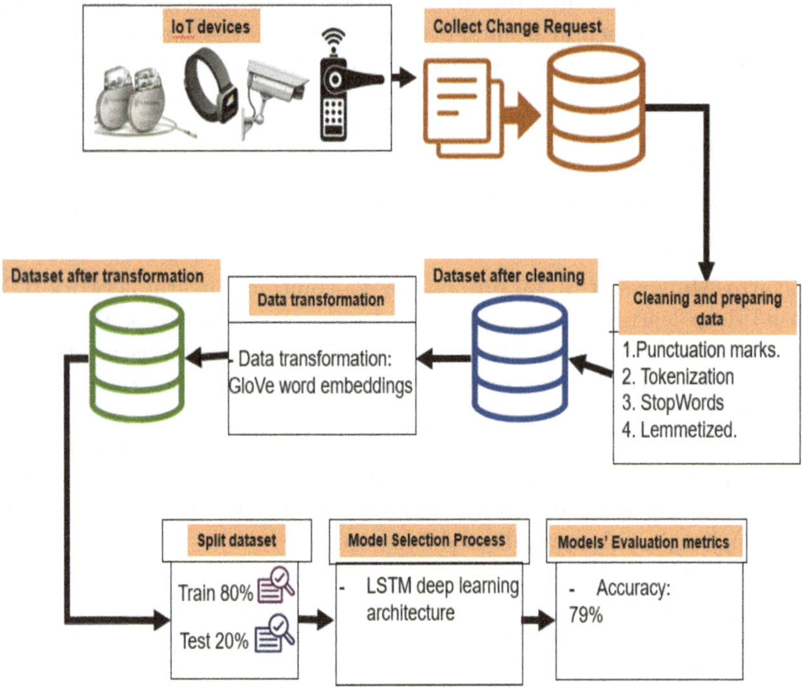


**Fig. 1.** Research Methodology

### 3.1 Research Questions

In our study, we address the following research questions:

1. How can we help testers select and prioritize TC appropriately?
2. How well can we classify security-related change requests based on ISO/IEC 25010?

### 3.2 Collecting Change Requests

The dataset was collected from IoT application developers using Google Forms. Some of the data are collected from websites describing IoT applications. All the data collected are used for securing IoT devices. The dataset contains 1100 security change requests for IoT devices such as security cameras, smart locks, medical pacemakers, fitness trackers, smart consumer drones, in-car entertainment systems, smart TVs, smart refrigerators, smartwatches, smart beds, and MRI machines. We generated 100 security change requests for each device.

Each row in the dataset includes the type of IoT device, the specific security change request (CR), and the security principle assigned to the CR. The security principles were generated based on the security goals of confidentiality, integrity, authenticity, and availability. The security change requests cover various security aspects, including data protection, access control, software updates, physical security, network security, and encryption. Examples of security change requests in the dataset include implementing end-to-end encryption to protect transmitted data, enforcing strong password policies to prevent brute force attacks, implementing intrusion detection systems, disabling unused services and ports, and regularly updating the firmware to fix vulnerabilities and bugs.

To verify the validity of the dataset, we asked a cybersecurity expert to evaluate the security change requests and the assigned security principles.

### 3.3 Cleaning and Preparing data

Preprocessing is an essential step in natural language processing (NLP) tasks, including text classification. The goal of preprocessing is to transform the raw input text into a format that is easier to work with and more suitable for downstream tasks. In this research project, we applied a preprocessing function to the security change requests dataset before training the classification model. The preprocessing function, shown in the code snippet above, involves several steps:

1. Converting the text to lowercase to ensure consistency and to reduce the number of unique tokens.
2. Removing the punctuation marks using the translate() method from the string module. This step helps in eliminating unnecessary characters that do not add meaning to the text.
3. Tokenizing the text into individual words using the word_tokenize() method from the Natural Language Toolkit (NLTK) library.

4. Removing stopwords using the set() from the NLTK library. Stopwords are common words that do not add meaning to the text, such as "the," "and," "is," etc
5. Converting the tokens into their base or dictionary form (*i.e.*, lemmatization). This step helps reduce the number of unique tokens and normalize the text. We used the WordNetLemmatizer from the NLTK library to perform lemmatization.
6. Changing the "-ing" verbs to their root form using the same lemmatizer. This step helps in capturing the meaning of the text and reduces the sparsity of the data.

After applying all the listed preprocessing steps, the tokens are joined back into a string and returned as the cleaned text (tokens).

### 3.4 Data Transformation

In this step, we perform the data transformation step that is related to word vectorization. This step is required for training our model on the corresponding textual data, which must be converted into numerical format. Word Vectorization is the process of converting phrases or words from an alphabetical list to their corresponding vector of real numbers, which can then be used for determining word forecasts and word similarities/semantics [7].

In this step, we used the Keras Tokenizer. The numerical sequences are then padded to ensure that all sequences have the same length. The embedding layer then maps the numerical sequences to their corresponding pre-trained GloVe word embeddings [8]. We use the embedding_dict object to load the pre-trained GloVe word embeddings, which we have downloaded beforehand. The embedding matrix is then initialized with the GloVe embeddings for each word in the tokenizer's vocabulary. The input layer of the model takes the numerical sequences as its input, and the embedding layer maps each input sequence to its corresponding pre-trained GloVe word embedding [8]. The spatial dropout layer helps to prevent overfitting by randomly dropping out embeddings.

### 3.5 LSTM Deep Learning Construction and Evaluation

This step focuses on how to classify security-related change requests based on ISO/IEC 25010 using the LSTM deep learning model. For that purpose, collected data falls into training data (880 requirements) and testing data (220 requirements). Thereafter, we use the LSTM classifier to automatically classify data (i.e., security-related change requests) into five sub-classes; integrity, confidentiality, non-repudiation, accountability, and authenticity. To do so, we first set the LSTM-based model parameters. Then we choose four evaluation metrics to assess the effectiveness of our constructed model.

**Set Model's Parameters:** the main parameters taken into account when building the classification model are optimization, loss function, and the number of epochs (See Table 1). Thereafter, the model is compiled using the Adam optimizer with a learning rate of 1e-3, and the loss function is set to categorical_crossentropy. The model is then trained on the training set, with a batch size of 32 and for 50 epochs.

**Table 1.** Set model's parameters

| Parameters | Values |
|---|---|
| Optimization | Adam optimizer |
| Loss function | categorical_crossentropy |
| number of epochs | 50 |

**Evaluation Metrics.** In the evaluation phase, we assessed the performance of the model on the testing set. We assessed the constructed models using a variety of evaluation metrics, including accuracy, precision, recall, and F1 score, which were measured as follows:

$$Accuracy = \frac{TP + TN}{TP + TN + FP + FN} \tag{1}$$

$$Precision = \frac{TP}{TP + FP} \tag{2}$$

$$Recall = \frac{TP}{TP + FN} \tag{3}$$

$$F1 = \frac{2 * Precision * Recall}{Precision + Recall} = \frac{2 * TP}{2 * TP + FP + FN} \tag{4}$$

The evaluation results of the classification model when using the LSTM deep learning classifier are shown in Table 2.

**Table 2.** Frequency of Special Characters

| Parameter | Results |
|---|---|
| accuracy | 79% |
| precision | 87% |
| recall | 86 % |
| F1 score | 86 % |

In Table 2, we found an accuracy of 79% which suggests that our model has achieved a reasonable level of performance in classifying security change requests into security subcharacteristics.

## 4 Discussion and comparison

The implications of this study are oriented toward software organizations that focus on testing software security in IoT-based devices. This is whether the CR is considered as a new customer non-functional (*i.e.*, quality) requirement or a modification or deletion of an existing security requirement. Since non-functional requirements are perceived as important in practice, many researchers have discovered that unresolved and insufficient non-functional requirements classification leads to project failure or increases production costs which may affect software quality evolution through frequent CR [9]. As a result, early non-functional requirement detection ensures quality software development while also decreasing development costs [6]. In this research study, classifying security CR in IoT systems is primarily used to assist testers in appropriately selecting security CR with their associated Test Cases, resulting in the successful management of IoT software system projects.

## 5 Threats to validity

Although various evaluation methods were used to conduct an experimental comparison, the scope of this current study remains limited. Threats to the validity of the present study impacted internal validity, external validity, and, finally, construct validity.

*Internal Validity.* is related first to the size of the dataset, where a greater number of instances is required. Second, the number of the attributes of dataset must be ameliorated.

*External Validity.* is related to the degree of the findings' generalization. The threat of this study could be the choice of the dataset, where only one dataset, privately gathered, was used.

*Threats of Construct.* validity is related to the degree of reliability of the features used to classify the security CR as well as measurement validity. The new challenge in this study is the use of ISO 25010 quality standards when identifying the different sub-classes of the security CR.

## 6 Conclusion

The main purpose of the presented work is to investigate the applicability of deep learning techniques in the classification of security-related change requests based on ISO 25010. This paper also gave an overview of our approach that proposes to use LSTM for classifying security change requests for IoT-based devices. Our work will be useful for researchers to help testers in selecting and prioritizing appropriate test cases. We recommend future studies to improve the effectiveness of the proposed classification model and to develop new tools for IoT device security.

## References


1. Molotoks, A., Smith, P., Dawson, T.P.: Impacts of land use, population, and climate change on global food security. Food Energy Secur. **10**(1) (2021). https://doi.org/10.1002/fes3.261
2. Medhat, N., Moussa, S.M., Badr, N.L., Tolba, M.F.: A framework for continuous regression and integration testing in IoT systems based on deep learning and search-based techniques. IEEE Access **8** (2020). https://doi.org/10.1109/ACCESS.2020.3039931
3. Navarro-Almanza, R., Juarez-Ramirez, R., Licea, G.: Towards supporting software engineering using deep learning: a case of software requirements classification. In: 2017 5th International Conference in Software Engineering Research and Innovation (CONISOFT), pp. 116–120. IEEE (2017). https://doi.org/10.1109/CONISOFT.2017.00021
4. Baker, C., Deng, L., Chakraborty, S., Dehlinger, J.: Automatic multi-class non-functional software requirements classification using neural networks. In: 2019 IEEE 43rd Annual Computer Software and Applications Conference (COMPSAC), vol. 2, pp. 610–615. IEEE (2019). https://doi.org/10.1109/COMPSAC.2019.10275
5. Rahman, M.A., Haque, M.A., Tawhid, M.N.A., Siddik, M.S.: Classifying non-functional requirements using RNN variants for quality software development. In: Proceedings of the 3rd ACM SIGSOFT International Workshop on Machine Learning Techniques for Software Quality Evaluation, pp. 25–30. (2019). https://doi.org/10.1145/3340482.3342745
6. Kurtanović, Z. and Maalej, W.: Automatically classifying functional and non-functional requirements using supervised machine learning. In: 2017 IEEE 25th International Requirements Engineering Conference (RE), pp. 490–495. IEEE (2017). https://doi.org/10.1109/RE.2017.82
7. Yahya, A.E., Gharbi, A., Yafooz, W.M., Al-Dhaqm, A.: A novel hybrid deep learning model for detecting and classifying non-functional requirements of mobile apps issues. Electronics **12**(5) (2023). https://doi.org/10.3390/electronics12051258
8. Akbik, A., Bergmann, T., Blythe, D., Rasul, K., Schweter, S., Vollgraf, R.: FLAIR: an easy-to-use framework for state-of-the-art NLP. In: Proceedings of the 2019 Conference of the North American Chapter of the Association for Computational Linguistics (demonstrations), pp. 54–59 (2019). https://doi.org/10.18653/v1/N19-4010
9. AlDhafer, O., Ahmad, I., Mahmood, S.: An end-to-end deep learning system for requirements classification using recurrent neural networks. Inf. Softw. Technol. **147**, 106877 (2022). https://doi.org/10.1016/j.infsof.2022.106877

# IIFSDR: Intelligent Integration Framework for Scientific Document Recommendation

Gerard Deepak(✉) and Achyuth Samavedhi

Department of Computer Science and Engineering, Manipal Institute of Technology Bengaluru, Manipal Academy of Higher Education, Manipal, India
gerard.deepak.christuni@gmail.com

**Abstract.** There is a dire need for scientific document recommendation frameworks in the area of Web 3.0. This paper proposes a Web 3.0 compatible semantic-oriented learning how-to reasoning strategy that encompasses ontologies for scientific document recommendation. The proposed model encompasses ontologies using the dataset categories and improving the contextual terms of the dataset in Latent Dirichlet Allocation and ontology alignment using concept similarity with the metadata of the preprocessed query words, metadata generation with each of the aligned ontological entities, and is classified utilizing a powerful deep learning RNN classifier are some of the noble features of the proposed framework. We use the XGBoost classifier, a machine learning feature-controlled classifier to classify the dataset and Lion's optimization computation using SemantoSim measure and concept similarity helps in standard strategic reasoning through semantic oriented measures and Lion's optimization serves as a strong metaheuristic optimization paradigm to yield the indicated terms for recommending the scientific documents from the dataset. An overall precision percentage with the recall percentage and the false discovery rate has been attained and, the suggested framework is the best possible model when compared to all other baseline models.

**Keywords:** Document Recommendation · Ontology · Semantics · Resource Description Framework (RDF) · Web 3.0

## 1 Introduction

The Semantic Web, also known as Web 3.0, is the next stage in the development of the Web. Through the application of semantic technologies and structured data, it seeks to increase the significance and connectivity of digital information. To fully benefit from the improved online ecosystem in the online 3.0 era, document recommendation systems must adhere to these frameworks. Frameworks that adhere to Web 3.0 are made to comprehend the semantics of Web content. They encode and represent the meaning of information using ontologies, RDF (Resource Description Framework), and linked data concepts. In terms of document recommendations, this means that the system can more accurately understand the context and content of documents. In Web 3.0 frameworks, domain-specific ontologies are frequently used. These ontologies can be used by document recommendation systems to comprehend the connections between


A. Bajaj et al. (Eds.): HIS 2023, LNNS 1224, pp. 226–238, 2025.
https://doi.org/10.1007/978-3-031-78925-0_23

concepts and entities in documents. For instance, a system for recommending scientific publications might use ontologies to find related research topics and suggest papers in line with those subjects. Knowledge graphs, which show structured knowledge about a certain area, are encouraged by Web 3.0. By recommending documents for inclusion and the extraction of structured information, document recommendation systems can aid in the creation of these knowledge graphs. On the other hand, they can also take advantage of current knowledge graphs to offer more contextually informed recommendations. Real-time and dynamic data sources are emphasized in Web 3.0. As new data and documents become available, document recommendation systems must be able to adapt to these dynamic data environments and continuously update their recommendations.

*Motivation* The key motivation for the proposed scientific document recommendation paradigm which semantically inclines is because there are not many semantically driven models for document recommendation in the existing works. Most importantly the existing works only focus on standard and strategic document recommendation and document retrieval systems. Scientific document recommendations are rarely tackled or are neglected. Scientific document recommendation models are quite complex and cumbersome to build due to the need for a large quantity of scientific knowledge and the data that must be modeled. The ultimate motivation is the need for scientific document recommendation frameworks that are semantically oriented and knowledge centric, and which encompasses ontological models which is the need of and has been proposed in this framework.

*Contributions* The contributions in the proposed framework are as follows: Encompassment of ontology generation in the perspective of the dataset and to further subject it to ontology alignment with that of the preprocessed query words, topic modelling utilizing two different topic modelling paradigms, namely the Latent Semantic Indexing (LSI) and Latent Dirichlet Allocation (LDA), was performed. Some of the unique aspects of the suggested framework include the generation of metadata and the classification of it using a typical deep learning RNN classifier with the feature selection of Shannon's entropy. Classification of their dataset using the XGBoost Classifier by encompassing the feature selection using Shannon's Entropy. Semantic Similarity using SemantoSim measure and Cosine similarity measure helps in initial feasible solution generation which is further optimized using the dataset that was recorded in this to align the document with the best-in-class recommendation basis using the Lion's Optimization Algorithm. Then, relative to the other baseline models, f-measure and precision-recall accuracy are lowered and raised.

*Organization* The paper is structured as follows. Section 2 provides an overview of Related Literature, Sect. 3 discusses the Proposed System Architecture, Sect. 4 presents the Results and Performance Evaluation and outcomes, and Sect. 5 concludes with insights.

## 2 Literature Survey

Huynh et al. [1] have put forward the S2RSCS model which is a submission recommendation system that draws the essential data from each paper submission's title, abstract, and keyword list. Gupta et al. [2] have proposed the SDRTG model that discusses combining multimodal distributed representations to create a novel technique for selecting related scientific articles given a query article. Li et al. [3] describe topic regression, a fresh approach to matrix factorization called Matrix Factorization (tr-MF) for recommending scientific papers to members of an online community. Habibi et al. [4] have proposed the KECDR model which addresses the issue of conversation keyword extraction to retrieve a small number of possibly pertinent resources for each brief conversation fragment and recommend them to participants. Kim et al. [5] have suggested the ConvMF model, which offers a novel context-aware recommendation model that combines convolutional neural networks (CNN) and probabilistic matrix factorization (PMF). Popescul et al. [6] published a unified probabilistic paradigm for fusing collaborative and content-based recommendations. In the presence of sparse data, global probabilistic models, and conventional expectation maximization (EM) learning techniques frequently overfit significantly, which are common in recommendation applications. Guan et al. [7] developed a novel graph-based representation learning strategy for resource recommendation based on tagging data in the same semantic space, which is defined by the close proximity of related entities, users, tags, and documents. They only suggest documents that are close enough to the user in question. Xu et al. [8] carried out a large-scale log study of users' engagement behavior on one of the top cloud document platforms, office.com. Their findings point to the potential for improving justifications and, more broadly, the conception of systems that make document suggestions and justify them. Weng et al. [9] propose using ontology and the spreading activation approach to research paper recommendations in the hopes of improving the system's performance and addressing the inadequacies of current recommendation systems. Zhu et al. [10] propose a heterogeneous hypergraph model that tailors the impact of various relational kinds while fully utilizing high-order relational information in tagged data. In [11–20] several models in support of the proposed frameworks literature have been depicted.

## 3 Proposed System Architecture

Figure 1. depicts the proposed computer architecture for the scientific document recommendation framework which is semantically inclined ontology based as well as knowledge centric. The user input query is the driving force for the entire framework, preprocessing of the query input includes tokenization, lemmatization, stop word removal, and NER, to obtain the specific query words. The query words obtained after pre-processing input question words directly from the user are free of stop words and are in their base form rather than their inflectional form and these query words serve as the driving force for the entire model for recommending the scientific documents. Once the query words are generated, the second step is to generate the ontologies. The dataset is used to develop ontologies, which is the categorical dataset specific to a scientific domain. OntoCollab as a tool is used for generating the ontology.

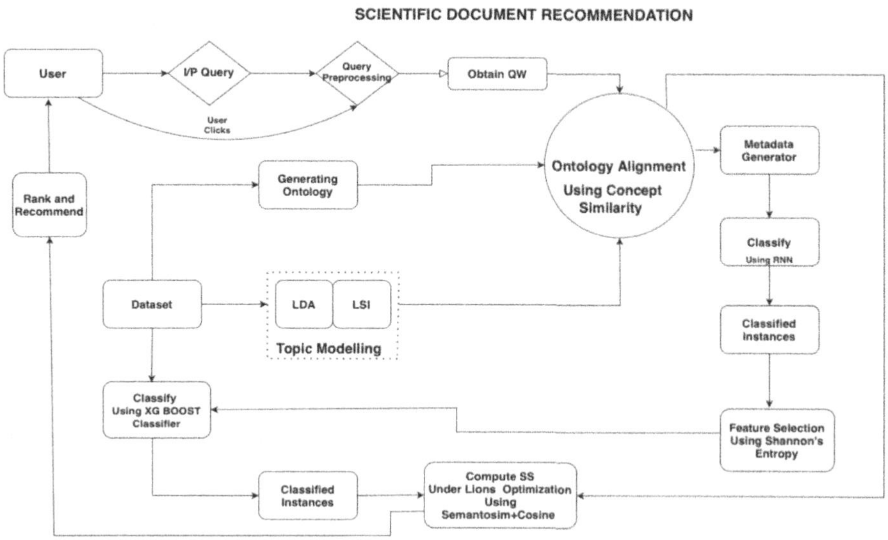


**Fig. 1.** Proposed System Architecture

The query words obtained are subjected to ontology alignment along with the generated ontology. The generated ontologies are highly subject-oriented ontology and ontology up to 7 levels are made use of. The generated ontology and the originally obtained query words are subjected to ontology alignment. Concept similarity is used in ontology alignment, with a threshold of 0.5. This is done because more entities need to be aligned at this step in order to develop an incremental ontology, which is why the threshold is set at 0.5. The domain is paleontology and fossils or categorically the dataset through paleontology and fossils made use of which is a document dataset that is categorically denatured the dataset's categories which are used to generate the ontology. OntoCollab is used as a tool to generate ontologies. Technical aspects for using RNN as a classifier: The functions used, the configurations used, if you have used tan sigmoid, all other functions or configurations, what are the parametric variations. XG Boost being a Machine Learning classifier, requires feature selection and feature controlling.

The system architecture for the user-centric, query-driven, semantically inclined, and knowledge-based recommendation of scientific information is shown in Fig. 1. The main input, the user query, is preprocessed. Pre-processing entails named entity recognition (NER), lemmatization, stop word elimination, and tokenization to obtain the individual query phrases. Once the individual query words are obtained, the query words are further used to drive the entire framework. Subsequently, the dataset is classified and to classify this dataset, the XG Boost classifier is used XG Boost classifier being a Machine Learning Classifier requires feature selection, since feature selection is not auto handcrafted or controlled, therefore the categories of the dataset are extracted, and this dataset is subjected to ontologies and generation of ontology is achieved with OntoCollab as the tool. OntoCollab generates ontologies, granular ontologies up to 7 levels, are retained and the rest within their vicinity are eliminated and these ontology instances are subjected to generation of metadata. Each of the ontology and their corresponding metadata is generated using the metadata generator Open Calais. Open Calais is used to generate the metadata. This metadata yields a lot of instances, and the metadata is generated from the extinct real-time World Wide Web and metadata increases the number of entities and henceforth increases the density of knowledge data and information thereby assuring a high entity density. The query words and the generated metadata are extensively large, Owing to the humungous size and the heterogeneity of the generated metadata, it cannot be handled rather it has to be classified, and to classify this we make use of the Recurrent Neural Network Classifier (RNN).

The RNN classifier, a deep learning classifier, because of the large size of the metadata, immediately classifies using auto-handcrafted feature selection. The classified instances then undergo feature selection using Shannon's entropy. Classified instances are yielded since they are of a large quantity Feature selection happens by computing Shannon's entropy. Shannon's entropy computes the information measure of each of the instances, Shannon's entropy of step deviance of 0.35 is used and the entities that are selected are passed through the XG Boost classifier as features to classify the dataset. Feature control mechanism takes place so that The XG Boost classifier's output is undefeated, and the dataset's classified instances of constitute the XG Boost classifier's output and are subjected to the computation of that of the semantic similarity. Initially, the query words and the generated ontology are subjected to ontology alignment using the concept similarity. These entities empirically provide informative examples with a concept similarity of at least 0.5, and these informative instances are fed into the development of metadata. These instances which are generated, the ontology generated and the query words, when computed concept similarity. These instances are put through computer semantic similarity comparison with the XG Boost classifier's classified instances using Lion's optimization algorithm. This time, semantic similarity is calculated using the SemantoSim measure with Cosine similarity both SemantoSim set to a threshold of 0.5 and Cosine similarity of 0.75. The reason for having this is to improve the variation heterogeneity of the instances aligned and Cosine similarity set to a threshold of 0.75 because it is not very strong however SemantoSim measure being a very strong measure is set to 0.5. Lion's optimization being a metaheuristic optimization algorithm that computes the initial solution uses the SemantoSim and Cosine similarity as the two objective functions and continually determines the ideal solution set, and this ideal solution set is the rank and suggested user-recommended increasing order of the SemantoSim measure. These are the phases that are produced, these phases are looped up with the dataset and all the documents which contain these phrases are subsequently and parallelly recommended along with the phrases, the document containing these phrases or terms are suggested to the user and they are all arranged in the SemantoSim measure's ascending order. If the user is satisfied, the search stops, but if they are not, they click on the phases that have been recorded and are included in the process, and this cycle continues until no more searches are made.

The Lion's Optimization Algorithm (LOA), a swarm intelligence optimization algorithm, draws its inspiration from how lions hunt. In LOA, the population of potential solutions is symbolized by a group of lions, with each lion's characteristics determined by where it now occupies the search space. The lions work together to hunt a candidate solution that represents the algorithm's current best answer during each iteration The lions work together to hunt the prey, which enables the algorithm to effectively scour the search space and avoid becoming snared in local optima: Flexible hunting tactics: The algorithm is flexible enough to tackle a variety of optimization issues since the lions change their hunting tactics according to the difficulty of the task. Randomness: LOA takes advantage of randomness to add variation to the lion population, enabling the algorithm to break free of local optima and explore various regions of the search space. When used to solve a variety of optimization problems, it has been shown that LOA performs better than other population-based optimization techniques, such as numerical optimization, combinatorial optimization, and global optimization.

Often employed for sequence modeling and classification tasks is an example of a neural network is the RNN (Recurrent Neural Network) classifier. RNNs, as opposed to feedforward neural networks, feature loops in their architecture that enable them to keep knowledge about prior inputs. Processing time series data or natural language for sequential data, can benefit especially from this, which are both examples of sequential data. In an RNN classifier, the network processes a series of inputs to create a series of outputs, with the last output frequently being used to make a prediction or classify data. At every step, the RNN's hidden state is changed, enabling it to consider data from earlier inputs. The Long Short-Term Memory (LSTM) network is a well-liked type of RNN classifier since it was developed to address the issue of disappearing gradients in traditional RNNs. LSTMs are particularly useful for long-term sequence modeling and classification problems because they employ gating mechanisms to selectively recall or forget information from prior inputs. The selection of activation functions for the hidden and output layers is frequently one of a Recurrent Neural Network's (RNN) technical considerations. The tanh (hyperbolic tangent) and sigmoid functions are frequently employed in RNNs as activation functions. The hyperbolic tangent (tanh) function, which maps its inputs to the range $[-1, 1]$, is a well-liked activation function for the hidden layer in RNNs, which can simulate both positive and negative values. The tanh function is defined in Eq. (1).

$$\mathrm{f}(x) = tanh(x) = \frac{2}{\left(1 + e^{(-2x)}\right)} - -1 \tag{1}$$

Another popular activation function that translates its inputs to the range $[0, 1]$ is the sigmoid function. The activation function for the output layer of RNNs commonly uses the sigmoid function, especially for binary classification issues. It is said that the sigmoid function is depicted in Eq. (2).

$$f(x) = \frac{1}{\left(1 + e^{(-x)}\right)} \tag{2}$$

The RNN's performance will be impacted by the activation function selection, and the best activation function will be determined by the nature of the data and the particular problem being handled. In addition to the activation functions, the number of recurrent units in the hidden layer is one of the other hyperparameters that must be changed when configuring an RNN, the sequence length, the learning rate, and the regularization method. Through a process of testing and cross-validation, the appropriate values for these hyperparameters can be found. In conclusion, choosing activation functions like tanh and sigmoid functions, as well as configuring other hyperparameters like the number of recurrent units, the length of the sequence, the learning rate, and the regularization technique are frequently included in the technical elements of an RNN. The individual problem being solved, and the properties of the data will determine the right settings for these hyperparameters.

A normalized semantic measure, the SemantoSim measure assesses semantic similarity and is derived from the Pointwise Mutual Information measure. When there are two terms in a question, they are paired as (x, y). If there are three of the three pairs of query phrases, permutations of those three are considered. Semantic similarity between the phrase and its most closely related semantically relevant term is taken into account if the query only contains one term. The search terms are the tokenized and stemmed query words. The SemantoSim (x, y) generates the semantic relatedness between two phrases, which is specified in Eq. (3). Equation (4) uses the equation to calculate the PMI (x, y). The likelihood of the terms x and y occurring together is shown in the equation p(x, y).p(x, y) is the likelihood that the phrase y will appear alongside x. The probabilities that the terms x and y will appear are p(x) and p(y), respectively. (The Bayes theorem states that the last two expressions are identical to the first). The expected value of the PMI is the mutual information (MI) of the random variables X and Y. (over all possible outcomes). Eq. (5) defines the equation for Shannon's Entropy, where N is the total number of observed occurrences and $p_i$ is the probability that the I event will occur. Think of a lexical database that contains two and two FCA concepts ($E_1$, $I_1$) and ($E_2$, $I_2$) of the same (or different) context(s), along with a lexical database for the English nouns E. Then, the concept similarity (Sim) between $E_1, I_1)(E_2, I_2)$ $Sim((E_1\ I_1), (E_2, I_2))$, is defined in Eq. (7) where ($I_1$, $I_2$) is defined as above, r, is the greatest between the cardinalities of the set $E_1$, $E_2$, and $I_1$, $I_2$, respectively. Finally, w is a weight that can be set by the user to increase the method's adaptability such that 0 < = w < = 1. It is based on the definition of the concept noun n probability, p(n), in Eq. (8). The frequency of n is estimated using noun frequencies from sizable text corpora using the formula *freq*(*n*). Similarity of information content (ICSs). Provided are two nouns $n_1$, $n_2 \in$ N, the weighted ISA hierarchy HE, and a noun dictionary for English $\varepsilon = (N, f(N), R, SynSet$. The following is a definition of the information content similarity $n_1$, $n_2$, indicated $asics(n_1, n_2)$, is defined in Eq. (9) and Eq. (10). If $n_1 = n_2$ or $n_1, n_2 \epsilon B_k \epsilon$ SynSet, for some k then Eq. (9) else Eq. (10). Where in Eq. (11) n' is a concept noun sharing the most information with $n_1$, $n_2$.

$$SemantoSim(x, v) = \frac{pmi(x, y) + P(x, y)\log\left[p(x, y)\right]}{\left[p(x).p(y)\right] + \log\left[p(y, x)\right]} \tag{3}$$

$$pmi(x; y) \equiv \log_2 \frac{p(x, y)}{P(x)p(y)} = \log_2 \frac{p(x|y)}{p(x)} = \log_2 \frac{p(y|x)}{p(y)} \tag{4}$$

$$H = -\frac{1}{logN} \sum_i p_i \log p_i \tag{5}$$

$$cosinesimilarityS_C(A, B) := \cos\theta = \frac{A.B}{\|A\|\|B\|} = \frac{\sum_{i=1}^{n} A_i B_i}{\sqrt{\sum_{i=1}^{n} A_i^2}\sqrt{\sum_{i=1}^{n} B_i^2}} \tag{6}$$

$$Sim((E_1, I_1), (E_2, I_2)) = \frac{\left|(E_1 \cap E_2)\right|}{r} * w + \frac{\mathcal{M}(I_1, I_2)}{m} * (1 - w) \tag{7}$$

$$P(n) = \frac{freq(n)}{M} \tag{8}$$

$$ics(n_1, n_2) = 1, \tag{9}$$

$$ics(n_1, n_2) = \frac{2\log p(n')}{\log p(n_1) + \log p(n_2)}, \tag{10}$$

$$-\log p(n') = \max_{n \epsilon S(n_1, n_2)} [-\log p(n)] \tag{11}$$

## 4 Performance Evaluation Results

The implementation was conducted using Python3 with Google Collaborators the preferred IDE. Python's NLDK library was used to conduct all the preprocessing and NLP tasks. Topic modeling was achieved using standard libraries for LDA and LSI implemented in Python. The deep learning model was realized using tensor flow. XGBOOST classifier was used in its native form Python for feature selection Shannon's entropy was completed using a customized agent subsequently the semantic similarity was computed using Lion's optimization using SemantoSim measure and Cosine similarity for which an agent was developed using JADE. Ontology alignment was achieved using concept similarity by using an agent-based model and ontology was generated using OntoCollab as a tool. Metadata was generated using Open Calais as a tool of choice. The dataset on which the experimentation was conducted is a single large integrated dataset comprising three main datasets namely the Dataset [21], Dataset [22] and Dataset [23]. But these three datasets will not be used as it is, rather the key terms in these datasets were extracted using a customized crawler, the Web's current structure was crawled for the documents pertinent to these datasets, and the dataset document category was formed for all the three independent datasets which comprise of large document corpora and prioritizations were done based on the categories, categorical terms, and the experimentations was covered into a single large dataset. F-Measure percentages are employed as prospective measures to evaluate the performance of the proposed Intelligent Integration Framework for Scientific Document Recommendation (IIFSDR), and another statistic used to evaluate the error is the false discovery rate (FDR). The proposed IIFSDR produces the highest precision-recall percentages of 97.74 and 99.01, the highest accuracy measure of 98.375, and the highest f-measure percentage of 98.370 with the FDR of 0.03 as shown in Table 1.

In order to assess and compare the effectiveness of the proposed IIFSDR framework, baseline models such as the SR, KECDR, SDRTG, and S2RSCS are used as baseline models. The SR model produces average overall precision of 87.39, average overall recall of 89.07, average overall accuracy of 88.23, average overall f-measure percentage of 88.22, and average overall FDR of 0.13. With an FDR of 0.11, the KECDR generates average precision percentages of 89.42, recall percentages of 90.14, accuracy percentages of 89.78, and f-measure percentages of 89.77. The SDRTG results in an overall average of 92.21% precision, 93.83% recall, 93.02 percent accuracy, and

**Table 1.** Comparison of Performance of the proposed IIFSDR with other approaches

| Model | Average Precision % | Average Recall % | Average Accuracy % (P + R)/2 | Average F-Measure % 2*P*R/ (P + R) | FDR |
|---|---|---|---|---|---|
| SR [1] | 87.39 | 89.07 | 88.23 | 88.22 | 0.13 |
| KECDR [2] | 89.42 | 90.14 | 89.78 | 89.77 | 0.11 |
| SDRTG [3] | 92.21 | 93.83 | 93.02 | 93.01 | 0.08 |
| S2RSCS [4] | 94.12 | 95.84 | 94.98 | 94.97 | 0.06 |
| **Proposed IIFSDR** | **97.74** | **99.01** | **98.375** | **98.370** | **0.03** |

93.01% f-measure with an FDR of 0.08. The S2RSCS produces average overall precision of 94.12, average overall recall of 95.84, average overall accuracy of 94.98, and average overall f-measure percentage of 94.97 with an FDR of 0.06. The precision, recall, accuracy, and f measures are the chosen potential metrics since they compute the relevant results whereas the quantity FDR denotes the error rate.

Importantly, optimization using Lion's optimization by using SemantoSim and Cosine similarity as two distinct semantic similarity models with differential thresholds ensures the most optimal solution set of the framework which is used to select the instances to yield the documents. All these factors contribute to the proposed scientific document recommendation framework IIFSDR's superior performance compared to all baseline models. All existing baseline frameworks are outperformed by the proposed IIFSDR framework because the IIFSDR framework is a knowledge-centric framework in which a large quantity of derivation-based knowledge along with semantic similarity frameworks and learning paradigms have been infused. Most importantly the query word enrichment takes place using standard ontologies which is a knowledge description model, which are lightweight knowledge description model, where the ontologies are generated in the purview, or the prospect of the dataset and the query words are enriched using ontology alignment using concept similarity. Subsequently, for each of the ontologies, metadata is generated, which adds to the auxiliary knowledge of the point exponential wave and handles intelligently by classifying it with RNN to yield classified encapsulated knowledge. Most importantly, the framework classifies the dataset using a lightweight machine learning, yet a strong XG Boost classifier in which a feature is selected using Shannon's entropy from the classified metadata, so a feature-controlled strategy is obtained, which prevents deviation conflicts and further deviation from the domain.

The reason why the SR model, the Scientific Article Recommendation model does not execute as anticipated is because topic-repulsive matrix factorization is the core strategy. This strategy is quite a conventional approach, but the matrix factorization works well if there are unique instances discovered and strategic knowledge which are relevant to the model is disseminated with all aspects. In the SR model, the knowledge is limited to auxiliary knowledge and there is no strong learning framework in the model and no strong alert completion mechanism consequently, the SR model does not execute as predicted. In comparison to the proposed IIFSDR, the KECDR model lags and performs below expectations because it makes use of keyword extraction and clustering. Keywords are informative terms, but they are not the best informative terms; there is no auxiliary knowledge; there is no robust relevance computation mechanism; the model only relies on clustering, which is a conventional method; there is no diversity of informative terms; auxiliary knowledge is embedded in the model; and the process of deriving the auxiliary knowledge is also not very concrete; as a result, the KECDR performs less well than expected when compared to the proposed framework. The SDRTG framework suggests that scientific documents or articles use distributed representations of texts and graphs, which produce semantic knowledge or auxiliary knowledge but are limited in nature and point shallow because they are limited representations since the distances are limited so there is no extensive density of knowledge. This is why the SDRTG framework does not execute as expected in comparison to the proposed framework. As a result, when compared to the suggested framework, the SDRTG model does not perform as predicted (Fig. 2).

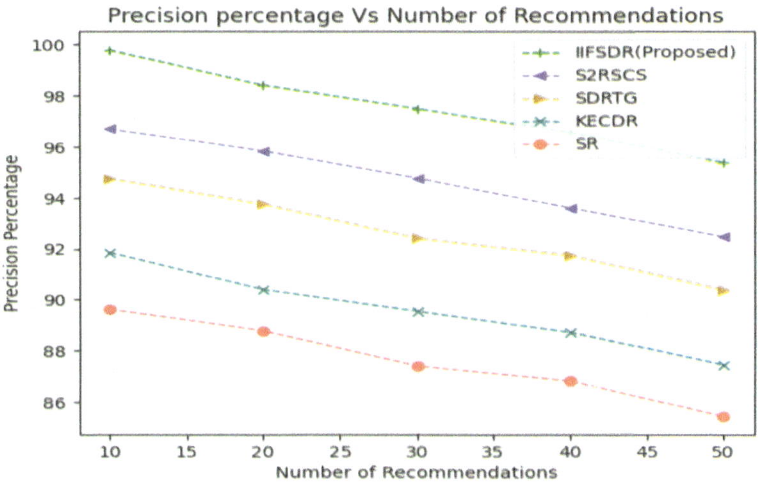


**Fig. 2.** Precision Percentage vs Number of recommendations

The S2RSCS model does not execute as anticipated in comparison to the suggested framework because it is domain-specific, which is good, but TF-IDF and chi-square statistics along with logistic regression also perform Multilayer positrons use a strong learning model but for this learning model there is very shallow knowledge, there is knowledge that is given but there is no knowledge derivation mechanism to handle the proper staging of knowledge. Moreover, the learning screens do not find a proper model-level substrate to yield knowledge. Because of this, this framework's S2RSCS model performs less well than the suggested framework. The proposed IIFSDR model outperforms all baseline models for all of the aforementioned reasons because they have noble topic derivation in terms of topic modelling, topic modelling refers to say LDA, LSI which enhances the probability of discovery of new topics thereby increases the rate of semantic similarity, generation of ontology to provide concrete knowledge models and alignment of the topics discovered to topic modelling and knowledge along with query words or using concept similarity to enrich the query words and generation metadata with handling of its classification using RNN and feature selection using the Shannon's entropy from the metadata utilizing a powerful machine learning classifier to categorize the dataset encourages strong learning models, strong topic models, strong knowledge models and metadata enhances our knowledge density and mostly the strong system of SemantoSim, Cosine similarity and concept similarity in the framework encompasses a good ecosystem of strong semantic similarity-based relevance computation techniques and lion's optimization performs at well on the substrate of the knowledge discovered and gives the best in class recommendations that is why all of the baseline models are outperformed by the suggested framework.

## 5 Conclusion

This paper proposes a strategic standard paradigm for scientific document recommendation the hybridization ofnna strong deep learning implicit feature selection base RNN classifier and a powerful machine learning feature control classifier but the machine learning classifier used is the XGBoost classifier which is used to classify the dataset and the RNN classifier which is used to categorize the intermediate metadata generated via the ontological entities which are common with that of the query words and the generated ontology. The model generates an ontology that provides a strong initial knowledge of the model, Topic modeling through LDA and LSI has been increasing the overall context of the terms in the framework. The proposition's knowledge density increases significantly because of the metadata, moving it into a more localized framework. Semantic similarity is calculated using the SemantoSim measure with Cosine similarity, differential thresholds, and step deviance measures, which produces the initial solution. Lion's optimization helps to yield the most optimal solution set from the initial feasible solution set to produce the fundamentals or indicated terms that are highly relevant to the user query, which is further classified with that of the dataset to recommend the scientific documents. Strong reasoning and an ecosystem for fact-based verification of the model's entities are provided by the presence of a strong learning model or strong knowledge generation paradigm. The proposed framework has an overall recall of 99.01%, a f measure of 98.370%, and an accuracy of 98.375%, which is the best baseline model for scientific document recommendation.

## References

1. Huynh, S. T., Huynh, P. T., Nguyen, D. H., Cuong, D. V., & Nguyen, B. T. (2020). S2rscs: an efficient scientific submission recommendation system for computer science. In Trends in Artificial Intelligence Theory and Applications. Artificial Intelligence Practices: 33rd International Conference on Industrial, Engineering and Other Applications of Applied Intelligent Systems, IEA/AIE 2020, Kitakyushu, Japan, September 22–25, 2020, Proceedings 33 (pp. 186–198). Springer International Publishing

2. Gupta, S., Varma, V.: Scientific article recommendation by using distributed representations of text and graphs. In: Proceedings of the 26th International Conference on World Wide Web Companion, pp. 1267–1268 (2017)
3. Li, Y., Yang, M., Zhang, Z.: Scientific articles recommendation. In: Proceedings of the 22nd ACM International Conference on Information & Knowledge Management, pp. 1147–1156 (2013)
4. Habibi, M., Popescu-Belis, A.: Keyword extraction and clustering for document recommendation in conversations. IEEE/ACM Trans. Audio Speech Lang. Process. **23**(4), 746–759 (2015)
5. Kim, D., Park, C., Oh, J., Lee, S., Yu, H.: Convolutional matrix factorization for document context-aware recommendation. In: Proceedings of the 10th ACM Conference on Recommender Systems, pp. 233–240 (2016)
6. Popescul, A., Ungar, L.H., Pennock, D.M., & Lawrence, S. (2013). Probabilistic models for unified collaborative and content-based recommendation in sparse-data environments
7. Guan, Z., Wang, C., Bu, J., Chen, C., Yang, K., Cai, D., He, X.: Document recommendation in social tagging services. In: Proc. of the 19th Int. Conf. on WWW, pp. 391–400 (2010)
8. Xu, X., Hassan Awadallah, A., Dumais, T.S., Omar, F., Popp, B., Rounthwaite, R., Jahanbakhsh, F.: Understanding user behavior for document recommendation. In: Proceedings of the Web Conference 2020, pp. 3012–3018 (2020)
9. Weng, S.S., Chang, H.L.: Using ontology network analysis for research document recommendation. Expert Syst. Appl. **34**(3), 1857–1869 (2008)
10. Zhu, Y., Guan, Z., Tan, S., Liu, H., Cai, D., He, X.: Heterogeneous hypergraph embedding for document recommendation. Neurocomputing. **216**, 150–162 (2016)
11. Surya, D., Palvannan, S., Deepak, G.: Bi-CSem: a semantically inclined bi-classification framework for web service recommendation. In: International Conference on Innovations in Bio-Inspired Computing and Applications, pp. 425–438. Springer Nature Switzerland, Cham (2022)
12. Jagan, N.D., Deepak, G., Santhanavijayan, A.: MPTR: a metadata-driven prospective tag recommendation model based on integrative classification. In: Advances in Data and Information Sciences: Proceedings of ICDIS 2022, pp. 291–301 (2022)
13. Deepak, G., Santhanavijayan, A.: OntoBestFit: a best-fit occurrence estimation strategy for RDF driven faceted semantic search. Comput. Commun. **160**, 284–298 (2020)
14. Pushpa, C.N., Deepak, G., Thriveni, J., Venugopal, K.R.: Onto Collab: strategic review oriented collaborative knowledge modeling using ontologies. In: 2015 Seventh International Conference on Advanced Computing (ICoAC), pp. 1–7. IEEE (2015)
15. Deepak, G., Rooban, S., Santhanavijayan, A.: A knowledge centric hybridized approach for crime classification incorporating deep bi-LSTM neural net-work. Multimed. Tools Appl. **80**(18), 28061–28085 (2021)
16. Rithish, H., Deepak, G., Santhanavijayan, A.: Automated assessment of question quality on online community forums. In: International Conference on Digital Technologies and Applications, pp. 791–800. Springer International Publishing, Cham (2021)
17. Kumar, N., Deepak, G., Santhanavijayan, A.: A novel semantic approach for intelligent response generation using emotion detection incorporating NPMI measure. Procedia Comput. Sci. **167**, 571–579 (2020)
18. Krishnan, N., Deepak, G.: KnowCrawler: AI classification cloud-driven framework for web crawling using collective knowledge. In: European, Asian, Middle Eastern, North African Conference on Management & Information Systems, pp. 371–382. Springer International Publishing, Cham (2021)
19. Deepak, G., Ahmed, A., Skanda, B.: An intelligent inventive system for personalised webpage recommendation based on ontology semantics. Int. J. Intell. Syst. Technol. Appl. **18**(1–2), 115–132 (2019)

20. Shaw, H., Deepak, G.: MetaBlog: a metadata driven semantics aware approach for blog tagging. In: International Conference on Digital Technologies and Applications, pp. 406–414. Springer International Publishing, Cham (2022)
21. Minerals management service Gulf of Mexico Region Paleontological Data CD-ROM [Dataset]
22. Lepore, T.J., Lundgren, L., Lawver, D.: The impact of field experiences in paleontology on high school learners [dataset] and James Lehane. A. A. Ekdale. **2023** (2016)
23. Morphometric analysis of graphoglytid trace fossils in two dimensions: implications for behavioral evolution in the deep sea [Dataset]

# Towards an Efficient and Intelligent Graph Model-Based Cooperative Intersection Collision Avoidance Systems

Safa Batita(✉), Achraf Makni, and Ikram Amous

University of Sfax, MIRACL Laboratory, Airport Road, km 4, B.P. 1088, 3018 Sfax, Tunisia
safabatita@gmail.com

**Abstract.** Traffic congestion and road safety continue to be critical challenges in urban environments, especially at intersections. Cooperative Intersection Collision Avoidance (CICA) systems have emerged as promising solutions to address these challenges. CICA systems leverage Vehicle-To-Vehicle (V2V) and Vehicle-To-Infrastructure (V2I) communications to enhance safety and optimize traffic flow at intersections.

Our research paper presents a novel CICA system specifically designed for unsignalized intersections. It utilizes a graph-based data model, which is particularly effective for encapsulating the complexities of Intelligent Transportation Systems (ITS) data. This system particularly focuses on rapid query processing, which is vital for effective and timely decision-making. Our method's primary component is an algorithmic framework that analyzes these graph-based representations to identify and prevent potential traffic issues. With an emphasis on query execution time, we have thoroughly evaluated the system. This paves the way for improving road safety and reducing urban traffic congestion.

**Keywords:** cooperative intersection collision avoidance system · graph databases · intelligent transportation systems · real-time

## 1 Introduction

In today's evolving transportation landscape, safety and efficiency play a central role, driven by the convergence of technology and innovation. Intelligent Transportation Systems (ITS) are pivotal in revolutionizing traffic flow, especially at intersections. While these intersections are integral to traffic fluidity, they also present significant challenges in collision avoidance, often becoming focal points for accidents.

Recognizing these issues, researchers worldwide are committed to enhancing intersection safety [1,2]. A key innovation in this domain is the CICA systems,


A. Bajaj et al. (Eds.): HIS 2023, LNNS 1224, pp. 239–250, 2025.
https://doi.org/10.1007/978-3-031-78925-0_24

which leverage real-time communication and sensor technologies to prevent collisions. These systems have evolved to provide a range of functionalities, including intersection collision prevention, collision risk alerts, overtaking assistance, and warnings for hazardous zones. However, the increasing complexity and volume of data managed by these systems pose challenges in data handling, potentially leading to data loss and serious consequences.

To address these challenges, recent contributions have focused on integrating and adapting Relational DataBases (RDB) for comprehensive data representation, aiming to augment the capabilities of these system's information and enhance existing systems [3–5]. However, a certain incompatibility has been noted between the relational model and the nuanced representation of real-world scenarios. Emerging as a viable solution, Graph DataBases (GDB) are a specialized type of non-relational databases that emphasize relationships and are known for their efficiency and scalability. Especially in ITS, GDB are seen as a natural and effective way to represent data. By accurately modeling intricate road networks and relationships among intersections and vehicles, GDB empower the CICA system with enhanced collision prediction and avoidance tactics. Though still in the developmental stages, GDB hold immense promise for revolutionizing intersection management and enhancing safety [6,7].

This study's primary goal is to develop a Cooperative Advanced Driver Assistance System (C-ADAS) designed to avoid collisions at unsignalized intersections named Intelligent-Graph model based-Cooperative Intersection Collision Avoidance Systems (IG-CICAS). This system utilizes GDB to analyze traffic scenarios, detect collision risks, and take necessary actions to prevent accidents.

The paper is organized as follows. Section 2 presents a literature review of related work. Section 3 describes the proposed IG-CICAS. Section 4 details the experimental study. Finally, Sect. 5 concludes this paper and discusses directions for future research.

## 2 Related Work

This section will provide an overview of research on cooperative collision avoidance systems at intersections. Additionally, it will delve deeper into the broader field of intelligent transportation research, with a particular focus on integrating GDB to improve various ITS. These studies collectively advance intersection safety and demonstrate the versatility of GBD to improve ITS.

### 2.1 Cooperative Collision Avoidance Systems at Intersections

[3–5] discuss a system that incorporates V2V and V2I communications with a Real-Time DataBase (RT-DB), focusing primarily on mitigating intersection collisions, addressing diverse road challenges, and overtaking. Similarly, [8] contributes to this field by creating an urban intersection collision avoidance system using onboard sensors and V2V communication, demonstrating collision frequency and severity. Broadening the scope of communication types, [9] introduced a system that employs Cellular Vehicle-to-Everything (C-V2X) technology. This system uses Received Signal Strength (RSS) to estimate intersection proximity and links collision avoidance probability to factors such as speed, traffic density, and urban road density. Further innovation is evident in [10], where a novel algorithm employing both V2V and V2I communications acts as a pre-warning system, utilizing a Dynamic Bayesian Networks (DBN) model to evaluate various vehicular parameters and predict potential risks. Additionally, Rules-based Intersection Collision Avoidance (RICA) system, introduced in [11], is a system that uses GPS and emphasizes V2X safety. It aims to enhance road safety by calculating intersection distance and collision risk. The system has shown promising results through simulation tests and experiments.

### 2.2 Integration of Graph Databases to Improve Various ITS

Graph databases have been used in many different ITS. Here we present some improved systems using GDB. For instance, [12] introduced a model for multi-modal transport relationships, implemented using Neo4j and Cypher. Similarly, [13] proposed a new algorithm called GraphRQI to classify driver behaviors using traffic graph spectral analysis. Moreover, [14] innovatively modeled Autonomous Transportation Systems (ATS) using a network knowledge graph, leveraging a Neo4j graph database. In a similar vein, [15] emphasized the implementation of a Local Dynamic Map (LDM) as a key component in Cooperative Intelligent Transport Systems (CITS). The authors introduced a new graph database, named iLDM, based on LDM, prioritizing interoperability and extensibility via standard interfaces.

### 2.3 Discussion

Most proposed CICA systems rely on computational methods rather than leveraging databases [8–11]. This heavy reliance on complex calculations and formulas creates complexity that can slow down real-time processing and limit adaptability. To overcome these limitations, other contributions propose to integrate RDB into these systems [3–5]. Although this transition brings some benefits, such as improved results and the simplification of formulas, it's not without disadvantages. Indeed, a key challenge emerged from the innate difference between the natural structure of relational databases and the multifaceted and interconnected reality of traffic scenarios. [12–15], which are capable of capturing complex relationships and dynamic patterns, offering a more comprehensive and accurate representation of real-world traffic situations. Based on this premise, the main goal of this research is to enhance the integration of GDB in the ITS field. This study will focus on applications related to road safety, known as Advanced Driver Assistance Systems (ADAS). Among these, this research will specifically highlight the domain of cooperative intersection collision avoidance and prevention applications.

# 3 Intelligent Graph Model-Based Cooperative Intersection Collision Avoidance System

In this section, we present the Intelligent Graph model-based Cooperative Intersection Collision Avoidance System (IG-CICAS), an innovative approach for enhancing safety at unsignalized intersections. Our method integrates a detailed characterization of intersections using parameters such as central point, danger zone, and control zone, along with a real-time tracking system for vehicle status within these zones. IG-CICAS leverages a graph database model to analyze and manage the complex interactions of vehicles as they approach intersections. This model aids in forecasting possible collision locations using Virtual Path Points (VPP) and Virtual Collision Points (VCP). Additionally, it allows vehicles to communicate their movement plans and make decisions based on current data. This part of the research is structured into four detailed subsections: the characterization of the intersection, the status of vehicles, an overview of IG-CICAS, and the graph database model, with each subsection being thoroughly explained in the following content.

## 3.1 Characterization of the Intersection

This research centers on a four-way intersection, illustrated in Fig. 1, where each incoming and outgoing street comprises two lanes.

The intersection features four connected streets, all following the same direction of travel. Additionally, each street provides three possible routes: turning left, proceeding straight, or turning right.

The intersection's characteristics are further defined by several key parameters:

- Central point (C): This point is defined by two coordinates on a Cartesian map, similar to the GPS coordinates inside the vehicle. In Fig. 1, the black dot's location is denoted by CX for the X coordinate and CY for the Y coordinate.
- Danger zone (D): This parameter indicates the distance from the intersection's center to the beginning of the intersection, plus a safety margin, and is depicted by the red square in Fig. 1.
- Control zone (E): This parameter represents the distance from where vehicles are controlled to the area where they begin to be actively controlled and is represented by the green square in Fig. 1.

These parameters are designed to describe both four-way and T-type intersections.

For this specific study, we propose a Control Zone (E) extending up to 100 m from the intersection and a Danger Zone (D) spanning 10 m.

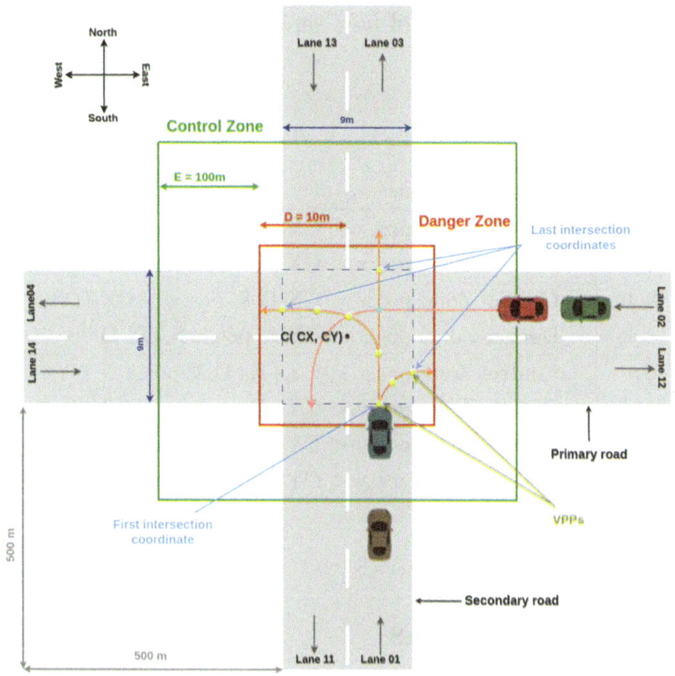


**Fig. 1.** Four-way intersection description

### 3.2 Vehicle Status

The 'vehicle_status' attribute is used to distinguish the real-time status of the vehicle approaching the intersection. This parameter can take one of eight distinct values:

- RightCZ: Signifies that the vehicle is located in the right section of the control zone.
- LeftCZ: Denotes that the vehicle is positioned in the left section of the control zone.
- DownCZ: Indicates that the vehicle is situated in the lower part of the control zone.
- UpCZ: Specifies that the vehicle is situated in the upper part of the control zone.
- RightDZ: Designates that the vehicle is in the right section of the danger zone.
- LeftDZ: Specifies that the vehicle is in the left section of the danger zone.
- DownDZ: Reveals that the vehicle is positioned in the lower part of the danger zone.
- UpDZ: Indicates that the vehicle is located in the upper part of the danger zone.

Table 1 provides an overview of how we can differentiate the status of each vehicle.

**Table 1.** The status of a vehicle approaching the intersection

| Direction | Zone Conditions |
|---|---|
| Right | Control Zone: pos_x < CX+D+E & pos_x > CX+D |
| | Danger Zone: pos_x < CX+D & pos_x > CX |
| Left | Control Zone: pos_x < CX-D & pos_x > CX-(D+E) |
| | Danger Zone: pos_x < CX & pos_x > CX-D |
| Up | Control Zone: pos_y < CY+E+D & pos_y > CY+E |
| | Danger Zone: pos_y < CY+E & pos_y > CY |
| Down | Control Zone: pos_y < CY-E & pos_y > CY-(E+D) |
| | Danger Zone: pos_y < CY & pos_y > CY-E |

### 3.3 Description of IG-CICAS

In this research, we introduce our Intelligent Graph model-based Cooperative Intersection Collision Avoidance System (IG-CICAS) to improve the management of unsignalized intersections. IG-CICAS allows drivers approaching intersections to assess the current situation by combining infrastructure-based and vehicle-based technology. It helps drivers avoid violating traffic control devices and safely navigate cross-traffic.

We focus on vehicles approaching intersections, where they have to make decisions that take into account the seriousness of the situation. Specifically, they must determine whether another vehicle will enter the same dangerous area within the intersection during their passage. To achieve this, a vehicle considers not only its movement intention but also the movement intention of other vehicles, captured by the 'mov_intention' parameter, which can be ToRight, ToLeft, or Forward. For a clearer illustration, consider a scenario in which a red vehicle, V1, as shown in Fig. 1, enters a control zone and navigates to head towards South Street. Its IG-CICAS creates a unique list for each vehicle_status using its local database. Notably, when several vehicles share the same vehicle_status, V1 prioritizes them. This strategy effectively reflects real-life situations and streamlines the necessary assessments. For example, in the scenario shown, V1 will only evaluate the risk of a collision with blue vehicle V2, which aims to head towards North Street. When vehicles enter critical areas, they share information about their movement intentions and store this data in their RT-DB, utilizing GDB for efficient data representation.

To define each vehicle's trajectory through the intersection, we adopt the proposition presented in [4], which utilizes Virtual Path Points (VPP). The number of VPP varies based on factors such as the vehicle's intent, intersection type, and road dimensions. For example, as shown in Fig. 1, a right turn may have three VPP, a left turn five, and going straight four, with one at the beginning and one at the end of the intersection zone. Vehicles calculate VPP coordinates for vehicles in the danger zone based on their movement intention stored in their RT-DB. The safe distance between consecutive vehicles is monitored in the danger zone.

Figure 1 illustrates a scenario in which IG-CICAS evaluates potential collisions. When a vehicle $V_M$ enters the control zone, its IG-CICAS creates lists of vehicles with different statuses based on its local database. The system evaluates the intersection's severity by checking for non-empty lists. In this scenario, there is a non-empty list, and IG-CICAS identifies potential collisions by estimating distances and times between vehicles.

V1 and V2 share the same VPP, which is called the Virtual Collision Point (VCP). The Distance to VCP (DVCP) has been calculated in previous research using the Euclidean distance formula, shown in formula (1). However, in our study, Neo4j provides this functionality. It not only automates calculations but also performs them faster than traditional formula-based calculations.

$$DVCP = \sqrt{(X_M - X_{VCP})^2 + (Y_M - Y_{VCP})^2} \tag{1}$$

The Time to Virtual Collision Point ($TVCP_{M,j}$) estimates the time required for vehicle $V_M$ to reach a specific VCP denoted as $VCP_j$, as depicted in formula (2). This time is calculated by taking the ratio of the distance between $V_M$ and $VCP_j$ to the current speed of $V_M$ ($CSV_M$).

$$TVCP_{M,j} = \frac{VCP - D_{M,j}}{CSV_M} \tag{2}$$

### 3.4 Graph Database Model

Formally, a graph denoted as G = (V,E) consists of a set V = $\{V_1, V_2,.., V_{|N|}\}$ of nodes and a set E=$\{E_{i,j}\}$ of edges connecting these nodes. In this context, the notation (i, j) ∈ E signifies the existence of an edge from node i ∈ V to node j ∈ V. Both nodes and edges can contain a list of attributes denoted as L[$V_i$] and L[$E_j$], respectively [16].

To facilitate a better understanding of the graphical representation structure, Fig. 2 shows how a real-world scenario (presented in Fig. 1) is converted into a graph format. The graph we have, denoted as G, is characterized as follows:

G = (V, E)
V = {Vehicle, Speed, Position, Direction, VPP, VCP, Intersection, Lane}
E = {HAS, LEADER, NEXT, BELONGS TO, STARTS WITH, ENDS WITH, OPPOSITE DIRECTION, FRONTAL ACCIDENT RISK}

Each node type is associated with a set of attributes:

L [Vehicle] = {vehicleID, model, serial_number, immatricule_number, lane, acceldecel, leaderID, leadergap, leaderspeed, mov_intention, vehicle_status}

L [Position] = {pos_x, pos_y, $d_{timestamp}$, $d_{duration}$}
L [Speed] = {speed, $d_{timestamp}$, $d_{duration}$}
L [Direction] = {direction, $d_{timestamp}$, $d_{duration}$}
L [VPP] = {pos_x, pos_y}
L [VCP] = {pos_x, pos_y}
L [Intersection] = {pos_x, pos_y}
L [Lane] = {LaneId}

Note that the attributes position, direction, and speed are real-time information. Therefore, we propose to adopt the data model proposed by [3,4] which makes it possible to maintain and share the current state of vehicles.

Thus, the structure of this real-time data will be represented by a triplet d = ($d_{value}$, $d_{timestamp}$, $d_{duration}$), where $d_{value}$ represents the value of the data, $d_{timestamp}$ indicates the instant when this value is measured, and $d_{duration}$ represents the validity period of this data.

This representation provides a structured and comprehensive way to model and manage data in a graph database system, especially suitable for applications in the ITS domain.

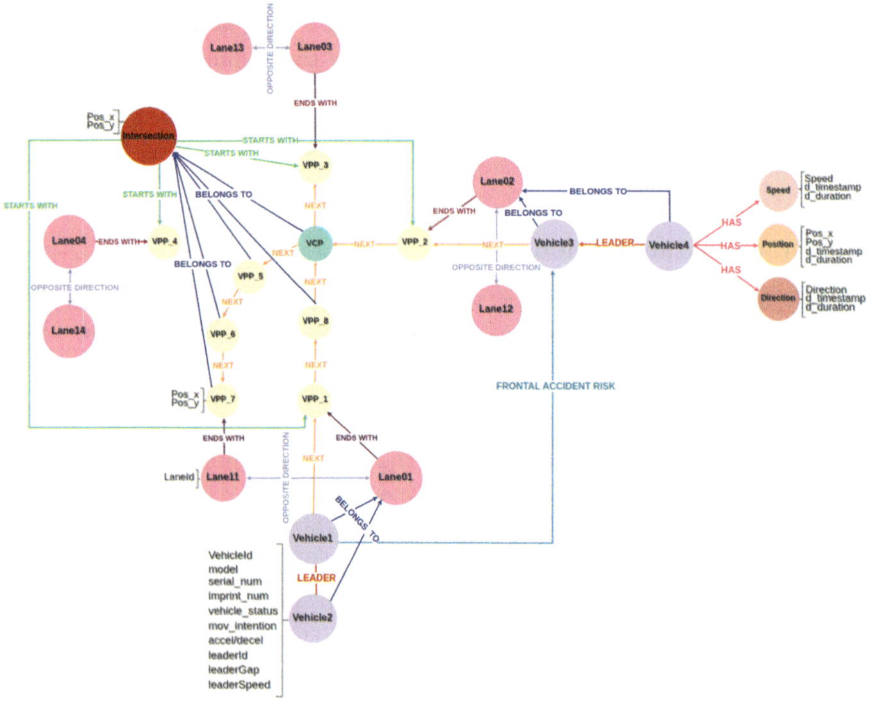


**Fig. 2.** Graph model of IG-CICAS

As shown in Table 2, the relations between various nodes are outlined with their corresponding semantic locations.

**Table 2.** Relations Between nodes

| Relations between nodes | Semantic location |
|---|---|
| FROM Vehicle TO Vehicle | LEADER/FRONTAL ACCIDENT RISK |
| FROM Vehicle TO Lane | BELONGS TO |
| FROM Vehicle TO VPP | NEXT |
| FROM Vehicle TO Direction | HAS |
| FROM Vehicle TO Speed | HAS |
| FROM Vehicle TO Position | HAS |
| FROM Lane TO Lane | OPPOSITE DIRECTION |
| FROM Lane TO VPP | ENDS WITH |
| FROM VPP TO VPP | NEXT |
| FROM VPP TO VCP | NEXT |
| FROM VCP TO VPP | NEXT |
| FROM Intersection TO VPP | STARTS WITH/BELONGS TO |

## 4 Experimental Study

Deploying our system in a real-world setting necessitates suitable infrastructure, sophisticated resources, and substantial expenses. Therefore, like most studies in this domain, we opted to assess its efficacy through a simulation-based approach. We relied on SUMO (Simulation of Urban MObility) to develop our road network, and concurrently, we implemented the RT-DB in IG-CICAS using Neo4j. During the IG-CICAS simulation, we applied two scenarios. Both have the same intersection, but vehicles take different routes and in different positions. To further evaluate the system, we varied the number of vehicles between 4 and 304 and measured the query execution time related to various critical queries concerning vehicles at intersections.

In an elaborate road situation detailed in Sect. 3.3, vehicles v1 (red) and v2 (blue) are on paths that could lead to a potential collision. such as:

Q1: List of vehicles preceding v1
Q2: The vehicle preceding v1
Q3: Distance between v1 and the leader of v1
Q4: List of vehicles in the opposite direction to v1
Q5: First vehicle in the opposite direction to v1
Q6: List of common VPPs between v1 and v3
Q7: Time to reach the common VPP for v1

The execution time for these queries proved notably efficient, ranging between 2 and 21 milliseconds, as illustrated in Fig. 3. This is significantly faster than the average reaction time of a human driver, which is typically estimated at one second [17]. This efficiency underscores the system's potential to enhance response time in real-world driving situations.

This implies that IG-CICAS can make potentially life-saving decisions faster than human drivers, especially to prevent intersection-related collisions. Furthermore, the system exhibited scalability and robustness as the number of vehicles grew, as depicted in Fig. 3. It maintained an acceptable response time without performance degradation, in contrast to systems based on relational database systems. This efficiency, particularly in handling complex and dynamic scenarios with multiple vehicles, highlights the system's suitability for real-world applications in ITS, improving safety and vehicular control at intersections.

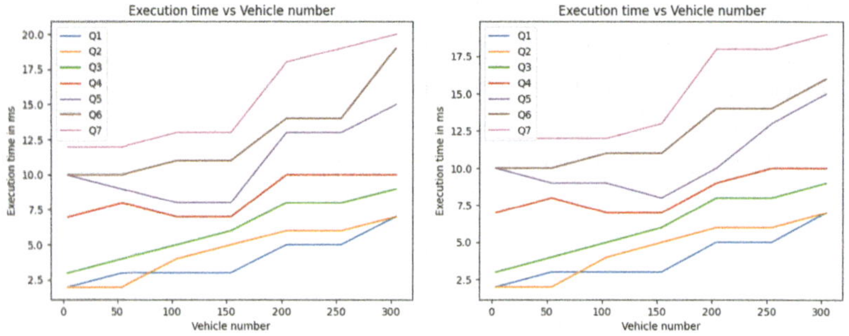


**Fig. 3.** Performance measurement of the 7 queries for the two scenarios

## 5 Conclusion

In this research, we have presented the IG-CICAS, a novel Cooperative Intersection Collision Avoidance System that uniquely harnesses the principles of graph theory. This innovative approach has not only opened new avenues in the realm of CICA systems but also demonstrated substantial potential in practical applications, especially with its efficient query execution time. Our findings underscore the pivotal role such systems can play in enhancing traffic safety, marking a critical advancement in the development of intelligent transportation systems aimed at reducing traffic incidents.

Looking ahead, our ambitions consist of integrating BDG into ADAS and other ITS, like road traffic management applications. Additionally, we aim to explore machine learning techniques to improve the evaluation of road situations and facilitate more efficient automated decision-making in autonomous vehicles.

## References


1. Oda, T., Tajima, T.: Driving safety support system for senior society. In: Proceedings Of 6th World Congress On Intelligent Transport Systems (ITS), Held Toronto, Canada, November 8-12, 1999. (1999)
2. Wang, J., Knipling, R.: Problem Size Assessment and Statistical Description. NHTSA Technical Report, Intersection Crossing Path Crashes (1994)
3. Elleuch, I., Makni, A., Bouaziz, R.: Cooperative overtaking assistance system based on V2V communications and RTDB. Comput. J. **62**, 1426–1449 (2019). https://doi.org/10.1093/comjnl/bxz026
4. Elleuch, I., Makni, A., Bouaziz, R.: CICAPS: a cooperative intersection collision avoidance persistent system for cooperative intersection ADAS. J. Supercomput. **79**, 6087–6114 (2023). https://doi.org/10.1007/s11227-022-04849-x
5. Elleuch, I., Makni, A., Bouaziz, R.: An intelligent and efficient safe driving system. Int. Conf. Hybrid Intell. Syst. 181–193 (2021). https://doi.org/10.1007/978-3-030-96305-7_17
6. Vela, B., Cavero, J., Cáceres, P., Sierra-Alonso, A., Cuesta, C.: Using a NoSQL graph oriented database to store accessible transport routes. EDBT/ICDT Workshops, pp. 62–66 (2018)
7. Tamilmani, R., Stefanakis, E.: Modelling and analysis of semantically enriched simplified trajectories using graph databases. Advances In Cartography And GIScience Of The ICA. 1, pp. 20 (2019). https://doi.org/10.5194/ica-adv-1-20-2019
8. Rashdan, I., Schmidhammer, M., Ponte Mueller, F., Sand, S.: Performance evaluation of vehicle-to-vehicle communication for cooperative collision avoidance at urban intersections. In: 2017 IEEE 86th Vehicular Technology Conference (VTC-Fall), pp. 1–5 (2017). https://doi.org/10.1109/VTCFall.2017.8288274
9. Wu, Q., Zhou, S., Pan, C., Tan, G., Zhang, Z., Zhan, J.: Performance analysis of cooperative intersection collision avoidance with C-V2X communications. In: 2020 IEEE 20th International Conference On Communication Technology (ICCT), pp. 757–762 (2020). https://doi.org/10.1109/ICCT50939.2020.9295949
10. Fu, Y., Li, C., Luan, T., Zhang, Y., Mao, G.: Infrastructure-cooperative algorithm for effective intersection collision avoidance. Transp. Res. Part C: Emerg. Technol. **89** 188–204 (2018). https://doi.org/10.1016/j.trc.2018.02.003
11. Yu, T., Wynn, T., Oo, M.: Rules based intersection collision avoidance system for V2X safety. Int. J. Adv. Sci. Res. Eng. **5**, 211–220 (2019). https://doi.org/10.31695/IJASRE.2019.33562
12. Wirawan, P., Riyanto, D., Nugraheni, D., Yasmin, Y.: Graph database schema for multimodal transportation in Semarang. J. Inf. Syst. Eng. Bus. Intell. **5**, 163–170 (2019). https://doi.org/10.20473/jisebi.5.2.163-170
13. Chandra, R., Bhattacharya, U., Mittal, T., Li, X., Bera, A., Manocha, D.: GraphRQI: classifying driver behaviors using graph spectrums. In: 2020 IEEE International Conference on Robotics and Automation (ICRA), pp. 4350–4357 (2020). https://doi.org/10.48550/arXiv.1910.00049
14. Zhang, L., Jiang, S., Huang, K., Xiao, Y., You, L., Cai, M.: Knowledge graph-based network analysis on the elements of autonomous transportation system. In: 2021 IEEE 21st International Conference On Software Quality, Reliability And Security Companion (QRS-C), pp. 536–542 (2021). https://doi.org/10.1109/QRS-C55045.2021.00082
15. García, M., Urbieta, I., Nieto, M., Mendibil, J., Otaegui, O.: iLDM: an interoperable graph-based local dynamic map. Vehicles **4**, 42–59 (2022). https://doi.org/10.3390/vehicles4010003

16. Czerepicki, A.: Application of graph databases for transport purposes. Bull. Pol. Acad. Sci. Tech. Sci. **64**, 457–466 (2016). https://doi.org/10.1515/bpasts-2016-0051
17. Ma, X., Andréasson, I.: Driver reaction delay estimation from real data and its application in gm-type model evaluation. Transp. Res. Record. 130–141 (2006)

# ABRHI: An Approach for Blog Recommendation Integrating Dual Classification and Hybrid Intelligence

Gerard Deepak[1,2](✉) and T. S. Amruth Mandappa[1,2]

[1] Department of Computer Science and Engineering, Manipal Institute of Technology Bengaluru, Manipal Academy of Higher Education, Bengaluru, India
gerard.deepak.christuni@gmail.com

[2] Department of Computer Science and Engineering, Indian Institute of Information Technology Dharwad, Dharwad, India

**Abstract.** A framework which is complaint to the Web 3.0, in which learning-inferencing through models for deep learning and machine learning along with Semantics Intelligence techniques has been proposed. This paper suggests a data set enrichment strategy for the key words extracted from the data set through structured topic modelling and the addition of terms from Wikidata and knowledge repositories. Subsequently the query is also enriched using meta data which is generated from the dataset which is further classified using transformers and formalized into a knowledge graph which uses Google Knowledge Graph API to be enhanced further. So the strategic encompassment and outgrowing of the dataset through topic modelling and auxiliary knowledge store repositories and knowledge graph through the meta data which ensures strong degree of knowledge density into the model with semantic relatedness computation using normalized compression distance, normalized information distance and Morisita's overlap index along with Pearsons correlation coefficient helps in semantics rigidness measure based reasoning and all baseline models are outperformed by the suggested framework. and achieves a overall Precision of 93.45% and F- measure of 189.30% and having a lowest FDR of 0.07, which model is the best in its class for microblog tagging in a highly dense Semantic Web environment.

**Keywords:** Micro-blog · Metadata · Web 3.0 · Semantic Web · Semantic Relatedness

## 1 Introduction

Semantic Web, commonly referred to as Web 3.0, is a crucial turning point in the development of the web. It moves beyond its function as a data store to become a dynamic platform that understands, combines, and links data from many sources. This virtual community has developed into a major hub for engagement, cooperation, and communication on a worldwide scale. The web links individuals and organizations across languages and cultures, fostering a vibrant interchange of information. This is done


A. Bajaj et al. (Eds.): HIS 2023, LNNS 1224, pp. 251–261, 2025.
https://doi.org/10.1007/978-3-031-78925-0_25

through social media, online communities, and collaborative technology. Despite this dynamic expansion, it can be difficult to effectively navigate the vast amount of internet information. The dearth of consistent annotations and helpful tags on many online sites, which impedes efficient information exchange and discovery, is a major problem. Blogs and microblogs are an essential part of this digital environment. While blogs offer a forum for in-depth inquiry through longer-form material, microblogs, represented by platforms like Twitter, enable fast transmission of succinct information.

There is a need for a competitive strategic microblog tag recommendation framework in the era of Web 3.0, where the conventional Web 2.0 is transforming into a highly data dense knowledge intensive Semantic Web. So, the existing models are not semantics oriented and are neither complained to a highly dense environment like Semantic Web or the Web 3.0. So, this paper proposes a microblog tag recommendation framework which is operable to Semantic Web which incrementally increases the knowledge and finds the confirmation with that of the queries in order to recommend the tags.

The primary contribution of the proposed framework is as follows. A strategic incrementation of knowledge through generation of metadata from keywords discovered from dataset and automating it by classifying it using a strong deep learning transformer classifier and formulating it into a knowledge graph using an intrinsic agent and then enriching the knowledge graph using Google Knowledge Graph API which is community contributed semantic knowledge repository, moreover the keywords from the microblog dataset are also enriched through structural topic modelling and Wikidata API by discovering entities for contextualization and to increase the information content. Strategic semantic similarity model with empirically decided threshold and step deviance measures brings about semantic reasoning strong infrastructure of classifiers like the transformers to transform the meta data and the AdaBoost machine learning feature control classifier to categorize the microblog data set are the suggested models' main contributions. Increased accuracy, recall, precision, and F-measure FDR are lower than other baseline models.

The paper is structured as follows. Section 2 provides an overview of Related Literature, Sect. 3 discusses the Proposed System Architecture, Sect. 4 presents the Results and Performance Evaluation and outcomes, and Sect. 5 concludes with insights.

## 2 Related Works

A model for contrasting retrieval techniques has been put forth by Arguello et al. [1] in which the blog or each post on the blog is viewed as the fundamental unit of retrieval, additionally, at the level of query development, where a user's original query is expanded in a novel way by using Wikipedia1's links and anchor text. We create two complimentary blog retrieval models that function at roughly the same levels of Accuracy and Recall. A model for a method of suggesting microblog topics that might give readers useful advice has been put out by Zhang et al. [2]. Users' similarity is calculated employing both the User Topic Feature Vector along with User Topic Feature Matrix in order to accomplish user clustering using the hierarchical clustering approach. The recommendation technique is provided to achieve customized microblogging topic recommendations inside user clustering sets by using subject heat degree and user authority. Experiments. A novel boosted inductive matrix completion technique (BIMC) for blog recommendation has been put out by Shin et al. [3]. In one component of the two-component additive low-rank BIMC model for user-blog preferences. A proactive suggestion model using the social, semantic, and value framework of the blogger has been put out by Tsai et al. [4]. Gao et al. [5] hybridized microblog recommendation technique is based on a deep neural network architecture and takes a range of inputs into account. Two new recommendation approaches were originally constructed in relation to the expanded user-interest tags along with themes. To gather the potential microblogs for the final suggestion, these two methodologies are combined with collaborative filtering. A deep neural network is later created to predict and rate the microblogs, including numerous hidden layers. A subject-rank recommendation model based on user preference and the topic's value to the user has been put out by Bao et al. [6]. They first used the term frequency-inverse document frequency (TFIDF) with latent Dirichlet allocation (LDA) approaches to develop a topic-screening model. A user influence model has also been provided (Chaoyi et al. [7]). The stereoscopic user model is first represented as a multi-dimensional vector. Later, they suggested the enhanced K-means technique, which was also used to obtain suggested microblog texts. In [11–17] several frameworks in support of the literature of the proposed work have been depicted.

## 3 Proposed System Architecture

Blog recommendation is the process of providing relevant and customized blog material to users based on their interests, preferences, and browsing history. It strives to improve the user experience by delivering articles, blogs, or stories that are relevant to their unique areas of interest. Blog recommendation systems examine user behaviour and content features to provide accurate, tailored recommendations utilizing data mining, natural language processing, and machine learning approaches. A blog recommendation system collects and analyses user data such as click patterns, reading history, and feedback. This information is used to generate user profiles, which include preferences, subjects of interest, and reading patterns. To match the qualities of the user's profile with the properties of accessible blog entries, content-based filtering approaches are used. This involves examining text characteristics, tags, categories, and other blog-related metadata. Collaborative filtering techniques may also be used to compare a user's preferences with those of similar users, allowing for suggestions based on a larger community's interests.

A recommendation system is a technique for information filtration that makes suggestions to consumers about products based on their preferences, interests, or previous actions. It is a piece of software or an algorithmic method created to examine user data and provide customized recommendations that meet specific demands and improve the user experience. The proposed system architecture of the is shown in Fig. 1. Microblog recommendation framework, which is knowledge-centric and semantically driven. So, this model is a query-initiated model. However, the dataset drives the entire model as it generates a lot of ancillary information. From the microblog data set which is categorical in nature is subjected to extraction of key words using NLTK tools and once the keywords are extracted from the data set it is enriched using two pipelines namely the structural topic modelling and the Wikidata. These enriched STM and Wikidata, both are enriched data set terms and both yield enriched data set terms and these are further used in the pipeline Also, the keywords extracted the production of metadata is applied to the microblog dataset, and metadata is generated using the following library and tool, namely "meta extractor software and my meta maker tool," and these meta data are generated from the current structure of the external Web 3.0. Large volumes of meta data are generated, which cannot be handled as they are; henceforth, they require classification, which is achieved using the transformers, which is a strong deep learning classifier. Transformers are preferred mainly for the reason that they do not require external features to be fed in and they work on large-scale data sets very well. Due to the strength of transformers, they are used, and a meta dataset is classified to give the classified instances, and these classified instances are formulated into a knowledge graph by computing their nodal Shannon's entropy using an agent designed using AgentSpeak.

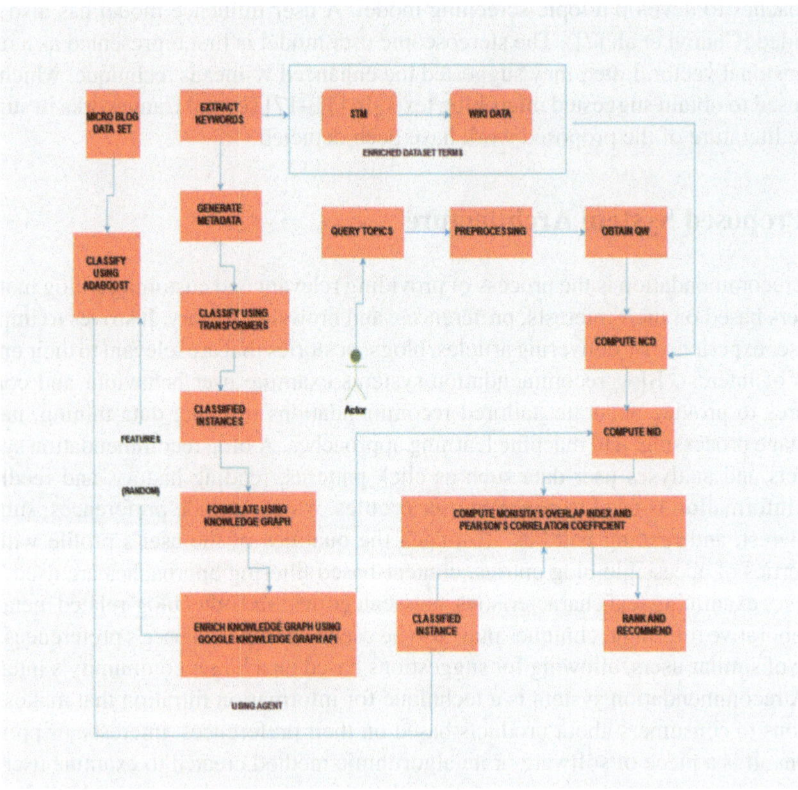


**Fig. 1.** Proposed system architecture of the Microblog recommendation framework

A key idea in information theory, Shannon's entropy measures the degree of unpredictability or uncertainty in a dataset. Numerous industries, including data compression, cryptography, and signal processing, use it. Shannon's entropy, which measures the randomness of data, enables effective data compression techniques and secure communication protocols, improving the dependability and effectiveness of different information systems. It is given by Eq. (1), where, H represents Shannon's Entropy, $p_i$ is the fraction of population composed of a single species $i$, $n$ is the number of species encountered.

$$H = -\sum_{i=1}^{n} p_i \log_2 p_i \tag{1}$$

Normalized Information Distance (NID) is a metric used to quantify the similarity between two pieces of information while considering their complexity. It accounts for both the size and structure of data, enabling effective comparisons of diverse datasets such as texts, images, and more. NID finds applications in fields like data compression, pattern recognition, and information retrieval, facilitating efficient analysis and classification of complex information. Assume that $x$ is a digital object, such as a binary string. The shortest binary program that produces $x$ has a length $(x)$, which is known as the Kolmogorov complexity of $x$. Additionally, the smallest binary program that calculates $x$ from $y$ and vice versa has a length equal to the information distance between $x$ and $y$, $E(x, y)$, The mathematical representation of the highest absolute value between the two variables $x$ and $y$, is provided by Eq. (2). Equation (3) depicts the Normalized Information Distance (NID).

$$E(x, y) = \max\{K(x|y), K(y|x)\} \tag{2}$$

$$NID(x, y) = \frac{\max\{K(x|y), K(y|x)\}}{\max\{K(x), K(y)\}}. \tag{3}$$

The Normalized Compression Distance (NCD) is a metric used to measure the similarity between two sequences or datasets based on their compression. It quantifies the amount of information shared between the sequences, making it useful for tasks like data clustering, NCD provides insights into their structural similarities and dissimilarities, aiding in various data analysis and information retrieval applications. It is depicted in Eq. (4), where $C$ (x) ($C$ (y)) denotes the joint compressed x and y size (in bits), and $C(xy)$ denotes the size of the individual compressed x and y.

$$NCD(x, y) = \frac{C(xy) - \min\{(Cx), C(y)\}}{\max\{(Cx), C(y)\}} \tag{4}$$

Morisita's overlap index is a quantitative measure used to assess the similarity between two ecological communities. It is depicted in Eq. (5), where, $Dx$ and $Dy$ are the Simpson's Index values for the x and y samples, and $x_i$ is the number of times an item is represented in the sum of X from one sample, and $y_i$ is the same. When it comes to species, $C_D$ equals 0 if there is no overlap between the two samples, and $C_D$ equals 1 if there is.

$$C_D = \frac{2\sum_{i=1}^{s} x_i y_i}{(Dx + Dy)XY} \tag{5}$$

Pearson's Correlation Coefficient is a statistical measure that assesses the linear relationship between two variables. In microblog recommendation systems, it helps quantify the similarity between users' behaviours and preferences. By analysing correlations in user interactions, such as likes and retweets, the system can suggest relevant microblogs to users based on their similarity to others with similar preferences, enhancing personalized content recommendations. It is depicted in Eq. (6), where $xy$ is the total of the products of the paired stocks, $x$ is the sum of the $x$ scores, $y$ is the sum of the $y$ scores, $x^2$ is the sum of the squared $x$ scores, and $y^2$ is the sum of the squared y scores, where $r$ is the Pearson coefficient.

$$r = \frac{n(\Sigma xy) - (\Sigma_x)(\sum y)}{\sqrt{\left[n\Sigma x^2 - \left(\sum x\right)^2\right]\left[n\Sigma y^2 - \left(\sum y\right)^2\right]}} \tag{6}$$

Once the knowledge graph is formulated, this knowledge graph is enriched by passing only the terminal nodes of the knowledge graph into Google's Knowledge Graph API. The partial knowledge subgraph and deep knowledge graph are harvested from Google's Knowledge Graph API, and this is further aggregated into the knowledge graph generated, and the large knowledge graph is formalized. However, the enriched knowledge graph that comes out of the whole knowledge graph API is filled with features at random by the AdaBoost classifier, which is a light-weight, strong machine learning classifier to classify the data set. The cause for preferring a strong machine learning classifiers are a result of its lightweightness compared to the deep learning model, and moreover, it is a feature control because it implicitly takes features, and we do not want domain deviation to take place. To minimize the domain deviance percentage, the AdaBoost classifier is preferred over the deep learning classifier, and it is also lightweight, which is advantageous. The classifier instances that come out of the AdaBoost classifier are then further used in response to the user query. The user query or the topic of choice for which the user requires a microblog is obtained from the user and subjected to preprocessing, which incorporates tokenization, lemmatization, stock word removal, and Named Entity Recommendation (NER). On preprocessing the query words into the query or topics, the inclusion query words "q" and suffix "w" are acquired, and these search terms are subjected to the computation of normalized compression distance (NCD) with a median threshold of 0.5, with that of the enriched data set term that comes out of the STM Wikidata pipeline. The reason for choosing the median threshold is because of the strength of normalized compression distance (NCD) and because we require a larger number of terms in the anchor, and the entities that come out of this NCD pipeline are subjected to computation of symmetric similarity with that of the enriched knowledge graph, which comes under Google's Knowledge Graph API. At this point, the symmetric similarity manager used or preferred is the Normalized Information Distance (NID), but with a strong threshold of 0.75. Although it is very strong, because the Google Knowledge Graph produces such a vast number of items, the NID is computed with a very high threshold of 0.8. The NIDs that come out of this pipeline are subjected to the computation of Morisita's Overlap Index and Pearson's Correlation Coefficient compared it to the classification of the AdaBoost classifier's output examples. The Morisitas's Overlap Index and Pearson's Correlation Coefficient are both set towards a step deviance of 0.255 and 0.25, which is considered because we do not want it to mean very stringent, and moreover, since two different indices are used for computing the step deviance, which is set at 0.25 empirically, and the entities that come out are ranked in ascending order of the Pearson's Correlation Coefficient measure and recommended to the user as the topics of choice, and the user clicks on these topics. The microblog realm containing these categorical terms is directly furnished to the user from the data set. The calls made here satisfy the user, and if the individual is not happy, the search proceeds until no more clicks from the user are logged.

## 4 Implementations, Results and Performance Evaluation

The effectiveness of the ABRHI model, a framework for blog recommendation that combines dual classification and hybrid intelligence, is evaluated using various metrics. These metrics include Accuracy, Recall, Precision, and F-Measure percentages, which are considered primary measures. Additionally, the False Discovery Rate (FDR) serves as a supplementary metric for assessing performance. In Table 1, we utilize Accuracy, Recall, Precision, and F-Measure percentages as they provide a quantifiable assessment of result relevance, while FDR serves to measure model error rates. The ABHRI model proposed in this study demonstrates outstanding performance with the highest average Precision percentage at 93.45%, the highest average Recall percentage at 95.88%, the highest average Accuracy percentage at 94.66%, and an impressive average F-Measure of 189.30, all while maintaining a low FDR of 0.07. To assess the efficacy of the ABRHI model, we conducted experiments involving 2482 blog recommendation queries, comparing it against similar models such as DRQBR, MTBUC, BRBIM, and PBRSSM, all evaluated on the same dataset and query set for a comprehensive analysis. From the data presented in Table 1, we can observe the performance of different models. DRQBR, for instance, demonstrates an average Precision of 85.46%, an average Recall of 87.22%, an average Accuracy of 86.34%, an average F-Measure of 172.66%, and an FDR of 0.15. Similarly, MTBUC exhibits an average Precision of 86.23%, an average Recall of 88.74%, an average Accuracy of 87.6%, an average F-Measure of 174.7%, and an FDR of 0.14. BRBIM, on the other hand, showcases an even higher average Precision of 89.91%, an average Recall of 90.78%, an average Accuracy of 90.3%, an average F-Measure of 180.57%, and an FDR of 0.11. Lastly, PBRSSM boasts an average Precision of 91.15%, an average Recall of 93.15%, an average Accuracy of 92.15%, an average F-Measure of 184.27%, and an FDR of 0.09. It is evident that, in comparison, our proposed ABRHI model surpasses all other models with the highest average Accuracy, Recall, Precision, and F-Measure percentages, while simultaneously maintaining the lowest FDR value.

The ABRHI model we propose achieves remarkable performance with the highest average values for Accuracy, Recall, Precision, and F-Measure percentages, while simultaneously achieving the lowest FDR. This outstanding performance can be attributed to the model's unique approach, which involves aggregating knowledge from established sources such as Wikidata and the Google Knowledge Graph API. Additionally, the incorporation of structural topic modeling aids in discovering relevant topics, thereby increasing the density of auxiliary knowledge. Notably, the model leverages keywords extracted from the micro-blog dataset to generate metadata, expanding beyond the confines of the current World Wide Web structure. To handle this wealth of information, we employ a robust deep learning transformer classifier. Moreover, classified instances undergo Knowledge Graph formalization with the assistance of an agent. The Knowledge Graph is further enriched through integration with the Google Knowledge Graph

**Table 1.** Comparison of Performance of the proposed SCRF with other approaches

| Model | Average Precision % | Average Recall % | Average Accuracy % | Average F-Measure % | FDR |
|---|---|---|---|---|---|
| DRQBR [] | 85.46 | 87.22 | 86.34 | 172.66 | 0.15 |
| MTBUC [] | 86.23 | 88.74 | 87.6 | 174.7 | 0.14 |
| BRBIM [] | 89.81 | 90.78 | 90.30 | 180.57 | 0.11 |
| PBRSSM [] | 91.15 | 93.15 | 92.15 | 184.27 | 0.09 |
| **Proposed ABRHI** | 93.45 | 95.88 | 94.66 | 189.30 | 0.07 |

API, which in turn serves as randomly selected features for dataset classification. This classification process is facilitated by a lightweight yet powerful AdaBoost classifier, enhancing the model's efficiency and effectiveness. Apart from this the query words were also enriched by computing a normalized compression distance with that of the enriched data set terms that come out of wikidata pipeline and normalized information distance with the outcome of the enriched Google Knowledge Graph API. And most importantly Morisita's Overlap Index and Pearson's Correlation Coefficient are used for final recommendation by computing the outcome of the NID and that of the AdaBoost classifier. Going through such strong relevance computation measures in terms of Morisita's Overlap Index and Pearson's Correlation Coefficient, Normalized Information Distance and Normalized Compression Distance aggregating knowledge from a strong repository namely the Wikidata API for enriching the knowledge and metadata to discover and generate knowledge and topic modelling the form of structural topic models and the presence of AdaBoost transformer classifier at two different ends to handle the dataset and the metadata separately and independently makes the proposed ABRHI model which outperform all the baseline model so this is the best in class model for microblog recommendation.

The limitations observed in the DRQBR model stem from several factors. Firstly, despite its implementation of query expansion and documentation mechanisms, it falls short due to the absence of robust relevance computation. Furthermore, the model lacks a strategic preferential learning approach, making it less effective in optimizing recommendations. Additionally, the DRQBR model does not incorporate metadata generation and handling, which limits its capacity to tap into valuable auxiliary information. Similarly, the MTBUC model faces shortcomings in its approach. While it leverages user clustering, this feature alone does not provide a strong foundation for recommendation accuracy. Although the model delves into modelling the topic feature graph and user microblog features, it lacks comprehensive integration between topic user feature vectors and topic feature matrices. Moreover, the computation of user authority is managed separately, leading to inefficiencies in the recommendation process. Although the algorithm arrangement is very strong, the target knowledge is absent, it works only for the data set which makes it under-fitted and clustering is weak and there are no strong learning models. The BRBIM model exhibits certain limitations, primarily attributable to its utilization of boosted inductive matrix completion, which indeed constitutes a robust mechanism for relevance computation. However, these advantages are somewhat overshadowed by the model's reliance on latent structures derived from side information. While this inclusion contributes to the model's overall complexity, it may not fully harness the potential of this auxiliary data, warranting further exploration for enhanced performance and also deep learning model are also with symmetrically rich features of image and text is used but there is not strong relevance computation mechanism, there is no attenuation of knowledge, permeability of knowledge density is not very high when compared to proposed model, knowledge permeability in the model and its regulation mechanism is absolutely unhandled. The PBRSSM model falls behind in comparison to the proposed framework is because, although it uses value semantics and social model, this model lags because several specific interactive like inclination, topic semantics and some other auxiliary social interactions are used in the model, but the learning model is quite weak, there is only absolute knowledge, there is no aggregation of current knowledge from the existing web and is also not compartment to web 3.0. Since the proposed model semantically inclined has strong learning ecosystems in terms of AdaBoost and transformers, generation if metadata and several knowledge repository topic modelling, and strong relevance mechanisms, the ABRHI model, as introduced in this study, surpasses all the baseline models and stands as the top-performing model in its class.

Figure 2 presents the distribution curves illustrating Precision percentages concerning the number of recommendations. Because ABRHI is a knowledge aggregation paradigm that uses traditional sources like Wikidata and the Google Knowledge Graph API to gather knowledge, it sits at the top of the hierarchy. To further increase the auxiliary knowledge density, structural topic modelling is also used for subject discovery. The runner-up position within the hierarchy is held by the PBRSSM model, which lags behind ABRHI because it uses inclination, topic semantics, and other auxiliary social interactions, but its learning model is quite deficient because it only incorporates absolute knowledge and does not aggregate current knowledge from the existing web or incorporate web 3.0. BRBIM occupies the middle region because it uses boosted inductive matrix completion, a powerful mechanism for computing relevance, latent structures using side information, and deep learning models with symmetrically rich features for both image and text. In contrast to our proposed model, the MTBUC model lacks a robust relevance computation mechanism, suffers from limited knowledge attenuation, and exhibits relatively lower knowledge permeability. Its position second to last in the hierarchy can be attributed to its modeling approach, which involves the subject feature graph, user microblog feature, topic user feature vector, and topic feature matrix. Additionally, the distinct method employed for computing user authority contributes to its placement in the hierarchy. Clustering together is a weak feature since it relies on user clustering. Although the algorithm layout is very robust, there are several drawbacks, including a lack of target knowledge, the fact that it only applies to the given data set, which makes it under-fitted, poor clustering, a lack of robust learning models, and the fact that DRQBR is in the last position in the hierarchy due to the absence of compelling significance. Strategic favoured learning, computing, and the production and processing of metadata are absent.

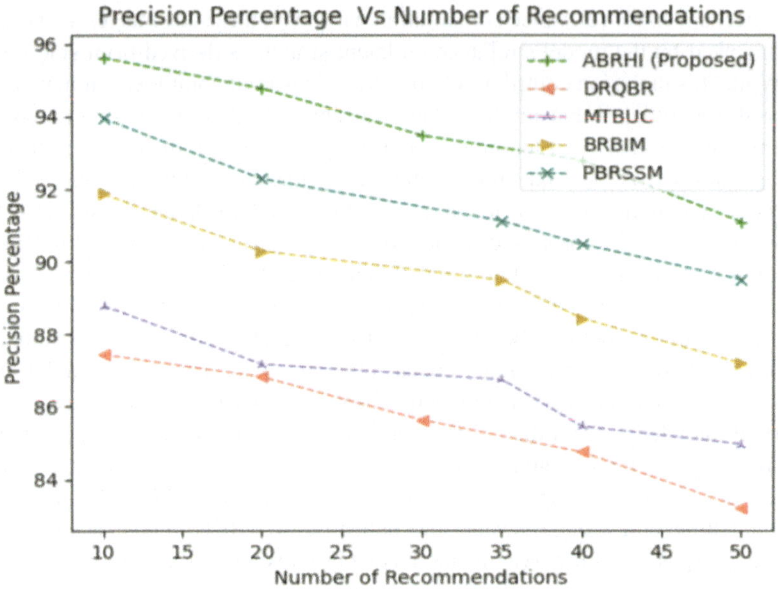


**Fig. 2.** Precision % Number of recommendations of the proposed ABRHI and other baseline models

## 5 Conclusion

This research paper presents a method for blog recommending integrating dual classification and hybrid intelligence strategies for tagging microblogs over a highly dense environment Web 3.0. The model has very strong mad rich knowledge derivation strategy through structural data modelling and Wikidata dependant entity extraction, generation of metadata through the keywords in the data set and handling it by classification using the transformers and formalization of the knowledge graph along with enrichment of the knowledge graph using the Google's Knowledge Graph API. This produces a very strong incremental knowledge infrastructure. Semantics reasoning is achieved using normalized compression distance, normalized information distance and Morisita's overlap index along with Pearsons's correlation coefficient with empirically decided threshold and step deviation measures in order to enrich the queries and to find the common instances between the enriched meta data driven terms. A robust learning framework incorporating advanced transformers for metadata classification and the AdaBoost algorithm to maintain domain diversity within the dataset has been employed. This comprehensive approach has yielded exceptional results, with an overall Accuracy of 94.66%, an impressive F-Measure of 189.30%, and a minimal FDR of 0.07%.

## References

1. Arguello, J., et al.: Document representation and query expansion models for blog recommendation. In: Proceedings of the International AAAI Conference on Web and Social Media, vol. 2, p. 1 (2008)
2. Zhang, S., et al.: The recommendation system of micro-blog topic based on user clustering. Mob. Netw. Appl. **22**, 228–239 (2017)
3. Shin, D., et al.: Tumblr blog recommendation with boosted inductive matrix completion. In: Proceedings of the 24th ACM International on Conference on Information and Knowledge Management (2015)
4. Tsai, T.-M., Shih, C.-C., Chou, S.-C.T.: Personalized blog recommendation using the value, semantic, and social model. In: 2006 Innovations in Information Technology, Dubai, United Arab Emirates, vol. 2006, pp. 1–5
5. Gao, J., et al.: Hybrid microblog recommendation with heterogeneous features using deep neural network. Expert Syst. Appl. **167**, 114191 (2021)
6. Bao, F., et al.: A topic-rank recommendation model based on microblog topic relevance & user preference analysis. Hum.-Cent. Comput. Inf. Sci. **12**, 10 (2022)
7. Li, C., Zhang, Y.: A personalized recommendation algorithm based on large-scale real microblog data. Neural Comput. & Applic. **32**, 11245–11252 (2020)
8. Ma, H., et al.: Combining tag correlation and user social relation for microblog recommendation. Inf. Sci. **385**, 325–337 (2017)
9. Yu, Y., Wan, X., Zhou, X.: User embedding for scholarly microblog recommendation. In: Proceedings of the 54th Annual Meeting of the Association for Computational Linguistics (Volume 2: Short Papers) (2016)
10. Zhao, F., et al.: A personalized hashtag recommendation approach using LDA-based topic model in microblog environment. Futur. Gener. Comput. Syst. **65**, 196–206 (2016)
11. Surya, D., Palvannan, S., Deepak, G.: Bi-CSem: a semantically inclined bi-classification framework for web service recommendation. In: International Conference on Innovations in Bio-Inspired Computing and Applications, pp. 425–438. Springer Nature Switzerland, Cham (2022)
12. Jagan, N.D., Deepak, G., Santhanavijayan, A.: MPTR: a metadata-driven prospective tag recommendation model based on integrative classification. In: Advances in Data and Information Sciences: Proceedings of ICDIS 2022, pp. 291–301. Springer Nature Singapore, Singapore (2022)
13. Deepak, G., Santhanavijayan, A.: OntoBestFit: a best-fit occurrence estimation strategy for RDF driven faceted semantic search. Comput. Commun. **160**, 284–298 (2020)
14. Pushpa, C.N., Deepak, G., Thriveni, J., Venugopal, K.R.: Onto Collab: strategic review oriented collaborative knowledge modeling using ontologies. In: 2015 Seventh International Conference on Advanced Computing (ICoAC), pp. 1–7. IEEE (2015)
15. Deepak, G., Rooban, S., Santhanavijayan, A.: A knowledge centric hybridized approach for crime classification incorporating deep bi-LSTM neural net-work. Multimed. Tools Appl. **80**(18), 28061–28085 (2021)
16. Rithish, H., Deepak, G., Santhanavijayan, A.: Automated assessment of question quality on online community forums. In: International Conference on Digital Technologies and Applications, pp. 791–800. Springer International Publishing, Cham (2021)
17. Kumar, N., Deepak, G., Santhanavijayan, A.: A novel semantic approach for intelligent response generation using emotion detection incorporating NPMI measure. Procedia Comput. Sci. **167**, 571–579 (2020)

# Date Varieties Identification Using DenseNet Model with GAN-Based Data Augmentation

Ines Neji[1(✉)], Najib Ben Aoun[2,3], Noureddine Boujnah[4], Hammadi Hamza[5], and Ridha Ejbali[1]

[1] Research Team in Intelligent Machines (RTIM), Engineering School of Gabes (ENIG), University of Gabes, Gabes, Tunisia
nejiines1997@gmail.com, ridha_ejbali@ieee.org

[2] Faculty of Computing and Information, Al-Baha University, Al Bahah 65779-7738, Saudi Arabia
nbenaoun@bu.edu.sa

[3] REGIM-Lab: Research Groups in Intelligent Machines, National School of Engineers of Sfax (ENIS), University of Sfax, 3038 Sfax, Tunisia

[4] Walton Institute, Waterford Institute of Technology, Waterford, Ireland
noureddine.boujnah@waltoninstitute.ie

[5] Arid Area Institute of Kébeli, University of Gabes, Gabes, Tunisia
hamza.hammadi@ira.rnrt.tn

**Abstract.** Date palm cultivation is a vital component of agriculture which contributes significantly to the global food supply. With the increasing demand for date fruits, date customer and traders face a big problem about differentiating between data varieties. Indeed, some date varieties are rare and expensive while they are very similar to other cheaper varieties. So, ensuring accurate classification and identification of date fruit varieties has become crucial. However, existing literature on date varieties identification remains limited, emphasizing the need for advanced systems capable of precise and automated classification. In this study, we propose an enhanced classification framework utilizing deep learning-based method trained with a date image dataset augmented using different data augmentation techniques such as Generative Adversarial Network (GAN) to accurately differentiate between 29 types of Tunisian date fruit varieties. By leveraging state-of-the-art Convolutional Neural Network (CNN) architectures, including MobileNet, VGG, Inception, and DenseNet, we aim at improving the efficiency and the accuracy of date fruit classification. Through an extensive comparative analysis with these architectures, we demonstrated the efficacy of our proposed approach in addressing the challenges posed by date fruit misclassification, thereby contributing to the advancement of sustainable agricultural practices and effective crop management.


A. Bajaj et al. (Eds.): HIS 2023, LNNS 1224, pp. 262–271, 2025.
https://doi.org/10.1007/978-3-031-78925-0_26

# 1 Introduction

The date fruit, renowned for its historical significance in the Middle East and North Africa for millennia, remains a vital agricultural and economic asset. Nonetheless, the misclassification of date fruit varieties continues to challenge the industry, resulting in significant production waste and market deceit. Efforts are underway to address this issue. A crucial factor in this endeavor is the accurate identification of various date fruit varieties to uphold market transparency. Moreover, ensuring precise date maturity classification, especially during pre-harvesting, is pivotal in preventing premature harvesting and reducing wastage. It's worth noting that distinguishing between different fruits, which often share similar characteristics, adds to the complexity of the situation.

In the domain of date palm research, the application of machine learning and deep learning techniques has gained traction, particularly in disease detection [2], harvesting optimization [4,5], and varietal identification [1,11,15]. Indeed, the emerging research in deep learning demonstrates promising advancements, showcasing enhanced accuracy and efficacy in classifying image data [6–9,17,18] for several applications and addressing the challenges of date palm cultivation and management.

The objective of this paper is focuses on enhancing date fruit variety identification through the utilization of advanced deep learning techniques. We employed a comprehensive set of CNN pretrained models, each offering unique design principles and capabilities. The VGG, developed by Simonyan and Zisserman [19] in 2014, features a 16-layer architecture and accepts images of size $224 \times 224$ pixels, with the final layer dedicated to the classification of over 1000 ImageNet objects. Notably, variations such as VGG16 and VGG19 have been extensively employed in image recognition tasks. The MobileNet, introduced by Howard et al. [12] in 2017, stands out for its lightweight and flexible design, making it particularly suitable for resource-constrained environments. This architecture operates on the principle of separable convolutional depths and encompasses three versions: MobileNetV1, MobileNetV2, and MobileNetV3. The InceptionV3, a successor to InceptionV1 and InceptionV2, was presented by Szegedy et al. [20] in 2016, featuring significant improvements including the use of the RMSProp optimizer, convolution factorization ($7 \times 7$), and a depth of 48 layers, with an input picture size of $299 \times 299$ pixels [21]. And the DenseNet, developed by Huang et al. [13] in 2016, emphasizes densely interconnected layers and employs the concept of direct connections. Notable variations include DenseNet121, DenseNet169, and DenseNet201.

We aim to address the challenges posed by limited data availability and the complex variability present in date cultivars by leveraging the capabilities of GAN to generate synthetic data, thereby enriching the dataset. Within the use of different data augmentation methods, DenseNet architecture performed consistently well across all approaches.

This paper is organized as follows. Section 2 provides an overview of the related works in date varieties identification. Section 3 describes our methodology. The results and analysis of our findings are presented in Sect. 4, followed by a comprehensive discussion in Sect. 5. Finally, Sect. 6 concludes the paper with a summary of the key findings and recommendations for future research.

## 2 Related Works

Within the domain of date fruit production, researchers have actively pursued the development of innovative systems aiming to automate key processes. Table 1 summarize the existing works in the field of date varieties identification.

**Table 1.** Summary of the related works on date varieties identification

| Paper | Methodology | Key Findings | Dataset | Accuracy (%) |
|---|---|---|---|---|
| Adnan et al. [1] (2020) | Attribute-based classification using SVM | Classification of 4 date fruit types based on color, size, and texture | No publicly available dataset: contains 325 images divided into 4 classes (Mabroom: 92, Sukkary: 81, Sagai: 47 and Ajwa: 105). | 73.88 |
| Koklu et al. [14] (2021) | Extracting 34 morphological features (shape, and color) which have been classified by a combination of LR and ANN models | A stacking model combining LR and ANN leading. | A 898 image dataset of 7 date fruit classes (Sagai: 94, DegletNour: 98, Safawi: 199, Sukkary: 204, RotabMozafati: 72, Ruthana: 166, and Barhee: 65 samples). | 92.80 |
| Alhamdan et al. [3] (2021) | CNN models with/ without data augmentation. | Adding noise layers which achieves higher accuracy of 97.00%. | A Saudian 1685 image dataset of 9 classes (NabtatAli: 177, Sokari: 264, Ajwa: 175, Galaxy: 190, Sugaey: 168, Medjool: 135, Meneifi: 232, Rutab: 146 and Shaishe: 171). | 97.00 |

In 2020, Adnan et al. [1] devised a classification system that categorized four types of date fruit based on color, size, and texture attributes, achieving a 73.88% accuracy using the SVM algorithm. Cordeiro et al. In 2021, Koklu et al. [14] proposed a classification method of 7 date fruit types. The research employed a computer vision system to procure 898 images of the mentioned date fruit types. Image processing methods were subsequently utilized to extract 34 pertinent features encompassing morphological aspects, shape, and color from the images. Initial model development LR and ANN methodologies, they obtained an accuracy of 91.0% and 92.2%. A stacking model was subsequently constructed by combining these models, resulting in an accuracy of 92.8%. In addition, Alhamdan et al.[3] proposed an application based on CNN for the identification of images of date fruits among 9 distinct varieties. They achieve an accuracy of 97.00%. In 2023, Cordeiro et al. [10] conducted a comparative study, employing five learning algorithms and observing a variation in the F-Measure from 83% to 96% across three datasets, emphasizing the significance of algorithm selection and data preprocessing techniques in date fruit classification. Ghulam [15] introduced an automated classification system for four date fruit types in 2014, integrating the local binary pattern and SVM, resulting in a 99% accuracy. The study encountered challenges in distinguishing between certain date types. Abdulhamid et al. [11] presented a classification system in 2012 that utilized 50 features and various classifiers, attaining accuracy ranging between 89% and 99%, with the artificial neural network emerging as the most accurate classifier.

To the best of our knowledge, the existing literature pertaining to the identification of date fruit varieties is notably scarce and offers limited insight into this specialized domain. In this study, we aim to address this gap by proposing an enhanced date palm varieties classification methodology. We compare various CNN architectures and leverage advanced deep learning techniques to accurately classify and differentiate 29 types of Tunisian date fruit cultivars.

## 3 Methodology

At first, our main goal was to tackle the difficulties caused by the small dataset. To get the greatest outcomes, our strategy concentrated on utilizing CNN architectures and hyperparameter optimization. We used data augmentation techniques to increase the size of the dataset, which gave us enough information for efficient learning. We sought to optimize pre-trained model expertise while customizing them for our dataset by utilizing transfer learning technology. Finding the ideal hyperparameters for the MobileNet, VGG19, InceptionV3, and DenseNet architecture was our main goal in order to guarantee reliable performance even with the restricted amount of data available.

### 3.1 Dataset

In this study, the raw data were collected at the Arid Area Institute of Kébéli, Tunisia. The obtained dataset encompasses 29 classes (KardEjmal, Kharrouby, KhaltAliEssassi, Tronja, GharsSouf, KhannegRouhah, GondaOuGondi, KhaletHmeid, Hissa, Hamra, KhaltDegloui, TezerzyetSafra, Horra, kessnarou, Yamyouli, KchedouAhmer, KchedouAbyadh, Zekri, Jeddadi, Fermla, Gosbi, Hloua, KhaltAbyadh, Tofli, Khadhra, KhaletGameh, Fezzani, Kenta, Gabsi) representing some of Tunisian date varieties, reflecting the rich diversity in this domain. The dataset initially consisted of 113 images, posing a significant challenge as each class was represented by only 3 or 4 samples. To overcome this limitation, we strategically employed data augmentation methods, effectively expanding the dataset to a substantial size of 1219 samples. Furthermore, the integration of the Generative Adversarial Network (GAN) allowed us to generate synthetic data, thereby enriching the dataset to a noteworthy 3200 samples.

Leveraging this combined dataset, which comprised the original, augmented, and GAN-generated data, significantly enhanced the robustness and generalization capabilities of our models, effectively overcoming the initial dataset constraints. The Fig. 1 depicts samples of original date fruit.

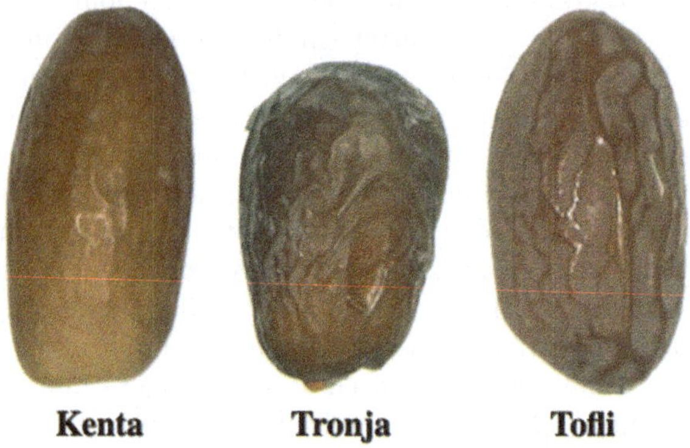


**Fig. 1.** Samples of initial data.

This paper provides an in-depth discussion of the qualitative and quantitative techniques used to augment the dataset, highlighting our efforts to address the limitations imposed by the dataset original size. Our goal is to offer valuable insights into the classification of Tunisian date palm varieties, all the while recognizing the inherent challenges posed by the initial dataset size.

### 3.2 Hyperparameters Selection

The hyperparameters utilized in our study were carefully selected to ensure optimal model performance and effective training. With 29 classes in consideration, the batch size was set to 32, enabling efficient processing while maintaining model stability. The training process was conducted over 100 epochs, allowing the models to sufficiently learn the intricate patterns and characteristics of the date fruit dataset. Moreover, the adoption of the Adam optimizer, coupled with a learning rate of 0.001, facilitated effective gradient-based optimization, contributing to the models robust convergence and accuracy. These hyperparameter configurations were integral to the successful development and training of our classification models.

### 3.3 Data Augmentation

One essential method for increasing the amount of data used for machine learning model training is data augmentation. In particular, deep learning models frequently require a large amount of training data in order to properly generalize and generate correct predictions. Nevertheless, it is not always possible to obtain a sufficiently large and diverse dataset. This problem can be solved by data augmentation, which increases the amount and diversity of the dataset by artificially adding variations of the existing data [16]. This procedure helps to achieve better predictive performance by providing models with broader information. In addition, the data augmentation methods can vary from basic ones like cropping and flipping to more complex ones like GAN [22].

* **Basic data augmentation**
  The data augmentation techniques employed in our study primarily involve image manipulation methods, which are detailed in Table 2. These fundamental image manipulations primarily revolve around image transformations, including operations such as rotation, flipping, cropping, and more. These techniques directly modify the images and are straightforward to implement [22]. The Fig. 2 shows some samples of data after fundamental data augmentation method.

**Table 2.** Basic data augmentation techniques

| Methods | Description |
|---|---|
| Rescale | Scales the pixel values to the range [0, 1] by dividing each pixel value by 255.0 |
| Rotation Range | Applies a random rotation in the range of [−20, 20] degrees |
| Width Shift Range | Shifts the image horizontally by a random fraction of its width within [−20%, 20%] |
| Height Shift Range | Shifts the image vertically by a random fraction of its height within [−20%, 20%] |
| Horizontal Flip | Randomly flips the image horizontally |

* **Advanced data augmentation (GAN)**
  adversarial networks are leveraged in data augmentation to generate synthetic data that closely resembles the original training data. By training a GAN on a dataset and then using the generator to create new data samples, it is possible to expand the dataset with realistic, artificially generated examples. Figure 3 shows some samples generated with this techniques. In our experiments, the GAN was configured with the following essential parameters: a batch size of 32, a latent dimension of 100, and a training duration of 100 epochs. Additionally, we set the generator learning rate and discriminator learning rate to 0.0002, respectively. These parameter values were crucial in ensuring the successful training and optimization of the GAN model for our study.

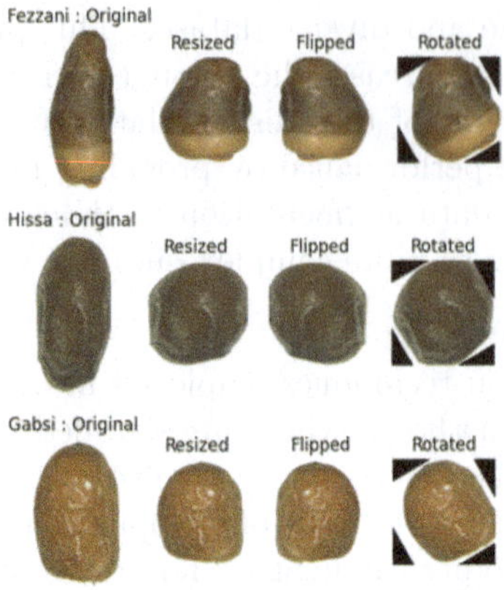


**Fig. 2.** Samples of date fruit images generated by basic data augmentation.

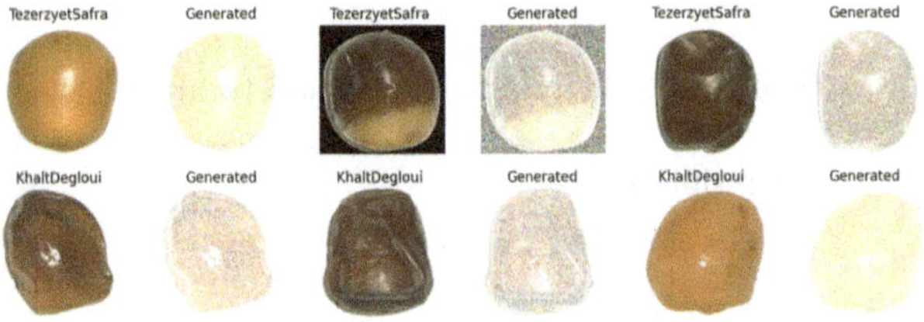


**Fig. 3.** Samples generated by GAN data augmentation.

## 4 Results

Our findings focus on the performance of each CNN architecture under various data augmentation strategies, with an emphasis on the impact of GAN-based augmentation and its combination with the basic data augmentation techniques.

**Table 3.** Summary of accuracy using different data augmentation techniques

| Data Augmentation | Architecture | | | |
|---|---|---|---|---|
| | MobileNet | VGG19 | InceptionV3 | DenseNet |
| Without data augmentation | 74.14 | 76.50 | 82.13 | 83.05 |
| Basic data augmentation | 93.02 | 65.03 | 64.02 | 77.59 |
| GAN data augmentation | 97.69 | 96.87 | 97.57 | 98.53 |
| Combined data augmentation | 99.57 | 80.06 | 97.22 | **99.89** |

The results depicted in Table 3 provide valuable insights into the performance of different architectures in our study. The MobileNet [12] architecture achieved a 99.57% accuracy with the combined data augmentation approach, showcasing its remarkable capabilities. On the other hand, the VGG19 [19] model demonstrated varying levels of accuracy across different methodologies, ranging from 65.03% with basic data augmentation to 96.87% with GAN data augmentation. Notably, the InceptionV3 [20] architecture showcased competitive performance, achieving 97.57% accuracy with GAN data augmentation. Furthermore, the DenseNet [13] model, presented in Fig. 4, performed consistently well across all approaches, reaching a peak accuracy of 99.89% with the combined data augmentation strategy. These findings emphasize the critical role of architecture selection and data augmentation techniques in optimizing the accuracy of date fruit classification models.

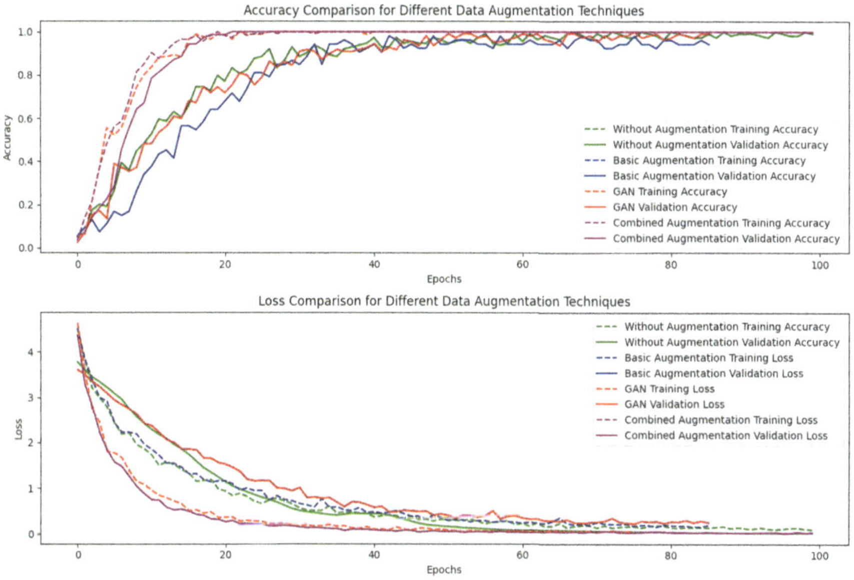


**Fig. 4.** DenseNet Accuracy and Loss Comparison for Different Data Augmentation Techniques.

## 5 Discussion

The results highlight the importance of selecting appropriate models and augmentation techniques for accurate identification. MobileNet showed robust performance across various augmentation strategies, handling the dataset complexity effectively. VGG19 varying performance indicated its sensitivity to data variations, requiring careful preprocessing.

InceptionV3 exhibited competitive performance, suited for intricate image recognition tasks. Similarly, DenseNet consistently captured complex patterns well. These findings emphasize leveraging advanced architectures and robust augmentation to enhance classification accuracy.

In summary, thoughtful model selection and effective augmentation are crucial for superior identification. Understanding each architecture strengths and weaknesses enables informed decisions, contributing to advancements in agricultural technology and computer vision.

## 6 Conclusion

Our research concludes by highlighting the significance of choosing appropriate architectures and efficient data augmentation techniques for precise classification tasks, particularly when distinguishing between various date fruit kinds. The findings underscore the significance of leveraging advanced models such as MobileNet, while also emphasizing the critical role of meticulous data preprocessing, as seen in the performance variations of the VGG19 model. The study also demonstrates how advanced architectures such as InceptionV3 and DenseNet can effectively capture complex patterns in the dataset. These observations offer insightful direction for future developments in agricultural technology and computer vision related to crop classification and analysis.

## References

1. Abi Sen, A.A., Bahbouh, N.M., Alkhodre, A.B., Aldhawi, A.M., Aldham, F.A., Aljabri, M.I.: A classification algorithm for date fruits. In: 2020 7th International Conference on Computing for Sustainable Global Development (INDIACom), pp. 235–239. IEEE (2020)
2. Al-Shalout, M., Mansour, K.: Detecting date palm diseases using convolutional neural networks. In: 2021 22nd International Arab Conference on Information Technology (ACIT), pp. 1–5. IEEE (2021)
3. Alhamdan, W., Howe, J.: Classification of Date Fruits in a Controlled Environment Using Convolutional Neural Networks, pp. 154–163. springer (2021)
4. Altaheri, H., Alsulaiman, M., Muhammad, G.: Date fruit classification for robotic harvesting in a natural environment using deep learning. IEEE Access **7**, 117115–117133 (2019)
5. Altaheri, H., Alsulaiman, M., Muhammad, G., Amin, S.U., Bencherif, M., Mekhtiche, M.: Date fruit dataset for intelligent harvesting. Data Brief **26**, 104514 (2019)
6. Ben Aoun, N., Mejdoub, M., Ben Amar, C.: Bag of sub-graphs for video event recognition. In: 39th IEEE International Conference on Acoustics, Speech, and Signal Processing (ICASSP'14), pp. 1566–1570 (2014). https://doi.org/10.1109/ICASSP.2014.6853857
7. Brahimi, S., Ben Aoun, N., Ben Amar, C.: Very deep recurrent convolutional neural network for object recognition. In: International Conference on Machine Vision (ICMV'2016). vol. 10341, p. 1034107 (2016). https://doi.org/10.1117/12.2268672
8. Brahimi, S., Ben Aoun, N., Ben Amar, C.: Improved very deep recurrent convolutional neural network for object recognition. In: IEEE International Conference on Systems, Man, and Cybernetics (SMC'2018), pp. 2497–2502 (2018). https://doi.org/10.1109/SMC.2018.00428

9. Brahimi, S., Ben Aoun, N., Ben Amar, C., Benoit, A., Lambert, P.: Multiscale fully convolutional DenseNet for semantic segmentation. J. WSCG **26**(2), 104–111 (2018)
10. Cordeiro, A.F., OliveiraJr, E., Costa, Y.M.: Date fruit classification using a wide range of classifiers. In: 2023 30th International Conference on Systems, Signals and Image Processing (IWSSIP), pp. 1–5. IEEE (2023)
11. Haidar, A., Dong, H., Mavridis, N.: Image-based date fruit classification. In: 2012 IV International Congress on Ultra Modern Telecommunications and Control Systems, pp. 357–363. IEEE (2012)
12. Howard, A.G., et al.: MobileNets: Efficient convolutional neural networks for mobile vision applications. arXiv preprint arXiv:1704.04861 (2017)
13. Huang, G., Liu, Z., Van Der Maaten, L., Weinberger, K.Q.: Densely connected convolutional networks. In: Proceedings of the IEEE Conference on Computer Vision and Pattern Recognition, pp. 4700–4708 (2017)
14. Koklu, M., Kursun, R., Taspinar, Y.S., Cinar, I.: Classification of date fruits into genetic varieties using image analysis. Mathematical Problems in Engineering (2021). https://api.semanticscholar.org/CorpusID:244063197
15. Muhammad, G.: Automatic date fruit classification by using local texture descriptors and shape-size features. In: 2014 European Modelling Symposium, pp. 174–179. IEEE (2014)
16. Mumuni, A., Mumuni, F.: Data augmentation: A comprehensive survey of modern approaches. Array p. 100258 (2022)
17. Nhidi, W., Ben Aoun, N., Ejbali, R.: Deep learning-based parasitic egg identification from a slender-billed gull's nest. IEEE Access **11**, 37194–37202 (2023). https://doi.org/10.1109/ACCESS.2023.3267083
18. Nhidi, W., Ben Aoun, N., Ejbali, R.: Ensemble machine learning-based egg parasitism identification for endangered bird conservation. In: 15th International Conference on Advances in Computational Collective Intelligence (ICCCI'2023). Communications in Computer and Information Science, vol. 1864, pp. 364–375 (2023)
19. Simonyan, K., Zisserman, A.: Very deep convolutional networks for large-scale image recognition. arXiv preprint arXiv:1409.1556 (2014)
20. Szegedy, C., et al.: Going deeper with convolutions. In: Proceedings of the IEEE Conference on Computer Vision and Pattern Recognition, pp. 1–9 (2015)
21. Szegedy, C., Vanhoucke, V., Ioffe, S., Shlens, J., Wojna, Z.: Rethinking the inception architecture for computer vision. In: Proceedings of the IEEE Conference on Computer Vision and Pattern Recognition, pp. 2818–2826 (2016)
22. Yang, S., Xiao, W., Zhang, M., Guo, S., Zhao, J., Shen, F.: Image data augmentation for deep learning: A survey. arXiv preprint arXiv:2204.08610 (2022)

# Intelligent System for Spectrum Detection and Selection in Cognitive Radio Networks

Daniel E. Asuquo[1,2(✉)], Uduak A. Umoh[1,2], Samuel A. Robinson[1], Emmanuel A. Dan[1], Samuel S. Udoh[1], and Kingsley F. Attai[3]

[1] Department of Computer Science, Faculty of Science, University of Uyo, Uyo, Nigeria
{danielasuquo,uduakumoh,samuelrobinson,emmanueldan, samueludoh}@uniuyo.edu.ng

[2] TETFund Center of Excellence in Computational Intelligence Research, UNIUYO, Uyo, Nigeria

[3] Department of Maths and Computer Science, Ritman University, Ikot Ekpene, Nigeria

**Abstract.** Most licensed spectrums in wireless communication systems are underutilized due to the fixed channel allocation policy. The limited availability of spectrum resource and great demand for spectrum access requires the deployment of techniques that enable efficient and timely utilization of available channels. With cognitive radio technology, secondary users (SUs) can use available spectrum holes of the primary users (PUs) without deteriorating the quality of service (QoS) of the PUs. Conventional spectrum holes detection and assignment methods consume a lot of time and energy power, which transpires to higher sensing error that increases spectrum handoff and negatively impacts data transmission. This paper develops a computational intelligence (CI)-based system that adaptively and dynamically manages spectrum resource in order to minimize spectrum handoff and sustain QoS. The developed system adopts Support Vector Machine (SVM) and Adaptive Neuro-Fuzzy Inference System (ANFIS) to predict spectrum holes availability and allocate best available channels to SUs. Then, a Java-based simulated network uses the output from the CI-based system and the interference temperature of selected channels in 2G, 3G and 4G spectrums to admit SUs' requests for diverse service demands. Results indicate that SVM achieves 98.8% accuracy in spectrum holes detection while ANFIS achieves 90.4% accuracy in channel allocation. The prediction accuracy and generalization capability of these models demonstrate ability to achieve significant improvements in spectrum utilization. Network operators can leverage on this result to maximize throughput and maintain QoS and fairness among users.

**Keywords:** Intelligent system · spectrum detection · channel allocation · cognitive radio network · machine leaning · java programming

## 1 Introduction

Spectrum is a scarce resource in wireless communications. However, the fixed channel allocation strategy prevents unlicensed users from accessing available frequency bands, causing the spectrum to be underutilized and sometimes unutilized by the licensed


A. Bajaj et al. (Eds.): HIS 2023, LNNS 1224, pp. 272–283, 2025.
https://doi.org/10.1007/978-3-031-78925-0_27

users. Nevertheless, the available frequency channels do not seem to meet the increasing users' demand [1, 2], for emerging versions of wireless applications [3, 4]. It has therefore become a challenge for researchers and telecommunication operators to determine mechanisms on how limited radio resources can be maximally utilized. One emerging technology to solving this problem is the cognitive radio (CR). This innovation uses software defined radio technology to intelligently adapt the dynamic radio environment, detecting which communication channels are occupied and which are not. This method allows CR users to access vacant channels for communication [5].

A cognitive radio network (CRN) includes a transceiver in which operating parameters such as frequency range, modulation type or maximum output power can be modified by software for dynamic spectrum allocation. Figure 1 shows the structure of a CRN indicating ad-hoc-based and infrastructure-based topologies with primary networks. The ad-hoc topology allows SUs to communicate with other users on a demand basis. The infrastructure-based topology allows communication through a cognitive base station. This technology permits SUs to access not just free channels in a specific spectrum but free channels of all spectrums contrary to known radio systems such as IEEE 802.11 wireless local area network, 2G, 3G, and 4G networks. Ideally, a CR system has two kinds of users namely, licensed or primary users (PUs) and unlicensed or secondary users (SUs). The system grants SUs the opportunity to either access or share unoccupied channels with a PU provided the SU's interference level does not exceed an acceptable threshold called interference temperature. The first access technique is called interweave. Here, the SU is asked to leave the channel to another available channel once a PU arrives. This method causes spectrum mobility with severe throughput degradation. The second access technique is called underlay. This scheme allows SUs to access any portion of the spectrum but restricts their transmission power from causing harmful interference to the PUs.

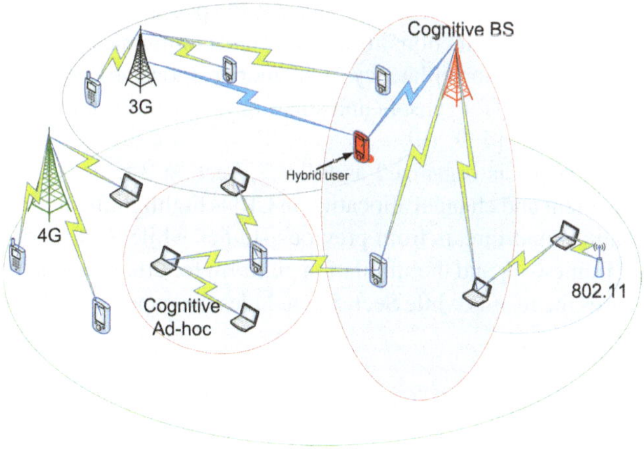


**Fig. 1.** A typical CRN architecture

However, existing spectrum detection techniques attempt to determine the occupancy status of PUs but come with numerous problems such as energy depletion, sensing delay, and unfairness. These schemes cannot guarantee seamless communication due to higher miss detection probability, incessant spectrum mobility, and service quality degradation. Spectrum sharing decision that aims to ensure efficient allocation of the available whitespaces to SUs within a given geographical location and at a given period of time becomes a challenge. Modeling real-world systems like radio access networks is difficult due to lack of exact and complete knowledge representation of the domain. Such systems are characterized with uncertainties, vagueness, nonlinearity and unanticipated changes in their environment. Hence, there is need for advanced decision making systems that can achieve certain acceptable performance in the midst of uncertainties and changes in system operating conditions. Although techniques for dealing with uncertainty in CR networks including models for good decision making and channel allocation exist [6–8], machine learning (ML) models that integrate artificial intelligence (AI) and computational intelligence (CI) techniques are widely accepted as they explore the construction of algorithms that can learn from and make predictions on data. Such algorithms operate by building a model from example inputs in order to make data-driven predictions or decisions, rather than following strictly static program instructions.

In order to mitigate the spectrum scarcity problem, this paper develops a computational intelligence (CI)-based framework for accurate spectrum holes detection and efficient channel allocation in CRNs. The essence is to facilitate efficient data transmission and minimize spectrum mobility including undesirable ping-pong effect. Other significant outcomes of this implementation include fairness among users, improved spectral efficiency and network throughput. The major contributions of this paper include to (i) develop a CI-based framework based on support vector machine (SVM) and adaptive neuro-fuzzy inference system (ANFIS) for accurate spectrum sensing and efficient spectrum decision functions in CRN; (ii) implement a simulated network in Java for the developed framework in (i); and (iii) evaluate the performance of the CI models and the CRN based on accuracy and error estimates as well as scenarios that assess fairness, spectral efficiency and spectrum handoff. This has the potential to sustain the assurance of 5G network to bring new functionalities to wireless communications, namely; greater speed (to move more data), lower latency (to be more responsive), and ability to connect a lot more devices at once (for sensors and smart devices) and suitability for Internet of Things (IoT) applications [4, 9, 10].

The rest of the paper is organized as follows. Section 2 reviews related works on spectrum management and channel allocation in CRNs highlighting the strengths, weaknesses, opportunities and threats from previous studies while Sect. 3 presents the proposed CI-based framework and the simulation platform for efficient spectrum allocation. Section 4 discusses the results while Sect. 5 concludes the paper with direction for future works.

## 2 Related Works

The widespread demand for real-time voice and data applications over mobile networks require that service providers guarantee a well-defined quality of service (QoS) for mobile users. Evaluating the QoS in CRNs to determine which primary network provides the best available channels for allocation to SUs is a challenging task, especially where an SU is not expected to significantly interfere with a PU's signal. The complex nature of mobile networks with multi-criteria and conflicting factors makes good decision making difficult [11]. CR systems are designed to coordinate the utilization of radio bands in a self-organizing manner. Essentially, a CR user is expected to recognize unused radio spectrum of the PU and opportunistically utilize the spectrum in an intelligent way. This new approach in wireless communications enhances radio resource management by detecting users' communication needs as a function of use context [12]. This intelligent communication system learns from its surrounding environment to adapt its internal states to dynamic changes in the radio frequency by adjusting certain operating parameters in order to achieve reliable communication and efficient utilization of radio spectrum [13].

Thus, instead of considering all the parameters that a wireless node or network can be aware of, spectrum-sensing CR detects the channels in the radio frequency spectrum. Sensing a spectrum at the physical layer implies adapting modulation schemes and other parameters for detecting and measuring the signals transmitted by PUs from different channels whereas media access control (MAC) layer sensing is responsible for scheduling time slots for SUs to search a particular channel for data transmission [14, 15]. Compared to the physical layer, MAC layer sensing has not received much attention and it is important for proper coordination and scheduling of SUs for optimal utilization of spectrum resource. This will facilitate spectrum holes detection, precise link estimation between nodes, fast and accurate frequency control, and power control that guarantees reliable communication between CR terminals including non-interference with the PUs. These functionalities require the CRN to be dynamically programmed according to changes in the radio environment. Apart from spectrum sensing, other important functions of a CR system include spectrum decision, spectrum sharing and spectrum handoff [16]. These spectrum management functions are contained in the CR terminal with possible transitions between them. Spectrum decision helps to select the best detected channel through the sensing phase, according to specific CR user's requirements (such as usage time, QoS, width of spectrum band). Spectrum sharing is the allocation of the selected spectrum band to the SU for coexistence with the PU. Spectrum handoff occurs when a CR terminal switches to another channel when agreement to share current channel with a PU terminates.

## 2.1 Spectrum Sensing Techniques

According to [4], the ability to identify PUs' presence is a cost-efficient way to detect unoccupied spectrum. The work in [17] grouped spectrum sensing techniques into three categories namely, non-cooperative detection, cooperative detection, and interference-based detection. The non-cooperative detection uses the transmitter detection techniques to sense the PU's signal which results in two sensing states, namely $Ho$ and $H_1$, where $Ho$ and $H_1$ signify the absence and presence of PU's signal, respectively. Signal specific and blind sensing methods are the two examples of non-cooperative technique where the former is known to cause energy depletion as the energy needed to operate the device is wasted for sensing functions thereby affecting the device's and network's lifetime. Also, higher missing and false detection probabilities are likely occurrences with this method. The later method does not permit the CR device to learn from the environment before taking channel switching decision. In cooperative sensing, multiple CRs work together to provide information to detect a PU and can be accomplished in a centralized or distributed arrangement [9]. Though it increases the probability of PU detection, it is not capable of detecting a second transmitting radio in the locality of the receiving radio due to shadowing or distance effects, resulting in the hidden terminal problem. Interference-based detection simply concentrates on measuring interference at the receiver using interference temperature model. This method sets the interference temperature as the tolerable limit of interference acceptable from SUs, accounting for cumulative radio frequency (RF) energy. This allows SUs to use a particular spectrum band as long as their transmissions do not exceed this limit. Nevertheless, [18] observed that apart from the interference temperature., the channel transmission range, channel transmission rank, transmission power and signal-to-interference ratio (SINR) of SUs are other important parameters to be considered when modeling spectrum management functions in CRNs.

The ability to control the transmission power permits the SU to simultaneously share the licensed channel for its transmission with acceptable QoS requirements. However, if no PU is present, the SU may transmit with maximum power to improve reception probability and improve the QoS for its transmissions. With SINR, the minimum value of transmit power required by the SU to reduce possible interference with the PU can be established. The transmission range significantly impacts mobility functions when an SU switches channel during spectrum handoff. Finally, the channel transmission rank provides an index indicating transmission channels sensitive to PU activity so that SUs can avoid them. This parameter makes available channels that are more stable, reduce unwanted interference with PUs, and minimize packet loss due to PU-SU collision. Numerous studies on spectrum sensing, channel allocation and sharing based on hidden Markov model, evolutionary algorithms, game theory, matching theory, and fuzzy logic methods abound [6, 19–22]. However, the opportunities and challenges posed by the exponential growth of interconnected objects and the increasing demand for wireless spectrum for data transmission with guaranteed QoS requirements, especially for IoT-enabled mission-critical applications and services has not yet been exhaustively explored. Tarek et al. [23] surveyed the problem of spectrum allocation and scheduling for CR-IoT. They concluded that IoT smart networks have to perform various functions and need integration of the CRN's capabilities in order to avoid the bandwidth scarcity issue. In [8], the greedy algorithm was used to improve channel allocation to priority based SUs. Here, the stable sub-bands are allocated to delay-sensitive multimedia traffic users. Simulated results indicate that the algorithm could reduce interference and maximize throughput through better allotment of unutilized licensed bands. Similar to the [24], the work in [20] proposed a priority-based SU call admission and channel allocation scheme aimed at reducing the blocking probability of higher-priority SU calls while maintaining a sufficient level of channel utilization. Markov chain model was used to estimate the arrival rates of SU calls of all priority classes and for dynamic channel reservation. Although this scheme performed better than the greedy algorithm, it could not achieve fairness among SUs as it increases the call dropping probability of low-priority SUs. Furthermore, [22] used Markov's secret mode to implement free channel identification and allocation in CRN-IoT. The authors aimed at using CRN to provide IoT connectivity to a huge number of linked IoT devices opportunistically, supporting low-energy IoT architecture for efficient spectrum utilization. The CR-IoT network was simulated in MATLAB, achieving lower switching delay.

[1] introduced a game theoretical-based model for efficient spectrum sharing among SUs using the Nash Equilibrium. The approach demonstrated that channel utilization with the SUs increased by 17.5%. Game-based solutions are complicated and more challenging as the objective of each CR user is to maximize its individual utility without considering fairness to others. A novel method based on particle swarm optimization for determining best position for CR base station and selection of optimal operating channel in IoT sensor networks was proposed in [19]. The method maximized the secondary system's capacity without deteriorating the QoS of the primary network compared to a random resource allocation algorithm. The work in [25] aimed to optimize spectrum management in CRNs using fuzzy logic. Their objective was to reduce interference and the number of channel switches while sustaining QoS. The fuzzy logic controller significantly managed the co-existence between PU and SU, minimizing interference, guaranteeing QoS, allowing the SU to select the best available channel and minimize the channel switching rate. The need to apply matching theory is due to the two-sided nature of CRNs in which licensed PUs own the channels that must be accessed by the unlicensed SUs. However, it is not easy to provide a centralized optimization solution in CRNs since PUs and SUs often belong to different operators and cannot be centrally controlled. Leshem et al. [26] presents one of the first works in this area, where a one-to-one matching problem is formulated between a number of SU and a number of PU channels. The preferences of both users and channels are based on the same utility function which primarily captures the rate of transmission.

The widespread adoption of ML and AI models for classification and prediction tasks in diverse disciplines informs their application in this study for spectrum holes prediction and allocation of best available channels to SUs for coordinated coexistence with PUs in CRNs. CI systems are intelligent computing solutions that have human-like capabilities to make decisions and future predictions with data cloaked with uncertainties and imprecision. They have the ability to learn and adapt, making them capable to dynamically change or evolve their parameters or structures in order to best meet their objectives [27]. A summary from reviewed literature reveals that a better optimization technique is needed to enhance accurate channel state prediction for a successful channel allocation in CRNs. The current paradigm does not guarantee equal and effective spectrum access that takes into account the CR-IoT multi-channel environment to ensure fairness among SUs and PUs.

# 3 Methodology

## 3.1 The Proposed System Architecture

The architectural design of the proposed CI-based spectrum detection and selection system is presented in Fig. 2. It has the following functional components, namely, Primary Network (PN), SU Database, Spectrum Hole Detection, Spectrum Selection and Channel Allocation, Performance Evaluation and Results Visualization.

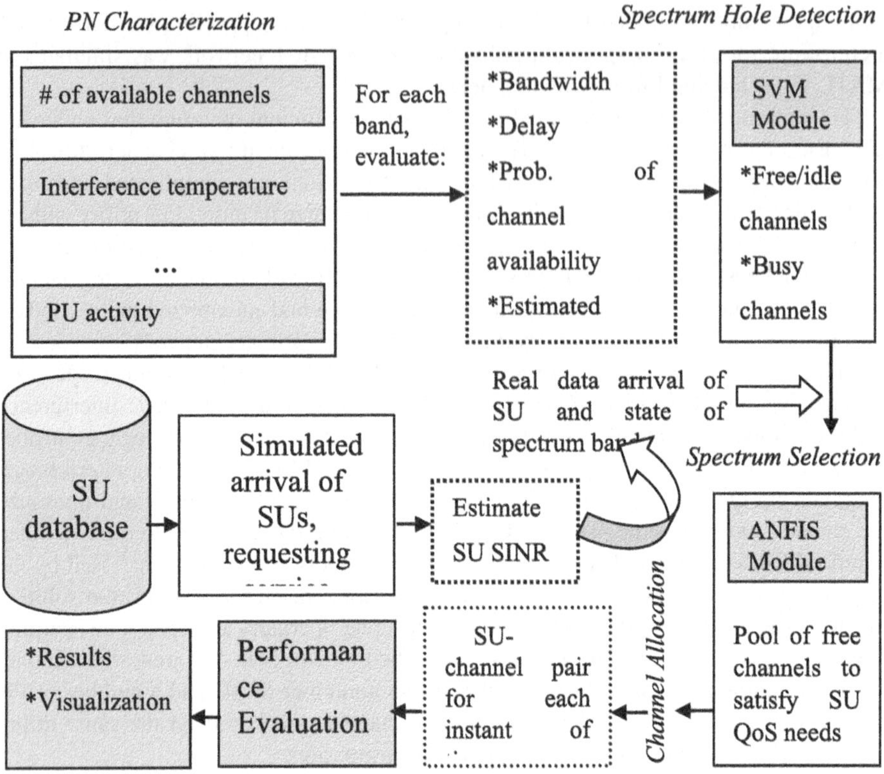


**Fig. 2.** Architecture of CI-based Spectrum Detection and Selection System

Initially, each spectrum band (2G, 3G, 4G) is characterized for the spectrum decision task based not only on local observations of CR users but also on statistical information of the primary networks. CR users exploit these local measurements, to estimate the channel conditions such as number of available channels (bandwidth), SINR, switching delay, probability of channel availability, and estimated available probability. In order to describe the dynamic nature of CRNs, a new parameter, PU activity, defined as the probability of the PU appearance during the CR user transmission is introduced. After the spectrum characterization and detection of whitespaces by SVM, the CRN chooses the best spectrum band through the ANFIS model using the multi-spectrum transmission technology. This decision process is modeled as an optimization problem. The proposed spectrum decision framework determines a set of spectrum bands by considering the QoS requirements of SUs' applications (contained in the SU database) as well as the dynamic and varying nature of the spectrum bands. Moreover, a dynamic admission control scheme is developed to decide on the spectrum bands adaptively dependent on the time-varying CRN's capacity. Spectrum handoff refers to a network-assisted procedure invoked on the CR users when they wish to transfer their connections to another spectrum band. This mobility scenario occurs when a PU appears or when current spectrum condition becomes worse.

The performance of these models was evaluated using error measures. Java with its software development toolkits was used to develop the system which was tested with data generated using different traffic intensities, representing the percentage utilization of the channels. Java runs on millions of personal computers worldwide including mobile devices, and is fast, secure, reliable, and portable.

The proposed approach is easy to implement and is not computationally complex due to its dependent on the interference temperature. It also facilitates the deployment of IoT applications on various use-cases. The SVM training process consists of finding a decision function capable of separating the spectrum bands making use of the optimization function [28]. The spectrum of frequencies with the greatest availability probability, estimated availability time and low delay are labeled as channels suitable for supporting the transmissions of real time (RT) applications (VoIP, streaming), while the rest of the points are labeled as channels for the transmission of best effort (BE) applications (messaging, web browsing, downloading). It is expected that the spectrum allocation process delivers the best quality channels to RT services as such SUs have higher priority than the BE SUs. However, if the range of frequencies in a category is exhausted, an SU may use the spectrum of the other category, subject to, in the worst case, a reduction in QoS.

The input parameters to the ANFIS model include available bandwidth (BW), availability probability (AP), estimated availability time (ET), delay (DL), and SINR while probability of channel selection (CS) is the output variable. The model adopts a five-layer ANFIS structure where the channel characteristics are found in the input layer; the membership functions are defined in the next layer (high and low) for each input; the third and fourth layers calculate the rules and establish the Takagi-Sugeno type inference; the fifth layer performs the summation of the outputs in order to obtain the scores for each channel and determine the ranking. The membership functions are defined in the same way for each input; the function that represents the low level has a maximum of 0, and the function that represents the high level has a maximum of 1, so that when the channel data are entered and then evaluated in the membership functions, coherent results are obtained. The decision was made to work with only two linguistic labels in order to limit the amount of computing resources required by the system, and thus reduce the algorithm's compilation time. The output error is used to adapt the premise parameters by means of a standard back-propagation algorithm to minimize the root mean squared-error cost function defined as:

$$\mathrm{RMSE} = \frac{1}{2}\sum \left\| t^i - a^i \right\|^2 \tag{1}$$

where, $t^i$and $a^i$ are the target output and the actual output, respectively. The hybrid learning algorithm is applied and structured by defining linear and nonlinear parameters illustrated at each iteration (epoch), where the gradient descent method updates the nonlinear parameters, while the least square method follows to identify the linear parameters.

## 4 Discussion of Results

A dataset comprising 700 data points was split into two where 70% was used for training the models while the remainder was used for testing. The SVM model yielded a satisfactory spectrum hole detection performance with an accuracy of 98.8%. The confusion matrix is represented in Table 1. Figure 3 indicates that ANFIS achieves a RMSE value of 0.095933 on the test data in the task of allocating best available channels to SUs. This shows that a prediction accuracy of 90.4% was obtained after learning from the trained data.

**Table 1.** Test performance of the SVM model

| | | | |
|---|---|---|---|
| Actual Class | 0 | 247 | 4 |
| | 1 | 2 | 237 |
| | | 0 | 1 |
| | | Predicted Class | |

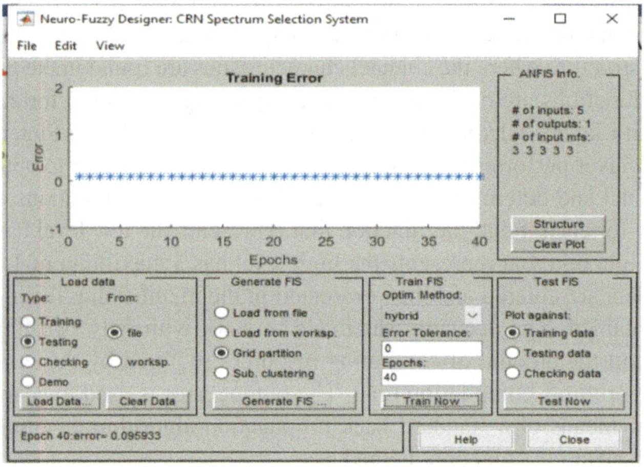


**Fig. 3.** Test performance of the ANFIS Model

### 4.1 Simulation Environment and Results

The simulated CRN presents 'START' and 'STOP' buttons for the user to click for commencement and ending of the experiment, simultaneously. After the 'STOP' button is pressed, the output is displayed with associated performance reports. The system logic has been programmed in Java object-oriented programming language to handle the SVM and ANFIS operations. Thus, a service request from an SU is accepted and channel allocated for the SU for a certain duration. The channel is released back to the system on service completion. However, the interference temperature must not be exceeded; otherwise the SU's request is rejected.

Figure 4 presents simulation results of the system for different observable scenarios. It indicates channel assignment different spectrums for VoIP, streaming, downloading and messaging services. This system is therefore able to intelligently and fairly allocate free frequency channels to SUs without interfering with PU traffic. Furthermore, it illustrates a case where an SU with identification SU7264702 having SINR greater than the interference temperature for 4G spectrum was rather assigned to 3G spectrum. Results indicate that the framework and the developed system can ensure fairness among SUs, reduce unnecessary interference, minimize delay and improve spectral efficiency and network throughput. A channel allocation accuracy of 90.4% implies that CR devices in 5G wireless communication networks can intelligently configure different parameters such as frequency band and transmit power based on the surrounding environment and consequently exploit under-utilized spectrum portions for different service demands, while avoiding annoying bottlenecks.

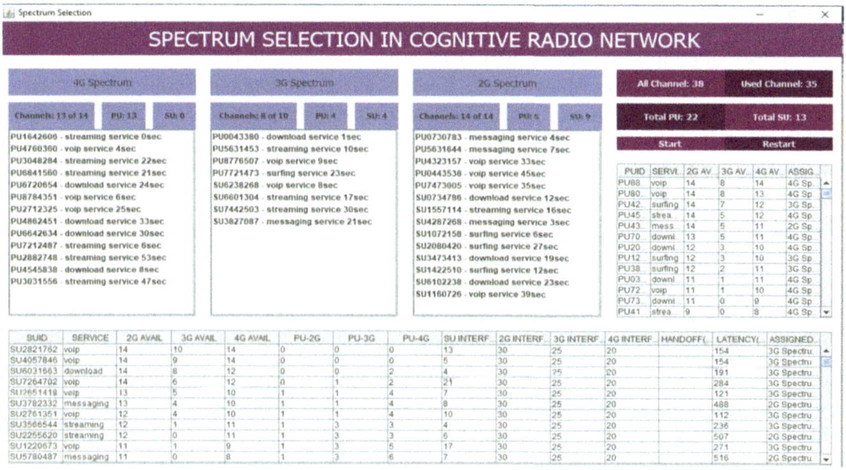


**Fig. 4.** Channel allocation when SU SINR > Interference temperature (21 > 20)

## 5 Conclusion and Future Directions

This paper presents a proactive approach for handling channel selection and allocation in CRNs. The proposed CI-based framework anticipates the arrival of SUs so that channels to be assigned in future are reserved in advance, using a spectrum classification system based on SVM and ANFIS methodologies. Data transmission was prioritized in 2G, 3G and 4G spectrums. This approach achieved probability of best channel selection with 90.4% accuracy, translating to the ability to improve CRN performance through maximum spectrum utilization and reduced spectrum handoff. It can be deployed for various IoT-based applications. Future works shall consider the simultaneous measurement of spectrum parameters in more than one geographical location within the same period of time to provide a wider view of spectrum occupancy estimates that can facilitate CRN implementations. The collation and analysis of data obtained from independent measurements will give a more accurate estimate that can be used to identify under-utilized bands with reference to frequency, time and spatial domains.

## References


1. Kumar, V., Dhurandher, S.K., Woungang, I., Gupta, S., Singh, S.: Channel allocation in cognitive radio networks: a game-theoretic approach. In: Barolli, L., Miwa, H., Enokido, T. (eds.) Advances in Network-Based Information Systems. Lecture Notes in Networks and Systems, vol. 526. Springer, Cham (2022). https://doi.org/10.1007/978-3-031-14314-4_18
2. Cai, B., Zhang, Q., Ge, J., Xie, W.: Resource allocation for cognitive leo satellite systems: facilitating iot communications. Sensors. **23**(3875), 1–19 (2023). https://doi.org/10.3390/s23083875
3. Giral, D., Hernández, C., Salgado, C.: Spectral decision for cognitive radio networks in a multi-user environment. Heliyon. **7**(5), 1–16 (2021)
4. Khalid, W., Yu, H.: Sum utilization of spectrum with spectrum handoff and imperfect sensing in interweave multi-channel cognitive radio networks. Sustain. For. **10**, 1–18 (2018)
5. Nathani, N., Khatri, V.K., Manna, G.C.: Cognitive radio networks-based internet of things: a survey on spectrum sharing or spectrum allocation schemes. Int. Res. J. Eng. Technol. **8**(5), 1082–1092 (2021)
6. Bharatula, S.: Efficient channel allocation schemes for cognitive radio networks. Int. J. Sci. Eng. Res. **5**(4), 23–42 (2014)
7. Bernal, C., Hernández, C.: Intelligent decision-making model for spectrum access in cognitive wireless networks. Contemp. Eng. Sci. **10**(15), 721–738 (2017)
8. Salahdine, F., Kaabouch, N., Ghazi, H.: Techniques for dealing with uncertainty in cognitive radio networks. In: 7th IEEE Annual Computing and Communication Workshop and Conference, Las Vegas, United States (2017)
9. Mi, Y., Lu, G., Li, Y., Bao, Z.: A novel semi-soft decision scheme for cooperative spectrum sensing in cognitive radio networks. Sensors. **19**, 1–12 (2019)
10. Ekpenyong, M.E., Asuquo, D.E., Udo, I.J., Robinson, S.A., Ijebu, F.F.: IPv6 routing protocol enhancements over low-power and lossy networks for iot applications: a systematic review. New Rev. Inf. Netw. **27**(1), 30–68 (2022). https://doi.org/10.1080/13614576.2022.2078396
11. Vithalani, A.A., Vithalani, C.H.: Application of combined TOPSIS and AHP method for spectrum selection in cognitive radio by channel characteristic evaluation. Int. J. Electron. Commun. Eng. **10**(2), 71–79 (2017)
12. Mitola, J.: Cognitive radio: an integrated agent architecture for software defined radio - Dissertation, Doctor of Technology, Royal Institute of Technology (2000)
13. Tabaković, Z.: Survey of cognitive radio systems, Croatian Post and Electronic Communications Agency, Jurišićeva 13, Zagreb, Croatia (2002)
14. Tlouyamma, J., Velempini, M.: Channel selection algorithm optimized for improved performance in cognitive radio networks. Wirel. Pers. Commun. **119**(4), 3161–3178 (2021). https://doi.org/10.1007/s11277-021-08392-5
15. Varalakshmi, S., Shanmugavel, S.: Adaptive MAC layer spectrum sensing algorithms for cognitive radio networks. Int. J. Adv. Inf. Sci. Technol. **30**(30), 45–52 (2014)
16. Xing, X., Jing, T., Cheng, W., Huo, Y., Cheng, X.: Spectrum prediction in cognitive radio networks. IEEE Trans. Wirel. Commun. **20**(2), 90–96 (2013)

17. Robinson, S.A., Asuquo, D.E.: Techniques for spectrum sensing and allocation in cognitive radio networks: a comprehensive review. J. Niger. Assoc. Math. Phys. **47**, 199–208 (2018)
18. Ali, A., Abbas, L., Shaq, M., Bashir, A.K., Afzal, M.K., Liaqat, H.B., Siddiqi, M.H., Kwak, K.S.: Hybrid fuzzy logic scheme for efficient channel utilization in cognitive radio networks. IEEE Access. **7**, 24463–24476 (2019)
19. Wen, J., Yang, Q., Yoo, S.-J.: Optimization of cognitive radio secondary information gathering station positioning and operating channel selection for IoT sensor networks. Mob. Inf. Syst. (2018). https://doi.org/10.1155/2018/4721956
20. Ali, A., Feng, L., Bashir, A.K., Shaker, S.H., Ahmed, S.H., Iqbal, M., Raja, G.: Quality of service provisioning for heterogeneous services in cognitive radio-enabled internet of things. IEEE Trans Netw Sci Eng. **7**(1), 328–342 (2020). https://doi.org/10.1109/TNSE.2018.287 7646
21. Asuquo, D., Ekpenyong, M., Udoh, S., Robinson, S., Attai, K.: Optimized channel allocation in emerging mobile cellular networks. Soft. Comput. (2020). https://doi.org/10.1007/s00500-020-04947-z
22. Leena, K., Hiremath, S.G.: Efficient channel allocation for cognitive radio internet of things. Turk. J. Comput. Math. Educ. **12**(10), 3476–3482 (2021)
23. Tarek, D., Benslimane, A., Darwish, M., Kotb, A.M.: Survey on spectrum sharing allocation for cognitive radio networks internet of things. Egypt. Inform. J. **21**, 231–239 (2020). https://doi.org/10.1016/j.eij.2020.02.003
24. Williams, E.E., Asuquo, D.E.: An efficient model for reducing soft blocking probability in wireless cellular networks. Int. J. Wirel. Mob. Netw. **6**(4), 85–99 (2014)
25. Mimbeu, Y., Matanga, J., Malong, Y., Ngalle, A., Doko, J.T., Essiben, J.F., Sévérin, N., Raphaël, O., Breunon, I.: Optimization of spectrum management in a cognitive radio network using fuzzy logic. Int. J. Eng. Res. Technol. **11**(6), 340–350 (2022)
26. Leshem, A., Zehavi, E., Yaffe, Y.: Multichannel opportunistic carrier sensing for stable channel access control in cognitive radio systems. IEEE J Sel Areas Commun. **30**(1), 82–95 (2012). https://doi.org/10.1109/JSAC.2012.120108
27. Xie, J., Song, Z., Li, Y., Zhang, Y., Yu, H., Zhan, J.: A survey on machine learning-based mobile big data analysis: challenges and applications. Wirel. Commun. Mob. Comput. **19** (2018)
28. Asuquo, D., Umoren, I., Osang, F., Attai, K.: A machine learning framework for length of stay minimization in healthcare emergency depart ment. Stud. Eng. Technol. J. **10**(1), 1–17 (2023). https://doi.org/10.11114/set.v10i1.6372

# A Comparison Between Traffic Classification Models for Bandwidth Management in Software-Defined Networks

Nilton J. Mocelin(✉), Rafael S. Parpinelli, and Adriano Fiorese

Department of Computer Science (DCC), Santa Catarina State University (UDESC), Santa Catarina, Brazil
nilton.junior@edu.udesc.br, {rafael.parpinelli,adriano.fiorese}@udesc.br

**Abstract.** Applications have different requirements in terms of bandwidth and delay to deliver the expected quality of service. For this and other reasons, new paradigms have emerged, such as software-defined networks, which allow the development of new applications for dynamic programming of routing devices in the network through a device called controller. In this work, five machine learning models for classifying network traffic are compared, which when implemented in the controller make it possible to discover the types of applications and their quality of service requirements. The obtained results demonstrate that the Random Forest model has an excellent performance in the classification of network traffic using few packets and achieves very high accuracy margins. On the other hand, it was noted that machine learning models based on streaming data are very sensitive to the dataset and the amount of packets used in the classification.

**Keywords:** Traffic Classification · Quality of Service · Machine Learning · Software-Defined Networks

## 1 Introduction

The term quality of service (QoS) can be defined by the characteristics of a telecommunications service that ensure the ability to meet the user's needs, both announced and implied requirements [8]. When these resources are not managed, their proper distribution cannot be guaranteed, leading to the detriment of some services.

Providing network resource management, especially bandwidth management, also brings various benefits to network administrators [7]. It can ensure that critical services are not affected when the network is congested. Additionally, bandwidth management helps network administrators allocate available network resources properly and avoid unnecessary infrastructure expenses.


A. Bajaj et al. (Eds.): HIS 2023, LNNS 1224, pp. 284–294, 2025.
https://doi.org/10.1007/978-3-031-78925-0_28

Conventional networks do not implement mechanisms to guarantee the quality of service requirements for application flows, resulting in competition for available network resources. The Internet is composed of independent systems [8]. Each autonomous system is responsible for traffic management and routing protocol usage within its domain. However, the interest in increasingly advanced applications and the emergence of new network paradigms like Software Defined Network (SDN) seek to change this landscape.

Thus, to address dynamic resource provisioning and the prioritized delivery needs of various applications, there are several frameworks like Flowing with Priority in SDN (FLOWPRI-SDN) [11]. This controller is an implementation of the Ryu SDN controller that aims to provide hierarchical resource reservation and traffic prioritization for flows that declare their requirements through a QoS contract. The framework can create specific forwarding rules to handle QoS flows within its domain, as well as discover other frameworks in other domains and distribute QoS contracts, being compatible with networks or domains that do not support it. However, FLOWPRI-SDN cannot identify flows and their requirements within its domain without the sender host announcing this information.

Therefore, this article aims to compare machine learning models to classify network traffic and identify bandwidth requirements. Data streams in application packets on the network can be interpreted as blocks or streams. Thus, there is a need to compare machine learning models that can work with blocks or flows in an SDN-like scenario. Learning models like Random Forest are based on decision trees and are used to classify blocks of data. On the other hand, Hoeffding Tree models (also a variation of decision trees) focus on streaming data. Then, the best classification models can be implemented in the FLOWPRI-SDN network traffic classification and QoS requirements identification processes.

The rest of the article is organized as follows. Section 2 presents related works and approaches used in network traffic classification. The methodology for obtaining and processing the datasets is presented in Sect. 3, and the machine learning models used are briefly introduced. Experiments and results are presented in Sect. 4, concluding with the work's conclusions in Sect. 5.

## 2 Related Works

Port-based classification was one of the most commonly used techniques in the past, as many applications used fixed port numbers assigned by the Internet Assigned Numbers Authority (IANA). However, over time, limitations of this approach began to appear. Numerous applications do not have registered port numbers, and many of them use dynamic port negotiation mechanisms to evade firewalls and network security tools. Moreover, the use of IP layer encryption has made discovering the original port numbers impossible [3].

The work presented in [15] evaluates the Support Vector Machine (SVM), GaussianNB, and Nearest Centroid models to identify HTTP, email, streaming, and VLC applications in SDN networks. However, they use the entire flow for classification and have a similar goal of adapting to a control framework. The work presented in [3] also aims to classify traffic using feature-based machine learning. Among the traffic classes are HTTP, YouTube, Skype, and Facebook. The models used were Random Forest, Stochastic Gradient Boosting Accuracy, and Extreme Gradient Boosting Accuracy, with Random Forest achieving the best results. The goal was also to use them in an SDN controller.

The work presented in [13] integrates a classification model for end-to-end encrypted traffic with a one-dimensional convolutional neural network model, which takes 784 bytes of each flow for classification. Likewise, several other recent works evaluate Deep Learning models for traffic classification, as presented in [9,16], however, these models are generally complex, require many computational resources and take a long time to train.

The related works provide some context for traffic classification. Optimizing a model and implementing it on an SDN controller is a common direction, but few define their objectives and use methods to generate adjusted subflow (valid flow) databases or define usage strategies for their models. Additionally, the works use either the entire flow or an excessive number of packets for classification, which hampers identifying short-duration flows and degrades network performance. To the best of the authors' knowledge, this work is the first to compare batch classification methods with methods adapted for streaming data in traffic classification, since network packets can be treated in both ways.

## 3 Development Methodology

In a TCP/IP protocol-based network, packets are responsible for communication between two devices [21]. A device sends data via packets to another device guided by the quintuplets: source and destination IP address, source and destination ports, and protocol type. Therefore, analyzing the characteristics and requirements of applications that are communicating over the network involves obtaining packets from these applications passing through a monitored path for a specific period of time. [21].

The ISCX-VPN-NonVPN-2016 dataset consists of captures of real network traffic packets (*pcap* files) with a wide variety of labeled application types [17]. This dataset was generated by capturing traffic from two computers, referred to as Bob and Alice, which were used to access various services such as Skype, Facebook, and others, as well as traffic over a Virtual Private Network (VPN).

In this work, only a portion of the generated dataset was used. Table 1 presents the capture files used and the rounded quantity of packets identified for each application type. The applications were grouped into service classes taking into account the nature of each application through searches on their websites, which makes processing easier.

Table 1 also summarizes a study of the bandwidth characteristics of service classes. Most applications do not clearly provide their requirements; however, other means of identifying this information exist. One way is by analyzing the network traffic of each application. Another way is to rely on information from

**Table 1.** Bandwidth and application label defined.

| Class of Service | Application | Flows | Pkts | Bandwidth | Label |
|---|---|---|---|---|---|
| Real-time Audio | VoipBuster, Skype, Hangouts | 5588 | 796.964 | 55–90kbps (codec) 100kbps (skype) | ar-100kbps |
| Static Audio | Spotify | 176 | 26.812 | 1-2Mbps | ae-2Mbps |
| Real-time Video | Skype, Hangouts | 580 | 978.291 | HD 1.5Mbps (skype) SD 300–500kbps (skype) | vr-1.5Mbps |
| Static Video | YouTube, Netflix | 657 | 370.356 | (Youtube) 1080p,30fps = 5 Mbps 720p,30fps = 2.5 Mbps 360p,30fps = 700kbps | ve-5Mbps |
| Chat | Facebook, Hangouts, ICQ, AIM, Skype, GmailChat | 4363 | 189.869 | chats in general 24–128kbps | chat-128kbps |
| Data Download | Bittorrent, ftps, scp | 758 | 589.583 | maximum (10Mbps) | down-max |
| Data Upload | ftps, sftp | 135 | 200.425 | maximum (10Mbps) | up-max |
| E-mail | Gmail | 1400 | 96.660 | 10kbps | email-10kbps |

Source: [6, 10, 19]

websites related to the applications or similar applications. Thus, the labels defined for each class represent the highest specified bandwidth by the services in the dataset.

### 3.1 Data Processing

To assemble the desired dataset from the *.pcap* files, a file manipulation library called Scapy [4] was used. This library allows implementing code in the Python language to extract packets, measure them and store them in another format if necessary. This way, the files can be processed to build a data set in a format similar to what FLOWPRI-SDN would manage in simulation to implement bandwidth reservation.

Data streams entering a domain managed by an SDN controller are forwarded by network devices for processing at the controller (when packet collection occurs). FLOWPRI-SDN handles these flows by defining forwarding rules with QoS on each forwarding device on the route. These rules expire if a flow is inactive for more than 10 s (subflow), requiring a new acceptance of the flow by the controller and creation of new rules.

Using the tool, we filter individual data streams from the capture files of each application, storing them in a new *.pcap* file. During flow filtering, some protocol flows unrelated to the quality of the desired applications were removed, such as DNS, LMNS and ARP, so as not to influence resource measurements. These types of flows are treated as flows without QoS requirements by FLOWPRI-SDN.

Finally, to extract valid subflows in this scenario, flow packets that have inter-arrival time (time elapsed between one packet and another) less than 10 s are extracted and stored in another *.pcap* file of the subflows. This technique is called [20] subflow packet sampling. Finally, short-lived subflows with less than 10 packets are also removed. This is done because the time required for the framework to create the rules on the forwarding devices is greater than the time of the active subflow.

### 3.2 Feature Extraction

After the separation of valid subflows, we have the *.pcap* files adjusted for each application, and it is possible to assemble the bases of features to train on machine learning models by extracting the information from the packets. Seeking to evaluate the influence of the number of packets used and the classification models, it was proposed to create four feature databases, two for batch data and two for streaming data.

The batch-type bases were created by extracting features from every 10 packets of a subflow, storing each result in a row of the features table of each application, the same for the base of 30 packets. To create the bases for streaming data, each packet is evaluated two-by-two, resulting in 10 lines in the table when evaluating 10 packets (or 30 lines for 30 packets). After the initial generation of feature bases for each application, the bases are grouped by class of service, balanced and randomized, and blocks with fewer packets than needed are discarded. For streaming databases, the feature blocks of the stream are randomized respecting the order of precedence of the inputs.

Table 2 presents the features calculated in each line of the feature dataset for batch or streaming packets. Among the information calculated are the size of the largest packet in bytes, the duration of the block, sum of packet sizes in blocks, time between packet arrivals (IAT), payload size (data load) of the packets and other information related to the TCP protocol header fields.

Due to the difference in the scales of the values collected from the features, the features with higher values have a greater influence on the classification in certain models. Thus, each feature was normalized to a calculated maximum size. Likewise, to avoid biases in models that benefit from a class, the feature bases were balanced, respecting the class with the least amount of feature data:

- Batch 10pkts: 1770 data per class (total: 14160).
- Batch 30pkts: 555 data per class (total: 4440).

**Table 2.** Table of selected features.

| protocol | duration | pkt_size_sum |
|---|---|---|
| pkt_biggest_size | pkt_smallest_size | pkts/sec |
| 1quartile_pkt_size | 3quartile_pkt_size | pkt_size_mean |
| pkt_size_median | pkt_size_std | pkts_size_smaller_mean |
| pkts_size_bigger_mean | payload_sum | payload_smallest |
| payload_biggest | payload_mean | payload_std |
| payload_smaller_128 | payload_between128_1024 | payload_bigger_1024 |
| 1quartile_payload_size | 3quartile_payload_size | IAT_menor |
| IAT_biggest | IAT_sum | IAT_mean |
| IAT_bigger_mean | IAT_smaller_media | IAT_std |
| 1quartile_IAT | 3quartile_IAT | tcp_push_flags |
| tcp_urg_flags | tcp_syn_flags | tcp_fin_flags |
| tcp_rst_flags | tcp_ack_flags | tcp_flags |
| tcp_windowsize_sum | tcp_windowsize_mean | tcp_windowsize_std |
| header_size_sum | header_size_mean | header_size_std |

Source: The authors, 2023

- Streams 10pkts: 18035 data per class (total: 144280).
- Streams 30pkts: 17763 data per class (total: 142110).

The features obtained in each dataset were analyzed using correlation matrices, which represent the links between the variables. Some variables have a strong correlation, while others appear to have a very weak correlation. Other methods for feature selection such as Recursive Feature Elimination (RFE) can help to reduce the base dimensionality as the ML model adapts, however, these tools are not easily implemented in data streaming-based learning models. That said, some features were manually removed:

- tcp_urg_flags: none of the packets used this flag.
- tcp_windowsize_std, tcp_windowsize_media: refer to network congestion control and do not exactly represent a characteristic of the flow.
- dport: It was removed because it influenced the classification too much (since it was traffic from only two computers).
- There are 41 features left.

### 3.3 Machine Learning Classification Models

ML models for classification for supervised (labeled) data types can be those that process data in batch or those that process data in streaming. Batch-type models group data features into a block to perform the classification, needing to store the data block. Models that handle streaming data, on the other hand, sort the data as it arrives, without the need to store and wait for data. Among the most common learning models in the literature are the decision tree algorithms and their variations, which generally have high accuracy rates, and classic models such as the SVM.

A Decision Tree is a tree-shaped learning model that works like a flowchart, where each node represents a feature, each branch represents a decision rule, and each leaf represents an output (classification) [12]. The Random Forest model is another machine learning model that can be used to perform block classifications on data. This algorithm randomly creates several Decision Trees and combines the results of all of them to arrive at the resulting classification.

The SVM model is one of the most popular classifiers for supervised learning. The SVM algorithm seeks to separate the $n$-dimensional space ($n$ represents the number of features) into classes by creating a division called a hyperplane. This hyperplane is constructed based on the support vectors, defined by the closest points to the separator. Thus, to separate two classes of data, the hyperplane found must be the one whose margin of distance between the support vectors is the largest [5].

The Hoeffding Tree model is a classification model for streaming data streams, where continuous data streams are categorized over time [18]. This model is also inspired by decision trees. It uses a coefficient to determine when a node should be split to generate a new tree. The trees are built incrementally, and it is assumed that if the data does not change over time, the results obtained are equivalent to those of batch methods. The main point highlighted in this model is that it does not need to store data for its operation.

The Adaptive Random Forest model is a modified version of the Random Forest to handle streaming data. One of the changes is that the learning base is modified trees of the Hoeffding Tree type. The algorithm starts a defined number of trees that receive data from the stream. When performance degradation is detected, backup trees are trained in parallel and can be used to replace an original tree [2].

## 4 Experiments and Results

The classification models described were evaluated in the experiments using the dataset division technique called cross validation. This method allows a better evaluation of the performance of machine learning models, as it allows partitioning the database into $K$ subsets with similar amounts of data. This allows you to perform $K$ training and testing iterations, iterating over which sets are used for testing and which are used for training [14].

In this way, we set $K$ to be 5 and divide the databases into blocks of equal size. Thus, in one iteration block 1 is used for testing while the others are used for training, in the next iteration block 2 is used for testing and the rest are used for training, continuing until each block has been used at least once for testing.

The datasets for streaming flows, as explained in the feature extraction process, were randomized, maintaining the order of precedence of the packets' features. Furthermore, to improve the performance of the streaming models, it was ensured that there were no more than 20 active streams at the same time, that is, the model does not receive mixed features from more than 20 streams. This causes models to improve learning power at the cost of limiting the type of input.

To run the tests, the machine used was a computer with Arch Linux 64bit operating system, AMD Ryzen 5 5600 processor and 20Gb of RAM, using only one thread for the tests. The batch-type machine learning models used were imported from the scikit-learn library, while the streaming-type machine learning models were imported from the scikit-multiflow library, which implements models based on the Massive Online Analysis (MOA) library.

The metrics selected for analysis of the models were precision, accuracy, recall and F1-score. Precision refers to how many data the classifier identified correctly, accuracy measures how often the classifier correctly identifies a data, recall represents the rate of how many real positive cases the model was able to correctly predict, and finally the F1-score is an average harmonic between precision and recall [1].

After running the 5 iterations with each dataset in the batch and streaming data classification models, it was possible to analyze the average values of the performance metrics obtained. Table 3, represents the comparison between the learning models for the feature databases of the subflows of 10 packets. As it is possible to identify, the Random Forest (batch) model obtained the best indicators and a low standard deviation in relation to the cross validation iterations. The Adaptive Random Forest model obtained better overall performance than its similar Hoeffding Tree in streaming databases.

**Table 3.** Comparison table between models (10pkts).

| ML Model | Precision | Accuracy | Recall | F1-Score | Standard deviation (precision) |
|---|---|---|---|---|---|
| Decision Tree | 0.97 | 0.97 | 0.97 | 0.97 | 0.0039 |
| Random Forest | 0.98 | 0.98 | 0.98 | 0.98 | 0.0045 |
| SVM | 0.9 | 0.9 | 0.9 | 0.9 | 0.0074 |
| Hoeffding Tree | 0.75 | 0.78 | 0.76 | 0.72 | 0.11 |
| Adaptive Random Forest | 0.88 | 0.85 | 0.86 | 0.85 | 0.024 |

Source: The authors, 2023.

**Table 4.** Comparison table between models (30pkts).

| ML Model | Precision | Accuracy | Recall | F1-Score | Standard deviation (precision) |
|---|---|---|---|---|---|
| Decision Tree | 0.95 | 0.95 | 0.95 | 0.95 | 0.013 |
| Random Forest | 0.98 | 0.98 | 0.97 | 0.98 | 0.0048 |
| SVM | 0.73 | 0.71 | 0.71 | 0.67 | 0.026 |
| Hoeffding Tree | 0.91 | 0.92 | 0.91 | 0.90 | 0.015 |
| Adaptive Random Forest | 0.91 | 0.92 | 0.92 | 0.91 | 0.025 |

Source: The authors, 2023.

Table 4, comprises the comparison between feature models for subflows of 30 packets. In these scenarios, batch-type databases shrink $3x$, as more packets are used to extract features, generating fewer data entries, but with more accurate values. Models for batch type data suffered a slight reduction in overall performance, in the cases of Decision Tree and SVM, while the Random Forest model did not change. In streaming-type models, the base size changes little, however, the packet sequences of each subflow increases by $3x$, causing more sets of features of the same flow to be analyzed. As a result, the models improved their overall performance, especially the Hoeffding Tree model, which increased accuracy by 16%.

The models for classifying data in batch train and test their bases practically instantly (less than 1 s), due to using only one base entry to perform the classification, as it represents the aggregate of 10 or 30 packets. For the base of subflows of 10 packets, the Hoeffding Tree model consumed an average of $9s$ to train the model in each cross validation iteration and $4.17s$ to test the model, on this same basis the Adaptive Random Forest consumed an average of $188.3s$ for training and $22.9s$ for the testing process. The database of 30 packets demanded an average of $7.9s$ for the Hoeffding Tree model to perform training in each iteration and $4.15s$ for the testing process, while the Adaptive Random Forest model consumed $176.3s$ for training and $23.56s$ for test.

Thus, based on the observations raised, it can be stated that the Random Forest model obtained the best performance in the classification of network traffic. By varying the number of packets, the models based on batch data were not affected, so it is possible to use a smaller number of samples to perform the classification. In this way, by collecting fewer packets from data streams, the impact on network performance is reduced, and fewer packets need to be forwarded to the SDN controller. For the record, until this stage, no fine-tuning of the parameters was carried out in any of the models.

## 5 Conclusion

In this work, the use of machine learning models for classifying network traffic was analyzed. The objective was to assess the viability of implementing these models in an SDN controller for the identification of application flows and their QoS requirements. The results indicated that Random Forest outperform the models based in data streaming in traffic classification using the dataset ISCX-VPN-NonVPN-2016.

The study focused on the classification of flows based on 10 or 30 packets so that an SDN controller can capture these packets without compromising the network performance at any time during packet transmission. As the tests demonstrated, some ML models achieved high precision rates for classifying packets over the QoS classes defined: real-time audio, static audio, real-time video, static video, chat, data download, data upload and e-mail. This way, this study demonstrates the potential of machine learning models in the context of network traffic classification, particularly in SDN environments, to enhance QoS management and flow identification.

In future work, the selected ML model can be implemented in the FLOWPRI-SDN framework and can be evaluated in complex network environments. In this way, it will be possible to observe the real overhead and the performance of the delivery rate and packet losses of the flows. Furthermore, more classes and bandwidth divisions can be associated and tested, making the classification more granular.

**Aknowledgments.** This work received financial support from the Coordination for the Improvement of Higher Education Personnel - CAPES - Brazil (PROAP/AUXPE).

## References

1. Agrawal, S.K.: Metrics to evaluate your classification model to take the right decisions. https://www.analyticsvidhya.com/blog/2021/07/metrics-to-evaluate-your-classification-model-to-take-the-right-decisions/ (2023)
2. Alkazaz, A., Saado Kharouki, M.: Evaluation of adaptive random forest algorithm for classification of evolving data stream (2020)
3. Amaral, P., Dinis, J., Pinto, P., Bernardo, L., Tavares, J., Mamede, H.S.: Machine learning in software defined networks: data collection and traffic classification. In: 2016 IEEE 24th International Conference on Network Protocols (ICNP), pp. 1–5 (2016)
4. Biondi, P.: Welcome to scapy's documentation! https://scapy.readthedocs.io/en/latest/ (2023)
5. Gandhi, R.: Support vector machine - introduction to machine learning algorithms. https://towardsdatascience.com/support-vector-machine-introduction-to-machine-learning-algorithms-934a444fca47 (2018)
6. Google: system requirements and supported devices for YouTube. https://support.google.com/youtube/answer/78358?hl=en (2023)
7. Huawei: overview of bandwidth managementt. https://support.huawei.com/enterprise/en/doc/ EDOC1000174073/1686bc c9/overview-of-bandwidth-management (2022)
8. ITU: End-user multimedia QoS categories. Int. Telecommun. Union (ITU) (2001)
9. Lotfollahi, M., Jafari Siavoshani, M., Shirali Hossein Zade, R., Saberian, M.: Deep packet: a novel approach for encrypted traffic classification using deep learning. Soft Comput. **24**(3), 1999–2012 (2020)
10. Microsoft: how much bandwidth does skype need?. https://support.skype.com/cn/faq/fa1417/how-much-bandwidth-does-skype-need (2023)
11. Mocelin Júnior, N.J., Fiorese, A.: FLOWPRI-SDN: a framework for bandwidth management for prioritary data flows applied to a smart city scenario. In: Barolli, L. (ed.) Advanced Information Networking and Applications: Proceedings of the 37th International Conference on Advanced Information Networking and Applications (AINA-2023), Volume 1, pp. 346–357. Springer International Publishing, Cham (2023). https://doi.org/10.1007/978-3-031-29056-5_31
12. Navlani, A.: Decision tree classification in Python tutorial. https://www.datacamp.com/tutorial/decision-tree-classification-python (2023)
13. Nguyen, T.T., Armitage, G.: A survey of techniques for internet traffic classification using machine learning. IEEE Commun. Surv. Tutorials **10**(4), 56–76 (2008)
14. Rabello, E.B.: Cross validation: avaliando seu modelo de machine learning. https://medium.com/@edubrazrabello/cross-validation-avaliando-seu-modelo-de-machine-learning-1fb70df15b78 (2017)
15. Raikar, M.M., Meena, S., Mulla, M.M., Shetti, N.S., Karanandi, M.: Data traffic classification in software defined networks (SDN) using supervised-learning. Procedia Comput. Sci. **171**, 2750–2759 (2020)
16. Rezaei, S., Liu, X.: Deep learning for encrypted traffic classification: an overview. IEEE Commun. Mag. **57**(5), 76–81 (2019)

17. UNB: VPN-nonVPN dataset (ISCXVPN2016). https://www.unb.ca/cic/datasets/vpn.html (2016)
18. Verma, Y.: A beginner's guide to Hoeffding tree with Python implementation. https://analyticsindiamag.com/a-beginners-guide-to-hoeffding-tree-with-python-implementation/ (2021)
19. Chen, Y., Toni Farley, N.Y.: QoS requirements of network applications on the internet. Syst. Manage. 4 (2004)
20. Zander, S., Nguyen, T., Armitage, G.: Sub-flow packet sampling for scalable ML classification of interactive traffic. In: 37th Annual IEEE Conference on Local Computer Networks, pp. 68–75 (2012)
21. Zhou, D., Yan, Z., Fu, Y., Yao, Z.: A survey on network data collection. J. Network Comput. Appl. **116**, 9–23 (2018). https://www.sciencedirect.com/science/article/pii/S1084804518301607

# A Reinforcement Learning Method for Integrated Production Scheduling and Distribution

Matheus de Freitas Araujo[1], Thiago Henrique Nogueira[2],
José Elias Claudio Arroyo[1(✉)], and Julio César Alves[3]

[1] Departamento de Informática, Universidade Federal de Viçosa, Av. Peter Henry Rolfs, s/n, 36570-900 Viçosa, MG, Brazil
{matheus.f.freitas,jarroyo}@ufv.br

[2] Universidade Federal de Viçosa, Engenharia de Produção, Rodovia MG-230 - Km 7, 38810-000 Rio Paranaíba, MG, Brazil
thiagoh.nogueira@ufv.br

[3] Universidade Federal de Lavras, Departamento de Computação Aplicada, Trevo Rotatório Professor Edmir Sá Santos, s/n, 37200-900 Lavras, MG, Brazil
juliocesar.alves@ufla.br

**Abstract.** This study addresses an integrate Parallel Machine Scheduling and Capacitated Vehicle Routing problem. In this problem, jobs must be processed efficiently on parallel machines before being distributed to customers through a fleet of vehicles. The main objective of this research is to optimize the machine production scheduling and the vehicle route selection minimizing the job's total weighted tardiness. Since the problem is NP-Hard, we introduce a novel approach using Hybrid Metaheuristics and Reinforcement Learning Algorithms, the PPO-VND, that combines strengths using a Variable Neighborhood Descent Algorithm (VND) and leverages using Proximal Policy Optimization (PPO). The PPO-VND method employs PPO to dynamically select the most effective local search strategy for each iteration within the VND algorithm, evaluated through computational experiments on several instances. The results demonstrate that the VND approach assisted by a Proximal Policy Optimization using a Reinforcement Learning Algorithm outperforms traditional methods such as the Random Variable Neighborhood Descent Algorithm (RVND) and the Mixed-Integer Linear Programming (MILP).

**Keywords:** Machine Learning · Reinforcement Learning · Parallel Machine Scheduling · Vehicle Routing Problem · Hybrid Metaheuristics

## 1 Introduction

Production and distribution methods have radically changed due to the Industry 4.0 emergence [9], requiring companies to operate based on parameters other than optimization processes or cost reduction. Therefore, this study aims to


A. Bajaj et al. (Eds.): HIS 2023, LNNS 1224, pp. 295–306, 2025.
https://doi.org/10.1007/978-3-031-78925-0_29

improve customer service quality rather than reduce costs. The authors attempt to solve the integrated production, planning, and distribution problems as efficiently and quickly as possible, adopting as a research problem the assemblage of products on unrelated parallel machines with sequence-dependent setup times that, once created, are sent to customers using a fleet of capable vehicles. Also, the research's general goal is to minimize the total weighted delay.

Mathematical models, heuristics, and metaheuristics commonly apply to combinatorial optimization problems, with several pieces of research proposing hybrid approaches. Hybrid algorithms do not purely follow the paradigm of a single traditional heuristic or metaheuristic, using distinct parts of different optimization techniques to find the best solution. For instance, Blum et al. [2] presented a Hybrid Metaheuristics (HM) survey investigating the complementary nature of different hybrid strategies.

Chen [3,4] provides an extensive literature review on integrated production scheduling and distribution problems. Also, Liu et al. [11] examined a single-machine scheduling and vehicle routing problem to minimize the totality of a job's delivery time using a Variable Neighborhood Search (VNS) heuristic, later comparing the VNS-produced results to those obtained with a Memetic Algorithm (MA) [13] and a Large Neighborhood Search Algorithm (LNS) [15].

Tamannaei et al. [17] proposed a mathematical model, a B&B algorithm, and a genetic algorithm to minimize the total weighted job delay and the transportation costs in a single-machine scheduling and capable vehicle routing problem. Likewise, Felix & Arroyo [6] proposed two hybrid metaheuristics based on ILS and RVND to solve the same problem, comparing their results with Tamannaei et al.'s.

Zhu et al. [19] proposed a Deep Reinforcement Learning (DRL) approach to solve a Flexible Flow Shop Scheduling Problem (FFSP) with makespan minimization to reduce the time required to complete a set of jobs on a series of identical parallel machines. The authors adopted Proximal Policy Optimization (PPO) to evaluate and optimize random and complex instances with different scales, including real-world instances. They established that the PPO always provided satisfactory solutions employing reasonable computational time.

Nahhas et al. [12] applied Deep Reinforcement Learning (DRL) techniques to optimize Hybrid Flow Shop (HFS) scheduling. The DRL techniques, Proximal Policy Optimization (PPO), and Asynchronous Advantage Actor-Critic (A3C) proved effective in learning appropriate policies for solving HFS scheduling and adjusting their strategies to different problems.

Kayhan and Yildiz [10] presented a comprehensive review of Reinforcement Learning (RL) applications for machine scheduling problems, analyzing algorithms, machine environments and characteristics, jobs, objectives, and reference methods. They indicated that the most frequently investigated topics are Job-Shop Scheduling Problems, Unrelated Parallel Machine Scheduling, and Single-Machine Scheduling. Their main contributions are thoroughly examining RL in machine scheduling, categorizing recurring problem types, objectives, and constraints, and divulging literature gaps.

In this study, we propose solving the abovementioned problem using a Mixed Integer Linear Programming Model (MILPM), a Constructive Heuristic, and a PPO-VND Hybrid Metaheuristic that combines VND with a Reinforcement Learning Algorithm and a Proximal Policy Optimization (PPO) algorithm, comparing the PPO-VND results with the MILPM and the Random Variable Neighborhood Search (RVND) metaheuristic, a variation of VND with a random neighborhood option.

This paper's organization is as follows: Sects. 2 and 3 state the research problem and describe the mathematical model, respectively. Section 4 presents the Constitutive Heuristics, the Proximal Policy Optimization, and the proposed Hybrid Metaheuristic. In Sect. 5, the authors demonstrate the computational experiments conducted for this research. Finally, Sect. 6 concludes the study.

## 2 Problem Definition

There is a set of jobs $I = \{1, 2, ..., n\}$ that must be processed, without preemption, on a set $M = \{1, 2, ..., m\}$ of $m$ unrelated parallel machines. Each job $j$ has a processing time $p_{ij}$ on each machine $i$, a due date $(d_j)$, a tardiness penalty $(w_j)$, and its size (space occupied in a vehicle) $(h_j)$. Also, there is a machine-dependent setup time $s_{ijk}$ for processing the job $k$, just after job $j$. In general, $s_{ijk} \neq s_{ikj}$.

After processing the jobs on the machines, they are split into batches, each delivered to the respective clients using a set $L = \{1, 2, ..., l\}$ of $l$ vehicles. For the routing problem, graph $G(V, A)$ displays a $V = \{0, 1, ..., n\}$ the set of nodes and an $A = \{(j, k) : 0 \geq j, k \geq n, j \neq k\}$ set of arcs. The node 0 represents the deposit, and each job is associated with a customer, being the set of customers $I = \{1, ..., n\}$. Furthermore, a travel time $t_{jk}$ is associated with each arc $(j, k)$. Each vehicle $v \in L$ has a capacity $(q_v)$ that must be respected. The problem assumes single-use vehicles and one visit per customer. The problem aims to determine the jobs' processing orders and the vehicles' delivery routes to minimize the jobs' total weighted tardiness (TWT).

Figure 1 illustrates the issue and presents a solution for an example with three machines ($m = 3$), six jobs ($n = 6$), and four vehicles ($v = 4$). Figure 1 shows each job's completion times, the delivery times for each job, and the vehicle routes. For instance, the completion and delivery time of job $I_4$ are 14 and 36, respectively. The vehicle $V_1$ starts at 14 and is responsible for delivering jobs $I_4$ and $I_5$, following the route: $\{I_0 \rightarrow I_4 \rightarrow I_5 \rightarrow I_0\}$, where $I_0$ represents the deposit.

## 3 Mixed Integer Linear Programming Model

The authors propose a Mixed Integer Linear Programming (MILP) model to solve the problem. This model uses the parameters in the problem description and the following decision variables: $X_{ijk}$ is a binary variable equal to '1' if job $j$ precedes job $k$ on machine $i$. The $Y_{ij}$ is a binary variable equal to '1' machine $i$ processes the job $j$. The $Z_{vjk}$ is a binary variable equal to '1' if the vehicle $v$

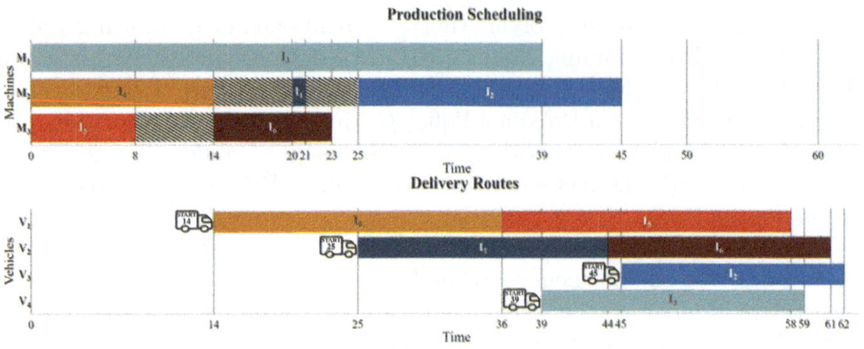


**Fig. 1.** The scheduling of jobs and delivery routes.

travels between points $j$ and $k$. The $U_v$ is a binary variable equal to '1' if the vehicle $v$ is used for deliveries. The $C_j$ is the job $j$'s completion time. The $S_v$ is the vehicle's start time $v$. The $T_j$ is the job $j$'s delivery delay, and $D_j$ is the job $j$'s delivery time.

The MILP model also requires the following parameters: $N = I \cup \{O\}$ represents a set containing the jobs and a dummy job starting from the origin, and $MG$ is a sufficiently large value defined by $\sum_{i \in M} \sum_{j \in N} \sum_{k \in N} t_{jk}\ p_{jk}\ s_{ijk}$. The mathematical model is presented below. The objective of the model is:

$$\min\ \text{TWT} = \sum_{j \in I} w_j T_j \tag{1}$$

$$\sum_{v \in L} \sum_{k \in N, k \neq j} Z_{vjk} = 1, \quad \forall j \in I \tag{2}$$

$$\sum_{j \in N} \sum_{k \in N, k \neq j} Z_{vjk} h_j \leq q_v U_v, \quad \forall v \in L \tag{3}$$

$$\sum_{k \in N} Z_{v0k} = U_v, \quad \forall v \in L \tag{4}$$

$$\sum_{j \in N} Z_{vjh} = \sum_{k \in N} Z_{vhk} \forall h \in N, \quad \forall v \in L \tag{5}$$

$$\sum_{k \in I} X_{i0k} \leq 1, \quad \forall i \in M \tag{6}$$

$$\sum_{i \in M} Y_{ij} = 1, \quad \forall j \in I \tag{7}$$

$$Y_{ij} = \sum_{k \in N, k \neq j} X_{ijk}, \quad \forall i \in M, \forall j \in I \tag{8}$$

$$Y_{ik} = \sum_{j \in N, k \neq j} X_{ijk}, \quad \forall i \in M, \forall k \in I \tag{9}$$

$$C_k - C_j \;\; MG(1 - X_{ijk}) \geq s_{ijk} \;\; p_{ik}, \quad \forall j \in N, \forall k \in I, j \neq k, \forall i \in M \tag{10}$$

$$S_v \geq C_k - MG(1 - \sum_{j \in N, j \neq k} Z_{vjk}), \quad \forall v \in L, \forall k \in N \tag{11}$$

$$D_k - S_v \geq t_{0k} - MG(1 - Z_{v0k}), \forall v \in L, \qquad \forall k \in N \tag{12}$$

$$D_k - D_j \geq t_{jk} - MG(1 - \sum_{v \in L} Z_{vjk}), \quad \forall j \in I, \forall k \in N, j \neq k \tag{13}$$

$$T_j \geq D_j - d_j, \quad \forall j \in I \tag{14}$$

The objective function, defined by Eq. (1), is designed to minimize the sum of the weighted job delays. The constraints (2) ensure that each customer is served only once and by a single vehicle. The restrictions (3) guarantee the maintenance of the vehicles' capacities, considering $h_0 = 0$. The constraints (4) and (5) assure that, when used, a vehicle must start and end its route at the depot. Constraints (6) ensure that only one job runs after each machine's dummy job $I_0$. Constraints (7) guarantee that each job runs on only one machine. Constraints (8) and (9) provide that each job has only one successor and one predecessor. Constraints (10) define the job completion time, considering $C_0 = 0$. The Constraints (11) define each travel's starting time. The vehicle arrival times are according to the constraints (12) and (13). Finally, constraints (14) define job delays based on due dates.

# 4 Proposed Algorithms

This section presents the Proximal Policy Optimization (PPO), Reinforcement Learning, and a Constructive Algorithm. Following, the authors propose the PPO-VND Hybrid Metaheuristic that incorporates the Variable Neighborhood Algorithm (VND) aided by the Reinforcement Learning Algorithm, PPO.

## 4.1 Initial Solution

To obtain an initial solution ($solution = \{s, r\}$), defining the job list $L1$ is necessary. The job list $L1$ ranks the jobs in descending order according to the sum of average processing and setup times. Subsequently, the NEH algorithm uses the job list $L1$ to define the machines' scheduling. The NEH algorithm sequentially inserts the jobs from the $L1$ list into the solution. At each algorithm iteration, a job $j \in L1$ is inserted into the machine's sequencing ($s$) to minimize the machine's completion time.

Defining the job list $L2$ is necessary to build the delivery routes since the $L2$ list comes from scheduling jobs using the Earliest Due Date (EDD) rule ($d_j - C_j$). In this case, jobs with more immediate due dates receive their routes first from the PIFH algorithm. The PIFH algorithm sequentially inserts the list of jobs $L2$ to minimize the weighted delay ($wt$). At the end of the algorithm, the vehicle routes $r$ are finally defined. Araujo et al.[7]'s research provides more information on NEH and PIFH heuristics.

### 4.2 The Proximal Policy Optimization

Reinforcement Learning (RL) commonly applies to sequential decision-making problems and consists of an agent interacting with a stochastic environment, learning to map states into actions and maximizing cumulative rewards. The RL problem can be modeled as a Markov Decision Process (MDP), a tuple $\{S, A, p, r(s,a)\}$, where $S$ is the set of states, $A$ is the set of actions, and $p$ is the transition function that represents the conditional probabilities of state change (between 0 and 1). Ultimately, $r(s,a)$ is the reward function that assigns a numerical value to the state-action pair. In these processes, the agent aims to find an approach that associates each state with an optimal action, maximizing the sum of rewards over time [1].

The Proximal Policy Optimization (PPO) algorithm, proposed by Schulman et al. [16], derives from the concept of Policy Gradients (PG) and attempts to directly optimize agents' policies through an on-policy approach, using a Deep Neural Network (DNN) to map actions to states. PPO trains the DRL agents to act in an environment, maximizing the cumulative reward and directly finding the best policy rather than learning the function value and to derive the best one.

Actor-critical are GP methods that approximate the policy and value function. The actor learns the policy, while the critic learns the value function to evaluate the stated policy. This approach speeds up the learning process by minimizing variation during training. Proximal Policy Optimization (PPO) balances implementation, parameterization, and sample complexity. Its main goal is to limit step size updates, revamping the policy with a new one similar to the current one. This process is important because large updates cause significant variation in the learning process, leading to poor performance [1]. The authors use the PPO algorithm from the stable-baselines3 library in Python [14]. The PPO pseudocode is presented in Algorithm 1.

**4.2.1 State, Action, and Reward Function** The state is the data that fully represents the current environment that contains the agent. The state's portrayal is a vector feature that describes the environment at a given time: a solution to the problem containing the number of machines ($m$), the number of jobs ($n$), the list of the jobs' completion times ($C_j, \forall j \in N$), the list of jobs' delivery dates ($D_j, \forall j \in N$), and the list of tardiness penalties ($w_j T_j, \forall j \in N$).

The action is the decision taken by the agent in a given state and represents the agent's next choice. In this case, the choice is the next neighborhood explored

**Algorithm 1.** Proximal Policy Optimization

Initialize $\Theta$ and $\Phi$ ▷ policy and value-function parameters, respectively
**for** $i = 0, 1, 2, \ldots$ **do**
  **for** $actor = 1, 2, \ldots, N_{actors}$ **do**
    Run policy $\pi(\Theta)$ for $T$ steps and collect the trajectories.
    Calculate advantage estimates $\hat{A}_1, \ldots, \hat{A}_T$ based on value-function $V_\Phi$.
  **end for**
  Form a *batch*, of size $NT$ , with collected trajectories and advantages.
  **for** $epoch = 1, 2, \ldots, K$ **do**
    Shuffle *batch* and split into *minibatches*
    **for all** *minibatches* **do**
      Update $\Theta$ wrt objective function via stochastic gradient.
      Update $\Phi$ wrt mean-squared error of value-function $V_\Phi$.
    **end for**
  **end for**
**end for**

by the local search. The model output is a $\sum_{j \in I} w_j T_j$ of solutions created from job lists $L1$ and $L2$ after the action taken by the agent.

The reward function comes from the objective function, calculated using the Eq. 15 and considering the lower bound ($LB$) and upper bound ($UB$). $LB$ is calculated using the Eqs. 16. To calculate $UB$, the authors assume that the jobs' distribution will be identical among machines and vehicles. The authors also consider that job $j$ will be the last executed on the machine and delivered on the route. Furthermore, all jobs have the longest possible processing, setup, and travel times.

$$\text{Reward Function} = -1 * (\sum_{j \in I} w_j T_j) - LB)/(UB - LB) * 100 \tag{15}$$

$$LB = \sum_{j \in I} w_j * Max((Min(p_{ij} \forall i \in M)\ t_{0j} - d_j), 0) \tag{16}$$

**4.2.2 The Environment Model** An agent's interaction with the environment demands an environmental model to describe the problem. The environment depicts the Integrated Production-Distribution Planning Problem. The model inputs consist of two job lists ($L1$ and $L2$) used to resolve said problem ($solution = \{s, r\}$). The model's output is the objective function that defines the criterion for evaluating the solution. The agent aims to find an approach that maximizes cumulative rewards over time, leading to better solutions for the integrated problem. This research problem required creating an environmental model for each group of instances with $n$ equal to 10 and 15, as the environmental model directly links to the number of jobs. The Environment Model appears in Algorithm 2.

**Algorithm 2.** Environment Model

```
solution = New solution{s, r}.
Sort the job sequence as L1 in in descending order of the sum of average processing
  and setup times.
for j ∈ L1 do
    Insert j in s to minimize the machine's completion time
end for
Sort the job sequence as L2 the earliest due date (EDD) rule (d_j − C_j).
for j ∈ L1 do
    Insert j in r to minimize the job's weighted delay (wt)
end for
Create solution with s and r
Return f(solution)
```

### 4.3 PPO - VND

The Variable Neighborhood Descent (VND) is a deterministic local search algorithm [8] that employs a series of neighborhood structures $V = V_1, V_2, ..., V_k$ until each solution $S$ becomes locally optimal. The PPO-VND hybrid metaheuristic is presented in the algorithm 3. A Proximal Policy Optimization Algorithm (PPO) predicts and defines the explored neighborhood at each algorithm iteration.

The algorithm starts with an initial solution, and at each algorithm iteration, a neighborhood structure $u$ is chosen by the PPO. Then, a local search occurs within the chosen neighborhood $u$, returning a local optimal solution. If the new solution is better than the "best_solution", the PPO updates it. Otherwise, the algorithm perturbates the solution. The process repeats until the number of non-improving iterations equals $n$ (number of instance jobs) or the solution found equals the Lower Bound (LB).

In this paper, the authors propose eight neighborhoods. The neighborhood methods use two types of actions: insertion and exchange with and without restarting. The movements occur within the job lists $L1$ and $L2$. The insertion implicates removing an element from the list and re-inserting it in a different position. The exchange movement involves exchanging two elements' positions and creating a new list. In addition, the neighborhoods with a restart option will restart the searches whenever a better solution occurs.

The solution's perturbation consists of randomly exchanging and inserting movements in the job lists. The perturbation happens in up to 10% of the jobs' list, i.e., 10%, with the list selection ($L1$, $L2$, or both) occurring randomly. Following, the Algorithm 2 uses these job lists ($L1$ and $L2$) to define a new solution, later employed in the algorithm.

## 5 Computational Experiments

In this section, the authors present the experiment to evaluate the performance of the proposed methods. The Mixed-Integer Linear Programming Model (MILP)

**Algorithm 3.** PPO - VND

```
Create L1 and L2,
Create solution using the NEH and PIFH Algorithms
best_solution ← solution
state ← initial state condition
while it < n or f(solution)! = LB do
    u ← PPO.predict(state) ▷ PPO chooses a neighborhood structure u for current state.
    solution ← Best Solution in N_u(solution)
    state ← update
    if F(solution) < F(best_solution) then
        best_solution ← solution
        it ← 0
    else
        it ← it 1
        perturbation the current solution
    end if
end while
return best_solution
```

was implemented in C++ and solved using the CPLEX solver [5]. The other methods were programmed in Python 3 and executed on an Intel Core i7-4790K (4.0 GHz) CPU with 32 GB of RAM. The algorithms presented in this paper are available at https://github.com/matheusinhofreitas/PPO-VND. The computational experiments occurred in a single thread with the MILP model processing time fixed at 180 s. The authors trained the agent for 500 thousand episodes with 36 randomly selected instances of each size and different complexities. The PPO hyperparameters were tuned using the RL Baselines3 Zoo library. Additionally, the authors developed a Random Variable Neighborhood Search algorithm (RVND) (similar to PPO-VND) that randomly selects the neighborhood structure at each algorithm iteration.

The solution comes from the best-known solution's relative percentage deviation (GAP). The $GAP = 100 \times \frac{(F_{method} - F_{best})}{F_{method}}$, where $F_{best}$ is the minimum TWT value obtained among all the compared methods and $F_{method}$ is the TWT obtained with a given method.

The authors employed the 360 instances proposed by Araujo et al. [7] to conduct the computational experiments. Each problem instance is characterized by two key parameters: the number of jobs denoted as $n \in \{10, 15\}$, and the number of machines represented by $m \in \{2, 4, 8\}$. For example, the instances "N_10_M_2" have 10 jobs and 2 machines.

## 5.1 Computational Results

Initially, the authors compare the MILP model to the PPO-VND and the RVND. For each instance, they run the PPO-VND and RVND Algorithms five times, classifying the average GAPs based on the number of jobs and machines. Table 1 represents the averages, the best GAPs, and the runtime across 60 instances. The smallest average GAPs are in bold.

**Table 1.** Average and best GAPs (%) and runtime for the compared methods.

| Instance | Average GAP (%) | | | | Best GAP(%) | | Time in seconds | | |
|---|---|---|---|---|---|---|---|---|---|
| | Initial Solution | MILPM | PPO-VND | RVND | PPO-VND | RVND | MILPM | PPO-VND | RVND |
| N_10_M_2 | 85% | 34% | **23%** | 46% | 12% | 25% | 173.22 | **1.46** | 1.53 |
| N_10_M_4 | 80% | **10%** | 34% | 38% | 26% | 22% | 173.22 | **1.45** | 1.54 |
| N_10_M_8 | 65% | **14%** | 18% | 17% | 12% | 5% | 173.20 | **1.44** | 1.54 |
| N_15_M_2 | 89% | 79% | **22%** | 53% | 9% | 31% | 173.19 | **1.43** | 1.54 |
| N_15_M_4 | 77% | 59% | **18%** | 31% | 8% | 10% | 172.95 | **1.44** | 1.55 |
| N_15_M_8 | 68% | 79% | **18%** | 19% | 7% | 5% | 172.94 | **1.45** | 1.55 |
| Average | 77% | 46% | **22%** | 34% | 12% | 17% | 173.12 | **1.45** | 1.54 |

Table 1 shows that the MILPM obtains the best average GAP for two groups out of 6 (N_10_M_4 and N_10_M_8 with the respective values: 10% and 14%), the PPO-VND produced the best GAP for 4 instance groups (N_10_M_2, N_15_M_2, N_15_M_4, N_15_M_8 with the respective values: 23%, 22%, 18% and 18%), and the RVND did not display the best GAP for any group. The average GAP for the MILPM, PPO-VND, and RVND methods is 46%, 22%, and 34%, respectively. Considering the overall average, the PPO-VND is the best method.

Considering all instances, the MILPM obtained the best GAP for 87 (24.1%) out of 360 instances. In 170 cases (47.2%), the PPO-VND found the best solutions. Finally, the RVND found the best GAP for 165 instances (45.8%). Furthermore, the MILPM managed to find the optimal solution for 18 instances.

Table 1 also shows that the execution time of the PPO-VND is similar to that of the RVND algorithm. The only drawback of the PPO-VND is that its algorithm requires prior training using the Deep Reinforcement Learning (DRL) model. However, it is a single-time training process; subsequently, the forecast time is instantaneous.

The authors performed an Analysis of Variance (ANOVA) [18] for a 5% significance level. The ANOVA yielded a p-value of 0.000102, indicating statistically significant differences between the experiments. Figure 2a shows the boxplot for the GAPs obtained using all methods. The constructive solution is limited when compared to other methods. Additionally, PPO outperforms RVND and rivals the mathematical model, providing equal or superior solutions. The boxplot illustrates each method in different colors, with the initial solution in blue, the MILPM in orange, the PPO-VND in green, and the RVND in yellow. Figure 2b shows the constant increase in the average reward throughout the model training, and the graph indicates a successful learning process for DRL agents to solve the adopted problem. The orange and blue lines represent the reward-changing process for instances with $n = 10$ and $n = 15$ sizes, respectively.

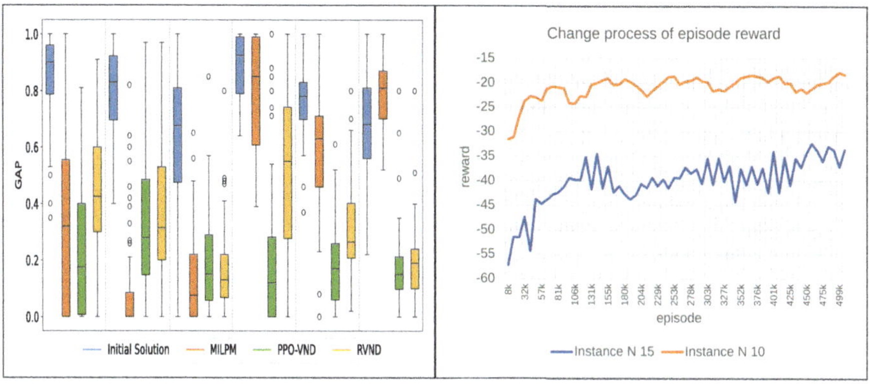


(a) BoxPlot of GAPs (%) for all methods

(b) Change process of episode reward for PPO

**Fig. 2.** General computational results for all methods.

## 6 Conclusions

This paper proposed a Mixed-Integer Linear Programming Model and the Hybrid Heuristics PPO-VND for an integrated production scheduling and distribution problem. The authors applied the Proximal Policy Optimization (PPO) Reinforcement Learning (RL) algorithm to predict the neighborhood structure for each PPO-VND interaction. The computational experiments revealed that the PPO-VND was more effective than the mathematical model and the RVND algorithm concerning execution time and solution quality. In this case, the authors disregarded the training time for the PPO since it only occurred once. It is essential to highlight that even in small instances with small search spaces, such as those used in this experiment, the PPO-VND algorithm obtained better results than the other resolution techniques presented. Compared with the other methods, PPO-VND results improved when the number of jobs increased from 10 to 15. The authors expect the PPO-VND algorithm to perform even better on larger instances and strive to apply the PPO-VND algorithm to solve such problems in future studies.

**Acknowledgments.** This work was supported by CAPES and CNPq.

## References

1. Alves, J.C., Mateus, G.R.: Deep reinforcement learning and optimization approach for multi-echelon supply chain with uncertain demands. In: International Conference on Computational Logistics, pp. 584–599. Springer, Cham (2020). https://doi.org/10.1007/978-3-030-59747-4_38
2. Blum, C., Puchinger, J., Raidl, G.R., Roli, A.: Hybrid metaheuristics in combinatorial optimization: a survey. Appl. Soft Comput. **11**(6), 4135–4151 (2011)

3. Chen, Z.L.: Integrated production and distribution operations. In: Handbook of Quantitative Supply Chain Analysis, pp. 711–745 (2004)
4. Chen, Z.L.: Integrated production and outbound distribution scheduling: review and extensions. Oper. Res. **58**(1), 130–148 (2010)
5. IBM ILOG CPLEX: V12. 1: User's manual for CPLEX. Int. Bus. Mach. Corporation **46**(53), 157 (2009)
6. Felix, G.P., Arroyo, J.E.C.: HeurÍsticas para o sequenciamento da produÇÃo e roteamento de veÍculos com frota heterogÊnea. LII Simpósio Brasileiro de Pesquisa Operacional (2020)
7. de Freitas Araujo, M., Arroyo, J.E.C., Nogueira, T.H.: Heuristics assisted by machine learning for the integrated production planning and distribution problem. In: Abraham, A., Pllana, S., Casalino, G., Ma, K., Bajaj, A. (eds.) Intelligent Systems Design and Applications: 22nd International Conference on Intelligent Systems Design and Applications (ISDA 2022) Held December 12-14, 2022 - Volume 4, pp. 120–131. Springer Nature Switzerland, Cham (2023). https://doi.org/10.1007/978-3-031-35510-3_13
8. Hansen, P., Mladenović, N.: Variable neighborhood search: principles and applications. Eur. J. Oper. Res. **130**(3), 449–467 (2001). https://doi.org/10.1016/S0377-2217(00)00100-4
9. Hofmann, E., Rüsch, M.: Industry 4.0 and the current status as well as future prospects on logistics. Comput. Ind. **89**, 23–34 (2017). https://doi.org/10.1016/j.compind.2017.04.002
10. Kayhan, B.M., Yildiz, G.: Reinforcement learning applications to machine scheduling problems: a comprehensive literature review. J. Intell. Manuf. **34**(3), 905–929 (2023). https://doi.org/10.1007/s10845-021-01847-3
11. Liu, L., Li, W., Li, K., Zou, X.: A coordinated production and transportation scheduling problem with minimum sum of order delivery times. J. Heuristics **26**(1), 33–58 (2020). https://doi.org/10.1007/s10732-019-09420-1
12. Nahhas, A., Kharitonov, A., Turowski, K.: Deep reinforcement learning techniques for solving hybrid flow shop scheduling problems: proximal policy optimization (PPO) and asynchronous advantage actor-critic (A3C) (2022)
13. Ngueveu, S.U., Prins, C., Calvo, R.W.: An effective memetic algorithm for the cumulative capacitated vehicle routing problem. Comput. Oper. Res. **37**(11), 1877–1885 (2010)
14. Raffin, A., Hill, A., Gleave, A., Kanervisto, A., Ernestus, M., Dormann, N.: Stable-baselines3: reliable reinforcement learning implementations. J. Mach. Learn. Res. **22**(268), 1–8 (2021). http://jmlr.org/papers/v22/20-1364.html
15. Ribeiro, G.M., Laporte, G.: An adaptive large neighborhood search heuristic for the cumulative capacitated vehicle routing problem. Comput. Oper. Res. **39**(3), 728–735 (2012)
16. Schulman, J., Wolski, F., Dhariwal, P., Radford, A., Klimov, O.: Proximal policy optimization algorithms. arXiv preprint arXiv:1707.06347 (2017)
17. Tamannaei, M., Rasti-Barzoki, M.: Mathematical programming and solution approaches for minimizing tardiness and transportation costs in the supply chain scheduling problem. Comput. Ind. Eng. **127**, 643–656 (2019)
18. Zar, J.H.: Biostatistical Analysis, 4th. New Jersey, USA, p. 929 (1999)
19. Zhu, J., Wang, H., Zhang, T.: A deep reinforcement learning approach to the flexible flowshop scheduling problem with makespan minimization. In: 2020 IEEE 9th Data Driven Control and Learning Systems Conference (DDCLS), pp. 1220–1225. IEEE (2020)

# Performance Evaluation of Metaheuristics Using TensorFlow Parallelism on GPU

Gustavo Gusmão Rocha[1], Fábio M. F. Lobato[1,3], and Omar Andres Carmona Cortes[1,2](✉)

[1] Programa de Pós-Graduação em Engenharia da Computação e Sistemas (PECS), Universidade Estadual do Maranhão (UEMA), São Luis, MA, Brazil
gustavorocha@outlook.com.br, fabio.lobato@ufopa.edu.br

[2] Instituto de Engenharia e Geociências (IEG), Universidade Federal do Oeste do Pará (UFOPA), Santarém, PA, Brazil
omar@ifma.edu.br

[3] Computing Department (DComp), Instituto Federal do Maranhão (IFMA), São Luis, MA, Brazil

**Abstract.** Parallel computing has become popular since multi-core advancements followed by multi-core architectures such as GPUs. Thus, it is essential to leverage these technologies when developing new applications to achieve better performance. In this context, this work investigates the parallel computing capabilities of TensorFlow to solve eight well-known optimization problems by applying metaheuristics such as Genetic Algorithm, Differential Evolution, Particle Swarm Optimization, and Cuckoo Search. Results show that TensorFlow can achieve a speedup of up to eight times in high-dimension benchmark functions with low programming effort using only one GPU. This work contributes to the body of knowledge by demonstrating and measuring how to get benefits from TensorFlow in metaheuristic implementations.

**Keywords:** Parallel Computing · Optimization · Metaheuristics · GPU · Performance

## 1 Introduction

Metaheuristics play an essential role in optimization problems because they can find approximate solutions to optimal solutions in complex, large-scale dimension applications in which deterministic approaches fail. Thus, metaheuristics have shown efficiency as the complexity of the search space increases. On the other hand, as dimensionality increases, its computational performance tends to deteriorate, sometimes exponentially increasing the execution time. In this context, Graphical Processing Units (GPUs) are an attractive solution for improving metaheuristic performance using high-parallel computing. The approach relies on a parallel architecture called Single Instruction Multiple Data (SIMD), in which hundreds of threads can be executed in thousands of cores simultaneously


A. Bajaj et al. (Eds.): HIS 2023, LNNS 1224, pp. 307–316, 2025.
https://doi.org/10.1007/978-3-031-78925-0_30

over different data, given the famous nickname so-called many-cores to GPU architectures.

To program GPUs, NVIDIA created the Computed Unified Device Architecture (CUDA), a framework to explore the parallel features in languages such as C and Fortran in several parallel programming models, mainly using a fine-based granularity. The issue with CUDA is that knowing the architecture details is required to extract an efficient parallelism; however, CUDA increases the programming learning curve.

Therefore, we propose using TensorFlow as a parallelism framework in this work. TensorFlow is a framework created by Google initially for scientific computation in 2015 but rapidly adopted by the Artificial Intelligence field as a framework for parallel training machine learning models. In our case, the main advantage of using TensorFlow is to state where to execute the code rather than how to parallelize it, providing a faster way to produce it. As might be expected, this kind of parallelization cannot reach the high level of CUDA code; however, the programming level is higher, allowing faster parallel code development.

Aiming to contribute to the body of knowledge, our experimental setup comprises four well-known metaheuristics: Genetic Algorithms (GA), Particle Swarm Optimization (PSO), Differential Evolution (DE), and Cuckoo Search (CS). To evaluate the parallel performance, we computed the speedup of eight benchmark functions: Discus, Rastrigin, Rosenbrock, Cigar, Elliptic, Ackley, Griewank, and Salomon, in high dimensionality (50, 100, and 500). The idea is to investigate parallelism; therefore, we are not seeking better solutions for metaheuristics. Instead, we are interested in better speedups.

The remainder of this work is organized as follows. Section 2 discusses some related works. Section 3 presents the parallelism strategy adopted in the present work. The benchmark functions, the runtime environment, and the results are detailed in Sect. 4. Finally, Sect. 5 gives the conclusions and future work directions.

## 2 Related Works

In the early days of GPU programming, GPU computing could be mentioned as a merely academic exercise [9] because GPUs were primarily employed as an efficient calculator, executing a predetermined set of fixed operations [3]. Since then, the programming models, tools, and GPU architectures have impressively evolved. For example, in June 2023, the fourth place in the Top 500 supercomputer list [1] is called Leonardo, being built with NVIDIA A100 GPUs totaling 1,824,768 cores with a performance of 238.70 PFlop/s. In fact, five supercomputers are GPU-based in the top 10 list. Additionally, GPUs became popular for game players, allowing non-academic people to program these kinds of applications in the so-called General Purpose GPU or GPGPU. Furthermore, cloud services such as Colab from Google, AWS from Amazon, and Azure from Microsoft make GPU programming available in many ways.

In terms of metaheuristics, Veronses [10] ported a metaheuristic named Particle Swarm Optimization (PSO) to a GPU platform in 2009; since then, researchers have been evolving parallel implementations of metaheuristics in different applications such as [5,6], [4,8,11], between many others. The main drawback of these implementations is that they are primarily based on CUDA Programming, which demands a good knowledge of the architecture being programmed.

In this context, this work proposes an alternative way of implementing parallel metaheuristic using the parallelism available in TensorFlow, exploring the semi-automatic parallelism by avoiding the CUDA programming details. In other words, we rely on the fact that in TensorFlow we only have to say where to execute code instead of how to execute them. To the best of our knowledge, no other works investigate the parallelism provided by TensorFlow in that way.

## 3 Parallelism and TensorFlow

In 2011, the team at [2] created a Google branch focused on research in artificial intelligence and developed a system called DistBelief, centered around neural networks and deep learning. Several Google subsidiaries started utilizing DistBelief in commercial and research applications, finding the system highly effective. As its usage proliferated within the company, Google assembled a team of computer scientists to redesign DistBelief, aiming to simplify and expedite its processes. This refinement ultimately resulted in the birth of TensorFlow.

TensorFlow is cross-platform and compatible with Windows, MacOS, and Linux. Moreover, it can operate on CPUs, GPUs (graphics cards that enhance processing), and even TPUs (specific integrated circuits developed by Google for accelerating machine learning tasks). TensorFlow APIs are available in multiple programming languages, including Python, JavaScript, C++, Java, Go, Swift, C sharp, Haskell, Julia, Ruby, Rust, and Scala [2].

We leveraged the TensorFlow API with metaheuristics, distributing training tasks across a GPU "/device" and a CPU "/device" using virtual machines implemented with just a few lines of code. This approach facilitated performance analysis with minimal adjustments. Utilizing TensorFlow's underlying component, *tf.distribute.Strategy*, we implemented a strategy-aware approach, effectively harnessing the power of GPU alongside a separate CPU.

**Algorithm 1.** - TensorFlow Algorithm for GPU

```
mirrored_strategy = tf.distribute.MirroredStrategy
(devices = [GPU:0],
cross_device_ops = tf.distribute.HierarchicalCopyAllReduce());
withtf.device(gpu:0) :;
tf.debugging.set_log_device_placement(True);
```

# 4 Experiments

In this section, we present the experimental framework adopted. First, the benchmark functions are described. Then, the runtime environment is detailed, followed by results and discussion.

## 4.1 Benchmark Functions

Benchmark functions are optimization problems presented as mathematical numerical functions [7]. These functions represent a set of optimization values with the most suitable parameters that help achieve the best solution, where $D$ represents the dimension of the problem. The best solution is among many optimal solutions spread across the entire problem scenario with different values. Any optimization algorithm, including metaheuristic algorithms, tends to search for the best solution, but there is no guarantee that the result expected by the function will always be achieved.

The efficiency of any metaheuristic is measured by its global search ability and local convergence ability. Algorithms with good global search capacity can explore a wide area of the solution space and are less likely to get stuck in local optima (local minima or maxima). Simultaneously, metaheuristics with efficient convergence ability can thoroughly explore a local area, minimizing the risk of missing the best solution. According to Hussain et al., [7], in the literature, different sets of benchmark functions have been introduced as configurations in new metaheuristic algorithms. Different experiments have determined specific dimensional parameters to propose any modified variant of an existing method.

That being said, [7] mentions that there is no fixed standard parameter, such as standard research settings, which leads new researchers to conduct experiments to validate their theses. These diversified approaches indicate the ability to solve specific problems, where a metaheuristic addresses a specific set of problems. However, most benchmark testing functions generally share a similar nature, featuring numerical optimization problems or, more specifically, problems with numerical constraints. Table 1 presents the list of benchmark function equations and their domains.

## 4.2 Runtime Environment

The local execution platform consisted of an Intel Xeon E3-1200 v6/7th i5-7200U Gen-Core notebook running a 64-bit Linux Operating System (Ubuntu) with 8GB of memory. Additionally, the Google Colab platform with TensorFlow version 2.8.0 was utilized. Two virtual machines were deployed for measuring the speedup metrics, one for CPU and another for GPU. The graphics card configuration included nvcc: NVIDIA(R) Cuda compiler driver, Copyright (c) 2005–2020 NVIDIA Corporation, Cuda compilation tools, release 11.1, V11.1.105. Linux operating system was pre-configured in Colab to enable parallel usage of both GPU and CPU.

**Table 1.** List of the benchmark functions used in the experiments

| Function | Equation | Domain |
|---|---|---|
| Elliptic | $\sum_{i=1}^{D}(10^6)^{\frac{i-1}{D-1}} x_i^2$ | -100,100 |
| Cigar | $x_i^2 + 10^6 \sum_{i=2}^{D} x_i^2$ | -100,100 |
| Discus | $10^6 x_i^2 + \sum_{i=2}^{D} x_i^2$ | -100,100 |
| Rosenbrock | $\sum_{i=1}^{D-1} (100(x_i^2 - x_{i+1})^2 + (x_i - 1)^2)$ | -5,10 |
| Ackley | $-20exp(-0.2\sqrt{\frac{1}{D}\sum_{i=1}^{D} x_i^2}) - exp(\frac{1}{D}\sum_{i=1}^{D} cos(2\pi x_i)) + 20 + e$ | -32,32 |
| Salomon | $1 - \cos(2\pi\sqrt{\sum inx_i^2}) + 0.1\sqrt{\sum inx_i^2}$ | -500,500 |
| Griewank | $\sum_{i=1}^{D-1} \frac{x_i^2}{4000} - \prod_{i=1}^{D} cos(\frac{x_i}{\sqrt{i}}) + 1$ | -600,600 |
| Rastrigin | $\sum_{i=1}^{D}(x_i^2 - 10cos(2\pi x_i) + 10)$ | -5,5 |

For this work, standard parameters were employed for each algorithm, including 10,000 interactions, a population size of 100 individuals, and dimensions of 50, 100, and 500. All execution processes were performed one at a time for each function, with memory clearance after each execution to prevent cached data that could bias the results. The GPU used in Colab was the T4 with 15GB of memory. Notably, there were limitations during execution, even in the Colab PRO version, due to access constraints for research testing.

### 4.3 Results and Discussions

As described in the previous Section, we focus on the speedup in terms of time. However, aiming to allow a fair comparison, we provided the benchmark function scores for dimension 500, as shown in Table 2. As previously stated, we are not looking for solution quality. Instead, we are looking for speedup.

Observing the results displayed in Table 2 it is possible to verify that PSO and GA got stuck in local minima, while DE and CS achieved better results for 10,000 interactions. Following, we analyze the speedup for each dimension considered in the experiments (50, 100, and 500), separating by metaheuristics to make it easy to follow. Tables 3 and 4 present the speedup results for PSO, GA, DE, and CS, respectively.

Analyzing the results given in Tables 3 and 4, it is possible to verify that, even for a lower dimensionality, it was possible to increase the speedup with GPU with a low programming effort. For instance, the TensorFlow implementation for PSO was 2.95 times faster than CPU on average. In GA, this difference is even higher, with the GPU being four times faster than the CPU implementation.

**Table 2.** Result function 500D and 10,000 iteration.

| | CPU/GPU | | CPU/GPU | | CPU/GPU | | CPU/GPU | |
|---|---|---|---|---|---|---|---|---|
| Function | **PSO** | | **GA** | | **DE** | | **CS** | |
| Discus | 46,781 | 47,741 | 56,841 | 61,666 | 14,513 | 19,348 | **10,666** | 11,001 |
| Rastrigin | 13,391 | 10,682 | 16,009 | 14,465 | **3,204** | 4,062 | 16,361 | 16,228 |
| Rosenbrock | 76,581 | 76,229 | 7,858 | 7,796 | 2,171 | **1,555** | 50,848 | 49,533 |
| Cigar | 43,973 | 42,850 | 10,959 | 10,934 | **4,461** | 59,354 | 11,090 | 11,259 |
| Elliptic | 34,114 | 47,345 | 37,564 | 35,596 | 32.0 | **12.0** | 67,065 | 62,876 |
| Ackley | 5.56 | 5.35 | 6.54 | 6.86 | **1.47** | 1.93 | 13.88 | 13.95 |
| Griewank | 3.0 | 27.53 | 103.06 | 99.05 | **0.007** | 5.47 | 3.79 | 3.72 |
| Salomon | 68.69 | 71.09 | 56.08 | 58.49 | **1.90** | 2.00 | 10.8 | 10.51 |

**Table 3.** Results - Speedup - 50D - PSO and GA

| | PSO | | | GA | | |
|---|---|---|---|---|---|---|
| **Function** | **CPU Time** | **GPU Time** | **Sp** | **CPU Time** | **GPU Time** | **Sp** |
| Discus | 198.62 | 67.83 | 2.93 | 550.11 | 63.85 | 8.62 |
| Rastrigin | 196.6 | 90.39 | 2.18 | 185.39 | 68.59 | 2.70 |
| Rosenbrock | 211.69 | 70.65 | 3.00 | 187.1 | 52.3 | 3.58 |
| Cigar | 210.86 | 77.12 | 2.73 | 186.33 | 54.33 | 3.43 |
| Elliptic | 211.85 | 71.23 | 2.97 | 189.15 | 58.7 | 3.22 |
| Ackley | 219.4 | 61.5 | 3.57 | 195.72 | 56.59 | 3.46 |
| Griewank | 209.91 | 64.07 | 3.28 | 186.42 | 58.48 | 3.19 |
| Salomon | 211.98 | 62.84 | 3.37 | 177.35 | 49.9 | 3.55 |

**Table 4.** Results - Speedup - 50D - DE and CS

| | DE | | | CS | | |
|---|---|---|---|---|---|---|
| **Function** | **CPU Time** | **GPU Time** | **Sp** | **CPU Time** | **GPU Time** | **Sp** |
| Discus | 544.06 | 67.6 | 8.05 | 359.27 | 76.48 | 4.7 |
| Rastrigin | 202.31 | 56.69 | 3.57 | 301.07 | 81.89 | 3.68 |
| Rosenbrock | 185.16 | 55.64 | 3.33 | 212.07 | 54.22 | 3.91 |
| Cigar | 185.71 | 59.85 | 3.10 | 202.31 | 83.15 | 2.43 |
| Elliptic | 192.28 | 81.67 | 2.35 | 214.22 | 60.78 | 3.52 |
| Ackley | 220.64 | 107.09 | 2.06 | 286.87 | 82.4 | 3.48 |
| Griewank | 187.77 | 81.26 | 2.31 | 225.12 | 76.01 | 2.96 |
| Salomon | 180.66 | 62.76 | 2.88 | 209.8 | 70.4 | 2.98 |

The same behavior is observed considering 100D, as presented in Tables 5 and 6. The PSO got a better GPU performance for the Elliptic function, almost six times faster; Discus presented the best results using GA (6.83x faster); and CS got 5.37x faster for Rosenbrock. The results for DE were consistent, with GPU being three times faster on average.

**Table 5.** Results - Speedup - 100D - PSO and GA

| | PSO | | | GA | | |
|---|---|---|---|---|---|---|
| **Function** | **CPU Time** | **GPU Time** | **Sp** | **CPU Time** | **GPU Time** | **Sp** |
| Discus | 203.78 | 68.00 | 3.00 | 503.69 | 73.7 | 6.83 |
| Rastrigin | 212.98 | 74.23 | 2.87 | 194.16 | 61.37 | 3.16 |
| Rosenbrock | 277.46 | 68.8 | 4.03 | 190.91 | 49.12 | 3.89 |
| Cigar | 197.79 | 67.09 | 2.95 | 188.16 | 51.52 | 3.65 |
| Elliptic | 351.52 | 61.5 | 5.72 | 189.65 | 54.38 | 3.49 |
| Ackley | 185.53 | 62.13 | 2.99 | 194.24 | 52.55 | 3.70 |
| Griewank | 187.93 | 70.7 | 2.66 | 196.4 | 52.44 | 3.75 |
| Salomon | 191.38 | 52.2 | 3.67 | 181.72 | 52.4 | 3.47 |

**Table 6.** Results - Speedup - 100D - DE and CS

| | DE | | | CS | | |
|---|---|---|---|---|---|---|
| **Function** | **CPU Time** | **GPU Time** | **Sp** | **CPU Time** | **GPU Time** | **Sp** |
| Discus | 212.72 | 84.82 | 2.51 | 302.6 | 71.62 | 4.23 |
| Rastrigin | 207.03 | 69.66 | 2.97 | 198.17 | 54.29 | 3.65 |
| Rosenbrock | 194.23 | 56.14 | 3.46 | 283.58 | 52.85 | 5.37 |
| Cigar | 183.43 | 53.74 | 3.41 | 184.4 | 50.56 | 3.65 |
| Elliptic | 230.8 | 69.32 | 3.33 | 199.94 | 49.71 | 4.02 |
| Ackley | 199.21 | 57.72 | 3.45 | 194.07 | 57.05 | 3.40 |
| Griewank | 203.2 | 71.49 | 2.84 | 188.41 | 63.52 | 2.97 |
| Salomon | 201.21 | 68.31 | 2.95 | 190.88 | 50.41 | 3.79 |

Finally, Tables 7 and 8 present the results for the dimension equal to 500. The speedup follows the same behavior seen for dimensions 50 and 100. Obviously, the time is higher considering the dimensionality (first two columns). We call attention to Table 7, in which GA had a prominent speedup of almost eight times faster in the Discuss function. Additionally, when examining the specific functions, Ackley, Griewank, and Salomon, considering all dimensions, it became evident that the use of GPU significantly impacted the results.

**Table 7.** Results - Speedup - 500D - PSO and GA

| | PSO | | | GA | | |
|---|---|---|---|---|---|---|
| **Function** | **CPU Time** | **GPU Time** | **Sp** | **CPU Time** | **GPU Time** | **Sp** |
| Discus | 199.22 | 61.72 | 3.23 | 720.72 | 94.38 | 7.64 |
| Rastrigin | 194.87 | 66.76 | 2.92 | 207.23 | 76.0 | 2.73 |
| Rosenbrock | 195.53 | 56.27 | 3.47 | 201.03 | 74.18 | 2.71 |
| Cigar | 213.75 | 60.38 | 3.54 | 519.49 | 81.21 | 6.40 |
| Elliptic | 193.57 | 63.83 | 3.03 | 188.9 | 54.99 | 3.44 |
| Ackley | 209.51 | 60.32 | 3.47 | 195.88 | 54.63 | 3.59 |
| Griewank | 195.24 | 64.25 | 3.04 | 527.34 | 83.6 | 6.31 |
| Salomon | 184.17 | 72.45 | 2.54 | 184.22 | 56.06 | 3.29 |

**Table 8.** Results - Speedup - 500D - DE and CS

| | DE | | | CS | | |
|---|---|---|---|---|---|---|
| **Function** | **CPU Time** | **GPU Time** | **Sp** | **CPU Time** | **GPU Time** | **Sp** |
| Discus | 410.02 | 65.5 | 6.26 | 489.35 | 79.64 | 6.14 |
| Rastrigin | 344.27 | 77.34 | 4.45 | 176.72 | 57.00 | 3.10 |
| Rosenbrock | 537.15 | 69.72 | 7.70 | 308.17 | 72.98 | 4.22 |
| Cigar | 188.34 | 69.37 | 2.72 | 189.47 | 53.84 | 3.52 |
| Elliptic | 198.66 | 69.11 | 2.87 | 188.7 | 55.39 | 3.41 |
| Ackley | 210.33 | 62.22 | 3.38 | 179.25 | 68.71 | 2.61 |
| Griewank | 482.25 | 76.44 | 6.31 | 262.27 | 80.01 | 3.28 |
| Salomon | 183.13 | 60.44 | 3.03 | 188.51 | 64.39 | 2.93 |

Analyzing all the dimensions considered, it is possible to conclude that the investigated approach can be eight times faster for high-dimension benchmark functions using only one GPU - this can be even higher using multiple GPUs. Moreover, it is noteworthy that it was achieved with low programming effort. The results indicated that for the Ackley, Griewank, and Salomon functions, the GPU outperformed the CPU in processing speed and provided high-quality solutions. This increase in efficiency is crucial for solving real-world problems.

The tests showed that the GPU excelled in speed even though the CPU exhibited excellent overall results for the functions. With low programming effort, these performance gains are crucial for complex optimization problems in industry, in which the engineers and data analysis teams have constraints in terms of time/resources - considering the CUDA learning curve - demonstrating the applicability of our research findings in real-world scenarios.

These findings underscore not only the importance of parallelism in computing architectures but also the specific relevance of the GPU when tackling challenging optimization functions. The results allow us to conclude that combining GPU with metaheuristics, especially when addressing intricate functions like Ackley, Griewank, and Salomon in multiple dimensions, offers a highly positive and practical approach to solving real-world problems. These results have significant implications for future optimization and parallel computing research, paving the way for faster and more precise solutions across various applied domains.

# 5 Conclusion

This work presented the evaluation of using TensorFlow parallelism for implementing metaheuristics. The experimental framework embraces eight well-known benchmark functions solved by four metaheuristics (Particle Swarm Optimization, Genetic Algorithm, Differential Evolution, and Cuckoo Search. The results demonstrate that TensorFlow parallelization can speed up to eight times in high-dimension benchmark functions with low programming effort.

Significant performance improvements, achieved with low programming effort in TensorFlow, are essential in addressing complex optimization challenges, especially in the industrial domain where time can be an expensive resource. These advancements are especially valuable when engineers and data analysis teams face limitations in terms of time and resources, particularly considering the learning curve associated with CUDA programming. Such results emphasize the practical applicability of our research findings in real-world scenarios.

In future work, we plan to enhance our experimental setup further, including implementing parallel hybrid metaheuristics that are more computing-intense. These expansions also include incorporating additional benchmark functions, conducting tests with multiple GPUs and diverse models, and exploring the potential of hybrid metaheuristics. These efforts aim to deepen our understanding and enlarge the scope of our research, ultimately advancing the state of the art in optimization techniques.

**Acknowledgments.** The authors acknowledge the support of the *Conselho Nacional de Desenvolvimento Científico e Tecnológico* (CNPq) - DT - 308334/2020; and the *Fundação Amazônia de Amparo a Estudos e Pesquisas* (FAPESPA) - PRONEM-FAPESPA/CNPq nº *045/2021.*

# References

1. Top 500 - June 2023 report. https://www.top500.org/lists/top500/2023/06/. Accessed 11 Nov 2023
2. Abadi, M., et al.: TensorFlow: large-scale machine learning on heterogeneous systems (2015). http://tensorflow.org/, "Software available from tensorflow.org"
3. Brodtkorb, A.R., Hagen, T.R., Sætra, M.L.: Graphics processing unit (GPU) programming strategies and trends in GPU computing. J. Parall. Distrib. Comput. **73**(1), 4–13 (2013). https://doi.org/10.1016/j.jpdc.2012.04.003, metaheuristics on GPUs

4. Charilogis, V., Tsoulos, I.G.: A parallel implementation of the differential evolution method. Analytics **2**(1), 17–30 (2023). https://doi.org/10.3390/analytics2010002
5. Fabris, F., Krohling, R.A.: A co-evolutionary differential evolution algorithm for solving min-max optimization problems implemented on GPU using C-CUDA. Expert Syst. Appl. **39**(12), 10324–10333 (2012)
6. Franco, M.A., Krasnogor, N., Bacardit, J.: Speeding up the evaluation of evolutionary learning systems using GPGPUs. In: Proceedings of the 12th Annual Conference on Genetic and Evolutionary Computation, pp. 1039–1046. GECCO '10, Association for Computing Machinery, New York, NY, USA (2010). https://doi.org/10.1145/1830483.1830672
7. Hussain, K., Salleh, M.N.M., Cheng, S., Naseem, R.: Common benchmark functions for metaheuristic evaluation: a review (2017)
8. Mane, S.U., Lokare, P.S., Gaikwad, H.R.: Overview and applications of GPGPU based parallel ant colony optimization (2022)
9. Owens, J.D., Houston, M., Luebke, D., Green, S., Stone, J.E., Phillips, J.C.: GPU computing. Proc. IEEE **96**(5), 879–899 (2008). https://doi.org/10.1109/JPROC.2008.917757
10. Veronese, L.d.P., Krohling, R.A.: Swarm's flight: accelerating the particles using C-CUDA. In: IEEE Congress on Evolutionary Computation, pp. 3264–3270 (2009). https://doi.org/10.1109/CEC.2009.4983358
11. Zhuo, Y., Zhang, T., Du, F., Liu, R.: A parallel particle swarm optimization algorithm based on GPU/CUDA. Appl. Soft Comput. **144**, 110499 (2023). https://doi.org/10.1016/j.asoc.2023.110499

# K-Nearest Neighborhood Based Hybrid Recommendation System Using Collaborative Filtering and Meta-Learning

Hrriday Agarwal, Lakshay Raghav, Krittin Nagar, and Seba Susan(✉)

Department of Information Technology, Delhi Technological University, Delhi, India
seba_406@yahoo.in

**Abstract.** Recommendation system is a niche area of research in the field of artificial intelligence where the inputs are user-item rating matrices. Collaborative filtering is the generally acknowledged solution for constructing recommendation systems. Matrix factorization is a memory-efficient collaborative filtering scheme that creates a latent feature space for users and items. In this paper, we propose a novel hybrid combination of two matrix factorization approaches employing stochastic gradient descent and alternating least squares, respectively, as learning algorithms, and an artificial neural network as the meta-learner. K-Nearest Neighborhood computation based on the cosine similarity metric is used for shortlisting the top-rated items under consideration for the meta-learning process. The primary task of the meta-learner is to learn parallelly from multiple collaborative filtering schemes which improves the accuracy of the recommendations. The efficacy of our approach is proved by the significantly reduced Root Mean Square Error (RMSE) of 0.6459 as compared to the baseline methods for the experiments conducted on the benchmark MovieLens-100 K dataset.

**Keywords:** Recommendation system · Collaborative filtering · Meta-learning · Matrix factorization · Alternating Least Squares · K-Nearest Neighbor

## 1 Introduction

Recommendation systems are an essential part of modern life, with an increasing number of people using this tool for discovering relevant products, movies, music tracks and friends in social networks [1]. In recent times, there has been observed an increasing involvement of Artificial Intelligence (AI) in designing efficient recommendation systems that recommend products/items to new users based on previous user ratings or reviews [2]. Simple K-Nearest Neighbor (KNN) based approaches [3, 4] have existed since long that rely on distance or similarity measures to find the closest user-item rating.

Collaborative filtering [5, 6] is considered to be the most traditional and popular tool for constructing recommendation systems. It is based on the collective user behavior, and the assumption that users would continue to prefer products based on the patterns


A. Bajaj et al. (Eds.): HIS 2023, LNNS 1224, pp. 317–323, 2025.
https://doi.org/10.1007/978-3-031-78925-0_31

observed in the past. Matrix factorization is a memory-efficient collaborative filtering scheme that maps users and items to a latent feature space. Korn et al. [7] introduced a recommender system using matrix factorization, considering user-item interactions, implicit feedback and auxiliary data. The authors in [8] proposed a deep neural network-architecture to learn the low dimensional matrix representations output from matrix factorization model; the system incorporates non-preference implicit feedback to improve the prediction accuracy. Stochastic gradient descent (SGD) and Alternating Least Squares (ALS) are the two popular learning algorithms for matrix factorization [7–9]. The ALS algorithm [9], as the name suggests, uses an alternative optimization scheme instead of the stochastic gradient optimization in traditional matrix factorization.

The advent of neural networks has revolutionized the field of AI in various domains due to the uncanny ability of neural networks to generalize for unseen examples [10]. The application of neural networks to construct efficient recommendation systems is a popular concept due to the well-known property of neural networks to generalize to unseen examples [11]. Hybrid models incorporating both Collaborative Filtering and neural network learning have been introduced recently [12, 13] to ensure that the goodness of both collaborative filtering and neural networks is brought together to construct efficient recommendation systems. The result is improved recommendation accuracy as compared to the traditional collaborative filtering. In this paper, we propose a hybrid combination of matrix factorization using SGD and ALS, with neural networks as the meta-learner that learns the predictions of the two collaborative filtering algorithms. The K-Nearest Neighborhood based on the cosine similarity metric is computed for generating the list of top-rated items which are then refined using the predictions of the proposed meta-learning model.

The paper is organized in the following manner. Section 2 reviews the related work in literature. Section 3 presents the proposed hybrid recommendation system and the methodology followed for the experiments. Section 4 discusses the experimentation and includes a discussion on the results, while Sect. 5 concludes the paper.

## 2 Background

Recommender system research in literature usually follow either a user-based or item-based approach; recent works suggest hybrid approaches. A recommender system is introduced in [12] that combines a graph-based model, user ratings similarity, and demographic/location information. Autoencoder feature extraction enhances the approach by creating new features from combined attributes. Recommender systems using Embarrassingly Shallow Autoencoders (EASE) are explored in [13], which are web technologies that suggest items based on user behavior and preferences. While previous evaluations focused on algorithm performance, recent research emphasizes user experience. This work surveys user experience in recommender systems, examining methods that improve user interaction, trust, and adoption. It covers three critical user-system interactions: preference elicitation, refinement, and recommendation presentation. Additionally, it discusses evaluation frameworks measuring overall system qualities and their impact on user intentions.

The KNN-based Collaborative Filtering Algorithm [14] addresses challenges of information overload in the digital age and its impact on the entertainment industry, specifically movies. To address this issue, the authors proposed a personalized movie recommendation system that leverages user behavior analysis. This system falls under the category of information filtering technology and combines various data mining algorithms to match users with their interests. The paper discusses common recommendation system types, emphasizing the effectiveness of hybrid approaches. The authors in [15] discuss recommender systems and the challenges they face in providing personalized recommendations for information, products, or services. They highlight the need for high-quality recommendations, efficient processing for large user and item databases, and addressing data sparsity issues. The paper explores item-based collaborative filtering techniques, analyzing different algorithms for generating recommendations.

Bayesian personalized ranking (BPR) [16] focuses on feedback information in recommender systems. This method extracts implicit feedback from non-observed items, and uses adversarial perturbations to reduce noise in embedding vectors through a minimax game. The authors in [17] review 85 recommendation papers, identify key evaluation factors, and establish benchmarks with standardized procedures. The top-N recommendation task is discussed in [18, 19], where the aim is to recommend the top-rated items without displaying predicted rating values. A hybrid recommendation system involving SGD and ALS [20] suggested training using ALS initially and then retraining with SGD.

## 3 Methodology

In this section, we describe the proposed hybrid recommendation system incorporating two collaborative filtering algorithms: - matrix factorization using SGD and ALS in a meta-learning framework. In matrix factorization, we break the user-item rating matrix into two smaller matrices of smaller dimensions. These matrices are often called 'Factor matrices' or Embedding layers that represent users and movies in a lower dimensional space. The process of factorization enables the system to predict user-personalized movie ratings. The objective of the matrix factorization algorithm is to reduce the error between the predicted and actual user ratings. The ALS algorithm also approximates the user-item interaction matrix as the product of two lower-dimensional matrices: user matrix and item matrix. However, ALS differs in the training procedure; the user matrix and item matrix are subject to the optimization procedure alternately while holding the other matrix fixed. The dimensions of these matrices are typically much smaller than the dimensions of the original interaction matrix.

The process flow of the proposed method containing the details of the proposed methodology is shown in Fig. 1. The predictions made by the two algorithms (SGD and ALS) are fed as input to an artificial neural network for predicting the final outcome. The artificial neural network is the meta-learner for the ensemble that learns the predicted ratings of the individual models and outputs one final prediction (user rating). Meta-learning is a well-tried concept in machine learning that aims to improve the overall prediction by learning the predictions of individual models in the ensemble [21, 22].

The K-Nearest Neighborhood computation is used amply in literature to distinguish and separate subsets of similar samples from the dissimilar ones [23]. As observed from the block diagram in Fig. 1, the K-Nearest Neighborhood is first computed (using the cosine similarity metric) for generating the list of top-rated items which are then refined

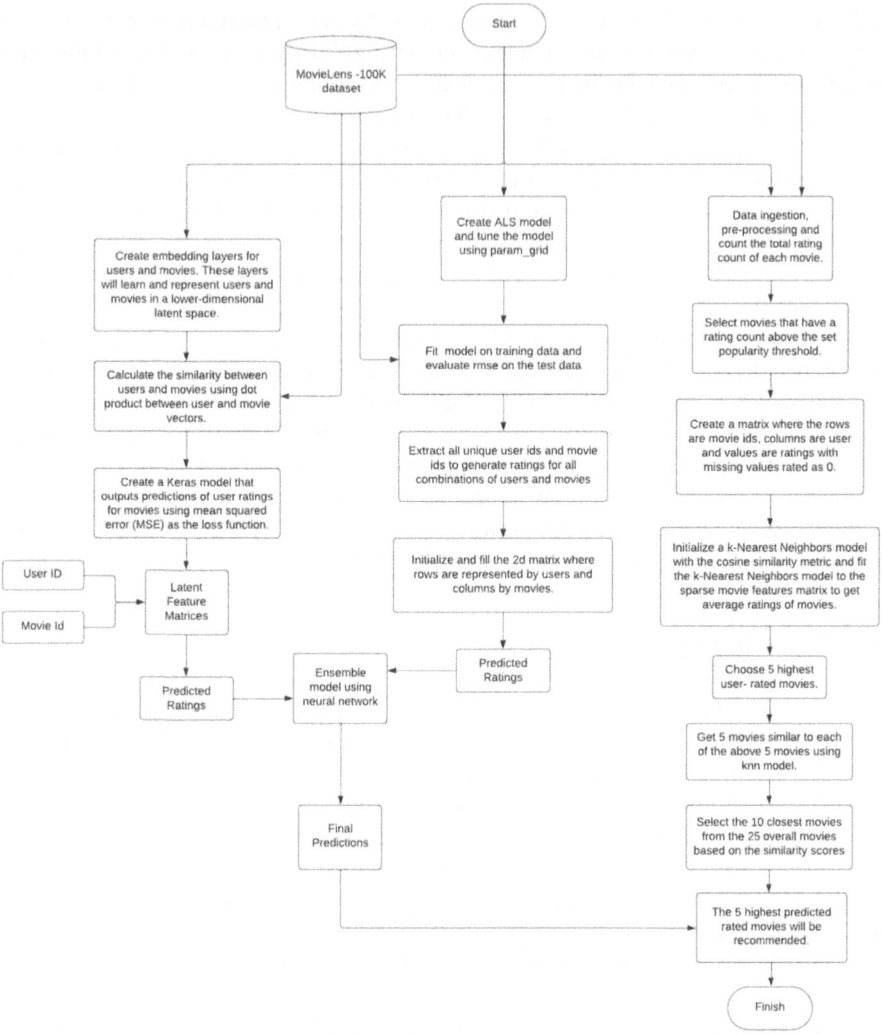


**Fig. 1.** Process flow of the proposed method.

using the neural network predictions of the proposed meta-learning model. The top-10 movies retrieved based on their similarity scores in the K-Nearest Neighborhood, are compared with respect to the predicted ratings output from the meta-learning module. The resulting top-5 rated movies are recommended to the user.

The methodology illustrated in Fig. 1 are summarized below.

- We take the prediction ratings from the SGD and ALS matrix factorization models and pass them to the meta-learner model which comprises of an artificial neural network.
- The neural network has (2,10,6,1) architecture.
- The final predicted ratings for all combinations of user ids and movie ids are obtained from the meta-learning stage.

- To make recommendations to the users, we first choose 5 highest user-rated movies.
- Now we retrieve 5 movies similar to each of the above 5 movies using the K-Nearest Neighbor algorithm.
- From the 25 movies, we choose 10 movies based on their similarity scores.
- We next compare the final predicted ratings of the 10 movies and choose 5 movies that have the highest ratings as predicted by the meta-learning model; these top-5 movies will be recommended to the user by the proposed recommendation system.

## 4 Experimentation and Results

The MovieLens-100 K dataset[1] was used for the experiments; it contains ratings of 9742 movies by 610 users. There is a minimum of 20 ratings per user. For the K-Nearest Neighbor computation, we retained only movies having more than 50 ratings (i.e. popularity threshold = 50). One sample of top-rated movies by a user is shown in Fig. 2.

| movieId | title | genres | Average Rating |
|---|---|---|---|
| 318 | Shawshank Redemption, The (1994) | Crime\|Drama | 4.49 |
| 356 | Forrest Gump (1994) | Comedy\|Drama\|Romance\|War | 4.05 |
| 1197 | Princess Bride, The (1987) | Action\|Adventure\|Comedy\|Fantasy\|Romance | 4.21 |
| 2959 | Fight Club (1999) | Action\|Crime\|Drama\|Thriller | 4.18 |
| 3949 | Requiem for a Dream (2000) | Drama | 3.87 |

**Fig. 2.** User-item rating for the MovieLens-100 K dataset (user-rated movies).

All experiments were performed on Python software (3.9.1 version) on a 1.6GHz Intel Core i5 PC. The Keras library was used to implement the collaborative filtering algorithms. The number of latent factors was constrained to 64. The number of epochs were set to 50 though the results converged around 16 epochs only. While 80% of the dataset was used for training, the remaining 20% was used for testing. A five-fold cross-validation approach was used. We illustrate in Fig. 3 the hybrid model's predicted rating for the top-rated movies of the same user whose ratings were shown in Fig. 2. As observed from Fig. 2, the user has given higher ratings for movies belonging to the Crime/Thriller genre. A similar trend is observed from Fig. 3 where the movies belonging to the Crime/Thriller genre are assigned higher ratings by the proposed recommendation system as compared to other genres such as Comedy, Documentary, Animation etc. which were not rated high by the user (Fig. 2).

We compare our approach with that of matrix factorization using SGD [7] and ALS [9] in Table 1. The average RMSE values for the SGD and ALS are 1.06 and 0.89, respectively. The proposed method has an average RMSE of 0.6459 (over five folds) which is a significant reduction as compared to the two collaborative filtering schemes establishing the efficacy of our hybrid approach.

1 https://grouplens.org/datasets/movielens/100 k/

| movieId | title | genres | Average Rating |
|---|---|---|---|
| 50 | Usual Suspects, The (1995) | Crime\|Mystery\|Thriller | 4.37 |
| 858 | Godfather, The (1972) | Crime\|Drama | 4.49 |
| 2542 | Lock, Stock & Two Smoking Barrels (1998) | Comedy\|Crime\|Thriller | 4.13 |
| 4226 | Memento (2000) | Mystery\|Thriller | 4.23 |
| 6016 | City of God (Cidade de Deus) (2002) | Action\|Adventure\|Crime\|Drama\|Thriller | 4.30 |

**Fig. 3.** User-item rating predicted by the proposed recommendation system (unrated movies).

**Table 1.** Average RMSE of different methods on the MovieLens-100 K dataset.

| Method | Average RMSE |
|---|---|
| Matrix factorization using SGD [7] | 1.06 |
| Matrix factorization using ALS [9] | 0.89 |
| Ours | 0.6459 |

## 5 Conclusion

In this paper, we present a hybrid recommendation system that involves two collaborative filtering algorithms and an artificial neural network as the meta-learner. The K-Nearest Neighborhood was computed for shortlisting the likely items which are movies in our case. Collaborative filtering captures the relationships between two entities-users and items. The predictions of the two collaborative filtering algorithms (matrix factorization using SGD and ALS) are learnt using the neural network meta-learner. The experiments on the benchmark MovieLens-100 K dataset proves the efficacy of the proposed approach which achieves the lowest RMSE of 0.6459.

However, automated recommendation systems that are modeled on generic user choices and preferences are sensitive to bias, a phenomenon in which some users tend to give higher ratings than other users to all items. Future recommendation systems need to build around this problem, and more personalized recommendation engines need to be devised that suit individual user's choices.

## References

1. Maniktala, M., Sachdev, S., Bansal, N., Susan, S.: Finding the most informational friends in a social network based recommender system. In: 2015 Annual IEEE India Conference (INDICON), pp. 1–6. IEEE (2015)
2. Zangerle, E., Bauer, C.: Evaluating recommender systems: survey and framework. ACM Comput. Surv. **55**(8), 1–38 (2022)
3. Adeniyi, D.A., Wei, Z., Yongquan, Y.: Automated web usage data mining and recommendation system using K-nearest neighbor (KNN) classification method. Appl. Comput. Inf. **12**(1), 90–108 (2016)

4. Airen, S., Agrawal, J.: Movie recommender system using k-nearest neighbors variants. Natl. Acad. Sci. Lett. **45**(1), 75–82 (2022)
5. Ben Schafer, J., Frankowski, D., Herlocker, J., Sen, S.: Collaborative filtering recommender systems. In: The Adaptive Web: Methods and Strategies of Web Personalization, pp. 291–324. Springer Berlin Heidelberg, Berlin, Heidelberg (2007)
6. Ekstrand, M.D., Riedl, J.T., Konstan, J.A.: Collaborative filtering recommender systems. Found. Trends Hum.–Comput. Interact. **4**(2), 81–173 (2011)
7. Koren, Y., Bell, R., Volinsky, C.: Matrix factorization techniques for recommender systems. Computer. **42**(8), 30–37 (2009)
8. Xue, H.-J., Dai, X., Zhang, J., Huang, S., Chen, J.: Deep matrix factorization models for recommender systems. IJCAI. **17**, 3203–3209 (2017)
9. Takács, G., Tikk, D.: Alternating least squares for personalized ranking. In: Proceedings of the Sixth ACM Conference on Recommender Systems, pp. 83–90 (2012)
10. Susan, S., Ranjan, R., Taluja, U., Rai, S., Agarwal, P.: Neural net optimization by weight-entropy monitoring. In: Computational Intelligence: Theories, Applications and Future Directions-Volume II: ICCI-2017, pp. 201–213. Springer Singapore (2019)
11. Abiodun, O.I., Jantan, A., Omolara, A.E., Dada, K.V., Mohamed, N.A.E., Arshad, H.: State-of-the-art in artificial neural network applications: a survey. Heliyon. **4**, 11 (2018)
12. Darban, Z.Z., Valipour, M.H.: GHRS: graph-based hybrid recommendation system with application to movie recommendation. Expert Syst. Appl. **200**, 116850 (2022)
13. Pu, P., Chen, L., Rong, H.: Evaluating recommender systems from the user's perspective: survey of the state of the art. User Model. User-Adap. Inter. **22**, 317–355 (2012)
14. Cui, B.-B.: Design and implementation of movie recommendation system based on Knn collaborative filtering algorithm. In: ITM Web of Conferences, vol. 12, p. 04008. EDP Sciences (2017)
15. Sarwar, B., Karypis, G., Konstan, J., Riedl, J.: Item-based collaborative filtering recommendation algorithms. In: Proceedings of the 10th International Conference on World Wide Web, pp. 285–295 (2001)
16. Wang, J., Han, P.: Adversarial training-based mean Bayesian personalized ranking for recommender system. IEEE Access. **8**, 7958–7968 (2019)
17. Sun, Z., Yu, D., Fang, H., Yang, J., Xinghua, Q., Zhang, J., Geng, C.: Are we evaluating rigorously? Benchmarking recommendation for reproducible evaluation and fair comparison. In: Proceedings of the 14th ACM Conference on Recommender Systems, pp. 23–32 (2020)
18. Anelli, V.W., Bellogín, A., Di Noia, T., Jannach, D., Pomo, C.: Top-n recommendation algorithms: a quest for the state-of-the-art. In: Proceedings of the 30th ACM Conference on User Modeling, Adaptation and Personalization, pp. 121–131 (2022)
19. Cremonesi, P., Koren, Y., Turrin, R.: Performance of recommender algorithms on top-n recommendation tasks. In: Proceedings of the Fourth ACM Conference on Recommender Systems, pp. 39–46 (2010)
20. Rao, P., Rao, T.V., Kurumalla, S., Prakash, B.: Matrix factorization based recommendation system using hybrid optimization technique. EAI Endorsed Trans. Energy Web. **8**, 35 (2021)
21. Susan, S., Kumar, A., Jain, A.: Evaluating heterogeneous ensembles with boosting meta-learner. In: Inventive Communication and Computational Technologies: Proceedings of ICICCT 2020, pp. 699–710. Springer Singapore (2021)
22. Jain, A., Kumar, A., Susan, S.: Evaluating deep neural network ensembles by majority voting cum meta-learning scheme. In: Soft Computing and Signal Processing: Proceedings of 3rd ICSCSP 2020, Volume 2, pp. 29–37. Springer Singapore (2022)
23. Susan, S., Kumar, A.: DST-ML-EkNN: data space transformation with metric learning and elite k-nearest neighbor cluster formation for classification of imbalanced datasets. In: International Conference on Artificial Intelligence and Data Engineering, pp. 319–328. Springer Nature Singapore, Singapore (2019)

# Evaluating Collaborative Forecasting in Non-horizontal Federated Learning

Raúl Llasag Rosero(✉), Catarina Silva, and Bernardete Ribeiro

University of Coimbra, Polo II - Pinhal de Marrocos, 3030-290 Coimbra, Portugal
{rosero,catarina,bribeiro}@dei.uc.pt

**Abstract.** Federated Learning (FL) is changing the way we do Machine Learning (ML) by moving from standard centralized to decentralized and privacy-preserving approaches. A common FL strategy, the horizontal approach, consists in aggregating the weights of neural networks sharing the same structure, thereby enhancing the predictive abilities of the models in a collaborative manner. However, this approach doesn't work when the models from different parties analyze diverse input data. In these situations, evaluating the performance improvement of models with varying architectures among parties requires considering the distinctions among them. In this work, we focus on a collaborative regression prognostics problem: to predict the Remaining Useful Life (RUL) of turbofan engines measured by multiple stations using different sensor spaces. We use two ML architectures: Multilayer Perceptron followed by a Kalman Filter (MLP-KF) and Functional Multilayer Perceptron (FMLP). We propose a non-horizontal approach, i.e., an approach that does not assume that all parties share the input data. Each party catches a different subsample of sensor signals to solve the federated problem utilizing index synchronization and the Vertical Asynchronous Federated Learning (VAFL) algorithm. We create a simulated environment where each party assesses multiple models by repeating the experiment for each ML architecture. Then, we apply appropriate key metrics to compare prediction variability among parties and experiments. Results show that it is possible to use non-horizontal federated approaches to tackle scenarios where the input space is different among parties. The evaluation revealed that FMLP outperformed MLP-KF, providing more stable predictions.

**Keywords:** Federated Learning · Machine Learning · Vertical Federated Learning · Functional Multilayer Perceptron · Remaining Useful Life

## 1 Introduction

Federated Learning (FL) is privacy-preserving and decentralized Machine Learning (ML) approach that allows federated participants to improve their models' performance prediction while keeping data on their producers [1]. In developing federated solutions, engineers face the recurring challenge of selecting the


A. Bajaj et al. (Eds.): HIS 2023, LNNS 1224, pp. 324–333, 2025.
https://doi.org/10.1007/978-3-031-78925-0_32

most suitable ML architecture to address regression problems with diverse data partitioning settings, including horizontal, vertical, and hybrid configurations. This process becomes crucial for improving the forecasting performance of the federated solution.

Performance forecasting evaluations of ML architectures have been facilitated in Horizontal Federated Learning (HFL) [2]. This federated category refers to parties' datasets sharing the same input feature space, while its collaborative learning is based on aggregating models with the same architecture. Therefore, it requires just evaluating the prediction performance of the global AI model. However, performance evaluations in non-horizontal data partitioning approaches require considering the prediction variability among parties. This is because parties' models fit with different input spaces such as in Vertical (VFL) and Hybrid Federated Learning (Hybrid-FL) [3,4] forecast continuous values with an inevitable variability.

Selecting which ML architecture achieves better forecasts in non-horizontal data partitioning settings differs from horizontal ones. Even if participants agree to evaluate the federated solution using the same testing dataset, local forecasts vary because parties use different input features. Therefore, selecting an adequate ML architecture must consider evaluating the forecast variability over the federated network.

This work configured a collaborative prognostics problem whose main objective is to cooperatively enhance the estimation of machinery's useful lifetime to reduce over and under-maintenance. The federated experiment, configured in a vertical data partitioning, simulates that four Condition Monitoring Systems (CMSs) contribute to enhancing the Remaining Useful Life (RUL) Estimation of turbofan engines data [5]. A single experiment first required randomly distributing the input feature space of the fourth C-MAPSS set called FD004. Then, we solved the collaborative prognostics problem utilizing the Vertical Asynchronous Federated Learning (VAFL) algorithm [3] through two ML architectures: Multilayer Perceptron followed by a Kalman Filter (MLP-KF) [6] and Functional Multilayer Perceptron (FMLP) [7].

In the final step, we evaluated the forecasting variability among parties and federated experiments on MLP-KF and FMMLP by adopting appropriate key metrics and repeating the federated experiment ten times for each architecture. Reductions in average bias, average variability, and reproducibility closer to zero by moving from MLP-KL to FMLP demonstrate that FMLPs provide more stable and reproducible estimations.

The contributions of this paper can be summarized as follows:

- The implementation of an MLP-KF and FMLP to collaboratively estimate the Remaining Useful Life of turbofan engines shared in the following repository: https://github.com/rhllasag/VerticalFederatedLearning.
- The RUL estimation in a collaborative problem in which parties' models are trained with datasets that differ in feature space.

## 2 Non-horizontal Federated Learning

The Federated Learning principle predicts local models' performance improvement without sharing raw data to accomplish data-privacy policies such as General Data Protection Regulation (GDPR). In the federated optimization problem orchestrated by a central server, the optimization of a global AI model is performed in $K$ model aggregation iterations. A single $k$-th iteration comprises four steps: 1) global AI model forecasting; 2) local training; 3) local model uploading, and 4) global AI model computation.

Defining these steps has mainly guided the exploration of FL algorithms in the presence of nonindependent and nonidentically distributed data (non-i.i.d), unbalanced data and three data partitioning settings [8].

Federated Learning algorithms are categorized depending on the data partitioning approach supported, horizontal, vertical and hybrid [9]. In this sense, Vertical FL (VFL) and Hybrid FL are considered Non-Horizontal Federated Learning categories because they allow parties to differ in input feature space. Horizontal and Vertical FL are exceptional cases of Hybrid FL in which parties may vary in feature and sample space. However, research on VFL and Hybrid FL gained less interest due to model aggregation of multiple parties implies integrating knowledge from models whose architecture differs.

Collaborative problems aim to overcome the challenges posed by data distribution [1,8]. This distribution can negatively impact the final performance prediction of the global AI model compared to a model trained using all the data stored in a single server [9]. Therefore, the primary objective of FL is to bridge this gap and achieve a global AI model whose prediction performance closely resembles that of the hypothetical model trained with centralized data.

Evaluating parties' performance improvements were simplified in HFL algorithms compatible with Artificial Neural Networks (ANNs) because the input vectors are mapped to predictions using the same ANN broadcast by the server. However, it does not happen in VFL and Hybrid FL because the dimension of input vectors and model architecture of parties vary. Figure 1 illustrates a typical federated optimization problem solved using VFL algorithms based on sample index synchronization such as Vertical Asynchronous FL (VAFL) [3], Federated Block Coordinated Descent (FedBCD) [10] and Vertical Federated Deep Learning [11].

### Synchronous Vertical Federated Learning

The federated optimization problem of VFL considers 1) labels $Y$ stored by the central server or a single party, and 2) parties sharing the same sample identifiers [3,10,11]. Those considerations enable the application of the sample index synchronization strategy to guarantee parties using the $i$-th sample index with the same label $Y_i$ correctly. Doing so allows the coordinated Stochastic Gradient Descent (SGD) optimization to fit an ML model collaboratively.

Locally, the problem of party $j$ corresponds to finding the weights $w^j$ that minimize the local loss function $F^j(.)$ at the $i$-th input sample. Collaboratively, a

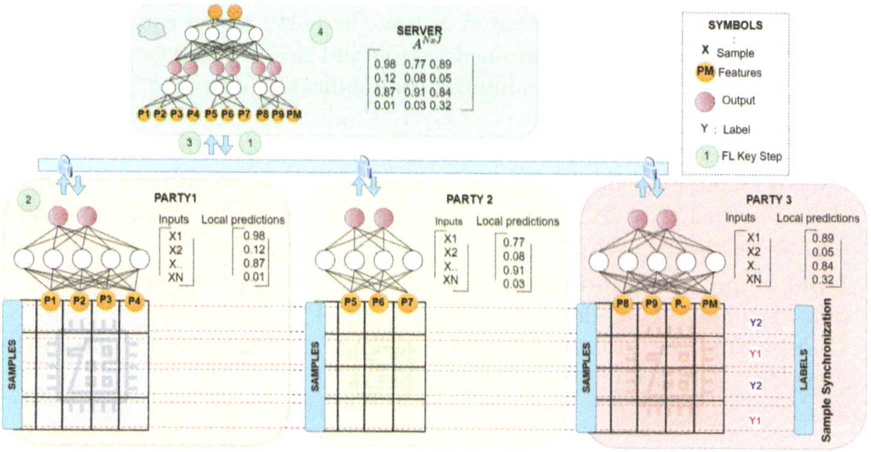


**Fig. 1.** Sample Synchronization for Vertical Federated Learning.

global loss function $F(.)$ evaluates how well the aggregation of $J$ local predictions weighted by $\psi$ predicts $Y_i$. In other words $\theta(\sum_{j=1}^{J} \psi \hat{Y_i^j}, Y_i)$ evaluates the distance between collaborative prediction $\sum_{j=1}^{J} \psi \hat{Y_i^j}$ and $Y_i$ by using the $i$-th sample.

Evaluating huge samples separately is computationally impractical. Therefore, the following global cost function $F(.)$ is calculated for a given batch $B$ of samples:

$$F(w) = \frac{1}{B} \sum_{i=1}^{B} \theta(\sum_{j=1}^{J} \psi \hat{Y_i^j},, Y_i). \tag{1}$$

where finding the optimal weights $w^*$ for the global AI model is also expressed as $w_* \triangleq \arg\min F(w)$.

The learning procedure of VFL algorithms starts by privately exchanging the local predictions with the server, which constantly updates a prediction matrix $A^{NxJ}$ of $N$ rows and $J$ columns. Figure 1 illustrates this matrix's update, a process performed via parties' pull and push requests to the server.

Considering a federated network with three parties, Fig. 1 illustrates how parties alight their samples to update $A$ using the same indexes. In vertical data partitioning, parties may differ in input features (yellow circles), while the architecture of their models differs. Therefore, the central server repeatedly ensembles the global AI model by joining parties' models. At a single model aggregation iteration, the following steps are conducted:

1. **Broadcasting the global AI model** is optional at zero time $\kappa = 0$ because the architecture of the parties' models differs. Models differing in architecture can not be initialized with the same weights even using the same random seed. Therefore, the architecture of the global AI model is defined after the first knowledge aggregation iteration $\kappa = 1$.

2. **Local training** refers to privately fitting the party model using input samples and predictions of determined sample indexes. Block-wise coordination support weight updating to minimize communication overhead by computing the local cost function defined in Eq. 1. Predictions of $A^{N \times J}$ matrix allow deriving the local loss function $F^j()$ w.r.t. the activation function of the last layer $L$ of the local model $\phi^L$. Then, the resulting gradient back-propagates the estimation error by applying the remaining chain rule. It allows calculating the weights $w^j_{\kappa+1}$ for the next global AI model computation $\kappa + 1$.
3. **Uploading local models** of parties and their local predictions $\hat{Y^j}$ to compute the global AI model and update matrix $A^{N \times J}$.
4. **Computing the global AI model** refers to creating a bigger model that joins local parameters $w^j_\kappa$ of $J$ parties.

## 3 Evaluation Metrics for Collaborative Forecasting

Evaluating how precise $M$ parties' models perform in forecasting horizon H of $N$ states (labels) is crucial for selecting the ML architecture that better solves a collaborative forecasting problem. Considering that parties agree on evaluating the variance prediction of those models using the same testing data composed of $N$ input samples, evaluation metrics presented in this section aim to determine how similar the forecasts of $M$ parties' models are.

Metrics such as Mean Error (ME) and Root Mean Square Error (RMSE) have commonly been used to evaluate the mean distance between the theoretical states $Y_i$ and predicted values $\hat{Y_i}$ on a single ML model $m$.

$$ME(m) = \frac{1}{N} \sum_{i=1}^{N} (Y_i - \hat{Y_i^m}) \tag{2}$$

$$RMSE(m) = \sqrt{\frac{1}{N} \sum_{i=1}^{N} (Y_i - \hat{Y_i^m})^2} \tag{3}$$

Those metrics were enough in HFL because the central server broadcasts the global model to parties. However, VFL and Hybrid FL require evaluating the prediction variability of their parties' models. Besides, federated applications often need to evaluate the experiments' reproducibility using a determined ML architecture. The use of the Overall Average Bias (OAB), Overall Average Variability (OAV) and Reproducibility (Rep) are considered for these purposes:

$$OAB = \frac{1}{M} \sum_{m=1}^{M} |ME(m)|, \tag{4}$$

$$OAV = \frac{1}{M} \sum_{m=1}^{M} RMSE(m), \tag{5}$$

$$Rep = \sqrt{\frac{2}{M(M-1)} \sum_{m=1}^{M} \sum_{b=m+1}^{M} [Mn(b) - Mn(m)]^2 + [Std(b) - Std(m)]^2}, \quad (6)$$

where $M$ refers to the number of parties' models.

These metrics were taken from Prognostics and Health Maintenance (PHM) theory because it is a well-established discipline that aims to enhance the maintenance processes by monitoring the health state of machinery [12]. In Predictive Maintenance, the machinery's useful time estimation has extended to collaborative prognostics due to the scarcity of valuable data at isolated parties [8,13].

## 4 Experimental Setup

In this study, we address a collaborative forecasting problem with vertical data partitioning using the Vertical Asynchronous Federated Learning (VAFL) algorithm. The experiments conducted in this work aim to select the ML architecture that better solves the collaborative problem using the evaluation metrics mentioned above.

The proposed collaborative prognostics problem simulates multiple Condition Monitoring Systems (CMSs) constructing and sharing AI models to collaboratively improve the Remaining Useful Life (RUL) Estimation of turbofan engines [5,13,14]. To reduce downtimes and minimize turbofans maintenance costs as Predictive Maintenance claims, we adopted two ML architectures: 1) a Multilayer Perceptron followed by a Kalman Filter (MLP-KF [6]), and 2) a Functional Multilayer Perceptron (FMLP [7]). Therefore, our study aims to identify which architecture better solves the collaborative prognostics problem.

We used C-MAPSS data from the Turbofan Engine Degradation Simulation Dataset because it comprises labeled data [5]. From the available FD00x sets on this dataset, we used FD004 sets due to presenting two failure modes $fm$ and six operating conditions $op$. FD004 training set shows labelled condition-monitoring data from 249 run-to-failure experiments as time-series data with 21 sensor signals and three operating settings (Altitude, Mach number and Throttle Angle Resolver). From the 24 input signals, we only used 16 of them (2, 3, 4, 6, 7, 8, 9, 10, 11, 12, 13, 14, 15, 17, 20, 21) because they better describe the health degradation of turbofans [15,16]. Then, those 16 input signals were randomly distributed among parties using a combination 13**C**16 to simulate a federated experiment with four parties $J = 4$.

We adopted the data pre-processing steps from [6] to tackle the collaborative prognostics problem, irrespective of the ML architecture employed. Those steps refer to:

- Data normalization per operating condition $N(feature, op)$,
- Label normalization s.t. Health Indicator or $Y \in [0, 1]$, and
- Failure degradation curve in which turbofan engines experience a constant RUL ($RC = 120$) until crossing a start-of-failure point $t_{SoF}$.

We evaluated the prediction variability among parties by considering a forecasting horizon H composed of 248 degradation cases available in the FD004 testing set. Finally, for each federated experiment solved with a particular ML architecture (MLP-KF or FMLP), we repeated the federated optimization ten times to analyze the reproducibility of our results.

### 4.1 Multilayer Perceptron and Kalman Filter

This architecture, used by [6,15], was used to estimate the RUL of turbofan engines in three stages locally. In the first stage, the $m$-th Multilayer Perceptron (MLP) of the $j$-th party estimates Health Indicators $\hat{Y^m} \in [0,1]$ via mapping function $\hat{Y_i^m} = F^j(X_i^j)$ using local vectors $X_{i \in N}^j$ with 13 values. In the second stage, the Kalman Filter (KL), also implemented locally, reduces prediction oscillations given by the MLP on a forecasting horizon H s.t $N > 1$. In the final stage, RUL denormalization refers to calculating $R\hat{U}L^m = RC * \hat{Y^m}$.

The mentioned MLP consists of a single hidden layer containing ten nodes and an output layer. The first layer is activated using the sigmoid activation function, while the second layer is activated using a linear function. For every federated experiment, local MLPs were trained by setting a step-size at 0.01, a bath size at 128 and adopting the RMSE as a cost function. To assess the collaborative forecast, we employed the KF on each run-to-failure state vector within the FD004 testing set by configuring:

- A degradation rate $Q = 1/209$ that corresponds to 209 operating cycles on average; and
- A initial conditions as $\hat{Y}_0 = 1$ because it corresponds to the normalized RUL with an imperceptible health degradation.

Once estimating $\hat{Y^m}$ for any run-to-failure case of the FD004 testing set, the estimated useful lifetime $R\hat{U}L^m$ for such given forecasting horizon was obtained by multiplying their estimated Health Indicators with $RC$.

### 4.2 Functional Multilayer Perceptron

In contrast to a conventional MLP, Functional Multilayer Perceptron (FMLP) from [7,17] eliminates the need for filtering noisy predictions. Therefore, FMLP estimates the RUL in two stages. In the first stage, FMLP predicts a Health Indicator $Y_i$ mapping $\hat{Y_i^m} = F^j(X_i^j)$, in which the input functional data $X_i^j$ is refers to a numerical matrix. This matrix of size $\mathbb{R}^{T \times 13}$ can be written as $[X_{i,1}^m(t); \ldots; X_{i,13}^m(t)]$ s.t. $t = 1, \ldots, T$ because it decomposes 13 curves of dimension $T$.

FMLP extends from the conventional MLP by adding a functional hidden layer previous to numerical hidden layers. The architecture adopted considers adding a functional hidden layer of 16 functional neurons before two numerical hidden layers with eight and four numerical neurons, respectively. We trained parties' models using the RMSE loss function and setting the Fractional Variance

Explained ($P_{FVE}$) at 0.85, the step-size at 0.01 and sigmoid activation functions. Finally, the useful lifetime $R\hat{U}L^m$ is assessed in the last step by multiplying $RC$ with the estimated Health Indicators $\hat{Y}^m$ of every run-to-failure testing case.

## 5 Results

The experiments of this work were conducted on a computer with AMD Ryzen 9 3900X 12-Core processor, 64 GB RAM, NVIDIA GeForce RTX 3080 GPU, Ubuntu 20.04 LTS 64-bit operating system and MATLAB R2021a.

The federated optimization solved using the VAFL algorithm was repeated ten times for each ML proposal, namely MLP-FK and FMLP. This is because the 13**C**16 selection criteria randomly distributed features among parties. We employed the Stochastic Gradient Descent optimization algorithm with early stopping criteria waiting for performance losses during eight model aggregations at each federated optimization, independently from the ML architecture.

Using the 248 run-to-failure cases ten times from the FD004 testing set, we collected the $\hat{Y^m}$ and $R\hat{U}L^m$ estimations of 10 global AI models and 40 local models ($m = 40$) trained with different feature spaces for each federated experiment and ML architecture. Those predictions evaluated the estimation bias of the global AI model and the collaborative forecast using $R\hat{U}L$ and $\hat{Y^m}$, respectively.

A reduction in Mean Absolute Error (MAE) and Root Mean Squared Error (RMSE) estimations from MLP-KF to FMLP in Table 1 suggests that FMLP presents more assertive RUL estimations on global AI models. This estimation error reduction is also evident in Fig. 2, in which the RUL is estimated using the 25th run-to-failure trajectory and adopting the FMLP and MLP-KF architectures. In addition, lower values for OAB, OAV, and Reproducibility adopting FMLPs suggest that parties forecast Health Indicators $\hat{Y}$ more similarly using FMLP instead of MLP-KF.

**Table 1.** Collaborative forecasting performance of MLP-KF and FMLP models in a Vertical Federated Learning problem solved using the VAFL algorithm.

| Models | Global AI model ($R\hat{U}L$) | | Collaborative forecasting ($\hat{Y}$) | | |
|---|---|---|---|---|---|
| | MAE | RMSE | OAB | OAV | Reproducibility |
| MLP-FL | 27.6966 | 29.8184 | 0.1380 | 0.2534 | $4.05e + 08$ |
| FMLP | 15.7356 | 19.4461 | 0.0954 | 0.1959 | $1.32e + 09$ |

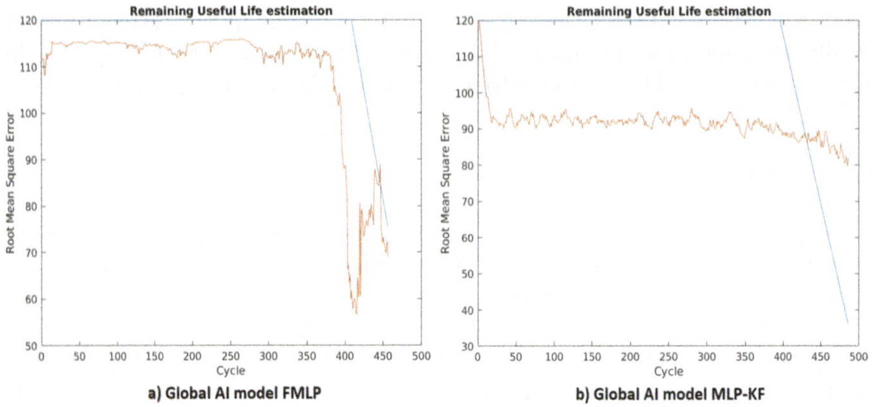


**Fig. 2.** Estimated RUL (orange curve) vs ground truth using (blue curve) using: a) FMLP and b) MLP-KF architectures. (Color figure online)

## 6 Conclusion and Future Works

In this work, collaborative forecasting was evaluated in a Vertical Federated Learning scenario where parties own AI models that differ in architecture because their private datasets present differences in feature space. In the context of the Remaining Useful Life Estimation of turbofan engines, collaborative forecasting evaluation metrics exhibited performance gains on parties' models using FMLP in comparison to MLP-KF architectures.

Lower values in overall average bias, overall average variability, and reproducibility metrics indicate that predictions made using FMLP are more stable and reproducible compared to MLP-KF. Despite collaboration among parties was assessed using Vertical Asynchronous Federated Algorithm (VAFL), the evaluation of the collaborative forecasting variability among parties in a prediction horizon H is extensible to horizontal and hybrid data partitioning settings. We considered extending our study to Hybrid Federated Learning applications because this federated category considers parties' datasets differing in feature and sample space.

## References

1. McMahan, H.B., Moore, E., Ramage, D., Agüera y Arcas, B.: Federated learning of deep networks using model averaging (2016). arXiv: 1602.05629
2. Oluwaseun, O.R., Mirsa, S., Maskeliunas, R., Samasevicius, R.: A review on federated learning and machine learning approaches: categorization, application areas, and blockchain technology. MDPI Inf. **13**(5), 263 (2022). https://doi.org/10.3390/info13050263
3. Tianyi, C., Xiao, J., Yuejiao, S., Wotao, Y.: VAFL: a method of vertical asynchronous federated learning (2020). arXiv: 2007.06081

4. Su, L., Lau Vincent, K.N.: Hierarchical federated learning for hybrid data partitioning across multitype sensors. IEEE Internet Things J. **13**(8), 10922–10939 (2021). https://doi.org/10.1109/JIOT.2021.3051382
5. Saxena, A., Goebel, K.: Phm08 challenge data set (Vol. Tech. Rep.). NASA Ames Prognostics Data Repository, NASA Ames Research Center, Moffett Field, CA (2008). https://www.nasa.gov/content/prognostics-center-of-excellence-data-set-repository
6. Rosero, R.L., Silva, C., Ribeiro, B.: Remaining useful life estimation in aircraft components with federated learning. Int. J. Prognost. Health Manag. **5**(1), 9–17 (2020). https://doi.org/10.36001/phme.2020.v5i1.1228
7. Rosero, R.L,, Silva, C., Ribeiro B.: Forecasting functional time series using federated learning. In: 24th International Conference on Engineering Applications of Neural Networks - EAAAI, vol. 1826 (2023). https://doi.org/10.1007/978-3-031-34204-2_40
8. Prathap Kumar, B., Chouksey, S., Amarlingam, M., Ashok, S.: Evaluation of federated learning strategies on industrial time series fault classification. In: Advances in Distributed Computing and Machine Learning, vol. 660 (2023). https://doi.org/10.1007/978-981-99-1203-2_36
9. Wen, J., Zhang, Z., Lan, Y., et al.: A survey on federated learning: challenges and applications. Int. J. Mach. Learn. Cybern. **14**, 513–535 (2023). https://doi.org/10.1007/s13042-022-01647-y
10. Novikova, E., Doynikova, E., Golubev, S.: Federated learning for intrusion detection in the critical infrastructures: vertically partitioned data use case. MDPI Algor. **15**(4), 263 (2022). https://doi.org/10.3390/a15040104
11. Mingjun, D., Annan, X., Qingwen, H., Zhonghao, Z., Xiaohui, L.: Vertical federated DNN training. Phys. Commun. **49** (2021). https://doi.org/10.1016/j.phycom.2021.101465
12. Yuxin W., Fashiar, R.Md., Honglun, X., Tzu-Liang, B.T.: Recent advances and trends of predictive maintenance from data-driven machine prognostics perspective. Measurement **187**, 110276 (2002). https://doi.org/10.1016/j.measurement.2021.110276
13. Dhada, M., Palau, A.S., Parlikad, A.K.: Federated learning for collaborative prognosis. In: International Conference on Precision, Meso, Micro, and Nano Engineering (COPEN 2019), vol. IIT Indore (2019). https://doi.org/10.17863/CAM.50577
14. Ajay, D.S., Krishnna, S., Jhajharia, K.: Predictive maintenance of NASA turbofan engines using traditional and ensemble machine learning techniques. In: Advances in Distributed Computing and Machine Learning, vol. 660 (2023). https://doi.org/10.1007/978-981-99-1203-2_31
15. Olivares, A., Gonzalez, A., Tovar, S., Gorrostieta, E.: Remaining useful life prediction for turbofan based on a multilayer perceptron and kalman filter. In: 16th International Conference on Electrical Engineering, Computing Science and Automatic Control - CCE (2019). https://doi.org/10.1109/ICEEE.2019.8884495
16. Sahu, A.K., Li, T., Sanjabi, S., et al.: On the convergence of federated optimization in heterogeneous networks (2019). arXiv: 1812.06127
17. Conan-Guez, B., Rossi, F.: Functional multi-layer perceptron: a nonlinear tool for functional data analysis (2017). arXiv: 0709.3642

# Analysing Metadata of Articles Connecting IOT to Agriculture and Climate Change

Luiz Felipe Conrado, Cristina Moll Huther, Flávio Silva Machado, Robisom Damasceno Calado, and Helder Gomes Costa(✉)

Universidade Federal Fluminense, Niterói, RJ, Brazil
{luizconrado,cristinahuther,flaviomachado,robisomcalado, heldergc}@id.uff.br

**Abstract.** A growing global need for food and the environmental impact of agriculture has led to an increase in the use of electronics, information technology, and telecommunications in this sector. To gain insights and consolidate knowledge on the intersection of IoT, agriculture, and climate change, a bibliometric analysis was conducted on 68 documents extracted from the Web of Science database, utilizing the bibliometrix and biblioshiny programs. It was evident through the analysis that there had been a notable surge in interest in the subject, as indicated by the increasing number of publications and research groups dedicated to the area. These findings shed light on the most significant contributions published to date within the interconnected realm of IoT, agriculture, and climate change.

## 1 Introduction

Improving agricultural production is a critical challenge of our time. The world's population will put a strain on the global food supply [5]. To meet the growing demand for food, we must find ways to improve agricultural production while also reducing the environmental impact of agriculture [10].

One way to improve agricultural production is to use technology to collect and analyze data about crop yields, soil conditions, and weather patterns. This data can be used to improve decisions about crop rotation, irrigation, and fertilization [17].

The Internet of Things (IoT) consists of a network of interconnected computing devices, mechanical and digital machines, objects, animals, or people that are identified by unique identifiers and are capable of transferring data over a network without human-to-human or human-to-computer interaction [12]. It is important to note that the IoT is a broad term that encompasses a broad range of technologies and applications, it has the potential to revolutionize a large number of industries and areas, making our lives easier, safer, and more efficient as the Internet of Things continues to develop [20].


A. Bajaj et al. (Eds.): HIS 2023, LNNS 1224, pp. 334–344, 2025.
https://doi.org/10.1007/978-3-031-78925-0_33

In addition to improving agricultural production, we must also reduce the environmental impact of agriculture. This can be done by using more efficient irrigation practices, reducing the use of pesticides and fertilizers, water consumption, and minimizing soil erosion [11]. By taking these steps, we can ensure that future generations have access to a reliable and sustainable food supply.

Considering the need to improve the efficiency of agricultural production while reducing the effects of climate change, the research area of the Internet of Things connected to agriculture and climate change, is urgent [18]. To understand IoT and agriculture based on the urgency to future-proofing for climate change, this article aims to provide an overview and synthesize the knowledge on the aforementioned topics. The main objective of this research consists of providing information concerning the metrics related to the research published so far on the concepts of IoT, agriculture, and climate change. There are three main specific objectives in this research:

- To know the topics included in the academic literature related to the concepts of IoT, agriculture, and climate change.
- To analyze areas of research related to this field.
- To provide future areas of research on this topic.

Therefore, the research questions to be answered in this paper are the following:

- What has been the historical evolution of the literature on IoT connected to agriculture and climate change?
- Which documents have mostly influenced the intellectual structure of the research topic?
- What are the hubs or groups working on the subject of IoT, agriculture, and climate change?
- The production per country as well as their collaborations;
- What research words regarding this matter are distributed around the globe?

## 2 Research Steps

This section will list the research steps to improve understanding and help researchers who want to reproduce this article.

- To select a database to be searched, looking for articles to be used in the research. We decided to use the databases indexed in the Core Collection of the Web of Science (WoS). a decision like this one contributes to mitigate the possibility of using grey literature.

- To search for articles adherent to the research objective, which were published in sources indexed in the database: a search was carried out on WoS databases looking for records having "IoT" AND ["agricult* OR agribusi*] in the topic fields (titles, abstracts, author keywords, and Keywords Plus). The outcome of this search was reduced by using the term "6.153 Climate Change" in the filter Meso Level Citation Topics.
- To analyze the metadata of the articles identified in the previous step. The metadata of the documents were exported from WoS to a BibTeX file which was then analyzed with the support of bibliometrix [2], which is an R [15] package developed by statisticians to perform bibliometric analysis. By running the following code into the RStudio IDE [16], it was possible to conduct the metadata analysis by accessing the bibliometirx functionalities through a web interface.

**Code for running bibliometrix as a web app**

```
library(bibliometrix)

biblioshiny()
```

## 3 Results

In this section, we describe the results from the analysis of the metadata of the 68 documents selected from the search described in the previous section. The full list of these documents can be found at https://doi.org/10.17632/B7VHYD982R.1.

### 3.1 Overview

Table 1 illustrates an overview of information about the 68 documents collected. analyzing this figure, one can observe that:

- The documents' period is only a few years, the starter paper in the set was published in 2017 which begins which means less than 10 years ago.
- There is an increasing interest in this subject, once the annual is 25.99%

**Table 1.** Bibliometric Overview.

| Exported data | | | |
|---|---|---|---|
| MAIN INFORMATION ABOUT DATA | | AUTHORS | |
| Timespan | 2017:2023 | Authors | 296 |
| Sources (Journals, Books, etc) | 56 | Authors of single-authored docs | 1 |
| Documents | 68 | AUTHORS COLLABORATION | |
| Annual Growth Rate % | 25.99 | Single-authored docs | 1 |
| Document Average Age | 1.74 | Co-Authors per Doc | 4.57 |
| Average citations per doc | 3.147 | International co-authorships % | 33.82 |
| References | 2651 | DOCUMENT TYPES | |
| DOCUMENT CONTENTS | | Article | 38 |
| Keywords Plus (ID) | | Article; early access | 3 |
| Author's Keywords (DE) | 292 | Proceedings paper | 22 |
| | | Review | 5 |

### 3.2 Production per Countries

As shown in Table 2, the country with the biggest contribution to the group is India, which produced 18 documents, with only one Multiple Country Publication (MCP). Pakistan produced a total of 6 documents, with only 2 being Single Country Publications (SCP). Followed by Italy and Spain, with 5 documents, and 60% SCP For both of them. Romania with 4 documents, 25% SCP.

**Table 2.** Author's Countries.

| Country | Articles | SCP | MCP | MCP Ratio |
|---|---|---|---|---|
| PAKISTAN | 6 | 2 | 4 | 67% |
| COLOMBIA | 2 | 1 | 1 | 50% |
| ITALY | 5 | 3 | 2 | 40% |
| SPAIN | 5 | 3 | 2 | 40% |
| U ARAB EMIRATES | 3 | 2 | 1 | 33% |
| USA | 3 | 2 | 1 | 33% |
| ROMANIA | 4 | 3 | 1 | 25% |
| INDIA | 18 | 17 | 1 | 6% |
| CHINA | 3 | 3 | 0 | 0% |
| PORTUGAL | 3 | 3 | 0 | 0% |

As displayed in Table 3, these results are linked to the number of citations made from documents from which country: India has 78 citations, a number 371.43% bigger than Pakistan in second place with 21.

**Table 3.** Countries Citations.

| Country | Total Citations | Average Article Citations |
|---|---|---|
| India | 78 | 4,30 |
| Pakistan | 21 | 3,50 |
| Spain | 19 | 3,80 |
| Romania | 14 | 3,50 |
| China | 12 | 4,00 |
| Italy | 12 | 2,40 |
| Colombia | 11 | 5,50 |
| Ireland | 8 | 8,00 |
| Panama | 7 | 7,00 |
| Portugal | 6 | 2,00 |

### 3.3 Affiliations

In this research a total of 149 affiliations were found, the ten more relevant are shown in Table 4. India only has one affiliation on the rank, which shows how diluted the studies about IoT, agriculture, and climate change are in the country.

**Table 4.** Most Relevant Affiliations.

| Affiliation | Articles |
|---|---|
| NATL TSING HUA UNIV | 9 |
| COMSATS UNIV ISLAMABAD | 6 |
| KALASALINGAM ACAD RES AND EDUC | 6 |
| AMER UNIV SHARJAH | 5 |
| UNIV ALMERIA | 5 |
| UNIV POLITEHN BUCURESTI | 5 |
| NATL YANG MING CHIAO TUNG UNIV | 4 |
| QASSIM UNIV | 4 |
| TAIF UNIV | 4 |
| UNIV COSTA | 4 |

### 3.4 Authors

Following the data presented in the overview in Fig. 1, it is possible to see some of the authors and their productions through time, starting in 2017 with a lower number of authors, to a bigger group of researchers in later years.

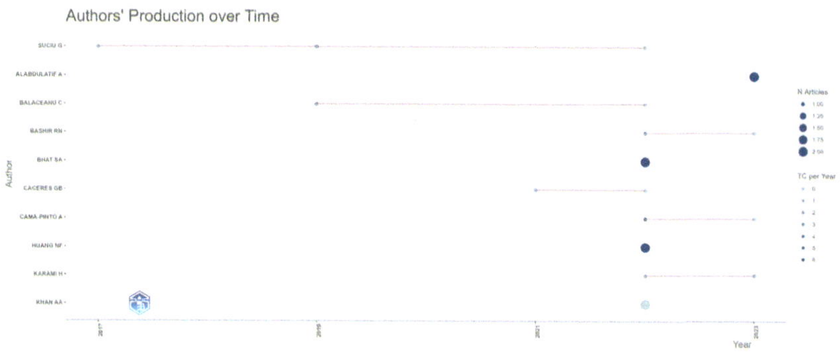


**Fig. 1.** Authors Production Overtime.

While the aforementioned graphic shows a timeline Fig. 2 gives a general view of hubs or research groups about the groups.

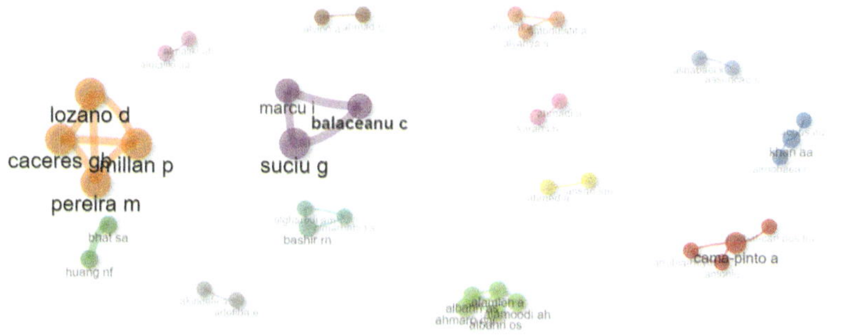


**Fig. 2.** Co occurrence Groups.

When examining both Fig. 1 and Fig. 2, it becomes apparent that a majority of authors have recently begun their work in this research field. Additionally, there has been little to no collaboration among the groups, as they have not been in contact with another one.

### 3.5 Documents Citation

The reference with more uses locally, which means references inside the group of 68 documents that were used, are shown in Table 5.

**Table 5.** Local or inside the group citations

| Cited References | Citations |
|---|---|
| GARCIA L, 2020, SENSORS-BASEL [6] | 8 |
| GOAP A, 2018, COMPUT ELECTRON AGR [7] | 7 |
| ALFRED R, 2021, IEEE ACCESS [1] | 6 |
| KAMIENSKI C, 2019, SENSORS-BASEL [9] | 6 |
| OJHA T, 2015, COMPUT ELECTRON AGR [14] | 6 |
| NAWANDAR NK, 2019, COMPUT ELECTRON AGR [13] | 5 |
| ZAMORA-IZQUIERDO MA, 2019, BIOSYST ENG [19] | 5 |
| BHAT SA, 2021, IEEE ACCESS [3] | 4 |
| BU FY, 2019, FUTURE GENER COMP SY [4] | 4 |
| GU Z, 2020, J IRRIG DRAIN ENG [8] | 4 |

This graphic shows the answers about documents that have mostly influenced the intellectual structure of the research topic where the work with the most local citation was about an overview of recent trends on sensors and IoT systems used in irrigation within precision agriculture [6].

### 3.6 Re-occurrences Maps

Figure 3 illustrates the relationship between words and their occurrence, which determines how the most common words are connected.

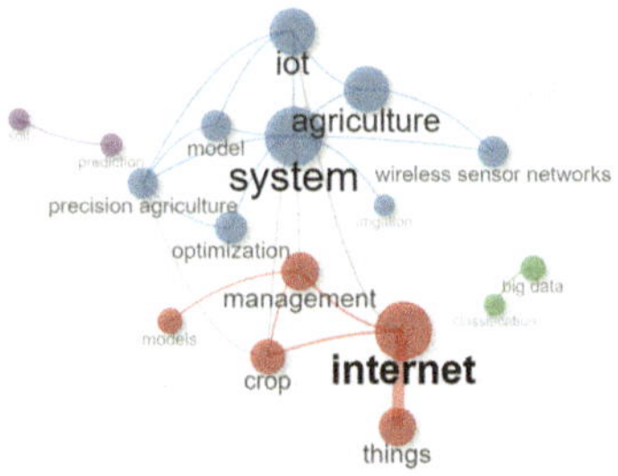


**Fig. 3.** Word's Network

Not only words do have frequency, but some Themes also repeat themselves in the search, making them hot topics. To understand and categorize them, the thematic map below in Fig. 4 has four quarters:

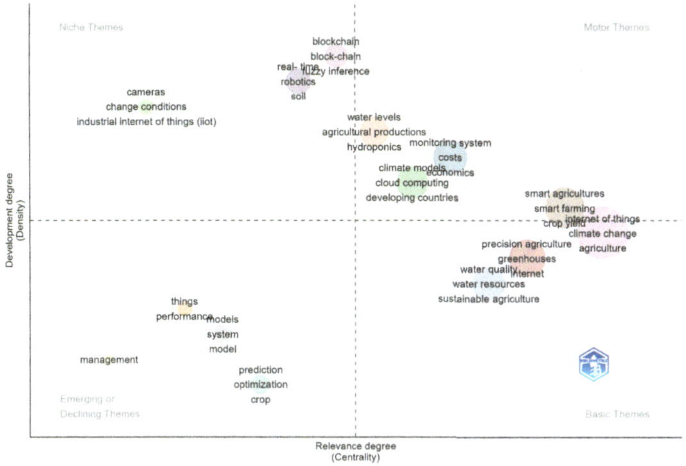


**Fig. 4.** Thematic Map

In *niche themes* are present themes with low Relevance (when compared to the others in the map) and with a high density of results such as smart irrigation, raspberry, machine, and learning.

While in *Basic themes* we deal with studies in machine learning itself, sensors, humidity, and random forest.

The *Emerging and Declining Themes* consists of themes with low relevance and development, which is a feature of too-new or too-old themes, in this study those themes were related to water, soil, and sustainability.

The biggest group of themes is the *Motor Themes*, here are presents themes with a higher density and relevance of publications they began in subjects such as precision agriculture/farming and smart farming, passing through automation to end in artificial intelligence.

In Fig. 5, also known as "tree map" the number and the percentage of re-occurrence of which word can be seen, different forms of writing IoT represents close to 20% of recurrences, leaving some topics such as challenges, performance, and quality with 5% combine, identifying good research possibilities.

With Information presented through the word cloud, thematic map, and tree map, was possible to map the research words regarding this matter around the globe, answering another one of this paper's questions. What research words regarding this matter are distributed around the globe? It seems that performance is becoming an increasingly relevant aspect, while management and optimization are consistently present in discussions surrounding this subject. This suggests that studies examining the correlation between these keywords and this theme may be on the rise.

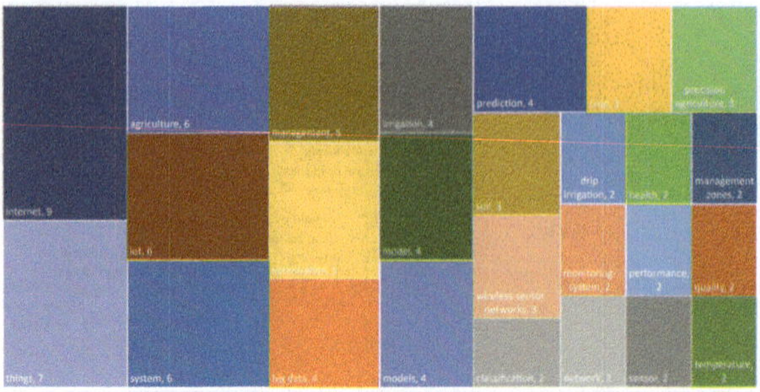


**Fig. 5.** Tree Map

## 4 Conclusion

The text explores a notable rise in interest at the intersection of IoT, agriculture, and climate change. It conveys insights on affiliations, countries, and impact, illuminating scientific production and author networks. In examining recurrence maps, we uncover the prevalent terminology used across diverse groups. Word clouds and tree maps present this information with clarity. The Thematic Map highlights the thematic focus of the study, emphasizing electronic components like sensors and micro-controllers, cloud computing, machine learning, data analytics, and artificial intelligence, all aimed at advancing 'precision farming' and 'smart farming.'

However, a notable limitation is the scarcity of relevant articles in this research domain. Expanding the volume of research may yield fresh perspectives and insights. Therefore, we recommend further exploration by future researchers to an understanding of the nexus between IoT, agriculture, and climate change.

**Acknowledgments.** This study was partially funded by: Ministério da Agricultura, Pecuária e Abastecimento do Brasil (TED 805/2022, number: 00001420220104-000805); Coordenação de Aperfeiçoamento de Pessoal de Nível Superior - (CAPES) - Finance Code 001; Conselho Nacional de Desenvolvimento Científico e Tecnológico - (CNPQ)-Grants 314953/2021-3 and 421779/2021-7; and, Fundação de Amparo a Pesquisa do Estado do Rio de Janeiro - (FAPERJ), Grant 200.974/2022).

## References

1. Alfred, R., Obit, J.H., Chin, C.P.Y., Haviluddin, H., Lim, Y.: Towards paddy rice smart farming: a review on big data, machine learning, and rice production tasks. IEEE Access **9**, 50358 - 50380 (2021). https://doi.org/10.1109/ACCESS.2021.3069449. https://www.scopus.com/inward/record.uri?eid=2-s2.0-85103778651&doi=10.1109%2fACCESS.2021.3069449&partnerID=40&md5=5b5001ecc864327b8c5deb8b13a8b30f

2. Aria, M., Cuccurullo, C.: bibliometrix: an r-tool for comprehensive science mapping analysis. J. Informetr. **11**(4), 959–975 (2017). https://doi.org/10.1016/j.joi.2017.08.007
3. Bhat, S.A., Huang, N.F.: Big data and ai revolution in precision agriculture: survey and challenges. IEEE Access **9**, 110209–110222 (2021). https://doi.org/10.1109/ACCESS.2021.3102227. https://www.scopus.com/inward/record.uri?eid=2-s2.0-85112664253&doi=10.1109%2fACCESS.2021.3102227&partnerID=40&md5=0693e27a7b14de0da6a7173bd8927876, cited by: 71; All Open Access, Gold Open Access
4. Bu, F., Wang, X.: A smart agriculture iot system based on deep reinforcement learning. Future Gener. Comput. Syst. **99**, 500–507 (2019). https://doi.org/10.1016/j.future.2019.04.041. https://www.scopus.com/inward/record.uri?eid=2-s2.0-85065541268&doi=10.1016%2fj.future.2019.04.041&partnerID=40&md5=8007b0429b5e64daed81371e4c039c68
5. Collado, E., Fossatti, A., Saez, Y.: Smart farming: a potential solution towards a modern and sustainable agriculture in panama. Aims Agric. Food **4**(2), 266–284 (2019). https://doi.org/10.3934/agrfood.2019.2.266
6. García, L., Parra, L., Jimenez, J.M., Lloret, J., Lorenz, P.: Iot-based smart irrigation systems: an overview on the recent trends on sensors and iot systems for irrigation in precision agriculture. Sensors (Switzerland) **20**(4) (2020). https://doi.org/10.3390/s20041042. https://l1nq.com/v3x3M
7. Goap, A., Sharma, D., Shukla, A., Rama Krishna, C.: An iot based smart irrigation management system using machine learning and open source technologies. Comput. Electron. Agric. **155**, 41–49 (2018). https://doi.org/10.1016/j.compag.2018.09.040. https://www.scopus.com/inward/record.uri?eid=2-s2.0-85054441693&doi=10.1016%2fj.compag.2018.09.040&partnerID=40&md5=c15c17285c03177967fb1a0f200188af, cited by: 340
8. Gu, Z., Qi, Z., Burghate, R., Yuan, S., Jiao, X., Xu, J.: Irrigation scheduling approaches and applications: a review. J. Irrigat. Drain. Eng. **146**(6) (2020). https://doi.org/10.1061/(ASCE)IR.1943-4774.0001464
9. Kamienski, C., et al.: Smart water management platform: Iot-based precision irrigation for agriculture. Sensors (Switzerland) **19**(2) (2019). https://doi.org/10.3390/s19020276. https://www.scopus.com/inward/record.uri?eid=2-s2.0-85060015568&doi=10.3390%2fs19020276&partnerID=40&md5=6859fae60c2cd47a547d1efc2cd90e4a
10. Kiani, F., et al.: A smart and mechanized agricultural application: from cultivation to harvest. Appl. Sci.-Basel **12**(12) (2022). https://doi.org/10.3390/app12126021
11. Kumar, P., Udayakumar, A., Kumar, A.A., Kannan, K.S., Krishnan, N.: Multi-parameter optimization system with dcnn in precision agriculture for advanced irrigation planning and scheduling based on soil moisture estimation. Environ. Monit. Assess. **195**(1) (2023). https://doi.org/10.1007/s10661-022-10529-3
12. Nagpure, S., Ingale, S., Pahurkar, S., Bobade, A.M., Ghosal, M., Dhope, T.: Smart agriculture using iot. Helix **9**(3), 5081–5083 (2019). https://doi.org/10.29042/2019-5081-5083
13. Nawandar, N.K., Satpute, V.R.: Iot based low cost and intelligent module for smart irrigation system. Comput. Electron. Agric. **162**, 979–990 (2019). https://doi.org/10.1016/j.compag.2019.05.027. https://www.scopus.com/inward/record.uri?eid=2-s2.0-85066236624&doi=10.1016%2fj.compag.2019.05.027&partnerID=40&md5=a558809fe27192b3568869a5ba18af31

14. Ojha, T., Misra, S., Raghuwanshi, N.S.: Wireless sensor networks for agriculture: the state-of-the-art in practice and future challenges. Comput. Electron. Agric. **118**, 66–84 (2015). https://doi.org/10.1016/j.compag.2015.08.011. https://www.scopus.com/inward/record.uri?eid=2-s2.0-84941123409&doi=10.1016%2fj.compag.2015.08.011&partnerID=40&md5=378c46a5115a771cb5bc55d5578b6cf2, cited by: 643
15. R Core Team: R: A Language and Environment for Statistical Computing. R Foundation for Statistical Computing, Vienna, Austria (2022). https://www.R-project.org/
16. RStudio Team: RStudio: Integrated Development Environment for R. RStudio, PBC., Boston, MA (2020). http://www.rstudio.com/
17. Veerachamy, R., Ramar, R., Balaji, S., Sharmila, L.: Autonomous application controls on smart irrigation. Comput. Electr. Eng. **100** (2022). https://doi.org/10.1016/j.compeleceng.2022.107855
18. Velani, A.F., Narwane, V.S., Gardas, B.B.: Contribution of internet of things in water supply chain management: a bibliometric and content analysis. J. Model. Manag. **18**(2), 549–577 (2023). https://doi.org/10.1108/JM2-04-2021-0090
19. Zamora-Izquierdo, M.A., Santa, J., Martínez, J.A., Martínez, V., Skarmeta, A.F.: Smart farming iot platform based on edge and cloud computing. Biosyst. Eng. **177**, 4 - 17 (2019). https://doi.org/10.1016/j.biosystemseng.2018.10.014. https://www.scopus.com/inward/record.uri?eid=2-s2.0-85056253107&doi=10.1016%2fj.biosystemseng.2018.10.014&partnerID=40&md5=0dbe0d88c3b930762bf7c8827b4094e9
20. Zulkifli, C.Z., et al.: Iot-based water monitoring systems: a systematic review. Water **14**(22) (2022). https://doi.org/10.3390/w14223621

# Graph Neural Network-Based Collaborative Filtering for Movie Recommendation

S. Adithya[1], S. Bhuvaneswari[1], and V. Subramaniyaswamy[2](✉)

[1] School of Computing, SASTRA Deemed University, Thanjavur 613401, India
[2] School of Computer Science and Engineering, Vellore Institute of Technology, Vellore 632014, India
subramaniyaswamy.v@vit.ac.in

**Abstract.** Customer satisfaction and engagement are two of the most important things companies look for to expand their businesses. Organisations have started using machine learning techniques to recommend items for users to try. If the user engages with the recommendation and likes it, they will come back to the company for more, which will then result in more revenue for the company. Movie recommendation systems have played a big part in the success of many multinational corporations in the entertainment industry. Previous works on movie recommendations have focussed on one of two primary Recommender system methods, namely Collaborative Filtering (CF) or Content-Based filtering (CBF). Both methods suffer from certain drawbacks, which sometimes hold them back from making meaningful recommendations. CF often faces challenges when dealing with sparse data, and CBF sometimes suffers due to over-reliance on user-defined preferences. One of the upcoming techniques widely being explored in recommender systems today is the usage of graphs or Graph Neural Networks (GNN) to make suggestions. This paper explores a method which plots the users as nodes of a graph, and each edge in the graph represents a common movie that the two users have rated. Then, a wide range of graph features are calculated and applied using GNN to make predictions and recommendations for particular user. Various evaluation metrics are then employed to test how well the model performs. This model is pretty advantageous as GNNs can understand user-item interactions and make good recommendations.

**Keywords:** Recommender system · Collaborative Filtering · Content-Based Filtering · Graph Neural Networks

## 1 Introduction

Recommendation systems have garnered significant attention due to their widespread applications, including e-commerce, social media, news articles and, notably, movie recommendations. These systems assist users in discovering new content based on their historical interactions, leading to increased user satisfaction and engagement. Movie recommendation systems, in particular, have become essential tools for enhancing user experiences and engagement in the entertainment industry.


A. Bajaj et al. (Eds.): HIS 2023, LNNS 1224, pp. 345–353, 2025.
https://doi.org/10.1007/978-3-031-78925-0_34

Recommender systems are broadly categorised into Collaborative Filtering (CF) or Content-Based filtering (CBF). CF operates under the principle that users who have exhibited similar behaviours or preferences in the past are bound to share similar predilection in the future. It also has the ability to offer recommendations without requiring explicit knowledge about the items being recommended [1]. CBF, on the other hand, provides recommendations based on intrinsic features of items and their alignment with a user's profile [2]. This approach contrasts with CF, which relies on the user-item interactions by focussing on the characteristics of the items linked with the user.

AI based models bring strength to recommendation tasks which leads to more accurate predictions and subsequently better recommendations [3]. Graph Neural Networks in particular have been gaining popularity in recent years because of their ability to understand and capture complex relationships in in graph-structured data. GNNs capture collaborative features among users and movies and use the structural information of the graph to enhance recommendation accuracy [4, 5]. For this reason, the graph neural network model is opted for in this work to provide strong movie recommendations.

As an initial step, a user-item interaction graph is constructed, and significant graph features are calculated, which help the GNN to make a precise prediction for movie recommendation. The explicit user ratings as well as the graph features help the model make diverse content-relevant suggestions. The rest of this paper is organised as follows. Section 2 contains information about existing implementations for movie recommendation systems. The next section elaborates on the techniques used to implement the model. Experimental Results and Discussion deep dives into the implementation and the experimental results of the proposed movie recommender system. The final section contains the conclusion and talks about why the proposed model performed as it did as well as a short note on possible future work ideas.

## 2 Related Works

This section presents existing research work which implemented a personalised movie recommender system. Since the chosen field is widespread and active, only some of the most popular and recent ones are mentioned.

Darban and Valipour, proposed a new hybrid recommender system that combines a graph-based model with side information such as users' demographic and location information. The graph-based model was used to capture the similarity between users, while the side information was used to extract new features that can improve the recommendation accuracy [4]. Muhan Zhang et al. propose a new inductive matrix completion model which is based on GNNs. The model is able to generalise the unseen users and items, even if they are not included in the training set. This is in contrast to traditional matrix completion methods, which cannot be generalised to unseen data. The model works by training a GNN on the subgraph around each user-item pair. The GNN learns to encode the relationships between the user, the item, and the neighbours in the subgraph. The encoded representations are then used to predict the rating of the user for the item [5]. Van den Berg et al. proposed a new matrix completion model called Graph Convolutional Matrix Completion, which leverages graph-structured data to improve the performance of matrix completion. The model works by constructing a graph from the data, where nodes represent users and items, and edges represent interactions between users and items. It then uses a Graph Convolutional Network (GCN) for learning the features of users and items from the graph data. The learned features are then used to predict the missing entries in the matrix [6].

He et al. proposed a new Graph Convolutional Network model for recommender systems called LightGCN. LightGCN is a simplified version of GCN that removes some of the less important components of the GCN architecture, such as the feature transformation layer and the nonlinear activation function [7]. Nesmaoui proposed a graph neural network (GNN)- based collaborative filtering movie recommendation system. The system works by constructing a graph from the rating data and then uses a GNN to learn the latent features of users and movies from graph data. The architecture is based on a graph convolutional network (GCN) with a novel attention mechanism. The attention mechanism allows the GNN to focus on the most important interactions in the graph when learning the features of users and movies [8].

Wang et al. proposed Neural Graph Collaborative Filtering (NGCF), a new CF framework that leverages GNNs to learn the features of users and items from the user-item graph. NCGF works by first constructing the user-item graph from the historical interaction data. Then, it uses GNN to propagate the features of users and items over the graph to predict the ratings of users for unrated items [9]. Sangeetha et al. proposed a GNN--based personalised recommendation model which first constructs a user-item graph from the historical interaction data. The model then uses GNN to learn the features of the users and items. The GNN uses a message-passing mechanism to aggregate information from the neighbours of each node in the graph [10].

Song et al. propose a deep GNN model with multi-graph attention fusion (MAF-GNN). Two feature graph attention modules and a multi-scale latent features module are created to generate better user and item latent features from input information. A dual-branch residual graph attention (DBRGA) module is used to extract similar features between neighbours from both user and item graphs. A hybrid fusion graph attention (HFGA) module is used to obtain collaborative information from user-item interaction graph. Finally, the whole MAF-GNN framework is optimized by a geometric factorized regularization loss [11]. Huang et al. propose a dual light graph convolutional network model called Dual-LightGCN which filters out items disliked by users. Dual-LightGCN divides the original user-item interaction graph into two bipartite subgraphs. One of the subgraphs are used to model the preferences between users and items and the other subgraph is used to model the dislike relationships between them and then used to make recommendations [12].

## 3 Proposed Methodology

This section shares an overview of the proposed system—the workflow diagram of the same is given in Fig. 1. The movie recommender system uses both the user's ratings for the movies as well as the graph features, which were calculated based on the user-item interactions to provide a personalised recommendation for the user. The user's ID is taken as input, and the movie IDs are shown as output for that specific user.

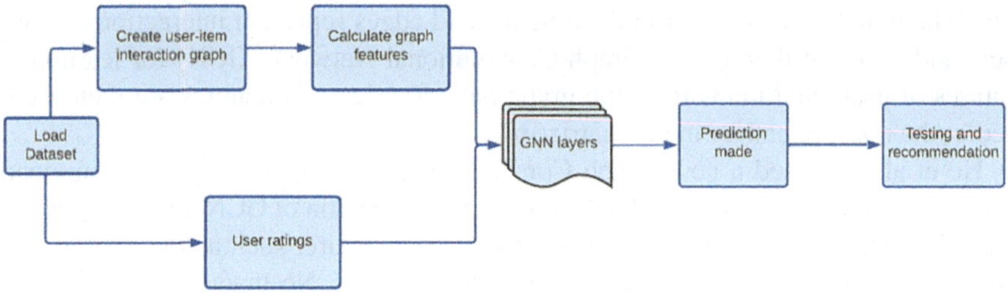


**Fig. 1.** Overall workflow of the recommender system

## 3.1 Data Collection

To train and evaluate our recommender system, the widely recognized and stable benchmark dataset, MovieLens 1 M is utilized. This dataset comprises a substantial collection of user-movie interactions, totalling over one million ratings from 6040 users across 4000 movies. It also contains useful side information of the users, which can be used as auxiliary information to improve recommendation accuracy.

## 3.2 Data Processing

The core of our data consists of user-movie interactions where each entry records a user's rating for a specific movie. This information is used to create a collaborative filtering model which relies on the users' ratings to make recommendations. The categorical data such as the user's age, gender and job information are converted into one-hot encodings so that they can be seamlessly integrated into our pipeline. For prediction, two ways of data are considered including explicit rating of the move and extracted graph features.

Collaborative Filtering is a widely used recommendation technique that leverages the collective preferences and behaviours of users. One of the core components of our Collaborative Filtering approach is the creation of a User-Item Matrix. This matrix is a two-dimensional structure where rows represent users, columns represent movies, and the values at the intersection store user ratings for movies.

Next, a user-item interaction graph is created which serves as a representation of user behaviour and movie preferences. The pairs of users are identified who have rated the same movies, which is then used to create the graph. The users are represented as nodes, and each edge in the graph represents a common movie that two users have rated. To better understand the role of users within the graph, various graph-based metrics are computed for each user. These metrics offer insights into user centrality and influence. The calculated metrics include PageRank, Degree Centrality, Closeness Centrality, Betweenness Centrality, Load Centrality and Average Neighbor Degree. These graph features are discussed along with their definitions below.

**Page Rank.** Page Rank is an algorithm used by search engines to rank web pages in their search engine results. It is based on the idea that the importance of a web page can be determined by the number of quality links pointing to it referring to Eq. 1.

$$\mathrm{PR(n)} = \sum\nolimits_{\mathrm{m}\in\mathrm{S_n}} \frac{\mathrm{PR(m)}}{\mathrm{L(m)}} \tag{1}$$

Where *PR(n)* denotes the Page Rank of node n and *PR(m)* denotes the Page Rank of node m. $S_n$ is the set containing all nodes which have a link to node n. *L(m)* denotes the total number of links from node m.

**Degree Centrality.** Degree Centrality is defined as the number of edges connected to a node, which is often referred to as the "degree" of the node. Below Eq. 2 is calculated as a ratio of the number of edges connected to the node to the total number of nodes in the network.

$$C_D(n) = \sum_{j} A_{ij}/(N - 1) \tag{2}$$

Where $C_D(n)$ represents the Degree Centrality of node $n$ and $A_{ij}$ represents the Adjacency Matrix of the given graph. $N$ is the total number of nodes in the network.

**Closeness Centrality.** Closeness Centrality is a measure that tells how close a node is to all other nodes in the network. It is calculated as the reciprocal of the sum of the length of the shortest paths between the node and all other nodes in the graph, shown in Eq. 3.

$$C_c(n) = \frac{N - 1}{\sum_m d(n, m)} \tag{3}$$

Where $C_C(n)$ represents the Closeness Centrality of node $n$ and *d(n, m)* is the length of shortest path between nodes n and m in the network. $N$ is the total number of nodes in the network.

**Betweenness Centrality.** Betweenness Centrality is a measure that quantifies the extent to which a node in a network lies on the shortest paths between other nodes. The shortest path between each pair of nodes is calculated and the number of such paths that pass through the target node is also calculated. This is then used to calculate the betweenness centrality of the node shown in Eq. 4.

$$C_B(n) = \sum_{i<j} \frac{\sigma_{ij}(n)}{\sigma_{ij}} \tag{4}$$

Where $C_B(n)$ represents the Betweenness Centrality of node $n$. $\sigma_{ij}(n)$ is the number of shortest paths between nodes $i$ and $j$ which pass through node $n$ and $\sigma_{ij}$ is the total number of shortest paths between nodes $i$ and $j$.

**Load Centrality.** Load Centrality is a measure that talks about the influence of a node in a network based on the concept of network traffic. It is calculated by considering the number of shortest paths that pass-through a given node as shown in Eq. 5.

$$C_L(n) = \sum \frac{1}{d(n, m)} \tag{5}$$

Where $C_L(n)$ represents the Load Centrality of node $n$ and *d(n, m)* is the shortest path between nodes $n$ and $m$.

**Average Neighbour Degree.** Average Neighbour Degree is a centrality measure used to assess the centrality and connectivity of a node within a network. It measures the average degree of a node's neighbours providing insights into local network structure.

$$\mathrm{AND(n)} = \sum_{\mathrm{m \in N(n)}} \mathrm{d(m)/N(n)} \tag{6}$$

In Eq. 6, *AND(n)* represents the Average Neighbour Degree of node *n*. *d(m)* represents the degree of nodes *m* which are neighbours of node *n*. *N(n)* is the number of neighbours of node *n*.

To ensure the fair and meaningful utilization of these graph metrics, they are normalized to a consistent range between 0 and 1. This normalization process helps us prevent any single metric from dominating the recommendation process and ensures that each metric contributes fairly to the overall recommendations.

Graph Neural Networks are a class of machine learning models well suited for graph-structured data. In our movie recommender system, GNN model is designed specifically for the user-item interaction graph which had been created. It learns user embeddings that encapsulate user preferences and movie embeddings that capture movie characteristics. The model is trained to predict user ratings for movies based on their interactions and connections in the graph.

Then the outcomes from Collaborative Filtering model and the GNN model are aggregated to make movie recommender system. The two approaches have been used complement each other: Collaborative Filtering excels at capturing user preferences from explicit ratings, while GNN excels at leveraging implicit interactions within the graph. The integration of these models allows for more accurate and personalised movie recommendations.

**Procedure 1: Graph Neural Network based Recommendation**

Input: User ID
Output: Top 10 Movie recommendations
Step 1: The data is preprocessed, and the user-item interaction graph is created.
Step 2: The graph-based features (PageRank, Degree Centrality, Closeness Centrality, Betweenness Centrality, Load Centrality and Average Neighbor Degree) are calculated for users based on their interactions.
Step 3: The user ratings as well as the calculated graph features are passed through the GNN layers, and a prediction is made for the user
Step 4: The predicted ratings are then used to make personalized recommendations for the user
Step 4.1: The dataset is split, and a part of the dataset is considered.
Step 4.2: Movie ratings are predicted for the users in the partitioned dataset.
Step 4.3: These are then tested with the actual ratings made by the user.
Step 4.4: Evaluation metrics are calculated, and the model is tested.
Step 5: The user ID is typed in by the user.
Step 6: The recommendations are displayed to the user.

## 4 Experimental Results and Discussion

The dataset used for evaluation is discussed below. The loss graph and some evaluation metrics like RMSE are also discussed.

### 4.1 Dataset

The Movielens 1 M stable benchmark dataset was used to test the efficiency of our model. The dataset consists of a little over a million ratings provided by 6400 users for around 3900 movies (Table 1).

**Table 1.** Details of dataset used for testing the model

| S.No | Dataset | Users | Movies | Total Ratings | Rating Range |
|---|---|---|---|---|---|
| 1 | MovieLens 1 M | 6040 | 3900 | 1000209 | 1–5 |

The accuracy for the rating prediction is calculated over 300 epochs with early stopping, which stopped training at the 265-epoch mark. The validation accuracy kept increasing steadily for 220 epochs, and then it started plateauing. The final accuracy after 265 epochs was 86.05% (Fig. 2).

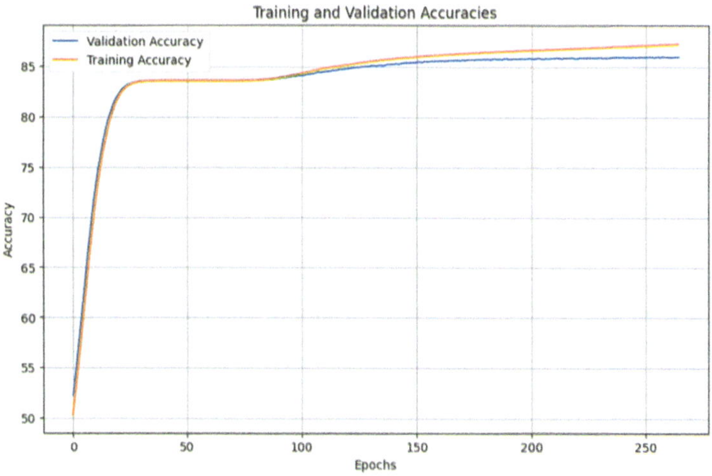


**Fig. 2.** Variation of Accuracy over Epochs

The training and validation loss over epochs has been shown in Fig. 3. It can be seen that the training and validation loss converges at around 220 epochs, and then the training loss goes lower than the validation loss. The training loss starts at 0.737 and goes down to 0.338 at epoch 265. Around the 220-epoch mark is also when the accuracy of the prediction starts plateauing, and the trend can be seen here. The RMSE of the proposed model is compared with a few other pre-existing recommendation models to show that our model performs better than these models on the MovieLens 1 M dataset (Table 2).

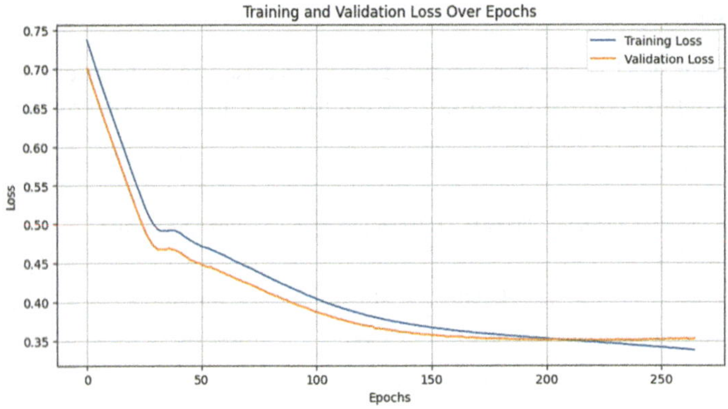


**Fig. 3.** Training and Validation Loss over Epochs

**Table 2.** RMSE comparison with pre-existing models

| Model | RMSE |
|---|---|
| IGC-MC [6] | 1.259 |
| Noise Correction Based RS [13] | 1.7 |
| Genetic Algorithm and gravitational emulation [14] | 1.087 |
| **GNN based Recommendation system** | **1.063** |

The comparison shows that the proposed model slightly improves upon the previous implementation's RMSE. This leads to better recommendations and increased satisfaction for the user. The result of the system cannot be compared with a standard metric, but this comparative study gives us an idea of the real-world performance of the model.

## 5 Conclusion

The goal of this work is to develop a movie recommendation system that provides personalised recommendations for users. A graph-based collaborative filtering model that uses graph neural networks was explored and tested on a benchmark dataset, the MovieLens 1 M, to check its real-time performance. This approach demonstrated the ability to harness the power of GNNs to make strong content-relevant recommendations. The proposed model incorporates user-item interactions and graph-based features derived from a user-user interaction graph. Our experiments and evaluations revealed promising results, as the model achieved competitive performance in predicting user ratings and providing movie recommendations. As the data size keeps increasing, a user-item interaction graph may not be able to capture all the dependencies among nodes. As future work, a hypergraph-based framework can be used for data representation which capture more features especially when dealing with complex data.

**Acknowledgements.** The authors also acknowledge SASTRA Deemed University, Thanjavur, for extending infrastructural support to carry out this research.

## References

1. Salehi, M.: Application of implicit and explicit attribute based collaborative filtering and BIDE for learning resource recommendation. Data Knowl. Eng. **87**, 130–145 (2013)
2. Lops, P., De Gemmis, M., Semeraro, G.: Content-based recommender systems: State of the art and trends. Recommender systems handbook, 73–105 (2011)
3. Çano, E., Morisio, M.: Hybrid recommender systems: A systematic literature review. Intelligent data analysis. **21**(6), 1487–1524 (2017)
4. Darban, Z.Z., Valipour, M.H.: GHRS: Graph-based hybrid recommendation system with application to movie recommendation. Expert Syst. Appl. **200**, 116850 (2022)
5. Zhang, M., & Chen, Y. (2019). Inductive matrix completion based on graph neural networks. arXiv preprint arXiv:1904.12058
6. Berg, R. V. D., Kipf, T. N., & Welling, M. (2017). Graph convolutional matrix completion. arXiv preprint arXiv:1706.02263
7. He, X., Deng, K., Wang, X., Li, Y., Zhang, Y., Wang, M.: Lightgcn: simplifying and powering graph convolution network for recommendation. In: Proceedings of the 43rd International ACM SIGIR Conference on Research and Development in Information Retrieval, pp. 639–648 (2020)
8. Nesmaoui, R., Louhichi, M., Lazaar, M.: A collaborative filtering movies recommendation system based on graph neural network. Procedia Computer Science. **220**, 456–461 (2023)
9. Wang, X., He, X., Wang, M., Feng, F., Chua, T.S.: Neural graph collaborative filtering. In: Proceedings of the 42nd International ACM SIGIR Conference on Research and Development in Information Retrieval, pp. 165–174 (2019, July)
10. Sangeetha, M., Manjuladevi, R., Sagana, C., Suruthi, B.S., Thejesvika, S.S., Vaibhav, R.: Predicting personalized recommendations using GNN. In: 2022 6th International Conference on Computing Methodologies and Communication (ICCMC), pp. 228–234. IEEE (2022)
11. Song, Y., Ye, H., Li, M., Cao, F.: Deep multi-graph neural networks with attention fusion for recommendation. Expert Syst. Appl. **191**, 116240 (2022)
12. Huang, W., Hao, F., Shang, J., Yu, W., Zeng, S., Bisogni, C., Loia, V.: Dual-LightGCN: Dual light graph convolutional network for discriminative recommendation. Comput. Commun. **204**, 89–100 (2023)
13. Bag, S., Kumar, S., Awasthi, A., Tiwari, M.K.: A noise correction-based approach to support a recommender system in a highly sparse rating environment. Decis. Support. Syst. **118**, 46–57 (2019)
14. Mohammadpour, T., Bidgoli, A.M., Enayatifar, R., Javadi, H.H.S.: Efficient clustering in collaborative filtering recommender system: Hybrid method based on genetic algorithm and gravitational emulation local search algorithm. Genomics. **111**(6), 1902–1912 (2019)

# SobdoKrom: An Unsupervised Bengali Keyword Extraction Model Using Pre-trained Large Language Model

Tabia Tanzin Prama(✉) and Md. Musfique Anwar

Jahangirnagar University, Savar 1342, Dhaka, Bangladesh
{I.prama.stu2017,II.manwar}@juniv.edu

**Abstract.** Keyword extraction is a Natural Language Processing task that aims to automatically identify significant terms representing the core themes and subjects covered in a document. This process entails providing a text to a computer, which then returns a set of keywords suggesting significant words related to the document's content. Keyword extraction methods find applications in various fields, particularly in information retrieval, where they help provide an overview of a document. This is particularly crucial as people often search for essential information based on keywords. While several techniques exist for English text, there is a notable scarcity of efforts in the domain of Bengali keyword extraction and context comprehension. In this article, we present an unsupervised approach 'SobdoKrom' (শব্দক্রোম), which utilizes POS tagger, Bengali pre-trained word embeddings, and cosine similarity to extract keywords from the "Prothom Alo" news portal dataset in Bengali. We utilized various pre-processing steps to derive candidate keyphrases from the documents. Our model achieves a precision of 81% for extracting 5 keywords and 66.67% for extracting 10 keywords. Compared to existing approaches, specifically the neural-based approach, our approach demonstrates a notable improvement of 22% and 19% for extracting 5 and 10 keywords, respectively.

**Keywords:** Bangla Keyword Extraction · Unsupervised Learning · Word Embedding · Cosine Similarity · POS tagger

## 1 Introduction

Keywords or keyphrases are significant words in a document that convey concrete ideas within the content. With the widespread use of the internet, keywords have become an essential tool for people to quickly comprehend the content of a document and understand its context [1]. Automated Keyphrase Extraction (KPE) is defined as the task of automatically identifying a set of words that best describe the subject of a document without any assistance from a human reader. When extracting keywords, these words are explicitly selected from the


A. Bajaj et al. (Eds.): HIS 2023, LNNS 1224, pp. 354–363, 2025.
https://doi.org/10.1007/978-3-031-78925-0_35

document. Keyphrases serve various purposes, such as summarization, categorization, indexing, enabling precise searching in search engines, and supporting data mining. The amount of content in the Bangla language on the internet is growing significantly day by day. To process or mine this vast amount of text and articles, there is a substantial demand for automated keyphrase extraction systems. Therefore, an efficient automated keyphrase extraction system for the Bengali language is an essential tool. Overall, the KPE methods described in the literature can be divided into two main categories: supervised and unsupervised, each with pros and cons of its own. Compared to unsupervised techniques, supervised procedures have a major limitation in that they depend on a substantial amount of training data that has been annotated by human experts [6]. The authors of [7] highlight that since unsupervised techniques are computationally efficient, they perform well in real-time applications. This research mainly focuses on unsupervised techniques for these reasons. Moreover, traditional and embeddings-based unsupervised KPE techniques are distinguished [8]. Compared to embeddings-based methods, classical techniques often overlook important semantic information in the text, which lowers their F1 scores. Nevertheless, a lot of modern embeddings-based KPE techniques don't take use of current developments in deep learning models, which have shown improved accuracy in various natural language processing (NLP) tasks.

In this paper, we propose an embedding-based approach that utilizes the Bangla POS tagger and cosine similarity to extract keywords. There has been limited research on unsupervised Bengali keyword extraction [17]. This presents a significant opportunity in the field of Bengali keyword extraction, as automated keyword extraction helps readers easily identify the category or type of documents, thereby reducing the reader's effort and time required to understand the document.

## 2 Related Work

In Bangla language processing, the majority of prior studies have concentrated on areas such as machine translation [9–11], reading compression [13,15], and sentiment analysis. Limited research has been dedicated to information retrieval aspects like relation extraction [12,14,16], where authors have explored the integration of knowledge bases.Research on keyword extraction from Bengali documents is limited, and we provide an overview of these studies along with relevant work below:

[18] introduced three machine learning-based approaches employing Decision Trees, Naïve Bayes, and Artificial Neural Networks for keyphrase extraction. These approaches were compared with each other and with the KPE system. Notably, their method did not incorporate lexical features in the candidate phrase extraction module. In [19], a neural network-based approach was presented for keyphrase extraction from scientific articles. This method used estimated class probabilities as confidence scores to classify phrases as positive or negative keyphrases. Their method employed five features, including TF*IDF, phrase position, phrase length, word length, and phrase connections. However,

this approach relied solely on supervised learning models.[20] outlined a comprehensive approach for automatic keyphrase extraction from scientific articles using Conditional Random Fields (CRF), a supervised tool. They trained the system on 144 scientific articles and tested it on 100 more. Eleven features were used, but this system couldn't extract multi-word keyphrases. A hybrid approach for automatic Bangla scientific document keyword extraction, mentioned in [23], combined a ranking model with supervised machine learning, specifically the Naïve Bayes classifier. This approach achieved a 58% precision rate for extracting 5 keywords, evaluated on a dataset of 100 scientific documents.

In an unsupervised context, [21] conducted an initial investigation into extracting keyphrases from Bengali documents. They utilized features such as TF*IDF and the first occurrence of phrases. Their prototype system involved extracting n-grams from source articles, identifying candidate keyphrases, and subsequently ranking them. The method proposed in [5] merges statistical techniques and machine learning for automatic keyword extraction to summarize text. It incorporates a Hidden Markov Model (HMM)-based POS tagger to identify parts of speech, followed by statistical methods to extract keywords. Each word receives a score for keyword extraction based on a learned probability distribution. The chosen keywords are then used to summarize the article, removing redundancies. The system also uses HMM-based POS tagging to identify POS data from a training corpus, particularly in newspaper headlines. By analyzing keyword information from the training corpus, it calculates the importance of each tagged word. This approach integrates both sentence-level and word-level features for extracting crucial sentences from articles [4]. An alternative strategy, as presented in [22], concentrated on Bangla keyphrase extraction using n-grams. This method encompassed steps like n-gram extraction, identifying candidate keyphrases, and assigning scores to these candidates. The author introduced a lightweight stemmer specifically for stemming candidate keyphrases. However, it's noteworthy that this system was assessed on a limited dataset of 41 documents. Another approach to keyword extraction involved the application of a Conditional Random Field (CRF) model, a contemporary sequence labeling model renowned for its efficiency in segmenting and labeling sequential data [2]. CRF, as a probabilistic undirected graph model, captures conditional probability distributions based on given attributes. The methodology employed a maximum entropy learning algorithm to train the CRF model, transforming documents into tagging sequences during training, encompassing sentence segmentation and POS tagging. Subsequently, features were extracted [3].

## 3 Proposed Approach

The multi-step technique of the suggested solution is shown in Fig. 1.

### 3.1 Data Collection and Preprocessing

For our research, we collected data from the prominent Bengali news portal 'Prothom Alo[1].' The dataset comprises 9,908 news articles spanning various

[1] https://www.prothomalo.com/.

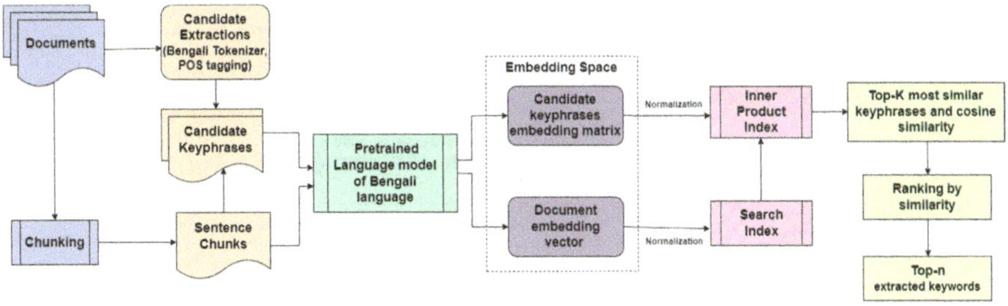


**Fig. 1.** Proposed workflow architecture for KPE approach

categories on lifestyle, travel, fashion, style, beauty, interior, recipe, shopping, horoscope, and international relations. After data preprocessing, the average number of keywords per article was reduced from 445 to 238. In our approach, we first extracted a candidate keyphrases list from the news content. During this extraction, we included keywords while excluding punctuation like commas. We excluded non-alphabetic words, numbers, Roman numerals, URLs, and irrelevant content from the dataset. Since Bengali words are not case-sensitive, there was no need for case normalization in our document processing. We acquired a stopwords[2] list from an online source, totaling 363 stopwords. Removal of stop words was performed through exact matches. The proportion of stop words in our training dataset is 18.60%. For stemming Bangla words, we employed a lightweight Bangla stemmer[3]. This stemmer adheres to Bangla grammar rules and effectively removes suffixes. It achieves an average accuracy of 88% in suffix removal.

### 3.2 Candidate Keywords Extractions

To generate a set of potential keyphrases from the text, we initiate the process by employing the Basic Tokenizer from the bnlp-toolkit[4] to tokenize the news content. Subsequently, we undertake the removal of duplicate tokens to filter out redundant keyphrases. Utilizing the FuzzyWuzzy toolkit[5], we assess similarity scores and establish a threshold of 90. This implies that keyphrases exhibiting more than 90% similarity in their characters, particularly within the largest common substring, are excluded from the candidate keyphrase list. Next, we utilize the Bengali Part-of-Speech (POS) tagger by the Bangla NLP toolkit[6] to filter out tokens categorized as prepositions, conjunctions, adverbs, pronouns, and interjections as these do not convey essential information about the content. Finally, we create the primary candidate keyword list. Figure 2 illustrates an example of candidate keywords extracted from the initial content.

[2] https://www.ranks.nl/stopwords/bengali.
[3] http://bit.ly/1N4ib1W.
[4] https://pypi.org/project/bnlp-toolkit/.
[5] https://github.com/seatgeek/fuzzywuzzy.
[6] https://github.com/Foysal87/sbnltk.

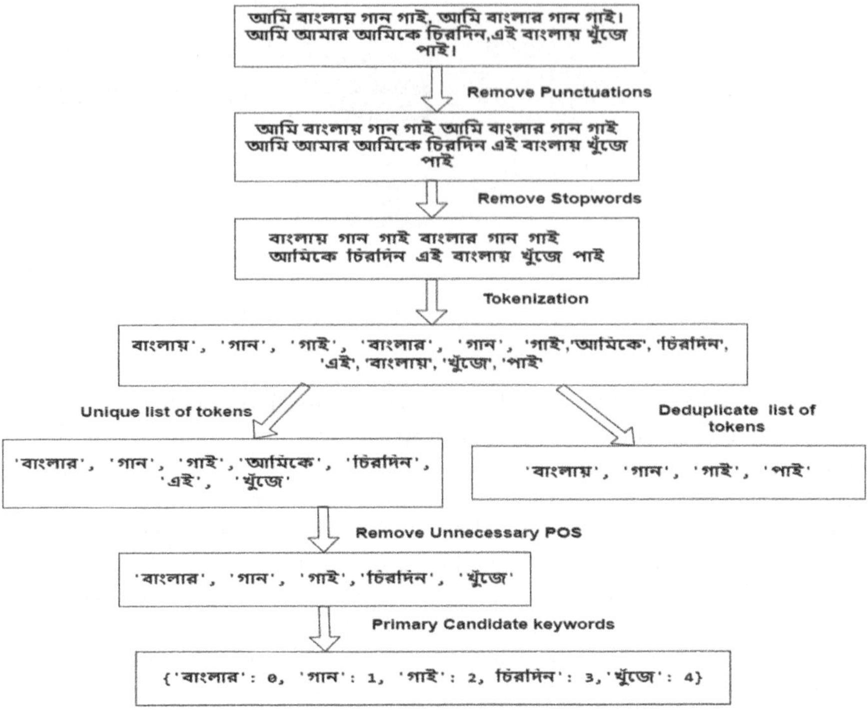


**Fig. 2.** Example of candidate keywords extraction

### 3.3 Embedding

The second phase of our approach involves computing embeddings for both candidate keyphrases and the content text, projecting them into a shared embedding space. Initially, we employ Bangla-Bert-Base[7], a trained Bengali language model using mask language modeling, in conjunction with the corresponding tokenizer. This process is initiated by tokenizing the input Bengali text into tokens and subsequently converting these tokens into unique input IDs. This leads to the creation of a tensor, with necessary adjustments made to data types as required. The tensor is then primed for a forward pass through the BERT model, enabling the generation of contextual embeddings for the input text. To ensure compliance with the model's specified input size, a series of checks are conducted. In cases where the document exceeds the model's input size, it is segmented into chunks, each of which adheres to the model's limitations. The divided segments undergo processing via the sentence transformers model. The document's embedding is derived by averaging the embeddings of all the individual chunks.

[7] https://github.com/sagorbrur/bangla-bert.

### 3.4 Candidate Keywords Ranking

Finally, we compute the cosine similarity between the document representation and candidate keyphrases. Initially, each candidate keyphrase's embedding vector undergoes normalization and is stored in an inner product index. This index facilitates cosine similarity calculations between the query vector (normalized document vector) and vectors in the index. Subsequently, the document vector is normalized, and a query is executed against the index to identify the top-k most similar vectors and their corresponding similarity scores. The top-k value corresponds to the length of the candidate keyphrase list and is employed to rank the entire list based on descending similarity scores. Algorithm 1 provides the pseudocode outlining the proposed KPE approach.

## 4 Results and Discussion

This section presents the outcomes and interpretations of our proposed approach.

### 4.1 Performance Evaluation

Initially, we provide an overview of our method's performance using the precision measure, which signifies the correctly classified keywords among the total keywords. This is expressed using the following equation:

$$\text{Precision} = \frac{TP}{TP + FP}$$

Here, $TP$ signifies correctly identified keywords extracted as true positives, while $FP$ denotes the extracted keywords that are unrelated to the topics and are classified as false positives.

To assess the performance of our model, we selected the top-k extracted keywords for evaluation, where we considered different values of k, namely 5, 7, 10, and 15 and precision is shown in Table 1. Among these configurations, the highest precision, reaching 81%, is achieved when considering k = 5. Conversely, the lowest precision, 60%, is observed when using k = 15, presented in Fig. 3. Figure 3 offers a comparison of precision between our proposed system and systems employing a neural-based approach with a multilayer perceptron. Notably, our approach demonstrates significant improvements, with precision enhancements of 22% and 19% for extracting 5 keywords and 10 keywords, respectively, compared to the multilayer perceptron approach.

### 4.2 Sample Output

We analyze the results generated by the proposed system using two sample test documents:

**Algorithm 1:** Enhanced Keyphrase Extraction Algorithm

```
Data: InputText, Params, NLPModel, SentTransModel
Result: TopNKeyphrases and their RankScores
candidatePhrases ← [];
nlpOutput ← NLPModel(InputText);
foreach chunk in nlpOutput do
    if phrase not in NLPModel.stopWords and phrase.pos_tag not in ['pronoun',
      'particle'] and phrase.length ¿ 2 and not phrase.starts_with_digit then
        if Params.keepNounAdjs then
            if chunk.consists_only_of_nouns_adjs then
                candidatePhrases.append((chunk, firstPos))
        else
            candidatePhrases.append((chunk, firstPos))
if Params.deduplicate then
    candidatePhrases ← lexSort(candidatePhrases);
    foreach phrase in candidatePhrases do
        matches ← getCloseMatches(phrase, candidatePhrases);
        candidatePhrases.remove(matches);
embeddings ← SentTransModel.encode(candidatePhrases);
docEmbedding ← SentTransModel.encode(chunk(InputText));
embeddings ← normalize(embeddings);
docEmbedding ← normalize(docEmbedding);
topK ← len(candidatePhrases);
index ← vectorIndex.innerProductFlatIndex(embeddings);
phrases, similarities ← index.query(docEmbedding, topK);
if Params.positionMetric then
    positionScores ← softmax(phrases.positions);
    foreach phrase, posScore in phrases, positionScores do
        phrase.score ← phrase.similarity · posScore;
    phrases ← lexSort(phrases, key=lambda x: x.score, reverse=True);
else
    foreach phrase in phrases do
        phrase.score ← phrase.similarity;
topNPhrasesScoresList ← phrases.get(Params.topN);
return topNPhrasesScoresList;
```

**Table 1.** Evaluation Outcomes Regarding Keyword Extraction Precision

| Top-K Keywords | Precision |
|---|---|
| k = 5 | 81% |
| k = 7 | 77.5% |
| k = 10 | 66.67% |
| k = 15 | 60% |

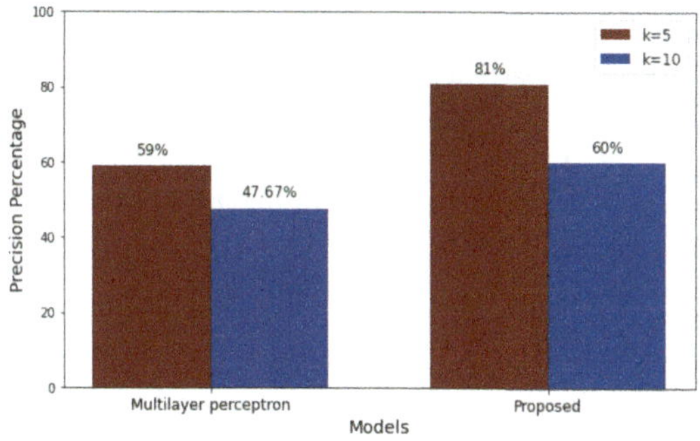


**Fig. 3.** Comparison of our approach with neural-based approach

**Test Case-1:** "অস্কারজয়ী অভিনেত্রী উইন্সলেটের অভিনয় একঘেয়েএটা বর্ষীয়ান হলিউড পরিচালক অ্যালেনের মন্তব্য হ্যালো ম্যাগাজিন সাক্ষাৎকারে জানা টাইটানিক ছবির তারকা কেটআগামী ডিসেম্বর মুক্তি অ্যালেনের ওয়ান্ড হুইল অভিনয় উইন্সলেট জানান ছবির শুটিংয় কেটের অভিনয় বোরিং দৃশ্যে অভিনয় মাঝে হঠাৎ থামা জানো তোমাকে শুনে পেয়েছি সহজে স্বীক সততা পছন্দ যতক্ষণ আঘাত বলেননিকেট উইন্সলেটের ওয়ান্ড হুইলএআরও মার্কিন গায়ক অভিনেতা জাস্টিন টিম্বারলেককেসত্র হিন্দুস্তান টাইমস"

**Test Case-2:** "চট্টগ্রাম চন্দনাইশ উপজেল দোহাজারী সোমব রোহিঙ্গা সশস্ত্র হামলা লুটপাট চালিয় পাহাড়ি গ্রাম হাতিয়াখোল পাঁচ ঘরেগভীর সশস্ত্র সন্ত্রাসীরা পাহাড় হামলা লুটপাট চালায় জানিয়েছেন স্থানীয় প্রত্যক্ষদর্শী জনপ্রতিনিধি সন্ত্রাসীরা ঘরবাড়ি ভাঙচুর করেছেস্থানীয় প্রত্যক্ষদর্শী ইউপি মোহামঞ্চদ ইস্কান্দর জানান সোমব মুখোশ একদল সশস্ত্র রোহিঙ্গা পাহাড় গ্রাম চম্পাতলী পাঁচ হামলা চালায় দরজা সবাইকে এলোপাতাড়ি মারধর লুটপাট চালায় সন্ত্রাসী হামলায় মহিলা পুরুষ মিলে কমপক্ষে হয়আহত ব্যক্তি দুজন গুরুতর দোহাজারী হাসপাতালে ভর্তি ব্যক্তি স্থানীয়ভাবে চিকিৎসা দেওয়া হাসপাতালে ভর্তি ব্যক্তি খুরশিদা বেগম মোহামঞ্চদ সেলিম হামল মালিকেরা মোহামঞ্চদ ফকির মাহাবুব কমলা বেগম আবদুল আজিজ কোরবান আলীদোহাজারী ইউনিয়ন পরিষ চেয়ারম্যান আবদুল্লাহ নোমান হামলা লুটপাট ঘটনা করেছেন চন্দনাইশ থানা ভারপ্রাপ্ত কর্মকর্তা মোহামঞ্চদ খালেদ হামলা লুটপাট থানায় পক্ষ অভিযোগ করেনি রোহিঙ্গা অপরাধ সংগঠিত স্বীক"

**Table 2.** Top-10 Keywords of Testcase-1

| Keyword | RankScore | Rank |
|---|---|---|
| অস্কারজয়ী | 0.58 | 1 |
| অভিনয় | 0.51 | 2 |
| সাক্ষাৎকারে | 0.48 | 3 |
| পছন্দ | 0.45 | 4 |
| অ্যালেন | 0.42 | 5 |
| শুটিং | 0.41 | 6 |
| হ্যালো | 0.40 | 7 |
| ডিসেম্বর | 0.40 | 8 |
| মন্তব্য | 0.39 | 9 |
| উইন্সলেট | 0.38 | 10 |

**Table 3.** Top-10 Keywords of Testcase-2

| Keyword | Rank Score | Rank |
|---|---|---|
| প্রত্যক্ষদর্শী | 0.61 | 1 |
| কর্মকর্তা | 0.56 | 2 |
| ইস্কান্দর | 0.49 | 3 |
| জনপ্রতিনিধি | 0.45 | 4 |
| চালায় | 0.44 | 5 |
| মোহামঞ্চদ | 0.42 | 6 |
| সশস্ত্র | . 0.41 | 7 |
| মাহাবুব | 0.40 | 8 |
| পরিষ | 0.39 | 9 |
| ইউনিয়ন | 0.39 | 10 |

The results presented in Tables 2 and 3 provide valuable insights into the performance of the proposed algorithm. In test case 1, we observe that the algorithm successfully identified relevant keywords, with the top-ranked keywords exhibiting high similarity scores to the document embedding. However, it is important to address the issue of suboptimal stemming performance, as highlighted by the inclusion of the keyword "সাক্ষাৎকারে" instead of the expected "সাক্ষাৎকার". Moving on to test case 2, we encounter an interesting challenge. The algorithm tends to generate individual keywords like "ইউনিয়ন" and "পরিষদ", which ideally should be combined into a single keyword, such as "ইউনিয়ন পরিষদ". This highlights an area for improvement in the model's keyword generation capabilities. However, there is room for refinement, particularly in handling stemming and generating combined keywords. These findings provide valuable insights for further enhancing the algorithm's performance.

## 5 Conclusion

Keyword extraction serves as a valuable tool for readers to expedite their comprehension of dataset contexts. In this research paper, we introduced an unsupervised keyword extraction approach that leverages POS tagging and the Bangla-Bert-Base model to generate contextual embeddings for both candidate keywords and documents. Subsequently, we employ cosine similarity to rank these keywords effectively. The keyword extraction process demonstrated high efficiency in identifying keywords, resulting in robust precision levels. While our current approach focuses on extracting individual keywords, we aspire to enhance our algorithm by incorporating more precise stemmers and POS taggers to enable keyphrase extraction in the future. Additionally, we envision the applicability of this method to micro-blogging datasets and its adaptability to other graph-based keyword extraction models tailored for the Bengali language.

## References

1. Matsuo, Y., Ishizuka, M.: Keyword extraction from a document using word co-occurrence statistical information. Trans. Jpn. Soc. Artif. Intell. **17**, 217–223 (2002). https://doi.org/10.1142/S0218213004001466
2. Lafferty, J., McCallum, A., Pereira, F.C.N.: Conditional random fields: probabilistic models for segmenting and labeling sequence data. In: Proceedings of the Eighteenth International Conference on Machine Learning (2001)
3. Zhang, C.: Automatic keyword extraction from documents using conditional random fields. J. Comput. Inf. Syst. **4**(3), 1169–1180 (2008)
4. Bharti, S.K., et al.: Konwar: automatic keyword extraction for text summarization in multi-document e-newspapers articles. Eur. J. Adv. Eng. Technol. **4**(6), 410–427 (2017)
5. Bharti, S., Vachha, B., Pradhan, R., Babu, K.S., Jena, S.: Sarcastic sentiment detection in tweets streamed in real time: a big data approach. Digital Commun. Netw. **2**(3), 108–121 (2016)

6. Hasan, S., Ng, V.: Conundrums in unsupervised keyphrase extraction: making sense of the state-of-the-art. In: Proceedings of Coling, Beijing, China, pp. 365–373 (2010)
7. Alami Merrouni, Z., Frikh, B., Ouhbi, B.: Automatic keyphrase extraction: a survey and trends. J. Intell. Inf. Syst. **54**(2), 391–424 (2019). https://doi.org/10.1007/s10844-019-00558-9
8. Papagiannopoulou, E., Tsoumakas, G.: A review of keyphrase extraction. WIREs Data Min. Knowl. Disc. **10**(2), e1339 (2020). https://doi.org/10.1002/widm.1339
9. Anwar, M.M., Anwar, M.Z., Bhuiyan, M.A.: Syntax analysis and machine translation of Bangla sentences. Int. J. Comput. Sci. Netw. Secur. **9**(8), 317–326 (2009)
10. Ashrafi, S.S., Kabir, M.H., Anwar, M.M., Noman, A.: English to Bangla machine translation system using context-free grammars. Int. J. Comput. Sci. Issues (IJCSI) **10**(3), 144–153 (2013)
11. Anwar, M.M.: Bangla to English machine translation using fuzzy logic. Int. J. Comput. Sci. Inf. Secur. (IJCSIS) **16**(11), 156–165 (2018)
12. Habib, R., Anwar, M.M.: Finding out noisy patterns for relation extraction of Bangla sentences. Int. J. Nat. Lang. Comput. (IJNLC) **9**(1), 09–20 (2020)
13. Aurpa, T.T., Rifat, R.K., Ahmed, M.S., Anwar, M.M., Ali, A.S.: Reading comprehension based question answering system in Bangla language with transformer-based learning. Heliyon **8**(10), e11052 (2022)
14. Mahfuz, T., Suha, T.F., Anwar, M.M.: Reducing wrong labels using conflict score in distant supervision for relation extraction in bangla language. In: IEEE Asia-Pacific Conference on Computer Science and Data Engineering (CSDE), pp. 1–6 (2020)
15. Aurpa, T.T., Ahmed, M.S., Rifat, R.K., Anwar, M.M., Ali, A.S.: UDDIPOK: a reading comprehension based question answering dataset in Bangla language. Data Brief **47**, 108933 (2023)
16. Habib, R., Ferdous, M., Musfique, A.M.: Creating Bengali freebase using wikidata. J. Comput. Commun. **11**(5), 151–160 (2023)
17. Showrov, M.I.H., Sobhan, M.: Keyword extraction from Bengali news. In: 5th International Conference on Advances in Electrical Engineering (ICAEE), Dhaka, Bangladesh (2019). https://doi.org/10.1109/ICAEE48372.2019.8941519.
18. Sarkar, K., Nasipuri, M., Ghose, S.: Machine learning based keyphrase extraction: comparing decision trees, naïve bayes, and artificial neural networks. J. Inf. Process. Syst. **8**(4), 693–712 (2012). https://doi.org/10.5121/ijci.2015.4304
19. Sarkar, K., Nasipuri, M., Ghose, S.: A new approach to keyphrase extraction using neural networks. Int. J. Comput. Sci. **7**(2) (2010)
20. Bhaskar, P., Nongmeikapam, K., Bandyopadhyay, S.: Keyphrase extraction in scientific articles: a supervised approach. In: COLING 2012: Demonstration Papers, pp. 17–24 (2012)
21. Sarkar, K.: Automatic keyphrase extraction from Bengali documents: a preliminary study. In: Second International Conference on Emerging Applications of Information Technology, pp. 125–128 (2011). https://doi.org/10.1109/EAIT.2011.35
22. Sarkar, K.: An N-for gram based method Bengali keyphrase extraction. In: Singh, C., Singh Lehal, G., Sengupta, J., Sharma, D.V., Goyal, V. (eds.) ICISIL 2011. CCIS, vol. 139, pp. 36–41. Springer, Heidelberg (2011). https://doi.org/10.1007/978-3-642-19403-0_6
23. Sazzad, K.H., Mumin, M.A., Azim, M.A.: Keyword extraction in Bangla scientific documents: a hybrid approach. Comput. Sci. Eng. Res. J. **11** (2018)

# IQOK: Intelligent Querying Using Ontology Prompting for Ontology-Knowledge Integration

A. Aravind Krishnan[1], Gerard Deepak[1(✉)], and A. Santhanavijayan[2]

[1] Department of Computer Science and Engineering, Manipal Institute of Technology Bengaluru, Manipal Academy of Higher Education, Manipal, India
gerard.deepak.christuni@gmail.com
[2] National Institute of Technology, Tiruchirappalli, India

**Abstract.** This paper proposes a strategic ontology knowledge integration framework for Web 3.0 through ontology-based prompting since Web 3.0 is shifting into a more organized era. This paper proposes a strategic ontology knowledge integration for political science and sports journalism as the domains of choice. Which incrementally increases the Auxiliary knowledge encompassment through topic modelling, metadata generation, and integrating entities through standard knowledge store repositories like the Google KG API. The model also formalizes ontologies from the categorical terms and from the classified instances of metadata which is achieved using the GANs classifier. Ontology shuffling is used to generate prompts from the bigram and trigram applied, pre-processed query words from the user query to increase the lateral assembly of the entities and thereby increase the overall diversity of querying in the model. The model strategically computes semantic similarities between the entities at different stages in the pipeline through Bell's isolation index, Jaccard's similarity, Jiang Conrath similarity Measure, and also Twitter Semantic Similarity. Reyni's Entropy helps in the synthesis of ontology, thereby organizing the knowledge accumulated in the model. Classification of the dataset through GANS and encompassment of semantic similarity measures to rank the documents helps in knowledge integration into the ontology which is optimizes using the Intelligent Water Drop Algorithm. This model can recommend documents as well as synthesize knowledge for political and sports journalism. This framework achieves a precision of 97.22%, an accuracy of 97.96%, and an FDR of 0.03

**Keywords:** Ontology · Knowledge Integration · Semantic Similarity · GANs · Intelligent Water Drop Algorithm.

## 1 Introduction

The creation of Web 3.0 is highly reliant on the creation of semantic understanding between entities on the web and the seamless dissemination of knowledge. The creation of ontologies and knowledge integration into these ontologies are of utmost importance to the creation of this highly interconnected semantic web. These ontologies are enhanced with the usage of metadata and topic modelling.


A. Bajaj et al. (Eds.): HIS 2023, LNNS 1224, pp. 364–373, 2025.
https://doi.org/10.1007/978-3-031-78925-0_36

Structured and standardized representations of knowledge and data are critical to the success of the Semantic Web and, in this regard, ontologies are crucial. Providing formal descriptions of concepts, entities, and their relationships in a specific domain allows machines to understand and relate information in a way that surpasses mere keyword matching. By integrating ontologies into the Semantic Web, data is enriched semantically, boosting precision and context-awareness in searches, integration, and interpretation. The outcome of such enrichment is basically the facilitation of interoperability among various databases and systems. This shrewd enhancement also holds enormous potentials for the automation of complex tasks such as data integration, question answering, and knowledge discovery.

**Motivation:** The primary motivation of the IQOK is to integrate ontological entities with the knowledge obtained from the Web 3.0 infrastructure to improve the overall standard associations and structure of the Web 3.0 framework. The encompassment of prompt engineering to re-formalize queries is a new concept that has to be incorporated into such frameworks to improve the relevance of the final queries that are yielded. Integrating ontological knowledge and thereby using the same facets for generating and recommending highly relevant documents is also the need of the hour. A very strong learning-reasoning framework is essential when constructing such frameworks in the present era of Web 3.0. Domains of political and sports journalism are quite unexplored by such frameworks and formal ontologies for these domains are required. The creation of the same would result in improving the overall structural knowledge present on the web.

**Contributions:** Synthesis of knowledge through ontology knowledge integration from the dataset and the metadata is a novel feature of the framework. Using this ontology knowledge synthesis via ontology prompting and thereby recommending documents that are highly specific to the domain is one of the new capabilities of the model. The diversification of query words through the application of bigram and trigram models not only helps in increasing the lateral ontological entity density but also the resulting prompt generation increases the overall diversity of results because ontology shuffling by using ontologies which are generated from the dataset is used. Optimization of the through the water drop optimization and Jiang Conrath similarity index along with measures such as Bell's isolation index, Jaccard's similarity, and Reyni's Entropy for semantic similarity induction into the model is also a novel concept utilized here. General Adversarial Networks for classifying the dataset and the metadata help in creating the best classification and provide the model with a very strong learning infrastructure.

**Organisation:** Section 1 depicts the Introduction, Section 2 contains the Related Works, Section 3 contains the Proposed System Architecture, Section 4 contains the Implementation, Performance Evaluation, and Results, Section 5 depicts a Conclusion, Section 6 depicts References and Datasets.

## 2 Related Works

Oprea et al., [1] have explored the application of graph theory to assess gene similarity between different species of plants using gene ontology and biomedical articles, aiming to enhance knowledge transfer in genetics. Pankowski et al., [2] have introduced a user-friendly method for facet ontology querying through a combination of faceted interface and query creation and an ontological database. It generates semantically equivalent queries to limited description entities. Cima et al., [3] have investigated the source query semantics through ontology-based abstractions and addressing limitations for a broader class of queries, monotone queries, by introducing a specialized query language.

Bhaveeasheshwar et al., [4] have highlighted the need for an automated ontology integration while maintaining knowledge representations and association for diverse domains using techniques such as LDA and Reyni Entropy to achieve high precision and recall. Koers-Stuiver et al., [5] have reviewed the research from the New Business Model Conference (NBMC), utilizing LDA and other text-mining techniques to analyze and create an understanding of the changing definitions and contextualization of new business models. Kim et al., [6] have used Generative adversarial networks (GANs), to simulate the molecular progression of tauopathy in a mouse model, identifying distinct early patterns of gene expression changes that emerge from the Gene Ontology enrichment analysis. Zhang et al., [7] have explored the scaling of biomedical ontologies at different levels and how they affect semantic similarity, demonstrating that similarity measures such as the Jiang-Conrath and Lin measures can reduce ontology size without losing important details, proving more stable and reliable. Shah-Hosseini et al., [8] have proposed a nature-inspired swarm-based optimization algorithm called the Intelligent Water Drop (IWD) algorithm, inspired by natural river behaviour to solve common computer science problems, demonstrating effectiveness in obtaining near-optimal or optimal solutions in these optimization tasks. Rafiq et al., [9] have dissected the querying and question data in e-learning systems, highlighting the importance of optimizing queries during online lectures. Deepak et al., [10–13] have proposed various frameworks and methodologies concerning ontologies and querying.

## 3 Proposed System Architecture

Fig. 1 Shows the architecture of the Intelligent Querying Framework using Ontology Prompting for Knowledge Ontology Integration in a Web 3.0 Environment. The dataset of political and sports journalism is the domain of choice for which ontology knowledge integration is facilitated in the proposed work. Moreover, recommendations also pertain to the domain of political and sports journalism as a prospective domain. The dataset of political and sports journalism is a document dataset and is subjected to the extraction of categories and terms. Categories are automatically extracted from the dataset's categories themselves. However, the terms from the sports and journalism documents are either extracted by querying using the TF-IDF model or by just shortlisting the keywords on the documents. This model employs TF-IDF. Following the strategic extraction of categories and terms from the dataset, they need to be contextualized. Contextualization is achieved using the Latent Dirichlet Allocation (LDA) which is a knowledge-topic modelling strategy from Web 3.0 which is used as the reference corpora. LDA has one major drawback which is that it only furnishes topics that are highly relevant and are very few. This poses a problem and must be subjected to enlargement. For this, the terms extracted from the LDA are sent to Google's Knowledge Graph API to load the knowledge graph and subgraphs. Google's KG API is a community-contributed, community-verified, and crowd-sourced knowledge repository, comprised of highly relevant and related content. The LDA and categorical terms are subjected to the production of metadata. Since this extensively large and voluminous, it cannot be used as is and therefore must be made permeable into the model for which they have to be classified. Due to the scale, diversity, and heterogeneity of the classification upon the metadata, needs a strong deep learning algorithm. Once the classified instances have been obtained, they are formalized as an ontology by computing the class similarity and the instance similarity using the Renyi entropy. The Renyi entropy at this juncture is set to an empirically evaluated value of step deviance of 0.15 to yield only the most relevant entities. Based on this information measure, the edges among the nodes are created

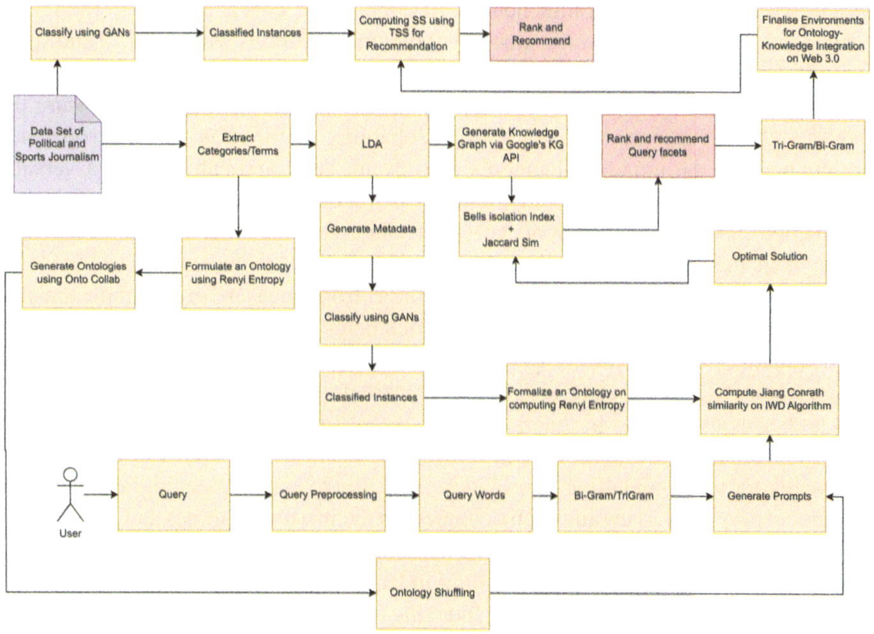


**Fig. 1.** Architecture of the IQOK framework

The user query which is one of the finer inputs is subject to preprocessing in the form of tokenization, lemmatization, word sense disambiguation, and entity recognition. This yields query words which are subjected to bigram and trigram models to generate prompts. This prompt generation, however, requires an ontology for which the entities and terms of the dataset are formulated as an ontology using Renyi entropy, yet again with a step deviance of 0.15. using this initial ontology, OntoCollab is used to generate larger ontologies up to 24 levels, exclusive of the individuals. These ontological entities are shuffled and added with bigram and trigram-generated words to generate larger prompts. These larger prompts are formed via a sentence formalization agent which is designed using AgentSpeak. These generated prompts are further subjected to the computation of the Jiang Conrath Similarity with that of the formalized ontologies that come out of the GANs Classifier. The Jiang Conrath Similarity is set to a threshold of 0.60 because the most relevant facets must be encompassed. This generates the initial solution. The initial solution may be extensively large which must be optimized using an optimization model like the intelligent water drop algorithm. This algorithm is run several times until the metadata is strategically depleted until a pseudo-optimal solution set is produced. This is further subjected to the computation of Bell's isolation index and Jaccard's similarity. Bell's isolation index is set to a step deviance of 0.10 and Jaccard's similarity is set to a threshold of 0.75. This yields the best possible query facets which are rearranged in the ascending order of Jaccard's similarity and ranked and the bigram and trigram model is run yet again, to generate a large number of queries which are then finalized based on a domain expert review. These finalized entities are subjected to the computation of Twitter semantic similarity with a threshold of 0.75 on the GANs classified dataset to yield the best-in-class recommendation of documents. This is a two-step model that recommends the queries as well as document recommendations with political and sports journalism as the domain of choice.

Renyi entropy like most entropy measures in information theory, is a measure that quantifies the uncertainty or disarray in a probability distribution. This entropy measure generalizes the popular Shannon's entropy measure and can provide a versatile framework for analyzing data. Renyi entropy is represented by the parameter $\alpha$ which when varied can represent different aspects of a dataset. One of the advantages of Renyi entropy is its flexibility. Shannon's entropy can be deduced from Renyi entropy when $\alpha$ equals 1. When $\alpha$ deviates from 1, however, the entropy measure grows more sensitive to extreme probabilities in the distribution which helps in dealing with skewed datasets. The adaptable nature of the Renyi entropy measure makes it suitable for this IQOK framework as it gives a greater insight into the structure and characteristics of data.

Another measure used in this framework is the Latent Dirichlet Allocation (LDA) which is a probabilistic model applicable in the case of NLP and learning designed specifically for topic modelling. It provides the ability to extract meaningful topics from a document corpus by assessing each document in the mixture of topics and each word in a document is attributed to a topic. LDA's flexibility and generative modelling approach make it a powerful tool for topic modelling in large corpora.

When compared to Latent Semantic Indexing (LSI), LDA has several advantages. LSI relies on singular value decomposition (SVD) to capture latent semantic relationships between words and documents. This is a traditional approach whereas LDA uses a generative probabilistic approach, making it more intuitive and interpretable. LDA not only identifies topics but also quantifies the distribution of topics within each document. LDA therefore, is better suited for handling the sparsity and high-dimensionality of text data since it is probabilistic. LDA has a clearer theoretical foundation as it explicitly models distribution, making it more suitable for capturing intricate relationships of textual data.

GANs, are primarily a type of neural network architecture used for generative tasks and it consists of two neural networks, a generator, and a discriminator, that engage in a competitive process to generate realistic data. The generator is responsible for the creation of data that cannot be differentiated from real data while the discriminator tries to fine tune and pick apart the differences. This process of training is continued iteratively until data is produced by the generator until a difference threshold is met.

TF-IDF is a metric that indicates the importance of a word inside a document collection/corpus. It serves as a weight factor in searches and text extraction/mining. TF-IDF is derived from two statistical measures: term frequency and inverse document frequency. The formula for calculating TF is given by Equation 1.

$$\mathrm{tf(t,\,d)} = \frac{f_{t,d}}{\sum_{t' \in d} f_{t,d}} \tag{1}$$

The value $f_{t,d}$ is represents the count of a term present in a document, indicating the frequency of term $t$ occurring in document $d$. The denominator represents the number of terms in document $d$. The formula for calculating IDF is given by Equation 2.

$$\mathrm{idf(t,\,D)} = \log\frac{N}{|\{d \in D : t \in d\}|} \tag{2}$$

Here, N represents the overall number of documents in the corpus, expressed as $N = |D|$. On the other hand, $|\{d \in D: t \in d\}|$ refers to the count of documents within the corpus where the term t is present. TF-IDF is generated by the product of the above-mentioned individual TF and IDF terms.

The Intelligent Water Drop Algorithm (IWD) is a metaheuristic optimization technique that is based on the foraging behaviour of water drops, which are tasked with finding the optimal path while flowing through a network of channels. In the context of optimization, the IWD Algorithm simulates the interaction and competition of intelligent water drops to find the best solution in the search space. For optimization problems, the Intelligent Water Drop Algorithm starts with an initial set of water drops (potential solutions) that are randomly distributed in the solution space. Each water drop evaluates the objective function for its current solution. The water drops compete with each other to find the best solution and the surviving water drops are adjusted towards better solutions based on their previous performance and the global best solution is found. This algorithm is run iteratively to repeat the competition and evaluation and improvement for a certain predefined number of iterations or until a convergence criterion is met.

## 4 Implementation, Performance Evaluation and Results

Implementation of the proposed IQOK framework was carried out using Python 3, with Google Colaboratory as the preferred IDE. Python's NLTK framework was used to perform common NLP tasks like lemmatization, stop word removal, etc. TensorFlow has been used to implement the GANs and other models such as the bigram and trigram. A customized annotator is designed with AgentSpeak.

The experimentations were conducted on 6 distinct datasets. To represent political journalism three distinct datasets were integrated to form a single dataset namely, the

2023 Political Communication Cultures in Western Europe dataset, the 2023 Politics and Journalism: Twitter domain shared by representatives and journalists in Spain Dataset, and the 2021 Digital Media Redefining the Modes of Political Communication; Disinformation, Misinformation & Fake News dataset. These three datasets are integrated strategically by their categories and by adding common annotations to a customized annotator. This is further integrated with another set of datasets for sports journalism. The sports journalism datasets are the Work With Data (2023) Sports Journalism dataset, the Work With Data (2023) Sports Journalism and coming out stories dataset, the Tomanek et al., and the Sports media management dataset. These are integrated with the other datasets using a custom annotator and they are reprioritized into a single large sports journalism dataset. The individual integrated political and sports journalism datasets are merged as is to form a single large dataset upon which experimentations were conducted.

The ontology knowledge integration framework for intelligent querying using ontology prompting over a Web 3.0 environment is evaluated using precision, recall, accuracy, and F-measure percentages in their average forms as the primary metrics and the false discovery (FDR) as the auxiliary metric.

Table 1 indicates the evaluation metrics of the IQOK framework in terms of these metrics. The primary metrics compute and quantify the relevance of results while the FDR quantifies the number of false positives furnished by the framework. Table 1 shows that the proposed IQOK framework yields the highest average percentages of precision, recall, accuracy, and F-measure of 97.22%, 98.71%, 97.96%, and 97.96% respectively, and a false positive rate of 0.03. The proposed IQOK is baselined with three models to benchmark the performance of the IQOK and to compare it. These baseline models are OBIM, ODFQ, and the MODM. The OBIM produces a precision of 88.84%, a recall of 90.22%, an accuracy of 89.35%, an F-measure of 89.70%, and an FDR of 0.12. The ODFQ shows a performance of 90.22%, 92.04%, 91.13%, 91.12%, and 0.10 for the same metrics respectively while the MODM has yielded a 91.46% rate of accuracy, 98.71% recall, 97.96% accuracy, 97.96% F-measure and an FDR of 0.09. The proposed IQOK produces the highest values of primary metrics and the lowest rate of false positives.

**Table 1.** IQOK metrics against other approaches

| Model | Average Precision % | Average Recall % | Average Accuracy % | Average F-Measure % | FDR |
|---|---|---|---|---|---|
| OIBM [1] | 88.48 | 90.22 | 89.35 | 89.70 | 0.12 |
| ODFQ [2] | 90.22 | 92.04 | 91.13 | 91.12 | 0.10 |
| MODM [3] | 91.46 | 93.07 | 92.26 | 92.26 | 0.09 |
| **Proposed IQOK** | 97.22 | 98.71 | 97.96 | 97.96 | 0.03 |

The reason for the better performance of the IQOK is that it encompasses incremental knowledge addition by extracting the terms and categories from the dataset about political and sports journalism as the domains of choice which are in-turn contextualized by topic modelling performed using the LDA. This model also includes the generation of metadata, generation of knowledge graphs, and strategic formalization of the ontology using the classified instances of the metadata which is classified using a very strong GANS classifier. The model synthesizes prompts from the user query after preprocessing and applying the bigram and trigram model which further enriches the auxiliary query addition and further generation of prompts through a prompting agent and ontology shuffling from the generated ontologies. The framework also encompasses very strong relevance computation methods in the form of Bell's Isolation Index, Jaccard's Similarity, Jiang Contract Similarity method, and Reyni entropy at different stages of the architectural pipeline. The Reyni entropy helps in synthesizing and organizing the ontology from the metadata and also formulates the ontology from the categorical terms and categories of the dataset. The intelligent water drop algorithm serves as a very strong metaheuristic optimization model that optimizes the initial solution set and transforms it into an optimal solution set. The Twitter Semantic Similarity measure is also used to link the ontology synthesized and integrated ontology with the recommendable entities from the original dataset and documents.

The OBIM is an ontology integration framework for biomedical data, specifically uses evolutional relationships among the entities, and gene similarity is computed. The similarity between genes and sequencing with a plethora of features helps the framework integrate the ontologies. Ontology integration here takes place between different formats primarily. Only the computation of lateral similarity is conducted rather than generating or synthesizing new knowledge which is its primary drawback. Additionally, the reasoning in the model is minimal as it depends on overlaps and similarities without thresholding the similarities. Uniform similarity measures are used which leads to a lack of heterogeneity which in turn causes it to underperform.

The ODFQ which is the ontology development with faceted queries generates faceted ontologies where the hierarchical view of nested facets of the ontology through a faceted interface is used. Ontology is converted to an ontological database where the axiom box is stored and further queried by formalizing queries that are semantically equivalent to SROIQ description logic. Weak reflexive cycles embedded in constrained rule sets are employed in this specific framework. Although it is an ontology querying model, the queries are of a specific type, which is again an assertional language querying and not a natural language querying. Ontology querying through integration should be as per the natural language querying rather than having an assertional query language suited to ontologies. This is one of the drawbacks of ODFQ. Most importantly, strong classification mechanisms are absent, only overlap measures and querying based on the assertions which again causes it to underperform in comparison.

The MODM is entirely based on monotone abstractions for ontology-based management of data. Monotone abstractions help in capturing the semantics of the query terms comprising the concepts and relations present in the ontology. The DL-Lite and unions of conjunctive logic are used through description logic-based querying. Abstractions over the description logic and the large classes through disjunctive dialogue over an epistemic operator are used which helps in reasoning. Epistemology-based reasoning is the only type of reasoning present. Reasoning through standard semantic similarity measures and quantitative reasoning is absent. Assertional reasoning and logic-based reasoning through description logic are alone made possible which cannot yield very strong semantics like qualitative semantic similarity measures. The model lacks new knowledge generation, creative emphasis on knowledge, and ontology knowledge integration. It only helps in the data management of existing ontological entities with very sparse addition of them. This makes this framework lag when compared to the proposed IQOK (Fig. 2).

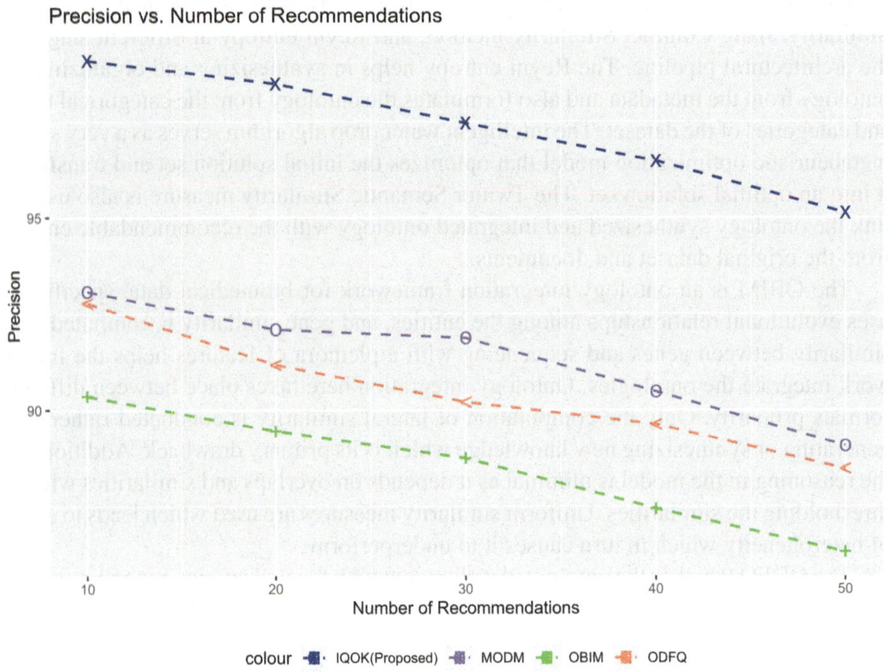


**Fig. 2.** Precision Vs No. of Recommendations Distribution

The line graph shows how the proposed IQOK framework occupies the highest position in the hierarchy of the precision against number of recommendations chart and outperforms all the other baseline models.

## 5 Conclusions

This paper proposes the IQOK framework which is a strategic ontology knowledge integration framework for Web 3.0 through ontology-based prompting. The framework takes in political science and sports journalism document datasets and strategically enhances the knowledge present through the use of community-contributed points obtained from Google's KG API. The model can encompass topics through LDA topic modelling and formalize ontologies for the mentioned domains of choice. Ontology shuffling is performed to generate prompts from the bigram and trigram models. The pre-processed query words from the user query are subjected to semantic similarity computation metrics such as Bell's isolation index, Jaccard's similarity, Jiang Conrath similarity, Twitter

Semantic Similarity, and Twitter Semantic Similarity. The Reyni Entropy measure helps in ontology synthesis. Ultimately, the GANs are used to encompass rank and recommend the knowledge integration into the ontology with the help of metaheuristic optimization techniques. The IQOK yields a final accuracy of 97.22%, a recall of 98.71%, and an FDR of 0.03.

## References


1. Oprea, A. (2023). Ontology integration for biomedical data
2. Pankowski, T.: Ontological databases with faceted queries. VLDB J. **32**(1), 103–121 (2023)
3. Cima, G., Console, M., Lenzerini, M., & Poggi, A. (2022). Monotone abstractions in ontology-based data management. In Proceedings of the AAAI Conference on Artificial Intelligence 36 5, pp. 5556-5563
4. Bhaveeasheshwar, E., Deepak, G.: OISHI: an ontology integration framework for domains of socio-humanitarian importance incorporating hybrid machine intelligence. In: International Conference on Electronic Governance with Emerging Technologies, pp. 155–167. Springer Nature Switzerland, Cham (2023)
5. Koers-Stuiver, D.M., Kamm, M.A.A., Schöller, K.: Uncovering the meaning of "new business models" and "multiple value creation". Maastricht University Press (2023)
6. Kim, H., Kim, Y., Lee, C.Y., Kim, D.G., Cheon, M.: Investigation of early molecular alterations in tauopathy with generative adversarial networks. Sci. Rep. **13**(1), 732 (2023)
7. Zhang, Q., Haglin, D.: Semantic similarity between ontologies at different scales. IEEE/CAA J. Automat. Sin. **3**(2), 132–140 (2016)
8. Shah-Hosseini, H.: The intelligent water drops algorithm: a nature-inspired swarm-based optimization algorithm. Int J Bio-inspired Comput. **1**(1-2), 71–79 (2009)
9. Rafiq, M.S., Jianshe, X., Arif, M., Barra, P.: Intelligent query optimization and course recommendation during online lectures in E-learning system. J. Ambient. Intell. Humaniz. Comput. **12**, 1–20 (2021)
10. Surya, D., Deepak, G., Santhanavijayan, A.: Ontology-based knowledge description model for climate change. In: International Conference on Intelligent Systems Design and Applications, pp. 1124–1133. Springer International Publishing, Cham (2020)
11. Deepak, G., Sheeba Priyadarshini, J.: A hybrid semantic algorithm for web image retrieval incorporating ontology classification and user-driven query expansion. In: Advances in big data and cloud computing, pp. 41–49. Springer Singapore (2018)
12. Deepak, G., Gulzar, Z., Leema, A.A.: An intelligent system for modeling and evaluation of domain ontologies for crystallography as a prospective domain with a focus on their retrieval. Comput. Electric. Eng. **96**, 107604 (2021)
13. Adithya, V., Deepak, G., Santhanavijayan, A.: OntoIntAIC: an approach for ontology integration using artificially intelligent cloud. In: Advances in Data Computing, Communication and Security: Proceedings of I3CS2021, pp. 3–13. Springer Nature Singapore, Singapore (2022)

# Predicting User Sentiments in Social Media with Machine Learning and Natural Language Processing Techniques

Kurmala Lakshmi Sathvika, Devineni Srujitha, Kavuru Tanya, Srilatha Tokala(✉), Murali Krishna Enduri, and Satish Anamalamudi

Algorithms and Complexity Theory Lab, Department of Computer Science and Engineering, SRM University-AP, Neerukonda, Andhra Pradesh, India
{sathvika_kurmala,srujitha_devineni,tanya_kavuru, srilatha_tokala,muralikrishna.e,satish.a}@srmap.edu.in

**Abstract.** Social media platforms have become an integral part of modern communication, offering individuals an unprecedented opportunity to express their thoughts, opinions, and emotions publicly. The primary objective of social media trends sentiment analysis is to analyze and interpret the sentiments associated with specific hashtags, topics, or viral content through the application of natural language processing and machine learning algorithms. To analyze the trends, we have used various ML algorithms such as CNN, Naive Bayes, SVM, Random-forest classifier, and Linear Regression. We have analyzed these algorithms by calculating various factors like f1 score, recall, precision, and accuracy. As social media plat-forms continue to evolve, this research field presents exciting opportunities for businesses, researchers, and policymakers to harness the collective voice of the online community and stay informed about the ever-changing public sentiment in the era of digital communication.

**Keywords:** Machine learning algorithms · Natural Language Processing · Text Mining · Twitter · Lexicon Based Methods

## 1 Introduction

Social media is one of the many platforms that involves and considers the emotions and opinions of people all over the world. Social media sentiment analysis plays a pivotal role in extracting sentiments or opinions out of the content posted on various platforms like Twitter, Facebook etc. [1]. The emergence of social media, access to opinions has become considerably more manageable and it's more critical than ever to measure social sentiment, as it changes frequently. Through this, we can judge whether the posts are positive, negative, or neutral. We can analyze these by collecting the datasets and using the Machine Learning algorithms accordingly to classify various parameters [2]. Using this technique, businesses, governments, and individuals tend to understand public opinion, identify ongoing trends, and make informed decisions. The study in this


A. Bajaj et al. (Eds.): HIS 2023, LNNS 1224, pp. 374–382, 2025.
https://doi.org/10.1007/978-3-031-78925-0_37

field is crucial due to several compelling reasons like it can be used in evaluating your respective brand's health, dealing with a crisis, understanding the competition, social listening and trend analysis, and customer feedback. Overall, sentiment analysis plays a pivotal role in extracting valuable insights from vast amounts of textual data generated on social media platforms or any other digital platforms [3]. It empowers researchers, businesses, and policymakers to make informed decisions, enhance user experiences, and engage with the public more efficiently.

ML and Deep Learning techniques offer valuable tools for exploring and understanding sentiment analysis. By leveraging diverse data sets, extracting meaningful features from the data, and utilizing various learning algorithms, these techniques can aid trend analysis, public opinion, customer service, and support [4]. Social media sentiment analysis is a specialized application of natural language processing (NLP) and machine learning techniques to analyze and interpret sentiments, emotions, and opinions. With the explosive growth of social media, this has become a crucial tool for understanding public perception, gauging customer satisfaction, and making data-driven decisions in diverse fields. Researchers have been using various methods such as lexicon-based approach, and hybrid approach, which is a way to get observations and draw conclusions [5]. Now in the modern era after the advent of ML and DL algorithms, numerous ways have evolved to efficiently collect, store, and analyze the data and perform complex computations. We are trying to solve this problem by identifying the patterns within the data extracting the pattern and calculating the trends. The study is important because it serves as a valuable resource for those seeking to lever-age sentiment analysis to gain insights into public sentiment and opinion in the era of social media. This study also delves into pre-processing techniques used to handle challenges specific to social media data. The overarching aim is to gain valuable insights into public perception, attitudes, and reactions towards various topics, brands, events, and trends in the digital realm. In sentiment analysis of social media, researchers and practitioners observe several knowledge patterns that emerge from the analysis of textual data and the application of machine learning techniques [6]. These knowledge patterns help in understanding the sentiments, emotions, and opinions expressed by users on social media platforms. Some of the common knowledge patterns observed in sentiment analysis of social media are viral sentiment cascades, Sentiment polarity distribution, Emotional analysis, and Contextual sentiment.

## 2 Related Works

Becker *et al.* proposed an innovative event identification method in social media using ensemble learning [7]. They assess text, time, and location, and introduce "URLs" and "bursty vocabularies" as new factors. Their evaluation of the upcoming dataset focused on Flickr photos preceding events demonstrates high accuracy. Challenges include prolonged inactivity in large-scale events and evolving bursty vocabulary, suggesting potential improvements with user participation and word frequency considerations. Gautam *et al.*, analysis for classifying customer reviews, which is useful for analyzing data in the form of the number of tweets with very illogical and either favorable or negative thoughts has been discussed [8]. The user's opinion has been classified into various sentiment classes which further help in the decision-making process. Sentamiselvan *et al.*, they analyzed the movie reviews given by people on various social media platforms through tweets, blogs, etc. using various techniques [9]. With this, they have identified the sentiment of an individual with respect to a given source of information. In the study by Ain *et al.*, enormous heaps of data from social networks, forums, and review sites have been gathered and analyzed to calculate the sentiment analysis [10]. To solve the problem of insufficient labeled data in the field of NLP, deep learning techniques are merged due to their automatic learning capability. It highlights studies that involve the implementation of convolutional neural networks, and deep neural networks. Jagdale *et al.*, A computer analysis of a person's buying interests and opinions about a company's business entity is also done in order to better business strategy and obtain insight into customer comments about their items [11].

In their investigation, Yue *et al.* focuses on providing standard approaches in the field of sentiment analysis from three distinct viewpoints [12]. Specifically, a variety of approaches and techniques are contrasted and arranged. However, a number of data formats and state-of-the-art research techniques are presented, along with their limitations. The identification and discussion of the most promising areas for sentiment analysis in the future are based on these resources.

The research by Nilekar *et al.* collected twitter data, focused on text, and used TF-IDF and machine learning to achieve 85% accuracy in trending tweet extraction, 80% in topic classification, and 75% in sentiment analysis for Arabic tweets [13]. Their approach advances trend analysis, offering the potential for business and organizational use in tracking opinions and trends. In this, Agboola *et al.* address challenges in spam filtering with a machine learning approach [14]. They use word embedding to convert words into vectors and apply various algorithms, achieving an impressive 96% accuracy on a spam message dataset. This underscores machine learning's effectiveness in adaptable spam detection, overcoming evolving spammer tactics. The next section describes about different machine learning methods that are used in our analysis.

## 3 Methodology

In this section, we outline the methods employed for conducting sentiment analysis on twitter dataset. The methodology encompasses data pre-processing, feature extraction, model building, and evaluation using a range of machine learning algorithms. The following steps were taken to achieve accurate sentiment classification as shown in Fig. 1.

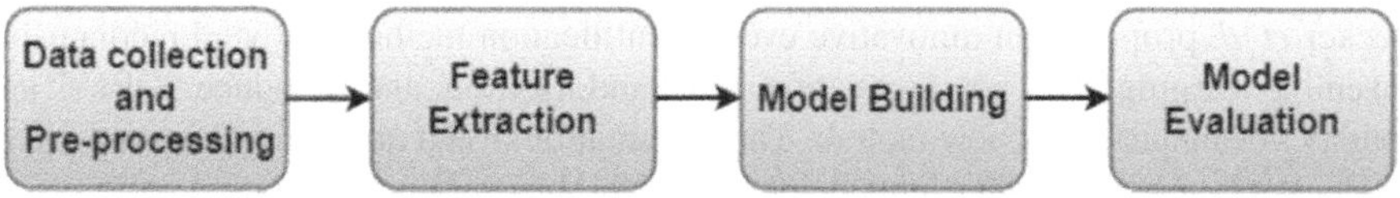


**Fig. 1.** Sentiment Analysis Process.

**Data Collection and Preprocessing:** We obtained twitter dataset from kaggle, which provided the necessary information for conducting sentiment analysis. We followed the subsequent steps for data pre-processing:

*Data Cleaning:* The datasets were cleaned to remove irrelevant information and focus on the essential attributes. Null values were either dropped or replaced based on the specific dataset and attribute.

*Tokenization and Stemming:* We employed the Porter Stemmer algorithm for stemming, which involved converting words into their root forms. Tokenization was performed to split tweets into individual words for analysis.

*Removal of User Handles:* Twitter user handles (starting with" @") were removed, as they were not informative for sentiment analysis.

**Feature Extraction:** We extracted features from the cleaned and processed tweets using various techniques. The following methods were employed:

*Bag-of-Words:* We applied the Bag-of-Words technique to convert the tokenized tweets into numerical vectors. Each word in the vocabulary was treated as a feature, and its frequency in each tweet was recorded [15].

*TF-IDF (Term Frequency-Inverse Document Frequency):* We used TF-IDF to calculate the importance of words in each tweet relative to the entire dataset. This technique considered both term frequency and the inverse document frequency to capture the significance of words.

**Model Building:** We employed a range of machine learning algorithms to build sentiment analysis models on the training dataset. The models were then evaluated on the testing dataset to assess their performance. The algorithms used were:

*SVM Algorithm:* In sentiment analysis, each text document is represented as a high-dimensional feature vector, where each feature represents a word or n-gram. SVM can handle this high-dimensional feature space efficiently. They are used for sentiment analysis because they are effective in handling high-dimensional text data, can capture non-linear relationships between features, are robust to overfitting, can handle imbalanced datasets, and have a solid theoretical foundation [16].

*KNN Algorithm:* It is not a common choice for sentiment analysis as it is mostly for regression tasks [17]. But is primarily a classification algorithm which makes it a good choice and also a complex choice for sentiment analysis. This is employed for sentiment analysis in the case of simple scenarios to understand its behavior and limitations when compared to other algorithms. If we ever need a quick and simple sentiment analysis for a small dataset, this algorithm could be implemented without extensive model building or for feature engineering.

*Logistic-Regression:* Logistic Regression is widely used, and also a common choice for sentiment analysis. The goal is to determine the sentiment expressed in a piece of text. This algorithm provides results that are interpretable. The coefficients of this model could be analyzed to understand which words and features contribute positive, negative, and neutral emotions to the sentiment prediction. This algorithm is also computationally efficient and it requires less computational resources, unlike more complex models. These algorithm coefficients help us to identify important features that contribute to a particular sentiment. This even works well with relatively small dataset, which are common in sentiment analysis tasks.

*Decision Tree Classifier:* This algorithm mainly consists of a tree structure with nodes, which is an important part of explaining this algorithm. Those are called tree nodes and root nodes, where each node represents the decision or a test on an input feature. These nodes are connected by branches. Root nodes are at the top of the tree structure and leaf nodes are the final nodes of the tree that represent output or prediction of a particular subset of data. This algorithm is known for its interpretable nature and can handle both categorical and numerical types of data. *Naive Bayes:* This is a probabilistic classifier that is most used in Natural Language Processing techniques like sentiment analysis. In Naive Bayes, probabilities are assigned to words or phrases, segregating them into different labels. Comparatively, this algorithm is much faster as it calculates the probabilities. Its easily scalable property makes it the most efficient algorithm.

*CNN (Convolutional Neutral Networks):* CNNs excel at capturing local patterns or features within the input data. In the context of sentiment analysis, local patterns can represent important clues about the sentiment expressed in a text. CNNs can learn hierarchical representations of the input data.

*XG Boost:* Basically, it's not a choice for sentiment analysis, as it's usually used for structured data problems [18]. But it's still possible to make XG Boost adapt for sentiment analysis tasks, and it will do its work efficiently. If the sentiment analysis task you provided involves both text data and structured features, XG Boost can handle the structured part efficiently while incorporating text features. XG Boost is widely known for use in hybrid models where you combine predictions from different models. For instance, the NLP model can be useful for text analysis and XG Boost for structured features as discussed, and then getting accuracy by combining their predictions.

**Model Evaluation:** The performance of each model was evaluated using key metrics, including accuracy, precision, recall, and F1-score.

The methodology outlined above facilitated the successful execution of sentiment analysis on Twitter datasets. By employing data pre-processing, feature extraction, and a diverse set of machine learning algorithms, we achieved accurate sentiment classification for the tweets. In this section, we will further discuss the characteristics of the datasets used in our study, shedding light on their relevance and significance. and the subsequent analysis of the sentiment patterns found in the Twitter data.

## 4 Dataset Statistics

In this study, the foundation of our research lies in three distinctive datasets meticulously gathered from kaggle:" EDA For Sentiment Analysis," "EDA Twitter Sentiments," and" Twitter Sentiment Dataset." These datasets form the bedrock of our exploration into sentiment analysis, providing the essential raw materials for our analytical journey. Let's delve into the specifics of each dataset (Table 1):

These datasets serve as the pillars of our study, providing a diverse range of textual expressions harnessed from social media platforms. Our analysis hinges on the comprehensive evaluation of these datasets, each contributing a unique dimension to our understanding of sentiment within various contextual frameworks. By dissecting the

**Table 1.** Dataset Statistics

| Dataset | No of rows | No of columns | Attributes | Text Sentiment |
|---|---|---|---|---|
| EDA Twitter Sentiments | 27,481 | 3 | Id, Text, Sentiment | Positive, Negative |
| Twitter Sentiments | 31,962 | 3 | Id, Label, Tweet | Positive, Neutral, Negative |
| EDA for Sentiment Analysis | 10,000 | 2 | Clean text, category | Positive, Neutral, Negative |

attributes and sentiments embedded within these datasets, we aim to construct robust sentiment analysis models capable of transcending the confines of any single dataset. This approach ensures that our research is adaptable and versatile, effectively addressing real-world sentiment dynamics across diverse scenarios.

## 5 Results

Having discussed the significance of the datasets, we will now delve into the out-comes of our sentiment analysis study. The performance of numerous machine learning algorithms across distinct datasets is examined in this study, illuminating their effectiveness in sentiment categorization. The evaluation encompassed accuracy, precision, recall and f1-score integral metrics that collectively illuminate the strengths and adaptability of these algorithms.

Let us consider M as True Positives - Instances correctly predicted as positive sentiments. N as False Positives - Instances incorrectly predicted as positive sentiments. O as False Negatives - Instances incorrectly predicted as negative sentiments when they were actually positive. P as True Negatives - Instances correctly predicted as negative sentiments.

**Accuracy:** Accuracy gauges the proportion of correct predictions in the entire dataset, reflecting how effectively models predict sentiment polarity from text data. It encompasses both true positive and true negative predictions and is calculated using the following formula:

$$Accuracy = \frac{m + p}{(m + n + o + p)}$$

**Precision:** Precision provides insights into the reliability of positive and negative sentiment predictions. It measures the proportion of true positive predictions (m) to all instances predicted as positive (m + n):

$$Precision = \frac{m}{(m + n)}$$

**Recall:** Recall measures a model's ability to correctly identify all instances of a particular sentiment class. It's the ratio of true positive predictions (m) to the total instances of that class (m + o):

$$Recall = \frac{m}{(m + o)}$$

**F1 Score:** The F1 score balances precision and recall, especially valuable for imbalanced datasets. It's defined as:

$$F1Score = \frac{2 * Recall * Precision}{Recall + Precision}$$

Figure 2 (a) steers our focus to the EDA Twitter Sentiments dataset, which encompasses neutral sentiments, we extend our analytical purview. Introduces a graphical dimension, presenting a visual depiction of algorithmic performances. Here, the XGBoost algorithm maintains its dominance, securing the highest accuracy of 0.947. Notably, the Logistic Regression and SVM method match closely with an almost impeccable accuracy of 0.945, reaffirming its robust standing. Nevertheless, in contrast, the Decision Tree algorithm surfaces with the lowest accuracy of 0.926. Expanding our analysis in Fig. 2(b) is the Twitter Sentiment dataset, aligning with characteristics of the prior dataset, we encounter a distinct algorithmic panorama. A visual representation further enhances our grasp of these algorithm's performances. Here, the Random forest and Decision tree algorithm emerges as a beacon of accuracy, boasting a pristine 1.0. Concurrently, the XGBoost method maintains exceptional accuracy at 0.909. Precision, recall, and f1-score harmoniously orbit accuracy. Nevertheless, in tandem with previous datasets, the KNN algorithm persistently reports the lowest accuracy of 0.610. Fig. 2(c) deals with the analysis of the EDA For Sentiment Analysis dataset ventured into a rich tapestry of XGBoost algorithm with higher accuracy, precision, recall, and f1-score of 1.0. As depicted in the figure, a visual representation adds depth to our understanding of each algorithm's performance on this dataset. Foremost among them, the Decision Tree and Random forest distinguished itself with the second highest accuracies with 0.99. In contrast, the KNN algorithm exhibited the dataset's lowest accuracy of 0.673.

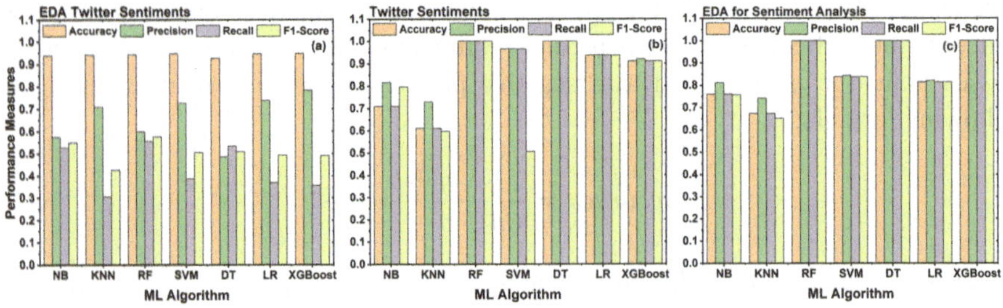


**Fig. 2.** Performance analysis on three datasets namely EDA Twitter Sentiments, Twitter Sentiment, and EDA For Sentiment Analysis.

In assessing algorithm performance across three sentiment analysis datasets, the XGBoost algorithm consistently emerged as the top performer, showcasing the highest accuracy, precision, recall, and F1-score. Notably, the decision tree and random forest algorithms exhibited their strength as the second-best performers, consistently achieving high accuracy, along with commendable precision, recall, and F1-scores. The SVM algorithm also demonstrated competitiveness, securing a solid third position in our evaluations. On the other hand, the k-nearest neighbors algorithm consistently reported the lowest accuracy, highlighting its limitations in sentiment analysis.

Our study goes beyond simply evaluating these algorithms; it serves as a valuable resource for understanding the intricate relationship between algorithms and datasets. This profound understanding equips decision-makers with the knowledge needed to make optimal algorithmic choices in diverse situations.

Comparatively, our approach excels in various critical aspects. Firstly, our research consistently achieves high accuracy in sentiment classification by utilizing advanced machine learning algorithms such as XGBoost and Random Forest. This precision extends to our detailed analysis of multiple datasets, including" EDA Twitter Sentiments," "Twitter Sentiments," and" EDA For Sentiment Analysis," enhancing the reliability of our results across diverse data sources.

Through effective utilization of machine learning algorithms, a comprehensive comparative analysis, and a strong emphasis on clarity and presentation, our work distinguishes itself in the field of sentiment analysis.

## 6 Conclusion and Future Work

In this sentiment analysis of the twitter datasets, we aimed to analyze and predict the overall sentiment expressed by users on the social media platform. We have utilized NLP techniques to interpret and summon a large number of tweets. We observed and calculated positive, negative, and neutral sentiments for the twitter datasets. As twitter consists of hashtags and mentions, we also have inspected and performed exploratory data analysis. With the help of ML algorithms, we have calculated the precision, accuracy, f1-score, and recall. With this, we were able to try and test various algorithms and conclude the best algorithms taking into count various parameters. Overall, the sentiment analysis of twitter or any social media helps us to provide insights into public opinion, and emotional trends which makes it a powerful tool for understanding and reviewing the sentiment of a diverse community on social media. When used responsibly and in conjunction with other research methods, sentiment analysis can offer meaningful and actionable insights for various fields, including market research, social studies, and public opinion analysis.

Currently, the study can lead to gaining the attention of the respective authorities that can address the issues which are obtained during the analysis, and in the future the obtained results can be used to deploy a perfect model. We are interested to explore and examine other datasets and try new social media platforms rather than twitter. Futurke work for this sentiment analysis can also focus on advancing the existing techniques and addressing the challenges and limitations observed in current approaches. Some potential areas of improvement and expansion may include fine-grained sentiment analysis in which we will enhance the sentiment analysis models by distinguishing various emotions such as joy, fear, sadness, etc., and multilingual sentiment analysis-this includes extending sentiment analysis models to support multiple languages. Twitter is a global platform with diverse users, and sentiment analysis should be adaptable to different languages, and sentiment over time which includes the evolution of sentiment trends over longer periods to identify patterns and understand how sentiment changes in response to global events and societal shifts.

## References


1. Pang, B., Lee, L., et al.: Opinion mining and sentiment analysis. Found. Trends Inf. Retr. **2**(1–2), 1–135 (2008)
2. Patel, V., Prabhu, G., Bhowmick, K.: A survey of opinion mining and sentiment analysis. Int. J. Comput. Appl. **131**(1), 24–27 (2015)
3. Philander, K., Zhong, Y.: Twitter sentiment analysis: Capturing sentiment from integrated resort tweets. Int. J. Hosp. Manag. **55**, 16–24 (2016)
4. Rout, J.K., Choo, K.-K.R., Dash, A.K., Bakshi, S., Jena, S.K., Williams, K.L.: A model for sentiment and emotion analysis of unstructured social media text. Electron. Commer. Res. **18**, 181–199 (2018)
5. Choudrie, J., Patil, S., Kotecha, K., Matta, N., Pappas, I.: Applying and understanding an advanced, novel deep learning approach: a covid 19, text based, emotions analysis study. Inf. Syst. Front. **23**, 1431–1465 (2021)
6. Ibrahim, N.F., Wang, X.: A text analytics approach for online retailing service improvement: evidence from twitter. Decis. Support. Syst. **121**, 37–50 (2019)
7. Becker, H., Naaman, M., Gravano, L.: Event identification in social media. In: WebDB (2009)
8. Gautam, G., Yadav, D.: Sentiment analysis of twitter data using machine learning approaches and semantic analysis. In: 2014 Seventh International Conference on Contemporary Computing (IC3), pp. 437–442. IEEE (2014)
9. Sentamilselvan, K., Aneri, D., Athithiya, A., Kumar, P.K.: Twitter sentiment analysis using machine learning techniques. Int. J. Eng. Adv. Technol. (IJEAT). **9**(3), 1–9 (2020)
10. Ain, Q.T., Ali, M., Riaz, A., Noureen, A., Kamran, M., Hayat, B., Rehman, A.: Sentiment analysis using deep learning techniques: a review. Int. J. Adv. Comput. Sci. Appl. **8**(6), 1 (2017)
11. Jagdale, R.S., Shirsat, V.S., Deshmukh, S.N.: Sentiment analysis on product reviews using machine learning techniques. In: Cognitive Informatics and Soft Computing: Proceeding of CISC 2017, pp. 639–647. Springer (2019)
12. Yue, L., Chen, W., Li, X., Zuo, W., Yin, M.: A survey of sentiment analysis in social media. Knowl. Inf. Syst. **60**, 617–663 (2019)
13. Nilekar, S., Rawat, S., Verma, R., Rahate, P.: Twitter trend analysis. Int. J. Sci. Res. Comput. Sci. Eng. Inf. Technol. (2020)
14. Agboola, O.: Spam detection using machine learning and deep learning. PhD thesis, Louisiana State University and Agricultural & Mechanical College (2022)
15. Neogi, A.S., Garg, K.A., Mishra, R.K., Dwivedi, Y.K.: Sentiment analysis and classification of indian farmers' protest using twitter data. Int. J. Inform. Manag. Data Insight. **1**(2), 100019 (2021)
16. Krawczyk, B., McInnes, B.T., Cano, A.: Sentiment classification from multi-class imbalanced twitter data using binarization. In: Hybrid Artificial Intelligent Systems: 12th International Conference, HAIS 2017, La Rioja, Spain, June 21–23, 2017, Proceedings 12, pp. 26–37. Springer (2017)
17. Imandoust, S.B., Bolandraftar, M., et al.: Application of k-nearest neighbor (knn) approach for predicting economic events: Theoretical background. Int. J. Eng. Res. Appl. **3**(5), 605–610 (2013)
18. Si, M., Du, K.: Development of a predictive emissions model using a gradient boosting machine learning method. Environ. Technol. Innov. **20**, 101028 (2020)

# Federated Learning for GDPR Compliance

Suyogita Singh, Satya Bhushan Verma(✉), Bineet Kumar Gupta, Monika Gupta, and Sandeep Dubey

Shri Ramswaroop Memorial University, Lucknow 225003, Uttar Pradesh, India
Satyabverma1@gmail.com

**Abstract.** Data privacy regulations, such as the General Data Protection Regulation (GDPR), have significantly impacted how organizations handle and process personal data. Compliance with GDPR requires organizations to protect individuals' privacy rights while still harnessing the benefits of data-driven decision-making. Federated Learning, a novel machine learning approach, offers a promising solution by enabling collaborative model training across decentralized data sources while preserving data privacy. This research paper delves into the intricacies of Federated Learning and its role in achieving GDPR compliance, highlighting its potential advantages, challenges, and future prospects.

## 1 Introduction

A period of unheard-of data collection, consolidation, and analysis has begun in the digital age. Businesses in a variety of industries have tapped into the power of data-driven decision-making to obtain knowledge, improve services, and spur innovation. However, this data-centric environment has also sparked serious worries about data security, individual privacy, and the moral implications of data processing. Regulations, most notably the General Data Protection Regulation (GDPR), have become effective tools for defending the rights and privacy of data subjects in response to these difficulties [1]. The European Union (EU) passed the GDPR in May 2018, and it has since established itself as the industry standard for privacy and data protection laws. Its guiding concepts and rules were created to guarantee that people have ownership over their personal data. While establishing stringent requirements for organizations that collect, process, or store this data. GDPR places a heavy emphasis on data minimization, purpose limitation, data subject rights, accountability, and transparency, challenging organizations to balance the benefits of data-driven insights with their responsibilities to protect individual privacy. One of the most pressing challenges in the context of GDPR is how organizations can leverage data for machine learning and analytics while remaining compliant with its rigorous standards. Traditional centralized machine learning approaches often involve the pooling of data into a central repository, a practice that may conflict with GDPR's principles of data minimization and data subject consent. Moreover, data aggregation can pose significant security and privacy risks, making it imperative to explore alternative methods that enable compliance without compromising the utility of data-driven models.


A. Bajaj et al. (Eds.): HIS 2023, LNNS 1224, pp. 383–392, 2025.
https://doi.org/10.1007/978-3-031-78925-0_38

Federated Learning is an innovative machine learning paradigm that offers a compelling solution to this conundrum. Federated Learning reimagines traditional centralized model training by enabling collaborative model development across decentralized data sources, often represented by diverse endpoints such as smartphones, IoT devices, or edge servers [2]. Unlike conventional machine learning, where data is centralized, Federated Learning trains models locally on client devices while sharing only model updates with a central server. This decentralized approach inherently preserves data privacy, minimizes data transfer, and aligns seamlessly with GDPR's principles.

The intersection of Federated Learning and GDPR compliance represents a compelling synergy—one that holds the promise of achieving data-driven insights while upholding the stringent standards set forth by GDPR. This research paper embarks on an exploration of federated Learning as a novel approach to GDPR compliance. It delves into the intricacies of federated Learning, its key components, advantages, and challenges, while elucidating how this innovative approach aligns with and enhances compliance with GDPR. By investigating the practical applications, limitations, and potential advancements in Federated Learning within the GDPR context, this paper seeks to provide a comprehensive roadmap for organizations navigating the complex terrain of data privacy regulations while pursuing the benefits of data-driven decision-making.

## 2 Federated Learning

A distributed machine learning method called federated learning enables several parties to build a machine learning model without sharing their data. Each party trains a local model on their own data to accomplish this, and the model updates are subsequently shared with the other parties. The individual data stays on the local devices, but the model updates are used to enhance the overall model. Federated learning is becoming increasingly popular as a way to protect user privacy [3]. This is because it does not require the data to be shared with a central server, which can be a target for cyberattacks. One example of how federated learning can be used is in the financial industry. Different financial institutions often have data that is siloed and cannot be shared due to privacy regulations. However, by using federated learning, these institutions can collaborate to train a machine learning model that can improve their lending decisions, fraud detection, and other applications [4]. There are a number of challenges to federated learning, such as the need to ensure that the model updates are accurate and that the data is not leaked. However, federated learning is a promising approach for protecting user privacy and enabling collaboration between different parties.

### 2.1 Key Components of Federated Learning

**Client Devices (Data Sources)**
Client devices are central to the Federated Learning ecosystem. These devices encompass a wide array of endpoints, including but not limited to smartphones, Internet of Things (IoT) devices, edge servers, and remote data centers. Each client device possesses its own unique local dataset, which can be thought of as a subset of the overall data relevant

to the machine learning task. These local datasets are often distributed across different geographic locations or organizational boundaries.

**Central Server**

In Federated Learning, a central server plays a pivotal role in orchestrating the collaborative model training process. The server's primary function is to coordinate the exchange of model updates between client devices, aggregate these updates, and iteratively improve the global model. Unlike traditional centralized machine learning, the central server does not have direct access to raw data but instead manages model parameters and facilitates communication between clients.

**Federated Learning Algorithm**

The choice of the Federated Learning Algorithm is critical to the success of the approach. Various algorithms, such as Federated Averaging and Secure Aggregation, have been developed to ensure efficient and secure model training in a federated setting [5]. These algorithms determine how model updates are communicated, aggregated, and incorporated into the global model while adhering to stringent privacy and security constraints.

## 2.2 Advantages of Federated Learning

**Privacy Preservation**

One of the most compelling advantages of Federated Learning is its inherent ability to preserve data privacy. Unlike traditional centralized approaches, federated learning does not necessitate the sharing of sensitive or personal data with a central entity. Instead, model updates are exchanged, allowing individual data sources to maintain control over their information.

**Decentralization**

Federated learning embraces the decentralized nature of data, making it particularly well-suited for scenarios where data cannot be easily centralized. This is crucial in cases where data resides on distributed devices, in multiple geographic locations, or within organizations bound by stringent data protection regulations [7].

**Efficiency and Scalability**

Federated learning can be highly efficient and scalable. Model updates are transmitted and processed locally on client devices, reducing the need for extensive data transfers. Additionally, this decentralized approach can lead to faster convergence and reduced computational overhead.

**Security**

Federated learning incorporates robust security measures to safeguard the model updates during transmission and aggregation. Techniques like differential privacy and cryptographic protocols help ensure the confidentiality and integrity of data and model parameters.

**Customization**
Organizations can customize Federated Learning to suit their specific needs. This includes defining the selection criteria for participating clients, setting communication frequency, and tailoring the aggregation mechanism to balance model accuracy and privacy requirements.

## 3 General Data Protection Regulation (GDPR)

The European Union (EU) passed the General Data Protection Policy (GDPR), a comprehensive data protection and privacy policy, in May 2018. Regardless of where the organization is physically located, it has broad ramifications for businesses that gather, handle, or keep the personal data of people living in the EU. The GDPR intends to give people more control over their personal data while placing strict requirements on corporations to protect the privacy and security of that data (Fig. 1).

**Fig. 1.** The General Data Protection Regulation (GDPR)

### 3.1 GDPR Principles

The cornerstone of the General Data Protection Regulation's (GDPR) approach to data protection and privacy is based on a number of key concepts. These principles serve as a guide for how businesses should handle personal data both inside and outside of the European Union, depending on the situation.

**Lawfulness, Fairness, and Transparency**
Organizations must process personal data in a way that is legal, just, and transparent. In other words, they should have a legitimate justification for processing data, let people know how their data will be used, and make sure processing is lawful, just, and reasonable.

**Purpose Limitation**
Only valid, explicit, and defined reasons should be used when collecting or processing personal data. Data should never be utilized for purposes other than those for which it was originally gathered.

**Data Minimization**
Businesses should only gather and use the minimal amount of personal data required to fulfil their stated objectives. Data that is excessive or irrelevant shouldn't be gathered or stored.

**Accuracy**
Organizations should make reasonable efforts to guarantee that erroneous personal data is promptly updated or deleted.

**Storage Restrictions**
Personal information shouldn't be stored any longer than is required to fulfil the original purposes for which it was obtained. Organizations should create data retention guidelines and get rid of outdated data [8].

# 4 Implications of GDPR in Federated Learning

## 4.1 Privacy Preservation

**Data Minimization**
GDPR mandates that organizations collect and process the personal data necessary for the intended purpose. In the context of federated learning, this aligns with the principle of data minimization. Federated Learning inherently supports this by allowing model training on decentralized, specific data subsets without centralizing all data. Organizations can thus comply with GDPR's data minimization requirements by minimizing the amount of data shared and processed centrally.

**Purpose Limitation**
The GDPR emphasizes the importance of processing personal data for legitimate, well-defined purposes. Federated learning can be tailored to meet these purposes efficiently. Organizations can design Federated Learning workflows with specific objectives in mind, ensuring that data is processed only for those predefined purposes, aligning with GDPR's principle of purpose limitation.

## 4.2 Consent Management

**Explicit Consent**
GDPR requires organizations to obtain explicit and informed consent from individuals before processing their personal data. In the context of federated learning, organizations should inform data subjects about their participation in collaborative model training, ensuring that individuals understand the nature and purpose of data processing. Organizations must obtain consent for data subjects' participation in such processes.

**Data Subject Rights**
GDPR grants individual's various rights, including the right to access, rectify, or erase their personal data. Federated learning implementations should provide mechanisms for data subjects to exercise these rights concerning their contributions to the federated training process. Organizations must be prepared to respond to such requests promptly.

### 4.3 Security and Data Integrity

**Data Security**
GDPR mandates that organizations implement appropriate technical and organizational measures to ensure data security. In federated learning, security measures are essential for safeguarding the confidentiality and integrity of model updates during transmission and aggregation. Techniques such as differential privacy and secure aggregation play a crucial role in ensuring GDPR compliance in this regard.

**Data Breach Notification**
GDPR requires organizations to report data breaches to the relevant authorities and affected data subjects promptly. Federated learning implementations should have mechanisms in place to detect and respond to any potential security breaches that could compromise data privacy.

### 4.4 Accountability and Transparency

**Accountability**
GDPR places an emphasis on accountability, requiring organizations to demonstrate compliance with its provisions. Organizations utilizing Federated Learning should maintain detailed records of data processing activities, consent management, and security measures to provide evidence of compliance.

**Transparency**
The GDPR mandates transparency in data processing. Organizations must inform data subjects about how their data is used. In the context of Federated Learning, organizations should provide clear and accessible information about the collaborative training process, including its purpose, the data used, and how data subjects can exercise their rights. GDPR has significant implications for Federated Learning, particularly in ensuring that data privacy, security, and individual rights are upheld throughout the collaborative model training process. Organizations must carefully consider GDPR's principles and requirements when implementing Federated Learning to ensure compliance and maintain trust with data subjects [9].

## 5 Design

### 5.1 Compliance Modes

We measure the strength of compliance along two dimensions, keeping in mind that GDPR compliance requires managing user requests to update or remove their data, including implied data [10].

**Frequency of Handling Modifications**

Modifications to Handling Frequency: This aspect considers how often the server processes updates to user data. The server will never reply to user notifications of changing data if "no" compliance is chosen. When the total number of requests reaches 10,000, 100, and 1, the server replies to user requests in "weak," "strong," and "strict" compliance modes, respectively. Each user request receives a prompt response from the server in strict accordance.

**Degree of Implied Data**

This dimension evaluates how certain global model weights are affected by a specific user's contribution. Implied data refers to data indirectly or directly influencing the global model. GDPR compliance, in this context, implies handling data associated with a user's request. Even if a user did not actively use affected data during training, compliance requires the server to update the global model if a user's data is deleted.

Our design allows for four compliance modes:

1. No Compliance: The model operates in this mode just like a typical Federated Machine Learning model.
2. Weak Compliance: Here, users can ask the Aggregator to update or delete any of their data that was used in earlier training sessions. Only the user's round participation information is affected by a server-side deletion or update.
3. Strong Compliance: In contrast to the weak compliance mode, the deletion or update requested by the user on the Aggregator side extends to all following rounds until the finish.
4. Strict Compliance: This mode is more stringent than strong compliance. User requests are handled immediately, including deep deletions and updates. This level of compliance prioritizes immediate privacy at the expense of computational costs.

## 5.2 Round

As previously mentioned, to accommodate various tasks and aggregation functions, our design employs functions as the unit elements within a round object [12]. To create a round, an engineer defines the following.

1. Data Function: A function that analyzes data points chosen for training and generates a list of data points in accordance with specifications laid down by engineers.
2. Training Function: On local devices, users can utilize this function to train on specific data and generate weight contributions that are delivered to the server.
3. Aggregation Function: This function, which uses the federated averaging algorithm or other techniques, is used by the Aggregator to integrate weight contributions and update the global model.

## 5.3 User

In contrast to previous work that only mentions users storing data on their devices, our design specifies components on the user side to achieve GDPR compliance [13].

1. Indicate whether specific data should be opted in for inference tasks by the central server.

2. Opt out their data from future training rounds by setting the "opt-in" variable associated with the data point to "false."
3. Apply local data selection functions sent by the Aggregator to determine which data is eligible for local training.
4. Associate participating data points with round information locally, allowing users to access this data through a front-end application.
5. Choose whether data updates or deletions should be local or propagated to previous training rounds. These requests are stored in a queue until the user requests the server to process them.

## 5.4 Aggregator

In our approach, the Aggregator serves as a go-between for users to get training information, for aggregating weight contributions, and for interacting with the Logger to store metadata data obtained from each round [14]. When the queue reaches a particular capacity, the aggregator maintains a queue of user requests to delete and update data, processing them as they come in.

**Logger**

The Logger stores essential information, including:

1. User IDs and Round IDs are mapped to actual users and rounds, making it simple to retrieve the right users and rounds for post-hoc training.
2. Weight contributions made by users in various rounds, allowing for quick adjustments in response to requests from users to delete or update their participation.
3. Global models, which are saved as mappings between round IDs and the assembled global models.

**Putting Everything Together**

To simulate various machine learning training techniques in a federated environment, we developed a tool. Users can run a Python script with specified parameters to create and manage user interactions, including data updates, training requests, and compliance modes. To implement GDPR compliance, users can add their data and choose its opt-in status. When the Aggregator sends training requests, users filter their data based on this opt-in status. Users can also request updates or deletions, and the server processes these based on the defined compliance mode. Concurrency is managed through producer-consumer queues, ensuring correct interaction between users and the server. GDPR-related requests from users are stored in a separate queue and processed according to the chosen compliance mode. Additionally, we have developed a front-end web application to visualize training statistics, user capabilities, and interactions with the server [15]. To experience full communication between the aggregator and Users, we recommend running the testers as described above, or users can run the web application with specific Python commands to explore and interact with the system.

Overall, our design addresses GDPR compliance while facilitating federated machine learning in a practical and efficient manner.

## 6 Conclusion

Federated learning offers a promising avenue for achieving GDPR compliance in the field of data-driven technologies. This innovative approach strikes a balance between the need for data privacy and the drive for technological advancement. By allowing models to be trained locally on user devices or within individual organizations, federated learning minimizes the need for centralized data storage and processing, thereby reducing the risk of data breaches and ensuring compliance with GDPR regulations. Federated learning's key strengths lie in its ability to empower individuals and organizations to retain control over their data while still benefiting from the collective knowledge gained through model updates. This collaborative approach aligns with the spirit of GDPR, which prioritizes user rights and data protection. However, challenges remain, including the need for standardized protocols, robust security measures, and ongoing education and awareness efforts. Organizations must invest in the development of secure federated learning frameworks and work closely with regulators to ensure that their implementations align with evolving privacy regulations. In the context of GDPR compliance, federated learning represents a significant step forward in addressing the complex intersection of data privacy and machine learning. As technology continues to advance and privacy regulations evolve, federated learning offers a flexible and privacy-preserving solution that holds the potential to reshape the way we approach data-driven innovation while safeguarding individual rights and data security. To fully harness this potential continued research, collaboration, and commitment to privacy principles are essential.

## References

1. Hardy, J.: Private federated learning on vertically partitioned data via entity resolution and additively homomorphic encryption. In: IEEE Transactions on Information Forensics and Security (2017)
2. Uribe, D.: Privacy Laws, Non-Fungible Tokens, and Genomics. J Br. Blockchain Assoc. **3**, 1–10 (2020)
3. Mir, S., Capretz, M.A., Grolinger, K., ElYamany, H.F., ElGayyar, M.M.: Blockchain-based federated identity and auditing. Int. J. Blockchains Crypt. **1**, 179 (2020)
4. Bonnah, E., Shiguang, J.: DecChain: a decentralized security approach in edge computing based on Blockchain. Futur. Gener. Comput. Syst. **113**, 363–379 (2020)
5. Blockchain and Artificial Intelligence ApplicationsSpecial Issue: Blockchain and Artificial Intelligence Applications. [Internet] (2021)
6. Bertino, E., Kundu, A., Sura, Z.: Data transparency with Blockchain and AI ethics. J. Data Inf. Qual. **11**(4), 1–8 (2019)
7. Zhao, Y., Zhao, J., Jiang, L., Tan, R., Niyato, D., Li, Z., Lyu, L.: Privacy-preserving blockchain-based federated learning for IoT devices. IEEE Int. Things J. **8**, 1817 (2021)
8. Yoon J., Democratic Senators Introduce the Consumer Online Privacy Rights Act. [Internet]. (2019)
9. Verma, S.B., Saravanan, C.: Performance Analysis of Various Fusion Methods in Multimodal Biometric International. IEEE (2018)
10. Verma, S.B., Yadav, A.: Detection of hard exudates in retinopathy images ADCAIJ: advances in distributed computing and artificial intelligence. J. Regul. Iss. **8**, 4 (2019)
11. Verma, S., Verma, S.B.: Data transmission in BPEL (Business Process Execution Language). ADCAIJ. **9**(3), 105–117 (2020) eISSN: 2255-2863

12. Verma, S.: Contactless palmprint verification system using 2-D Gabor filter and principal component analysis. Int. Arab J. Inf. Technol. **16**(1), 23–29 (2019)
13. Bernal, P.: The Right to be Forgotten in the post-Snowden era. PinG Privacy in Germany (2014)
14. Borgesius, F.Z.: Informed Consent: We Can Do Better to Defend Privacy. IEEE Secur. Privacy. **13**(2), 103–107 (2015)
15. Castelló Ferrer, E., Hardjono, T., Pentland, A.: Editorial: Proceedings of the First Symposium on Blockchain and Robotics. MIT Media Lab (2018)

# Spectral and Spatial Feature Extraction Techniques for Advanced Hyperspectral Image Classification

Kesava Sriram Kothamasu[1], Lekhana Golla[1], Priyusha Kilaru[1], Suresh Aala[2], and Sunil Chinnadurai[2](✉)

[1] Department of Computer Science and Engineering, SRM University – AP, Amaravati, Andhra Pradesh, India
{kesavasriram_k,lekhana_golla,priyusha_kilaru}@srmap.edu.in

[2] Department of Electronics and Communication Engineering, SRM University – AP, Amaravati, Andhra Pradesh, India
{suresh_aala,sunil.c}@srmap.edu.in

**Abstract.** Hyperspectral image processing, a critical domain in remote sensing and Earth observation, has witnessed exponential growth in feature extraction methods. In light of this, our study delves into the ongoing debate between spectral and spatial feature extraction strategies, a question of paramount importance. To address this, we embarked on a rigorous investigation using two eminent deep learning architectures, namely 3DCNN(Three dimensional convolutional neural network) and ResNet(Residual Neural Network) and employed two benchmark hyperspectral datasets: Indian Pines and Salinas. Our research was meticulously structured, encompassing spectral feature extraction methods such as mean, mode, standard deviation, and skewness, as well as spatial feature extraction techniques employing advanced convolutional operations. The spectral standard deviation feature extraction method using 3DCNN yields an overall accuracy of 98.71% for the Indian Pines dataset. In the Salinas dataset, the spectral and spatial mean feature extraction methods achieve accuracies of 99.78% and 99.89% respectively. When employing the ResNet architecture, the spectral skewness feature extraction method attains an accuracy of 99.83% for the Indian Pines dataset. Meanwhile, in the Salinas dataset, spectral standard deviation feature extraction excels with an accuracy of 99.99%. Analyzing computational times, the 3DCNN on the Indian Pines data set shows that spatial mode feature extraction has the shortest time of 0.33 s, while in the Salinas dataset, spatial skewness feature extraction requires 1.60 s. For ResNet, Indian Pines' spatial mode is the quickest at 2.10 s, and Salinas' spatial mode is 10.66 s. In the comparison of spectral and spatial feature extraction methods, the likelihood of achieving lower computational times favors spectral methods. Additionally, the probability of attaining higher accuracy is notably higher with spectral feature extraction methods. Therefore, spectral approaches emerge as a superior choice.

**Keywords:** Hyperspectral imaging · Spectral and Spatial Information · 3DCNN · ResNet


A. Bajaj et al. (Eds.): HIS 2023, LNNS 1224, pp. 393–404, 2025.
https://doi.org/10.1007/978-3-031-78925-0_39

# 1 Introduction

In the field of remote sensing, cutting-edge technology called hyperspectral imaging (HSI) captures information across a broad spectrum of wavelengths, ranging from visible light to infrared. Unlike traditional imaging techniques, HSI enables the acquisition of detailed spectral data for each pixel in an image, offering a wealth of information about the composition, properties, and conditions of the imaged objects or areas. This wealth of spectral data empowers scientists, researchers, and professionals to delve deep into various applications, such as environmental monitoring, agriculture, mineral exploration, medical diagnostics, and forensic sciences [1]. The ability to discriminate between materials based on their unique spectral signatures makes HSI a powerful tool for understanding complex landscapes and phenomena, making it indispensable in fields requiring precise and nuanced data analysis. Advancements in technology have driven a paradigm shift in the field, leading to an exponential surge in feature extraction methodologies. In HSI, interpreting the extensive spectral information from raw data poses a significant challenge. Addressing this challenge requires leveraging advanced methods, with feature extraction playing a crucial role among them. Feature extraction involves identifying and selecting relevant information from hyperspectral datasets, and transforming raw spectral data into meaningful, compact, and interpretable features. These features are the basis for subsequent analysis, classification, and interpretation. Feature extraction methods serve a multifaceted role, enhancing comprehension of the underlying data and enabling the development of precise and efficient algorithms for diverse applications. Additionally, they are applied for dimensionality reduction, optimizing computational efficiency, and simplifying complex datasets. These methods can be broadly categorized into spectral and spatial techniques. Spectral feature extraction is a data processing technique in which specific spectral characteristics, represented by patterns of light intensity at various wavelengths (bands), are identified and extracted from each pixel in a spectral image. Spatial feature extraction assesses relationships between neighboring pixels within the same band, capturing patterns based on pixel proximity and arrangement in the image. The scientific community is currently engaged in a spirited debate, driven by advancements in technology and the growing need for accuracy and efficiency. This discussion centers on choosing between spectral and spatial feature extraction strategies.

## 1.1 Spectral Information in HSI

In HSI, the spectral dimension refers to the range of wavelengths measured across each pixel in an image. Unlike traditional RGB (Red, Green, Blue) imaging, which records only three broad wavelength bands, HSI divides the spectrum into numerous narrow, contiguous spectral bands. This results in a rich dataset where each pixel not only provides spatial information but also reveals the unique spectral signature of the object or scene it represents. These spectral signatures are akin to fingerprints for materials, allowing researchers to identify and differentiate substances based on how they interact with light at various wavelengths.

Spectral information in HSI is a powerful tool for applications such as mineral identification, vegetation analysis, and environmental monitoring. Figure 1 (a) illustrates the spectral extraction method through a visual representation.

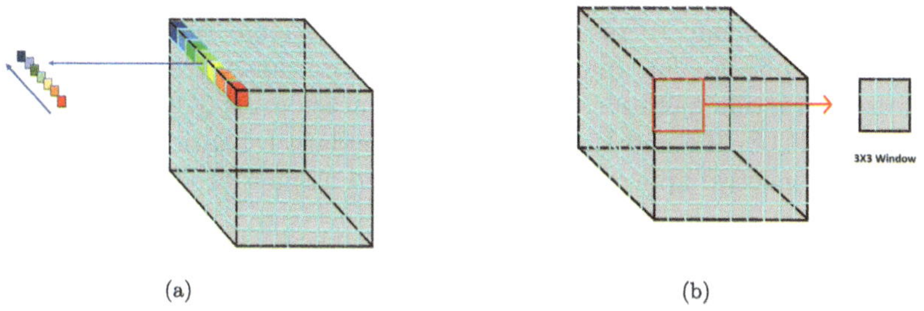


**Fig. 1.** Spectral and Spatial Extraction

### 1.2 Spatial Information in HSI

In addition to spectral information, HSI also captures detailed spatial information. The spatial dimension in HSI pertains to the arrangement and location of pixels within an image. Each pixel corresponds to a specific area on the ground or object being observed. Researchers can utilize spatial information to analyze the distribution of materials, study patterns in natural landscapes, and monitor changes in urban areas. Combining this spatial information with spectral data enhances the overall capability of HSI for various remote sensing applications. Figure 1 (b) demonstrates the spatial extraction technique through a detailed visual representation.

### 1.3 Distinguishing Spectral and Spatial Information

The debate centers on whether spectral or spatial feature extraction is superior in HSI, aiming to determine accuracy and computation speed via different architectural models. Spectral data deals with material signatures across wavelengths for identification, while spatial data focuses on pixel positions for object details and context, both vital for comprehensive scene analysis. The papers [2–5] provide comprehensive details on works related to spectral and spatial features, delving deeply into the intricacies of these aspects. The contributions from the paper are summarized as follows.

- To our knowledge, this study represents the first comprehensive analysis comparing spatial and spectral feature extraction techniques in remote sensing applications.

- We performed an in-depth analysis of spectral and spatial feature extraction methods, focusing on their accuracy and computational efficiency. Our research leveraged two distinct datasets and employed two separate architectural approaches, thereby ensuring a comprehensive evaluation.
- We compared spectral and spatial feature extraction methods, spectral methods are more likely to result in lower computational times. Moreover, the probability of achieving higher accuracy is significantly greater with spectral feature extraction methods, for instance, accuracy using 3DCNN-IP-Standard Deviation is 98.70% and ResNet-Salinas-Standard Deviation is 99.99%. Consequently, spectral approaches are considered a superior choice.

The rest of the paper is structured as follows. In Sect. 2, we explored several relevant papers where similar research has been conducted. Section 3 presents the methodology used in this work. Section 4 provides a detailed description of the experimental dataset, results, and a discussion of the obtained results. Finally, Sect. 5 concludes the paper.

## 2 Related Work

In the expansive field of hyperspectral image processing, a significant debate centers around the choice between spectral and spatial techniques for feature extraction. Our in-depth analysis of influential works has profoundly shaped our understanding, providing valuable insights into this matter. Hyperspectral data contains detailed information and is used in fields like environmental monitoring and mineral exploration. Understanding this complex data is a challenge, so researchers are developing new methods to make sense of it.

One approach involves classifying hyperspectral images by combining traditional techniques with modern methods using color and location data. Researchers like Xu and their team in 2021 have made significant progress by using advanced techniques to improve accuracy [6]. Another challenge is dealing with noise in hyperspectral data. Researchers like Maffei and their team in 2019 have developed efficient methods using deep learning to reduce this noise and improve the quality of the data [7]. Researchers are also focusing on the specific colors in the data. Deshpande and their team in 2021 have used advanced techniques to understand important color features in hyperspectral data [8]. Understanding hyperspectral data requires considering both color and location. Researchers like Kaufman and their team in 2014 emphasize the importance of understanding both aspects together to get a complete picture [9].

## 3 MEthodology

### 3.1 Datasets and Dimensionality Reduction

Two widely used datasets for classification, Indian Pines (IP) and Salinas (SA), were employed in this study. The IP dataset was collected using an AVIRIS

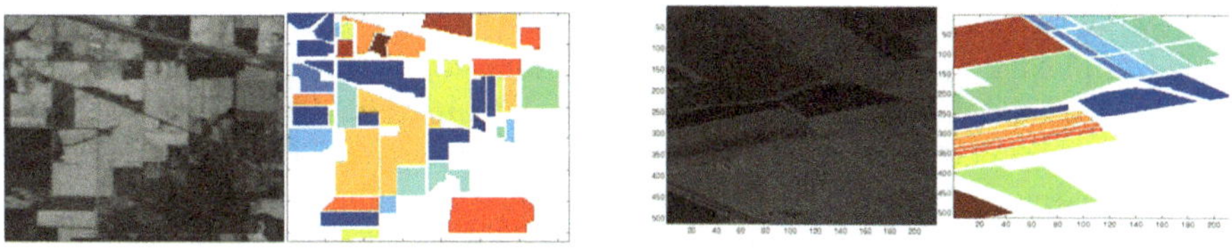

(a) (b)

**Fig. 2.** Indian pines and salinas dataset and its ground truth

sensor over the Indian Pines test site in northwestern Indiana, while the SA dataset was collected using an AVIRIS sensor over the Salinas Valley in California. The dimensions of this file are 145 × 145 pixels (224 bands) and 512 × 217 pixels (224 bands). Both IP and SA datasets comprise 16 classes. To reduce the dimensionality of these datasets, the Incremental Principal Component Analysis (IPCA) method was applied, which facilitates the extraction of the most crucial features from the datasets, and the resulting reduced datasets were then subjected to a feature extraction method for further analysis.

### 3.2 Analytical Framework For Feature Extraction

In our study, we utilized fundamental statistical techniques, such as mean, mode, standard deviation, and skewness, for feature extraction in both spectral and spatial contexts. We intentionally concentrated on these four methods to thoroughly assess their effectiveness within our analytical framework. While there are various other feature extraction methods available, our focused approach allowed us to gain a deep understanding of these core statistical measures and their applicability to our research objectives. This selection ensured a comprehensive assessment of the data's central tendencies, modes, variability, and skewness, making them indispensable for our analysis.

### 3.3 Deep Learning Architectures for Hyperspectral Image Processing

In the realm of hyperspectral image processing, our goal is clear i.e., to delve into the intricate domain of hyperspectral data and unveil its hidden valuable information. Our exploration is guided by two distinct deep learning techniques which are Three-dimensional Convolutional Neural Network (3DCNN) and the Residual Neural Network (ResNet). These choices aren't arbitrary; they are rooted in the specific challenges presented by hyperspectral data and the demonstrated success of these architectures in the field.

**Three-Dimensional Convolutional Neural Network (3DCNN).** The Three-dimensional Convolutional Neural Network(3DCNN) captures our attention due to its exceptional suitability for the spectral dimension of hyperspectral data. Unlike traditional two-dimensional Convolutional Neural Networks

(CNNs), which excel in conventional image-processing tasks, hyperspectral data adds complexity by introducing a spectral dimension. Every pixel contains more than just spatial details; it holds a diverse spectrum across various wavelengths, turning it into a valuable source of data. We opted for 3DCNN based on previous research accomplishments. Chang et al.'s seminal work exemplifies the effectiveness of 3DCNN in hyperspectral image classification [10]. Embracing 3DCNN empowers us to fully exploit the spectral dimension, a significant boost to hyperspectral data classification. Our decision isn't just a leap of faith; it's grounded in empirical evidence and a history of achievements in hyperspectral image processing.

**Residual Neural Network (ResNet).** In our research framework, the Residual Neural Network (ResNet) is a robust solution, purposefully crafted for feature extraction from hyperspectral data. ResNet stands out due to its deep architecture, fortified with residual learning, which empowers it to model intricate data structures with remarkable efficiency. This unique ability to capture subtle details in spectral information is highlighted by studies such as [11]. The subtle spectral characteristics play a crucial role in differentiating between materials or objects, requiring a profound understanding of the underlying spectral patterns. Furthermore, ResNet's architecture, featuring skip connections, adeptly conquers the vanishing gradient challenge common in deep neural network training. This skill equips us to extract highly discriminative features from hyperspectral data, adding significant value to the accuracy and robustness of our classification model.

## 4 Results and Analysis

### 4.1 Data Set and Experimental Details

Two datasets, Indian Pines and Salinas, each containing 16 distinct classes, have been selected for analysis. Figure 2(a) represents the Indian Pines dataset alongside its corresponding ground truth. Figure 2(b) represents the Salinas dataset depicted in conjunction with its ground truth.

### 4.2 Classification Results

To validate the CNN and resnet model in extraction, overall accuracy (OA), precision (P), recall (R), and F1 score (F1) were calculated during the model training with a learning rate of 0.001, determined from the training data. A total of 20 most informative data were selected using incremental principal component analysis(iPCA).

In the analysis of the IP dataset using 3D CNN, spectral mean accuracy (96.95%) trails slightly behind spatial mean accuracy (97.17%), with a marginal difference. Spectral mean processes data in 0.37 s (s), while spatial mean takes 0.47 s, indicating a slight advantage for spectral mean. Spectral mode accuracy

(98.48%) outperforms spatial mode accuracy (92.65%) with faster computations, clocking in at 0.37 s compared to 0.33 s for spatial mode. Furthermore, spectral standard deviation accuracy (98.70%) surpasses spatial accuracy (98.65%) and is computationally more efficient at 0.51 s, contrasting sharply with spatial standard deviation, which takes 1.75 s. Additionally, spectral skewness accuracy (98.53%) excels over spatial skewness (98.48%) and completes computations in 0.43 s, whereas spatial skewness takes 0.39 s, revealing spectral skewness' slight computational advantage. Figure 3 (a) and Fig. 3 (b) present the accuracy and computational time outcomes respectively for spectral and spatial methods applied to the IP dataset using the 3DCNN model. Table 1 and Table 2 contain accuracy and computational time information for feature extraction methods, specifically focusing on spectral and spatial aspects. The evaluation is conducted using the Indian Pines dataset and employs 3D CNN for analysis.

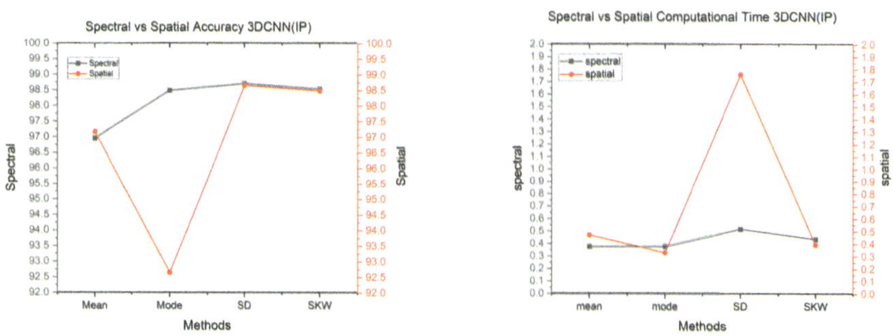


(a) (b)

**Fig. 3.** Accuracy and Computational Time Between Spectral methods with Spatial using 3D CNN for Indian pines Data Set

In the SA dataset analyzed with 3D CNN, spectral mean accuracy (99.77%) is slightly lower than spatial mean accuracy (97.88%). The spectral mean processes data in 203.17 s, while the spatial mean takes 203.14 s, indicating a marginal time difference with the spectral mean being slightly slower. Spectral mode accuracy (99.42%) surpasses spatial mode accuracy (94.97%), with spectral mode being computationally faster at 163.74 s compared to spatial mode's 168.58 s. Spectral

**Table 1.** Accuracy-IP-3DCNN

| Method | Mean | Mode | SD | Skewness |
|---|---|---|---|---|
| Spectral | 96.95 | 98.48 | 98.70 | 98.53 |
| Spatial | 97.17 | 92.65 | 98.65 | 98.48 |

**Table 2.** Computational Time IP-3DCNN

| Method | Mean | Mode | SD | Skewness |
|---|---|---|---|---|
| Spectral | 0.375 | 0.377 | 0.515 | 0.434 |
| Spatial | 0.473 | 0.331 | 1.759 | 0.394 |

standard deviation accuracy (98.92%) is higher than spatial accuracy (97.16%), but spectral standard deviation takes longer at 203.16 s, unlike spatial standard deviation, which processes data faster at 160.66 s. Additionally, spectral skewness accuracy (98.58%) outperforms spatial skewness (99.33%), with spectral skewness taking 202.99 s, while spatial skewness is slightly faster at 161.62 s, indicating a minor computational advantage for spatial skewness. Figure 4 (a) and Fig. 4 (b) present the accuracy and computational time outcomes respectively for spectral and spatial methods applied to the SA dataset using the 3DCNN model. Table 3 and Table 4 contain accuracy and computational time information for feature extraction methods, specifically focusing on spectral and spatial aspects. The evaluation is conducted using the Salinas dataset and employs 3D CNN for analysis.

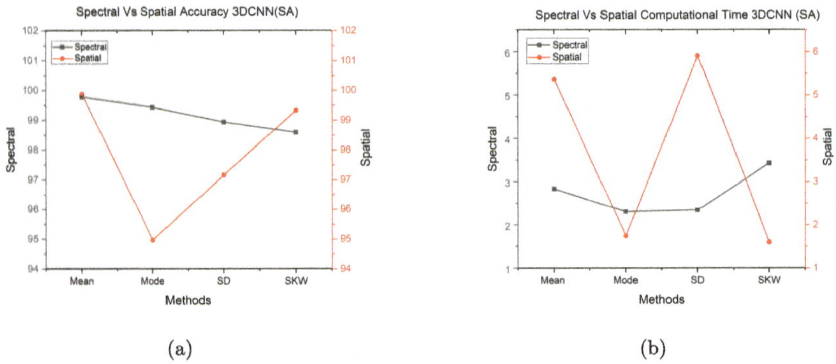


**Fig. 4.** Accuracy and Computational Time Between Spectral methods with Spatial using 3D CNN for Salinas Data Set

In the analysis of the IP dataset using ResNet, spectral mean accuracy (99.80%) is slightly higher than the spatial mean accuracy (99.04%). Spectral mean processes data in 2.13 s, while spatial mean takes 2.14 s, indicating a slight time advantage for spectral mean.

Spectral mode accuracy (99.68%) outperforms spatial mode accuracy (97.12%) and spatial mode processes data in 2.10 s, whereas the spectral mode takes 2.94 s, making the spatial mode more efficient in terms of computational time. Furthermore, spectral standard deviation accuracy (98.58%) is lesser than

**Table 3.** Accuracy-Salinas-3DCNN

| Method | Mean | Mode | SD | Skewness |
|---|---|---|---|---|
| Spectral | 99.77 | 99.42 | 98.92 | 98.58 |
| Spatial | 99.88 | 94.97 | 97.16 | 99.33 |

**Table 4.** Computational Salinas-3DCNN

| Method | Mean | Mode | SD | Skewness |
|---|---|---|---|---|
| Spectral | 2.832 | 2.299 | 2.339 | 3.423 |
| Spatial | 5.367 | 1.750 | 5.911 | 1.600 |

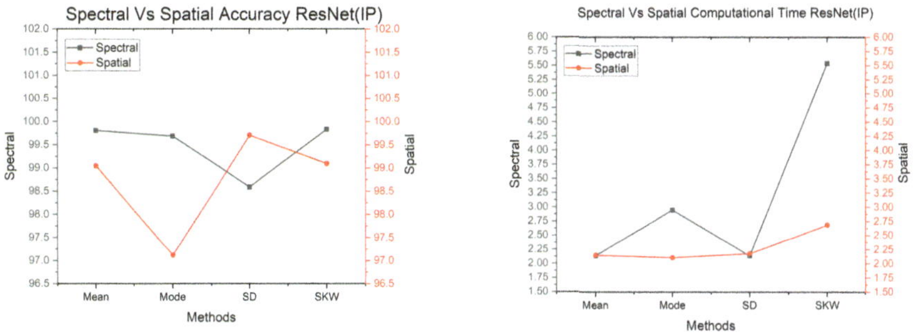


(a) (b)

**Fig. 5.** Accuracy and Computational Time Between Spectral methods with Spatial using ResNet for IndianPines Data Set

the spatial accuracy (99.70%) and spectral method is computationally more efficient at 2.13 s than spatial standard deviation, which takes 2.17 s. Additionally, spectral skewness accuracy (99.82%) excels over spatial skewness (99.09%) and completes computations in 5.53 s, whereas spatial skewness takes 2.68 s, revealing spatial skewness' slight computational advantage. Table 5 and Table 6 contain accuracy and computational time information for feature extraction methods, specifically focusing on spectral and spatial aspects. The evaluation is conducted using the Indian Pines dataset and employs ResNet for analysis. Figure 5 (a) and Fig. 5 (b) present the accuracy and computational time outcomes respectively for spectral and spatial methods applied to the IP dataset using the ResNet model.

In the SA dataset analyzed with ResNet, spectral mean accuracy (99.80%) is quite higher than the spatial mean accuracy (90.24%). The spectral mean processes data in 11.91 s, while the spatial mean takes 12.99 s, indicating a marginal time difference with the spatial mean being slower. Spectral mode accuracy (99.75%) is lower than the spatial mode accuracy (99.93%), with spectral mode being computationally slower at 13.63 s compared to spatial mode's 10.65 s. Spectral standard deviation accuracy (99.99%) is higher than spatial accuracy (99.84%), but spatial standard deviation takes longer at 11.74 s, unlike spectral standard deviation, which processes data faster at 10.67 s. Additionally, spectral skewness accuracy (99.36%) performs less than spatial skewness (99.65%), with

**Table 5.** Accuracy-IP-ResNet

| Method | Mean | Mode | SD | Skewness |
|---|---|---|---|---|
| Spectral | 99.80 | 99.68 | 98.58 | 99.82 |
| Spatial | 99.04 | 97.12 | 99.70 | 99.09 |

**Table 6.** Computational Time-IP-ResNet

| Method | Mean | Mode | SD | Skewness |
|---|---|---|---|---|
| Spectral | 2.137 | 2.946 | 2.137 | 5.538 |
| Spatial | 2.145 | 2.104 | 2.175 | 2.683 |

spectral skewness taking 10.84 s, while spatial skewness is slower at 11.45 s, indicating a minor computational advantage for spectral skewness. Figure 6 (a) and Fig. 6 (b) present the accuracy and computational time outcomes respectively for spectral and spatial methods applied to the SA dataset using the ResNet model. Table 7 and Table 8 contain accuracy and computational time information for feature extraction methods, specifically focusing on spectral and spatial aspects. The evaluation is conducted using the Salinas and employs ResNet for analysis.

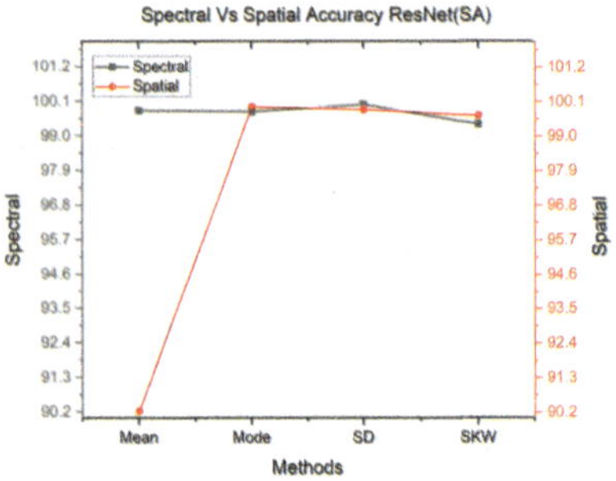


(a)

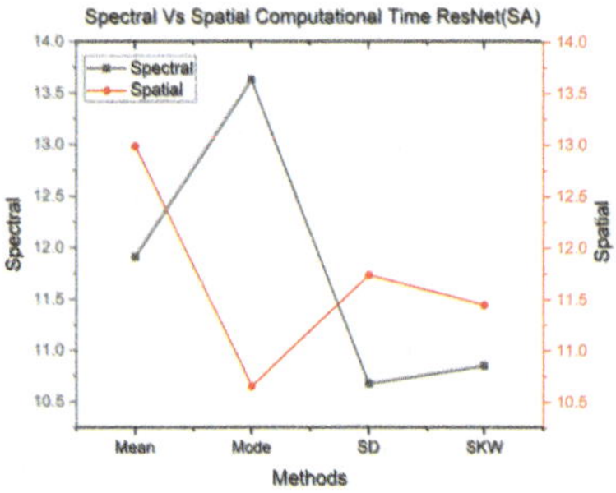


(b)

**Fig. 6.** Accuracy and Computational Time Between Spectral methods with Spatial using ResNet for Salinas Data Set

**Table 7.** Accuracy-Salinas-ResNet

| Method | Mean | Mode | SD | Skewness |
|---|---|---|---|---|
| Spectral | 99.80 | 99.75 | 99.99 | 99.36 |
| Spatial | 90.24 | 99.93 | 99.84 | 99.65 |

**Table 8.** Computational Time-Salinas-ResNet

| Method | Mean | Mode | SD | Skewness |
|---|---|---|---|---|
| Spectral | 11.911 | 13.631 | 10.676 | 10.846 |
| Spatial | 12.990 | 10.657 | 11.741 | 11.450 |

Figure 7 (a) and Fig. 7 (b) illustrate the training and validation loss curves for 3DCNN with spectral mean feature extraction on the Salinas dataset and ResNet with spatial standard deviation feature extraction on the IP dataset respectively.

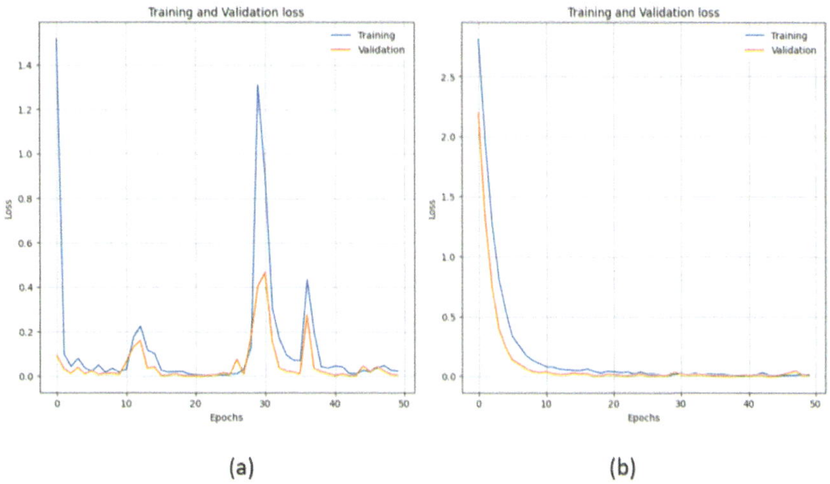


**Fig. 7.** a-3DCNN-Spectral Mean-SA and b-ResNet Spatial SD-IP

## 5 Conclusion

Our research paper dives into the realm of hyperspectral image processing, with a particular focus on the ongoing debate regarding spectral and spatial feature extraction strategies. To address this debate, we employed two advanced deep learning architectures, 3DCNN and ResNet, and applied them to two widely recognized hyperspectral datasets: Indian Pines and Salinas. The primary objective was to determine which feature extraction methods, whether spectral or spatial, would yield superior results in terms of accuracy and computational efficiency. Our findings highlight the dominance of spectral feature extraction methods, including mean, mode, standard deviation, and skewness, over their spatial counterparts. For instance, in the Indian Pines dataset, the 3DCNN-based spectral standard deviation method achieved an impressive accuracy of 98.71%, while in the Salinas dataset, both spectral and spatial mean methods excelled, with accuracies of 99.78% and 99.89%, respectively. Furthermore, when applying the ResNet architecture, the spectral skewness feature extraction method attained an accuracy of 99.83% for the Indian Pines dataset, and the spectral standard deviation method in the Salinas dataset reached an accuracy of 99.99%. In addition, our study demonstrated that spectral methods generally required less processing time compared to spatial methods, suggesting that spectral approaches are not only more computationally efficient but are also more likely to deliver higher classification accuracy. Consequently, we conclude that spectral feature extraction methods represent the superior choice for applications in remote sensing.

**Acknowledgment.** We extend our sincere gratitude to SRM University Andhra Pradesh for their exceptional support and state-of-the-art research facilities. The university's dedicated faculty, innovative atmosphere, and continuous guidance from the research support team significantly enriched our academic experience. In this paper, Kesava Sriram Kothamasu, Lekhana Golla, and Priyusha Kilaru contributed equally.

## References

1. Butt, M.H.F., Ayaz, H., Ahmad, M.: Fast and compact hybrid CNN for hyperspectral imaging-based bloodstain classification. In: Conference: 2022 IEEE Congress on Evolutionary Computation (CEC) (2022)
2. Maryam Imani, Hassan Ghassemian.An overview on spectral and spatial information fusion for hyperspectral image classification: Current trends and challenges( Volume:59 July 2020)
3. Vaddi, R., Manoharan, P.: Hyperspectral image classification using CNN with spectral and spatial features integration. Infrared Phys. Technol. **107**, 103296 (2020)
4. Li, L., Gao, J., Ge, H., Zhang, Y., Zhang, H.: An effective feature extraction method via spectral-spatial filter discrimination analysis for hyperspectral image. Multimedia Tools Appl. **81**(28), 40871–40904 (2022)
5. Zhou, Y., Wei, Y.: Learning hierarchical spectral–spatial features for hyperspectral image classification. IEEE Trans. Cybern. **46**(7), 1667–1678 (2015)
6. Xu, Z., Yu, H., Zheng, K., Gao, L., Song, M.: A novel classification framework for hyperspectral image classification based on multiscale spectral-spatial convolutional network. In: 2021 11th Workshop on Hyperspectral Imaging and Signal Processing: Evolution in Remote Sensing (WHISPERS). IEEE (2021)
7. Maffei, A., Haut, J.M., Paoletti, M.E., Plaza, J., Bruzone, L., Plaza, A.: Efficient convolutional neural network for spectral-spatial hyperspectral denoising. In: 2019 10th Workshop on Hyperspectral Imaging and Signal Processing: Evolution in Remote Sensing (WHISPERS). IEEE (2019)
8. Deshpande, S., Thakur, R., Balamuralidhar, P.: Learning deep spectral features for hyperspectral data using convolution over spectral signature shape. In: 2021 11th Workshop on Hyperspectral Imaging and Signal Processing: Evolution in Remote Sensing (WHISPERS). IEEE (2021)
9. Kaufman, J., Weinheimer, J.J., Celenk, M.: Spatial-spectral feature extraction on hyperspectral imagery. In: 2014 6th Workshop on Hyperspectral Image and Signal Processing: Evolution in Remote Sensing (WHISPERS). IEEE (2014)
10. Chang, C.I., et al.: Hyperspectral image classification and dimensionality reduction: an orthogonal subspace projection approach. IEEE Trans. Geosci. Remote Sens. **32**, 779–785 (1994)
11. Roy, S.K., Manna, S., Song, T., Bruzzone, L.: Attention-based adaptive spectral–spatial kernel ResNet for hyperspectral image classification. IEEE Trans. Geosci. Remote Sens. **59**(9), 7831–7843 (2020)

# Hyperspectral Image Classification with Deep Learning: Unleashed by Feature Selection and Extraction

Venkata Krishna Saadhvik Muddana[1], Pujitha Vadapalli[1], Dimple Kari[1], Suresh Aala[2], and Sunil Chinnadurai[2(✉)]

[1] Department of Computer Science and Engineering, SRM University – AP, Amaravati, Andhra Pradesh, India
{saadhvik_muddana,pujitha_vadapalli,dimple_kari}@srmap.edu.in
[2] Department of Electronics and Communication Engineering, SRM University – AP, Amaravati, Andhra Pradesh, India
{suresh_aala,sunil.c}@srmap.edu.in

**Abstract.** Hyperspectral Image (HSI) processing stands at the forefront of remote sensing technologies, offering a wealth of detailed information crucial for various environmental and agricultural applications. The complexity of HSI datasets demands sophisticated techniques for accurate and efficient analysis. In this research, we investigate the efficiency of feature selection and extraction methods for HSI classification using Fast-3D Convolutional Neural Networks (CNN) and Residual Neural Networks (ResNet). Our motivation stems from the necessity to enhance accuracy and computational efficiency in remote sensing applications. We conducted experiments on two prominent HSI datasets, Indian Pines and Salinas, leveraging feature selection techniques like Correlation-based Feature Selection (CFS), Mutual Information Feature Selection (MIFS), Linear Discriminant Analysis (LDA), and Recursive Feature Elimination (RFE), alongside feature extraction methods such as Principal Component Analysis (PCA), Incremental PCA (IPCA), Quadratic Discriminant Analysis (QDA), and Non-negative Matrix Factorization (NMF), implemented within both Fast-3D CNN and ResNet architectures. Our results indicate that in hyperspectral data classification for both Indian Pines and Salinas datasets, IPCA and QDA excelled in accuracy (99.68%, 99.14%) within the Fast-3D CNN architecture, with IPCA demonstrating speed (0.40 s, 1.40 s). CFS displayed notable efficiency (99.27%, 90.88%) with minimal processing time (0.56 s, 1.35 s) in the Fast-3D CNN architecture. RFE achieved high accuracy (99.88%, 96.39%) with reasonable processing times (2.28 s, 11.53 s) in the ResNet architecture. While IPCA led in extraction (99.90%, 99.12%), it required more time (21.17 s, 2.23 s) in the ResNet architecture. These results underscore the practical significance of feature extraction methods like IPCA for accuracy-focused tasks and the efficiency of feature selection methods like CFS in scenarios prioritizing computational time.


A. Bajaj et al. (Eds.): HIS 2023, LNNS 1224, pp. 405–416, 2025.
https://doi.org/10.1007/978-3-031-78925-0_40

**Keywords:** Hyperspectral Image Processing · Feature Selection · Feature Extraction · Fast-3D Fast-3D CNN · ResNet · Remote Sensing

# 1 Introduction

In the realm of remote sensing technologies, Hyperspectral Image (HSI) processing plays a pivotal role, offering essential information for environmental and agricultural applications. Addressing the complexity of HSI datasets, our research focuses on enhancing the efficiency of HSI classification. We delve into feature selection and extraction methods, employing cutting-edge technologies like Fast-3D Convolutional Neural Networks (CNN) and Residual Neural Networks (ResNet). The primary goal is to improve both accuracy and computational efficiency in remote sensing applications.

Feature selection is a critical step, aiming to choose relevant attributes from the original set. Methods like CFS, MIFS, LDA, and RFE evaluate feature importance using criteria such as correlation and mutual information. This streamlined dataset enhances interpretability and reduces overfitting. Simultaneously, feature extraction techniques like PCA and NMF transform original features into a reduced-dimensional representation, preserving vital patterns and improving model performance. Our research illuminates the balance between accuracy and computational time, offering practical insights for hyperspectral image analysis.

Our findings provide nuanced insights into the delicate balance between accuracy and computational time in hyperspectral image analysis. By exploring feature selection and extraction methods alongside advanced technologies like Fast-3D CNN and ResNet, our research significantly enhances the efficiency of HSI classification. Through rigorous experimentation, specific techniques have emerged as superior in distinct contexts, guiding researchers and practitioners in selecting appropriate methodologies tailored to individual HSI analysis tasks. These insights are not only academic but also pivotal for the practical applications of hyperspectral image analysis, illuminating the path for future advancements in the field.

## 1.1 Major Contributions of this Paper

- We Comprehensively evaluated various feature selection and feature extraction methods on hyperspectral remote sensing datasets, specifically Indian Pines and Salinas.
- Through our thorough analysis, IPCA exhibited outstanding accuracy, while CFS displayed remarkable computational efficiency.

## 1.2 Organisation of Paper

The paper is organized as follows. In Sect. 2, we delve into the cutting-edge research within the domain of Hyperspectral Image Classification with Deep

Learning, focusing on the advancements enabled by Feature Selection and Extraction. Section 3 provides a comprehensive overview of the methodology adopted in this study. Moving on to Sect. 4, showcasing the results derived from our experiments, and engaging in a meticulous discussion on the outcomes. Finally, Sect. 5 serves as the conclusion, summarizing the pivotal discoveries and their implications.

## 2 Related Work

In hyperspectral image classification, various feature extraction and selection techniques have been explored to tackle the challenges of high-dimensional data and limited training samples. One study examined [1] methods like Principal Component Analysis (PCA), Sparse Principal Component Analysis (SPCA), Feature Principal Component Analysis (FPCA), Kernel Principal Component Analysis (KPCA), and Kernel Entropy Component Analysis (KECA). It discussed the trade-offs between nonlinear techniques, such as KPCA and KECA, which offer better accuracy but demand more resources, and the memory-efficient FPCA, which provides comparable results. The paper also emphasized the significance of parameter settings and FPCA's ability to capture local features in complex data. Another paper [2] introduced an unsupervised feature selection technique using the Morisita Index to reduce redundancy and computational demands in hyperspectral land-cover classification. It highlighted hyperspectral imagery's superiority over traditional multispectral data for land-cover discrimination and advocated using these feature selection methods to improve efficiency and accuracy in hyperspectral imaging. A different study [3] focused on dealing with the complexities of hyperspectral imaging, such as high dimensionality and limited training samples. It emphasized the effectiveness of feature selection and extraction techniques like Minimum Redundancy and Maximum Relevance (mRMR) and PCA in reducing data dimensions. The hybrid approach of PCA-mRMR achieved an impressive 87.72% accuracy, showcasing its potential for dimensionality reduction in hyperspectral image classification. Yet another research [4] effort explored alternative dimensionality reduction techniques, such as Local Fisher Discriminant Analysis (LFDA) and two band selection methods, Independent Component Analysis based band selection (ICAbs)and modified version of ICAbs (mICAbs), as replacements for PCA. LFDA yielded the best classification accuracy, and mICAbs showed promise in classifying sparse representations. This study underscored the importance of feature extraction and selection techniques in enhancing hyperspectral image analysis. Together, these four research papers provide a comprehensive view of feature extraction, selection, and dimensionality reduction techniques in hyperspectral image classification, emphasizing the balance between methodology, accuracy, and computational efficiency in this challenging field.

Our research presents a distinct and improved approach to hyperspectral image classification, setting it apart from the previously mentioned papers Our focus lies in the evaluation of the efficiency of feature selection and extraction techniques within Fast 3D-CNN and ResNet, with the overarching aim of enhancing both accuracy and computational efficiency. Conducting experiments on the Indian Pines and Salinas hyperspectral datasets, we evaluate a comprehensive set of feature selection and extraction methods, including CFS, MIFS, LDA, RFE, PCA, IPCA, QDA and NMF. Our research provides a novel perspective on feature selection and extraction techniques. Through the introduction of a unique experimental setup and the analysis of various methods, our work contributes to advancing comprehension in the hyperspectral image classification domain and presents new possibilities for enhancing accuracy and computational efficiency.

# 3 Methodology

In this section, we describe the basic overview of the Architecture models, Dataset models, Feature selection, Feature extraction and Final analysis (Fig. 1).

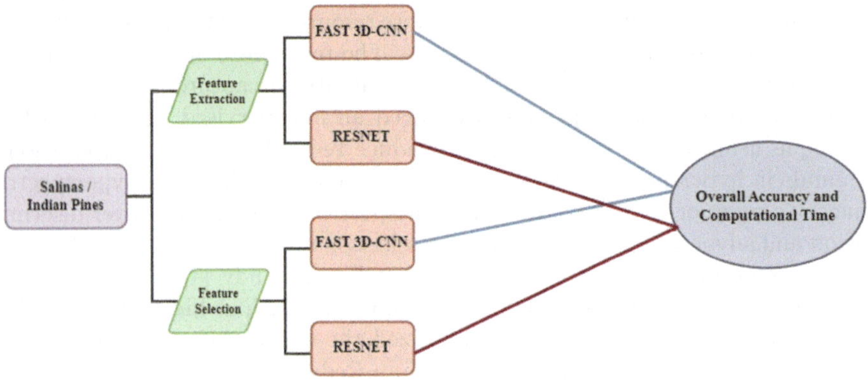


**Fig. 1.** Block diagram for Feature Selection and Extraction

## 3.1 Architectural Model

Fast 3D CNN [13] was chosen for its unique ability to process three-dimensional architectural data, aligning with the move towards 3D modelling in architecture. It leverages spatial and temporal information, providing a robust approach for dynamic architectural analysis. ResNet [14] was selected for its proven performance with complex visual data, including images, and its flexibility in handling deep architectures. Its residual connections excel at capturing intricate architectural details, making it an ideal choice for our study. Fast 3D CNN and ResNet is among the most widely used and open-source deep learning models, facilitating its integration into our research.

### 3.2 Dataset Models

The Indian Pines dataset is a widely used hyperspectral image dataset collected by the Airborne Visible/Infrared Imaging Spectrometer (AVIRIS) sensor over an agricultural area in Indiana, USA. It consists of $145 \times 145$ pixels, with 224 spectral bands.

The Salinas dataset is another widely recognized hyperspectral dataset captured by the AVIRIS sensor over an agricultural area in Salinas Valley, California, USA. It consists of $512 \times 217$ pixels, with 224 spectral bands. Indian Pines and Salinas is well-documented, widely accessible, and frequently used within the remote sensing community [14]. Its open nature promotes transparency and reproducibility in our research, enabling us to compare our results with other studies.

### 3.3 Feature Extraction and Feature Selection

In Feature Selection [12], we have chosen LDA, MIFS, CFS, and RFE due to their wide adoption and demonstrated effectiveness [9,10,15,16]. These methods have shown their capability to retain the most relevant features for architectural analysis, improving model performance and interpretability. Where as in feature extraction [8], we have chosen PCA, IPCA, NMF, and QDA [5–7,10]. These methods have consistently proven their proficiency in capturing and representing the crucial spectral information required for robust architectural analysis.

## 4 Results and Discussions

In our extensive study on hyperspectral image classification, we meticulously evaluated various feature selection and extraction methods using Indian Pines and Salinas datasets. Our findings consistently favored feature extraction techniques such as PCA, IPCA, QDA, and NMF, which demonstrated superior accuracy compared to feature selection methods like CFS, MIFS, LDA, and RFE, vice versa in terms of compuatational time.

Within the CNN architecture, IPCA stood out with outstanding accuracy (99.68% in Indian Pines and 98.37% in Salinas) while being remarkably efficient (0.40 s in Indian Pines and 1.40 s in Salinas). In contrast, CFS emerged as the most efficient feature selection method, achieving competitive accuracy (99.27% in Indian Pines and 90.88% in Salinas) with minimal processing time (0.56 s in Indian Pines and 1.35 s in Salinas). In the ResNet architecture, RFE demonstrated exceptional accuracy (99.88% in Salinas and 96.39% in Indian Pines) along with efficiency (2.28 s in Indian Pines and 11.53 s in Salinas). IPCA led in accuracy (99.90% in Salinas and 99.12% in Indian Pines) but required more processing time (21.17 s in Salinas and 2.23 s in Indian Pines).

In summary, feature extraction methods like IPCA offer higher accuracy but require more processing time, while feature selection methods like CFS and RFE maintain competitive accuracy with efficient computational times. Specifically,

CFS and LDA emerged as top choices for computational efficiency in hyperspectral image analysis across different architectures. This highlights the practical significance of IPCA for accuracy-focused tasks and the efficiency of CFS in scenarios prioritizing computational time.

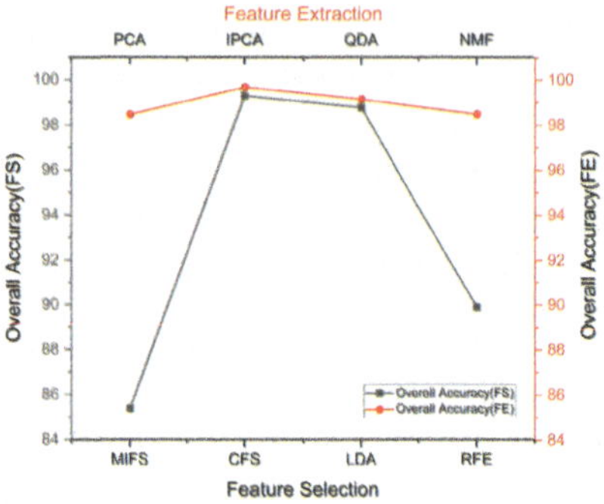


**Fig. 2.** Indian Pines (CNN) - Overall Accuracy

In Fig. 2, feature extraction methods, specifically IPCA and QDA, outperformed feature selection techniques. IPCA achieved an outstanding accuracy of 99.68%, emphasizing its capability in preserving essential spectral information for classification. Similarly, QDA demonstrated remarkable accuracy at 99.15%, highlighting the strength of feature extraction techniques in capturing critical data patterns. In contrast, feature selection methods like MIFS lagged behind with an accuracy of 85.39%, indicating their limitations in retaining crucial information compared to extraction techniques.

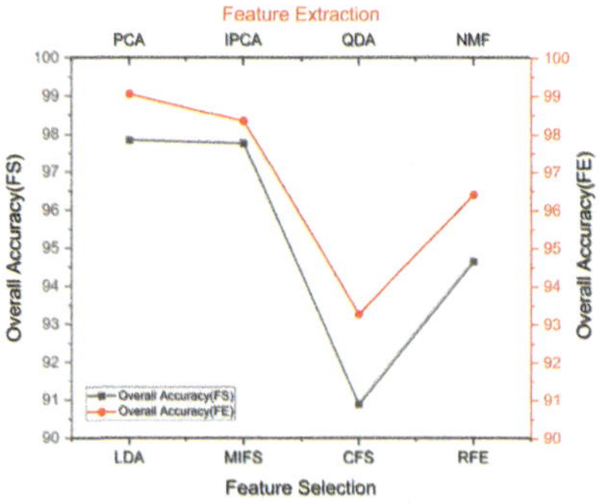


**Fig. 3.** Salinas (CNN) - Overall Accuracy

In Fig. 3, in terms of overall accuracy, feature extraction methods, particularly MIFS and QDA, showcased impressive performance. MIFS achieved a remarkable accuracy of 97.77%, underscoring its ability to capture essential spectral patterns for classification. Similarly, QDA demonstrated outstanding accuracy at 97.86%, emphasizing the strength of feature extraction techniques in preserving critical information. Feature selection methods like RFE, although competitive, fell slightly behind with an accuracy of 94.65%, indicating their limitations in capturing nuanced data patterns compared to extraction methods.

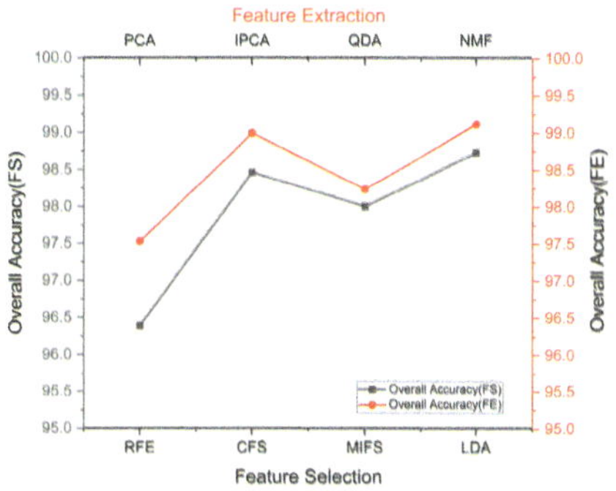


**Fig. 4.** Indian Pines (Resnet) - Overall Accuracy

In Fig. 4, feature extraction methods continued to dominate in terms of overall accuracy. IPCA and NMF displayed exceptional performance, achieving accuracies of 99.00% and 99.12%, respectively. These results underline the superiority of extraction techniques in capturing complex spectral features, leading to superior classification outcomes. On the other hand, feature selection methods like RFE exhibited relatively lower accuracy at 96.39%, indicating their limitations in capturing the intricate patterns present in hyperspectral data.

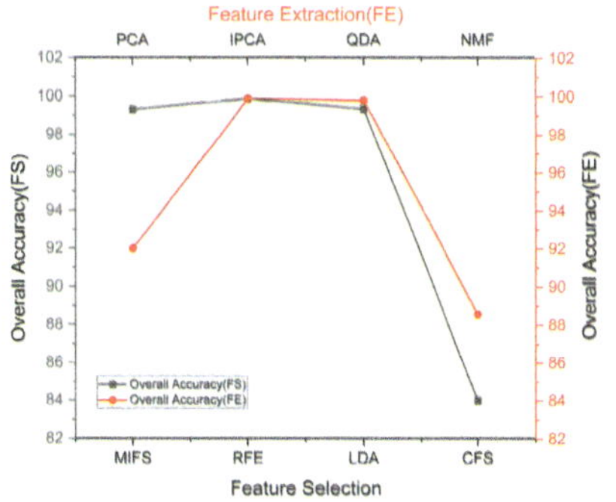


**Fig. 5.** Salinas (Resnet) - Overall Accuracy

In Fig. 5, feature extraction methods consistently outperformed feature selection techniques in terms of overall accuracy. Notably, IPCA and QDA exhibited exceptional accuracy scores, achieving 99.90% and 99.81%, respectively. These results underscore the superiority of extraction techniques in capturing intricate spectral features, leading to superior classification outcomes. Feature selection methods, although competitive, showed relatively lower accuracy, with CFS at 83.99%, indicating their limitations in capturing complex data patterns present in hyperspectral imagery.

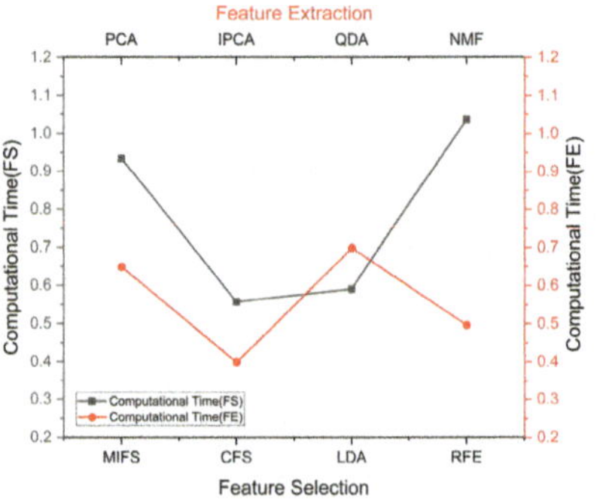


**Fig. 6.** Indian Pines (CNN) - Computational Time

In Fig. 6, in terms of computational time, CFS emerged as the fastest feature selection method, requiring only 0.56 s. However, for feature extraction, IPCA excelled with a mere 0.40 s, emphasizing its efficiency in quickly processing hyperspectral data. Overall, feature selection methods are much efficient in terms of computational time.

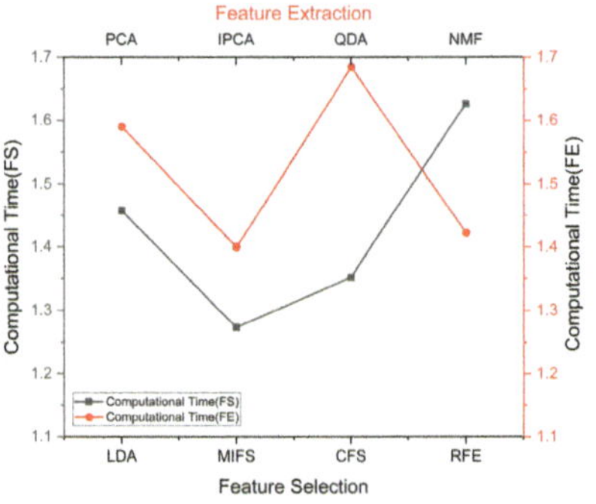


**Fig. 7.** Salinas (CNN) - Computational Time

In Fig. 7, regarding computational time, CFS emerged as the most time-efficient feature selection method, requiring only 1.35 s. Among feature extraction methods, IPCA demonstrated efficiency with a mere 1.40 s.

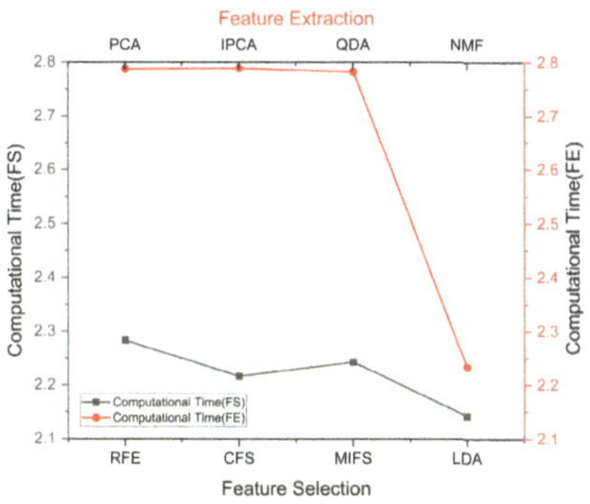


**Fig. 8.** Indian Pines (Resnet) - Computational Time

In Fig. 8, regarding computational time, CFS stood out among feature selection methods, requiring 2.22 s. In the extraction category, IPCA demonstrated efficiency with 2.79 s, but overall feature selection methods are much efficient in terms of computational time.

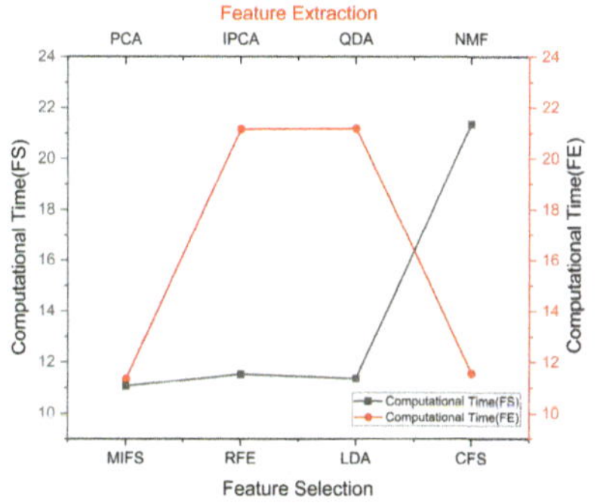


**Fig. 9.** Salinas (Resnet) - Computational Time

In Fig. 9, concerning computational time, CFS emerged as the fastest feature selection method, requiring only 21.36 s. In the extraction category, IPCA demonstrated efficiency with 21.17 s, aligning with its accuracy while highlighting the trade-off between precision and computational speed.

From Table 1, Table 2, Table 3, and Table 4, it can be observed that feature extraction techniques consistently outperform feature selection methods in terms of overall accuracy. Feature extraction methods such as IPCA and QDA consistently achieve higher accuracy rates compared to feature selection techniques across different datasets and neural network architectures. On the other hand, feature selection techniques, as seen in the cases of CFS and MIFS, demonstrate better performance in terms of computational time. These methods generally require less computational time compared to feature extraction techniques.

**Table 1.** Summary of Feature Selection and Extraction Methods - Indian Pines (CNN)

| Method | Feature Type | Overall Accuracy (%) | Comp. Time (s) |
|---|---|---|---|
| CFS | Sel. | 99.27 | 0.56 |
| MIFS | Sel. | 85.39 | 0.93 |
| LDA | Sel. | 98.76 | 0.59 |
| RFE | Sel. | 89.88 | 1.04 |
| PCA | Ext. | 98.49 | 0.65 |
| IPCA | Ext. | 99.68 | 0.40 |
| QDA | Ext. | 99.15 | 0.70 |
| NMF | Ext. | 98.49 | 0.50 |

**Table 2.** Summary of Feature Selection and Extraction Methods - Indian Pines (ResNet)

| Method | Feature Type | Overall Accuracy (%) | Comp. Time (s) |
|---|---|---|---|
| CFS | Sel. | 98.46 | 2.22 |
| MIFS | Sel. | 98.00 | 2.24 |
| LDA | Sel. | 98.73 | 2.14 |
| RFE | Sel. | 96.39 | 2.28 |
| PCA | Ext. | 97.54 | 2.79 |
| IPCA | Ext. | 99.00 | 2.79 |
| QDA | Ext. | 98.24 | 2.78 |
| NMF | Ext. | 99.12 | 2.23 |

**Table 3.** Summary of Feature Selection and Extraction Methods - Salinas (CNN)

| Method | Feature Type | Overall Accuracy (%) | Comp. Time (s) |
|---|---|---|---|
| CFS | Sel. | 90.88 | 1.35 |
| MIFS | Sel. | 97.77 | 1.27 |
| LDA | Sel. | 97.86 | 1.46 |
| RFE | Sel. | 94.65 | 1.63 |
| PCA | Ext. | 99.07 | 1.59 |
| IPCA | Ext. | 98.37 | 1.40 |
| QDA | Ext. | 93.28 | 1.69 |
| NMF | Ext. | 96.41 | 1.42 |

Therefore, the choice between feature extraction and feature selection depends on the specific requirements of the application: feature extraction for maximizing accuracy and feature selection for optimizing computational efficiency.

**Table 4.** Summary of Feature Selection and Extraction Methods - Salinas (ResNet)

| Method | Feature Type | Overall Accuracy (%) | Comp. Time (s) |
|---|---|---|---|
| CFS | Sel. | 83.99 | 21.36 |
| MIFS | Sel. | 99.31 | 11.07 |
| LDA | Sel. | 99.33 | 11.37 |
| RFE | Sel. | 99.88 | 11.53 |
| PCA | Ext. | 92.02 | 11.36 |
| IPCA | Ext. | 99.90 | 21.17 |
| QDA | Ext. | 99.81 | 21.19 |
| NMF | Ext. | 88.58 | 11.56 |

## 5 Conclusion and Future Work

In conclusion, our Research on hyperspectral image processing using Convolutional Neural Networks (CNN) and Residual Neural Networks (ResNet) demonstrates that feature extraction methods, specifically IPCA and QDA, consistently outperform feature selection techniques in terms of overall accuracy. IPCA achieved remarkable accuracy scores, reaching 99.68% in the Indian Pines dataset with CNN and 99.89% in Salinas with ResNet. Conversely, in terms of computational time, feature selection methods like CFS and LDA excelled, with CFS proving to be the fastest, processing the Indian Pines dataset in 0.56 s with CNN and 2.22 s with ResNet. These findings emphasize efficiency of feature extraction for accuracy-focused tasks and highlight the efficiency of feature selection methods for applications prioritizing computational time in hyperspectral image analysis.

We suggest considering IPCA for applications demanding high accuracy, such as medical imaging or mineral detection, due to its outstanding precision. In contrast, CFS could be a viable choice for real-time applications like environmental monitoring or precision agriculture.

**Acknowledgment.** I extend my sincere gratitude to SRM University Andhra Pradesh for providing an unparalleled environment of academic excellence and innovation. My heartfelt thanks go to the dedicated faculty members and supportive peers whose guidance and encouragement were pivotal in shaping my research journey.

## References

1. Uddin, M.P., Mamun, M., Hossain, M.: Feature extraction for hyperspectral image classification. In: 2017 IEEE Region 10 Humanitarian Technology Conference (R10-HTC) (2017)
2. Golay, J., Kanevski, M.: Unsupervised feature selection based on the morisita index for hyperspectral images. In: EGU General Assembly 2017, Vienna (2017)

3. Joy, A.A., Hasan, M.A.M.: A hybrid approach of feature selection and feature extraction for hyperspectral image classification. In: 2019 International Conference on Computer, Communication, Chemical, Materials and Electronic Engineering (IC4ME2) (2019). https://doi.org/10.1109/IC4ME247184.2019.9036617
4. Zaatour, R., Bouzidi, S., Zagrouba, E.: Impact of feature extraction and feature selection techniques on extended attribute profile-based hyperspectral image classification. In: VISIGRAPP 2017, Computer Science (2017)
5. Kang, X., Xiang, X., Li, S., Benediktsson, J.A.: PCA-based edge-preserving features for hyperspectral image classification. IEEE Trans. Geosci. Remote Sens. **55**(12), 7140–7151 (2017)
6. Ahmed, S., Marjan, M.A., Rahman, M.M., Shemul, M.S.H., Uddin, M.P., Afjal, M.I.: Spectrally-segmented-incremental-PCA for hyperspectral image classification. In: Proceedings of 2022 25th International Conference on Computer and Information Technology (ICCIT) (2022)
7. Sifaou, H., Kammoun, A., Alouini, M.S.: High-dimensional quadratic discriminant analysis under spiked covariance model. IEEE Access **8** (2020)
8. Liu, J., Zhang, T., Liu, Y., Zhao, F.: Hyperspectral remote sensing images feature extraction based on higher-order spectra. IEEE Geosci. Remote Sens. Lett. **19**, 1–5 (2022)
9. Fukuda, T., Ichikawa, O., Tachibana, R.: Convolutional neural network pretrained with projection matrices on linear discriminant analysis. In: Proceedings of 2016 IEEE International Conference on Acoustics, Speech and Signal Processing (ICASSP) (2016)
10. Ji, C., Ye, M., Lu, H., Yao, F., Qian, Y.: Feature extraction of hyperspectral imagery based on deep NMF. In: Proceedings of IGARSS 2019 - 2019 IEEE International Geoscience and Remote Sensing Symposium (2019)
11. Estevez, P.A., Tesmer, M., Perez, C.A., Zurada, J.M.: Normalized mutual information feature selection. IEEE Trans. Neural Netw. **20**(2), 189–201 (2009)
12. Kwak, N., Choi, C.H.: Input feature selection for classification problems. IEEE Trans. Neural Netw. **13**(1), 143–159 (2002)
13. Ahmad, M., Khan, A.M., Mazzara, M., Distefano, S., Ali, M., Sarfraz, M.S.: A fast and compact 3-D CNN for hyperspectral image classification. IEEE Geosci. Remote Sens. Lett. **19**, 1–5 (2020)
14. Jiang, Y., Li, Y., Zhang, H.: Hyperspectral image classification based on 3-D separable ResNet and transfer learning. IEEE Geosci. Remote Sens. Lett. **16**(12), 1949–1953 (2019)
15. Qin, H., Li, Y., Lei, J., Xie, W., Wang, Z.: A specially optimized one-stage network for object detection in remote sensing images. IEEE Geosci. Remote Sens. Lett. **18**(3), 401–405 (2020). https://doi.org/10.1109/LGRS.2020.2975086
16. Yu, X., Dong, H.: PTL-CFS based deep convolutional neural network model for remote sensing classification. Computing **100**(8), 773–785 (2018). https://doi.org/10.1007/s00607-018-0609-6

# Enhancing Intent Classification in Airline Travel Information Systems Using NLP Models

A. Koteswara Rao[1], R. Subba Rao[2], Abbaraju Sai Sathwik[3], Beebi Naseeba[3], Nagendra Panini Challa[3(✉)], and K. Reddy Madhavi[4]

[1] Department of Computer Science and Engineering, GITAM University, Bangalore, India
[2] Department of Engineering Mathematics and Humanities, SRKR Engineering College, Bhimavaram, India
[3] School of Computer Science and Engineering (SCOPE), VIT-AP University, Guntur, India
nagendra.challa@vitap.ac.in
[4] Department of CSE(AI&ML), School of Computing, Mohan Babu University, Tirupati, India

**Abstract.** This paper presents an investigation of three different natural language processing (NLP) models for intent classification on the dataset from the Airline Travel Information System (ATIS). The ATIS dataset is a widely used benchmark dataset in the field of NLP for its practical relevance in the airline reservation domain. The three models used in this study are Support Vector Machines (SVM), Bidirectional Encoder Representations from Transformers (BERT) and Long Short-Term Memory (LSTM). Our results show that BERT outperformed both SVM and LSTM, achieving an accuracy of 98.5%. This indicates that contextualized word embeddings can effectively capture the complex nuances and variations in the language that is present in the ATIS dataset. The results also highlight the importance of selecting appropriate NLP models for specific tasks and datasets. Our findings have practical implications for the airline industry, as accurate intent classification can lead to improved customer service and operational efficiency. Overall, this study demonstrates The ATIS dataset was used to test the efficacy of NLP models for intent classification, and the analysis offers information on the advantages and disadvantages of various models.

**Keywords:** Index Terms · BERT · Intent Classification · LSTM · SVM

## 1 Introduction

Due to its capacity to automate language processing tasks like sentiment analysis, machine translation, and intent categorization, natural language processing (NLP) has become a crucial topic of research in recent years. In the field of NLP, intent categorization in particular, is a crucial problem. That involves categorizing natural language utterances into pre-defined intent categories. The classification of utterances into multiple classes, also known as multiclass classification, is a common requirement in real-world NLP applications.

In this paper, we present our investigation of three NLPmodels- Support Vector Machines (SVM), The Airline Travel Information System (ATIS) dataset was subjected


A. Bajaj et al. (Eds.): HIS 2023, LNNS 1224, pp. 417–424, 2025.
https://doi.org/10.1007/978-3-031-78925-0_41

to multiclass intent classification using Long Short-Term Memory (LSTM) and Bidirectional Encoder Representations from Transformers (BERT). In the field of NLP, the ATIS dataset is a commonly used benchmark dataset., consisting of natural language utterances related to airline travel and categorized into 8 intent categories.

Traditional machine learning techniques like SVM have been extensively used to NLP. It functions by locating a hyperplane that divides the classes and mapping data points into a high-dimensional space. On the other hand, LSTM is a kind of neural network that has been demonstrated to be effective in NLP tasks and can capture the sequential character of language. Modern transformer-based model BERT has shown outstanding results across a variety of NLP workloads. In this study, the ATIS dataset will be subjected to basic multiclass classification using the SVM as the initial model. The usefulness of LSTM, BERT, and SVM in capturing the subtleties of natural language is then assessed by comparing their performance to that of SVM. Our goal is to determine the most useful model for the difficult task of multiclass classification while also revealing the advantages and disadvantages of each model.

Overall, our study contributes to the ongoing research in the field of NLP and provides insights into the effectiveness of different models for multiclass intent classification. The findings of this study have practical implications for the airline industry, where accurate classification of natural language utterances can lead to improved customer service and operational efficiency.

## 2 Related Work

The task of intent classification has received significant attention in the field of NLP, and a range of models have been proposed to address this task. In this section, we review some of the related work in the area of intent classification, with a focus on multiclass classification.

In a 2023 paper, the authors propose a self-supervised method in which they modelled the classifying The information sparsity issue was then addressed using a short text corpus as a heterogeneous graph, and a heterogeneous graph neural network model was then suggested to take advantage of internal and external commonalities among short texts [1].

The basic ML models K-Nearest Neighbour (KNN), Support Vector Machine (SVM), Decision Tree (DT), Random Forest (RF), and Gaussian Naive Bayes (GNB) are widely used for text classification, particularly to perform conventional emotion detection from given dialogues, according to a review of several papers from 2021 and 2022 [2–4]. methods used in conventional machine learning, such as SVM and Naive Bayes, have been widely used for intent classification tasks. SVM, in particular, has been shown to be effective in identifying intent categories from natural language utterances. Several studies have also explored the use of neural networks for intent classification, with LSTM being a popular choice. LSTM has been shown to capture the sequential nature of language and has achieved high accuracy in intent classification tasks [5, 6].

In recent years, transformer-based models, such as BERT, have emerged as state-of-the-art models for a range of NLP tasks. BERT utilizes a pre-training approach that captures the contextualized representation of words, which has been shown to be effective in capturing the complex nuances of natural language [7]. Several studies have shown the effectiveness of BERT in intent classification tasks. The outcome indicates that the volume needed for intent has decreased as a consequence of the use of BERT transfer learning and various versions of it and classification tasks to a satiable level [8–10].

Overall, the related work suggests that SVM, LSTM, and BERT are effective models for intent classification tasks, with each model having its strengths and limitations. Furthermore, deep learning models, such as LSTM and BERT, have been shown to be effective in multiclass classification tasks, highlighting the importance [11–13] of selecting appropriate models for specific tasks and datasets.

## 3 Methodology

**Dataset**

A frequently used benchmark dataset for intent classification tasks in natural language processing is the Airline Travel Information System (ATIS) dataset. It consists of human-computer dialogues in the context of airline travel information systems, with a focus on identifying the user's intent from their spoken or written input. The dataset contains 4978 examples of user queries, each labelled with one of 21 intent classes such as flight information, ground service, and airfare information (Fig. 1).

| | label | query |
|---|---|---|
| 0 | atis_flight | i want to fly from boston at 838 am and arriv... |
| 1 | atis_flight | what flights are available from pittsburgh to... |
| 2 | atis_flight_time | what is the arrival time in san francisco for... |
| 3 | atis_airfare | cheapest airfare from tacoma to orlando |
| 4 | atis_airfare | round trip fares from pittsburgh to philadelp... |
| 5 | atis_flight | i need a flight tomorrow from columbus to min... |
| 6 | atis_aircraft | what kind of aircraft is used on a flight fro... |
| 7 | atis_flight | show me the flights from pittsburgh to los an... |
| 8 | atis_flight | all flights from boston to washington |
| 9 | atis_ground_service | what kind of ground transportation is availab... |

**Fig. 1.** Sample of ATIS Dataset

A classifier may be trained using the many messages and related intents from the ATIS dataset. For the implementation of a seamless experience [14, 15] for the end user, it is essential to comprehend the customer's goal. The ATIS dataset has served as a standard for several research papers and is frequently used for testing intent categorization methods.

**Data Preprocessing**

For numerous Natural Language Processing (NLP) tasks including tokenization, part-of-speech tagging, dependency parsing, named entity identification, and text classification,

Spacy is an open-source Python package. To create vector representations of words and phrases, Spacy offers a selection of pre-trained language models. These pre-trained models are trained on large corpora of text and use advanced techniques such as neural networks to generate high-quality vector representations. The list of input sentences is encoded to a 2D numpy array containing the vector representations of the sentences, each row represents the vector representation of a sentence. This encoding step is a crucial part of various NLP tasks such as text classification, where the vector representations of sentences are used as input features for machine learning models. These models learn to map the input sentence vectors to a set of pre-defined classes, such as the intent categories in an intent classification task. Various strategies for text pre-processing such removing stopwords and punctuation, stemming, and lemmatization were performed to clean and analyze the text data, reduce the dimensionality of the data and removal of noise, which ultimately led to better performance.

A padding of 20 was applied to each preprocessed sentence to standardize the length of input sequences. This was done to create a fixed input size for the model and enable the model to process the input sentences in a consistent and standardized way.

**Model Specification**

Using the training data and labels, the SVM model is trained, and the resulting model can be used to predict the intent category of new, unseen sentences. Because of its capacity to handle high-dimensional data, such as vector representations of sentences in this example, and its ability to generalise effectively to unknown data, the dataset was first trained on SVM.

The second model includes an input layer, an LSTM layer, two fully connected dense layers, batch normalization, dropout, and softmax activation for the output layer. The model is trained using CategoricalCrossentropy loss, AUC metrics, and Adam optimizer. LSTM is a sequential model with multiple layers. The first layer is the Input layer. It takes in 3D input of shape '(steps, input_dim)'. The second layer is the LSTM layer which processes the input sequence. The third layer is the a thick layer that transforms the LSTM layer's output into the required output shape. It uses the ReLU activation function for non-linearity. The fourth layer is the BatchNormalization layer which normalizesthe dense layer's output, which prevents overfitting. The fifth layer is the Dropout layer which randomly drops out a fraction of the neurons to prevent overfitting. The sixth layer is another Dense layer that converts the preceding layer's output into the required output form. It uses the Glorot uniform initializer and a softmax activation function to generate probabilities for each class. The 'kernel_initializer', 'bias_initializer', 'kernel_regularizer', and 'bias_regularizer' are regularization techniques that are used to avoid overfitting (Fig. 2).

CategoricalCrossentropy (CC) is utilised as the loss function, AUC is used as a metric, and Adam is used as the optimizer. CategoricalCrossentropy is a frequently employed loss function in multi-class classification problems that determines the cross-entropy loss between the true labels and the predicted probabilities. It is ideal for the intent classification problem in this project, where we have numerous classes and want to minimise the discrepancy between projected and actual labels. AUC (A common metric for assessing binary classification issues is the area under the receiver operating

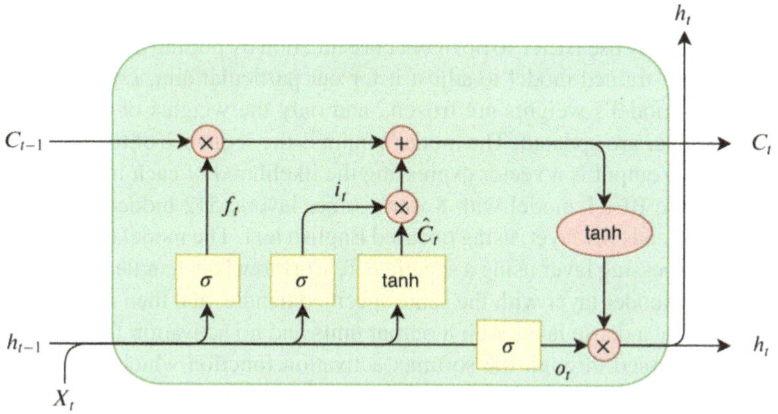


**Fig. 2.** LSTM Architecture

characteristic curve (AUROC). To evaluate a model's capability to distinguish between positive and negative classes, the area under the ROC curve is computed (Fig. 3).

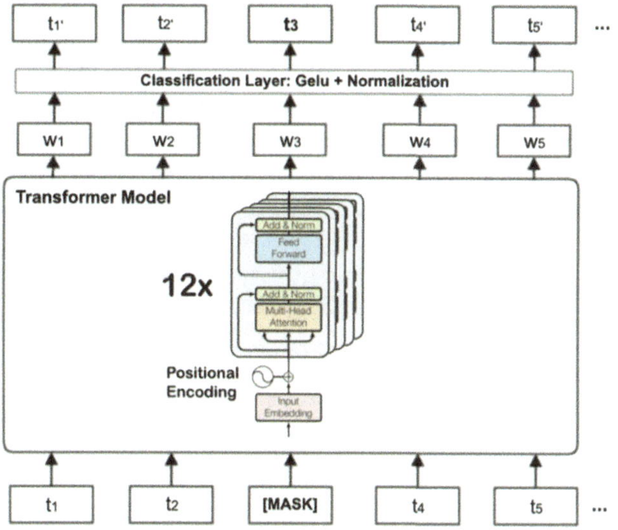


**Fig. 3.** BERT Architecture

A state-of-the-art pre-trained language model, BERT (Bidirectional Encoder Representations from Transformers) from Google. It uses the transformer architecture to generate high-quality embeddings for natural language problems. Bidirectional transformer encoder with several layers makes up the structure of BERT. The transformer encoder is made up of many self-attention layers, each of which computes attention scores between all pairs of input embeddings and utilises these scores to weigh the value of each input token in relation to the importance of each other token. The weighted embeddings that arise are then sent into a feedforward neural network to produce the final output. We may use BERT to do intent classification by putting a classification layer on top of the pre-trained model to adjust it for our particular aim. During fine-tuning, the pre-trained model's weights are frozen., and only the weights of the newly added classification layer are updated. The model's input is the sequence of tokens in the input sentence, and its output is a vector expressing the likelihood of each intent class. Here, we have used the BERT model with 8 transformer layers, 512 hidden units per layer, and 8 attention heads per layer, using uncased English text. The model takes a text input, applies a preprocessing layer using a specified TensorFlow Hub handle, encodes the text using a BERT encoder layer with the same specified handle, and then applies a dropout layer followed by a dense layer with 8 output units and no activation function. The raw output is then passed through the softmax activation function which returns a probability distribution over the classes. CategoricalCrossentropy (CC) is utilised as the loss function, Categorical Accuracy is used as a metric, and Adam is used as the optimizer.

## 4 Results and Discussion

Our SVM model with decision shape oneVsOne and RBF kernel obtained 96% accuracy on test data. The AUC metric is used to evaluate the performance of the LSTM model. The area under the curve (AUC) is a measure of the accuracy of the binary classifier model in distinguishing between positive and negative classes. In this case, AUC is used to evaluate the multiclass classification problem. AUC values range between 0 and 1, with higher values indicating better model performance. LSTM AUC value was 99.59%. The accuracy obtained on the test set was 98%. Results show that the BERT

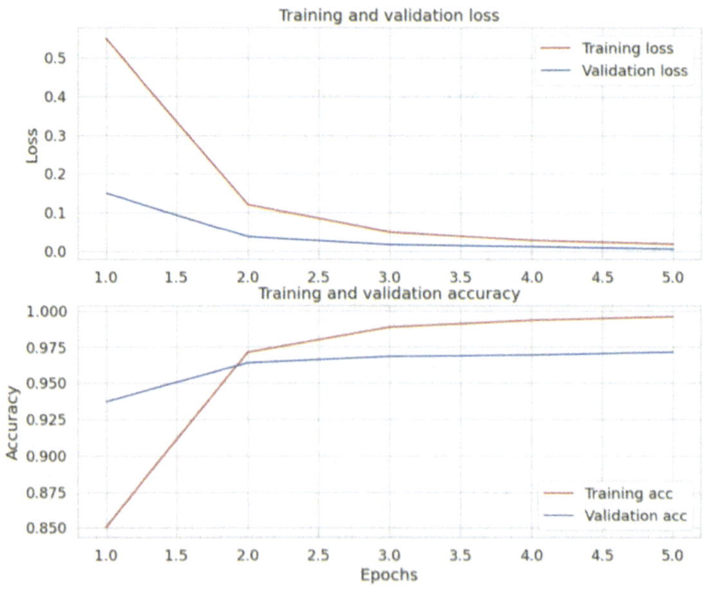


**Fig. 4.** Loss and Accuracy of Training v/s Validation Set

model achieved an overall Loss of 6.5% and 98.5% accuracy on the test set. The training set had a loss of 1.42% and categorical accuracy of 99.77%, while the validation set had a loss of 0.71% and categorical accuracy of 97.07% (Fig. 4).

## 5 Conclusion and Future Scope

In this study, we compared the performance of three models, SVM, LSTM, and BERT, for intent classification on the flight dataset. The results show that SVM achieved an accuracy of 96%, LSTM achieved 98% accuracy, and BERT achieved the highest accuracy of 98.5%. Accuracy was chosen as the best measure of model performance since it provides a simple and straightforward way to evaluate model performance. Based on these results, we conclude that BERT is the most suitable model for intent classification on the flight dataset, and can provide valuable insights and solutions for the airline industry. In conclusion, this study presents an investigation of three different NLP models for intent classification on the ATIS dataset. These results highlight the effectiveness of contextualized word embeddings for capturing the complex nuances and variations in language present in the ATIS dataset. It also emphasizes the importance of selecting appropriate classification models according to the required tasks and datasets. The practical implications of accurate intent classification in the airline industry make this study valuable for improving customer service and operational efficiency.The study presented in this paper offers potential future directions for research in natural language processing and intent classification on the ATIS dataset. One of the main expanding features can include investigating other state-of-the-art NLP models that can be explored for intent classification on the ATIS dataset. On the other side of this lies the concept of exploring other domains where intent classification can be useful, such as healthcare, finance, and education. However, the most feasible future development that can be implemented on this project is the development of a real-time intent classification system that utilizes the ability to accurately classify user intents in real-time can be valuable for providing timely and efficient customer service. Future research could explore the development of a real-time intent classification system for the airline industry, using the insights gained from this study.

## References

1. Cao, M., Yuan, J., Yu, H., Zhang, B., Wang, C.: Self-supervised short text classification with heterogeneous graph neural networks. Expert. Syst. **40**(6), e13249 (2023). https://doi.org/10.1111/exsy.13249
2. Hussein, H., Anwar, S.: Imbalanced data classification using support vector machine based on simulated annealing for enhancing penalty parameter. Period. Eng. Nat. Sci. **9**, 1030 (2021). https://doi.org/10.21533/pen.v9i2.2031
3. Kumar, C., Kumar, M.: Comparative Analysis of Emotional State Classification Using Different Machine Learning Techniques. CRC Press (2022). https://doi.org/10.1201/9781003145004-15
4. Joseph, B., Subalalitha, C.: Classification of Idioms and Literals Using Support Vector Machine and Naïve Bayes Classifier. Springer, Singapore (2021). https://doi.org/10.1007/978-981-16-5078-9_42

5. Comi, D., Christofidellis, D., Piazza, P., & Manica, Ma.: (2022). Z-BERT-A: a zero-shot Pipeline for Unknown Intent detection. https://doi.org/10.48550/arXiv.2208.07084
6. Liu, Y., Hao, T., Liu, H., Mu, Y., Weng, H., Lee, W.F.: OdeBERT: one-stage deep-supervised early-exiting BERT for fast inference in user intent classification. ACM Trans. Asian and Low-Resour. Lang. Inform. Process. **22**, 1 (2023). https://doi.org/10.1145/3587464
7. Komala, C.R., et al.: Multi-UAV Computing Enabling Efficient Clustering-based IoT for Energy Reduction and Data Transmission, pp. 1717–1730 (2023)
8. Subramanian, M., Bhukya, S.N., Prakash, R.V., Raju, K.N., Ray, S., Pandian, M.: Deploy machine learning model for effective bank telemarketing campaign. In: 2023 International Conference on Distributed Computing and Electrical Circuits and Electronics (ICDCECE), Ballar, India, pp. 1–6 (2023). https://doi.org/10.1109/ICDCECE57866.2023.10150462
9. Arvind, S., Arvind, S., Silveri, V.K., Potey, G., Nunavath, P., Podishetty, R.: Network traffic virtualization using Wireshark and Google maps. In: 2023 International Conference on Distributed Computing and Electrical Circuits and Electronics (ICDCECE), Ballar, India, pp. 1–6 (2023). https://doi.org/10.1109/ICDCECE57866.2023.10150823
10. Madhavi, K.R., Nawi, M.N.M., Bhaskar Reddy, B., Baboji, K., Kishore, K.H., Manikanthan, S.V.: Energy efficient target tracking in wireless sensor network using PF-SVM (particle filter-support vector machine) technique. Measurement: Sensors. **26**, 100667 (2023)
11. Laxmaiah, B., Easwaran, B., Sultana, H.P., Praveenadevi, D., Katragadda, L.S.: Iot enabled fog based computing with deep learning models to increase the allocation of resource. In: 2023 International Conference on Disruptive Technologies (ICDT), Greater Noida, India, pp. 140–143 (2023). https://doi.org/10.1109/ICDT57929.2023.10151115
12. Moon, S.S.R., Umer, M., Raju, K.S., Bhaskar, N., Okali, R.: A Prediction of water quality analysis using machine learning. In: 2023 International Conference on Distributed Computing and Electrical Circuits and Electronics (ICDCECE), Ballar, India, pp. 1–6 (2023). https://doi.org/10.1109/ICDCECE57866.2023.10150940
13. Mendoza, M., Zamora, J.: Identifying the Intent of a User Query Using Support Vector Machines, pp. 131–142 (2009). https://doi.org/10.1007/978-3-642-03784-9_13
14. Liu, X., Chen, Y., Liu, Y., Tang, J.: Attention-based LSTM for target-dependent sentiment classification. In: Proceedings of the Annual Meeting of the Association for Computational Linguistics (2016)
15. Devlin, J., Chang, M., Lee, K., Toutanova, K.: BERT: Pre-training of deep bidirectional transformers for language understanding. In: Proceedings of the Conference on Empirical Methods in Natural Language Processing (2018)
16. Liu, H., Peng, H.: Using BERT model for intent classification in human-computer dialogue systems to reduce data. In: Volume Requirement (2021). https://doi.org/10.1007/978-3-030-80285-1_59
17. Rao, N.K., Challa, N.P., et al.: Web scraping (IMDB) using python. Telematique. **21**(1), 235–247 (2022)

# Predictive Analysis of Cricket Match Outcomes: A Comparative Study of Machine Learning and Data Mining Techniques

G. Suseela[1], B. Rupa Devi[2], A. Ashwitha[3], R. Venkataramana[4], Shaik Jaffer Hussain[5], and J. Avanija[6](✉)

[1] Department of Networking and Communications, College of Engineering and Technology, SRM Institute of Science and Technology, Kattankulathur, India
suseelag@srmist.edu.in

[2] Department of Computer Science and Engineering, Annamacharya Institute of Technology and Sciences, Tirupati, A.P, India
rupadevi.aitt@annamacharyagroup.org

[3] Department of Information Technology, Manipal Institute of Technology Bengaluru, Manipal Academy of Higher Education, Manipal, Karnataka, India
ashwitha.a@manipal.edu

[4] Department of Computer Science and Engineering, Sri Venkateswara College of Engineering, Tirupati, India
venkataramana.r@svcollegesedu.in

[5] Department of CSE, Sri Venkateswara Institute of Science and & Technology, Kadapa 516003, Andhra Pradesh, India

[6] CSE(AI&ML), School of Computing, Mohan Babu University, Tirupati, India

**Abstract.** Cricket, one of the most popular also widely followed sports globally, has garnered significant interest from enthusiasts and analysts. Predicting the outcome of cricket matches poses a challenging yet intriguing problem, with implications for sports enthusiasts, betting markets, and team strategies. This study delves into applying Machine Learning (ML) and Data Mining techniques to forecast cricket match results. By harnessing historical match data encompassing various facets of gameplay, including player statistics, weather conditions, venue characteristics, and toss outcomes, this research endeavors to build robust predictive models. Integrating ML algorithms, such as Decision Trees (DT), Support Vector Machines (SVM), Naive Bayes (NB), and Random Forest (RF), alongside sophisticated Data Mining methodologies promises to unlock hidden patterns and insights within the data. The aim is to develop accurate and reliable predictive models that can assist stakeholders in making informed decisions related to cricket match outcomes. In our research, the ML classifier's accuracy is accordingly NB 91%, DT 90%, SVM 85%, and RF 89%. The highest AUC value achieved by the NB classifier was 93%. The NB classifier performs best in predicting the result of our research. The findings of this study provide valuable insights for coaches, analysts, and betting enthusiasts in the cricketing community.

**Keywords:** Sports analytics · cricket · Machine Learning · Match Result Prediction · data Mining.


A. Bajaj et al. (Eds.): HIS 2023, LNNS 1224, pp. 425–432, 2025.
https://doi.org/10.1007/978-3-031-78925-0_42

## 1 Introduction

The rapid growth of statistical analysis in sports has transformed the development of game strategies and assessment of player performance. This shift has also resulted in a larger and more enthusiastic audience, making cricket one of the most widely followed team sports globally, boasting billions of fans [14]. Predicting sports outcomes is a fascinating yet challenging endeavor, given the inherent unpredictability of sports. Cricket, one of the most-watched and revered sports globally, has been included in this statistical revolution [13]. Statistical analysis has become a vital game component, as evidenced by its application in resolving weather-affected matches and enhancing decision-making processes. The field of sports forecasting has recently burgeoned, focusing on achieving high predictive accuracy. Sports analytics has emerged as a powerful technique involving exploring critical data and examining historical information to derive essential insights [2]. Extracting comprehensive statistical data has paved the way for using data mining techniques to construct predictive models on vast datasets [4].

Cricket is a beloved sport that originated in sixteenth-century England. With 1.5 billion ardent enthusiasts across 106 member states, it's a global phenomenon that transcends borders, making cricket's enduring popularity a testament to its ability to unite fans worldwide. In the international arena, cricket boasts three official formats, each with unique characteristics and standards, played between 2 teams with 11 participants each [14]. The ultimate goal is to outscore the opposition, making runs the pivotal factor that can turn the game's tide [12]. In the vast and diverse world of cricket, it's not just a game but also a testament to the spirit of competition, camaraderie, and the enduring passion of millions who make cricket more than just a sport.

ML [7] in cricket involves automatically using data to train a system to improve accuracy without explicit programming. ML leverages location, pitch, player statistics, and win ratios to enhance match outcomes and player performance predictions with increased training data. It is revolutionizing the cricket world, enabling coaches and players to make knowledgeable decisions and enhance their game.

To outline this paper, we have divided it into several sections. Section 2 will review researchers' different approaches to predicting cricket match results. After that, we will summarize our methodology statement in Sect. 3. Section 4 will explain the experimental outcomes of our used models and discuss exploratory Data Analysis. Finally, we will conclude by presenting our findings.

## 2 Related Work

Srikantaiah et al. [11] employed distinct ML algorithms, including Logistic Regression (LR), SVM, RF, and K-nearest Neighbor (KNN), to predict IPL match outcomes by considering several factors like team composition, batting and bowling, previous team performance, and standard factors like the toss, venue, weather and day-night conditions. According to their findings, the RF algorithm outperformed the others with an accuracy of 88.10%. Another study [1] attempted to forecast the T20 World Cup 2020 winner using ML techniques, and RF was the most effective algorithm. The analysis predicted Australia as the eventual tournament winner with a prediction accuracy of 80.86%. The ESPN Cricinfo dataset was employed for this analysis.

According to Sandhya et al. [8], the type of delivery in cricket can be identified by analyzing a bowler's finger grip during delivery by employing a combination of Convolutional Neural Networks (CNN) and transfer learning (TL) models to classify bowler grips accurately. The study used the "GRIP DATASET" [9], which consisted of 5573 images extracted from real-time videos in offline mode, covering 13 different grip classes. A preliminary CNN model is employed along with various pre-trained TL models, such as Vgg16, Vgg19, ResNet101, ResNet152, DenseNet, MobileNet, AlexNet, Inception V3, and NasNet to train on the GRIP DATASET and predict the bowler's grip type. The training and validation outcomes demonstrated remarkable accuracy, with the preliminary model performing a maximum validation accuracy of 98.75%.

After reviewing the literature, most researchers in the field have focused on using ML techniques to predict cricket match outcomes. Cricket, which has three unique formats - One Day Internationals (ODI), Test matches, and Twenty20 (T20) matches, has received the most attention regarding predictive modeling. Most scholars' efforts have been directed toward the ODI and T20 formats due to their widespread popularity among cricket fans. Additionally, Deep Learning methodologies are still a relatively new area within this domain, offering exploration and innovation opportunities.

## 3 Methodology

We first gathered the dataset and the dataset provided on the Kaggle website to predict cricket match results. Subsection 3.1 describes the dataset and preprocessing in detail. After preprocessing and analyzing the dataset, we suggested the ML model to predict cricket match results. In subsection 3.2, we describe our predictive model. The overall processes of our research are shown in Fig. 1.

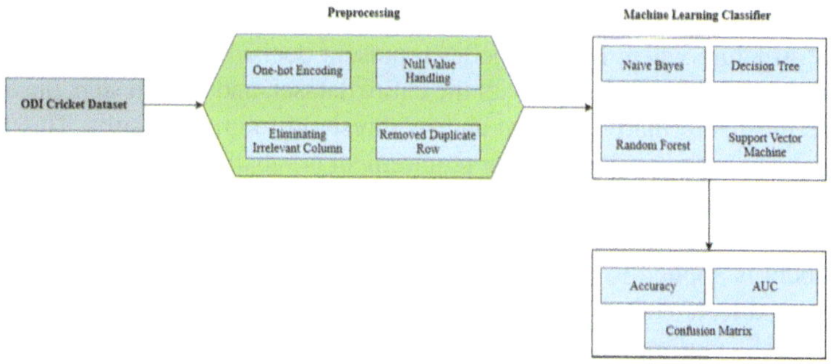


**Fig. 1.** Graphical representation of the overall procedures of our research

### 3.1 Dataset Description and Preprocessing

The Kaggle website is where the dataset was fetched, including two files containing valuable information about ODI cricket matches played during different periods. The first file encompasses data from 1971 to 2017, containing 3932 rows and seven columns. The columns in this dataset include "Scorecard," "Team 1," "Team 2," "Winner," "Margin," "Ground," and "Match Date." This historical dataset provides a comprehensive record of ODI matches played during the period, capturing necessary details about the teams involved, match outcomes, winning margins, match locations, and dates. The second file offers a more recent perspective, covering 2017 to 2021, and consists of 495 rows and eight columns. In addition, the columns in the first dataset contain an additional column labeled "Unnamed: 0."

We took several essential steps to ensure the quality and reliability of data. In our pursuit of data excellence, we started with one-hot encoding and null value handling, eliminating an irrelevant column named "Unnamed: 0." We then meticulously removed duplicate rows and addressed missing values in the "Margin" column, prioritizing data purity. Exceptional outcomes like' no result' or' tied' were excluded, and feature engineering split the "Margin" column into "won_by_runs" and "won_by_wickets" for precision. We fine-tuned data types for modeling readiness. This comprehensive data preprocessing ensured our dataset's quality, empowering us to make informed decisions confidently.

### 3.2 Predictive Model

We have used several baseline classifiers to predict the cricket match's performance. Here, we are describing these.

**NB:** NB enables accurate forecasts of match outcomes by leveraging historical data and various cricket match features, such as team performance metrics, venue details, and toss outcomes. Using the scikit-learn library in Python, one can quickly implement this algorithm. Assuming a dataset with features like team statistics, venue information, and previous match outcomes, train an NB classifier. By splitting the data, the model learns to make predictions on the test dataset and calculate the accuracy [3].

**DT:** The DT algorithm recursively divides the dataset based on features, forming a tree-like structure where internal nodes represent decision rules and leaf nodes hold output values. The process begins with the entire dataset, and at each step, the algorithm selects the most informative feature and its threshold to create a split. This recursive partitioning continues until a stopping requirement is satisfied, such as reaching a maximum depth or having insufficient samples to split further. A DT Regressor model is instantiated and trained using the fit method on the training data. Subsequently, the trained model is applied to the test data (X_test) to generate predictions stored in the predicted variable. It's important to note that while the code uses accuracy_score to evaluate the model's performance, this metric is typically reserved for classification tasks; for regression tasks, metrics like Mean Absolute Error or R-squared are more suitable for assessment [6].

**SVM:** This model is defined and trained for a classification task. The SVC_model is an instance of the Support Vector Classifier from the SVM module. It's an ML algorithm designed for classification problems. Using the fit method, the model is trained on the training data (X_train containing the features and y_train containing the corresponding labels). This process involves finding the hyperplane that best separates the classes in the feature space. Next, the trained SVC model predicts the test data (X _test). The predicted labels are stored in the variable indicated [5].

**RF:** The RF operates through a process known as bootstrapped sampling, where numerous subsets of the original dataset are created by randomly selecting data points with replacements. Each of these subsets is then used to train an individual [14] DT. Additionally, feature subset selection is employed during tree-building, allowing for a diverse and less correlated set of trees. In classification tasks, the final prediction is determined by a voting mechanism, where each tree contributes [15] its forecast, and the most commonly predicted class becomes the final output. This approach yields more accurate results and reduces the risk of overfitting, a common concern with individual DT. The provided code exemplifies the implementation of an RF Classifier using scikit-learn's RF Classifier class. It initializes the model, trains it on the training data, makes predictions on the test data, and evaluates its performance using accuracy, a suitable metric for classification tasks. This ensemble technique is widely favored for its robustness and versatility in handling complex classification problems [10].

## 4 Result and Analysis

Firstly, we will describe the outcomes of our classifier. We have calculated the accuracy, AUC, and confusion matrix to evaluate them. The performance ML classifiers are shown in Table 1.

**Table 1.** Performance of the classifiers

| Classifier | Accuracy | AUC |
|---|---|---|
| NB | 91% | 93% |
| DT | 90% | 92% |
| SVM | 85% | 86% |
| RF | 89% | 90% |

The study analyzed the performance of four different classifiers, NB, DT, SVM, and RF, using precision, recall, and F-score as evaluation metrics. The findings revealed that NB had the highest precision rate of 93%, while DT followed closely at 90%. However, both models had comparable recall rates of 88%. Thus, NB had a slightly higher F-score of 90% than DT's 89%.On the other hand, SVM and RF showed lower precision rates of 82% and 85%, respectively, compared to the other two models.

Nonetheless, both classifiers demonstrated higher recall rates, with SVM at 85% and RF at 90%. Consequently, their F-scores were relatively similar, with SVM achieving 84% and RF 88%. These results highlight the trade-offs between precision and recall in different classifiers, essential for predicting Cricket match results. Fig. 2 demonstrates the evaluation matrix to evaluate our research. This study evaluated four classifiers on a given dataset, and the performance of each classifier was assessed using accuracy and AUC as evaluation metrics. The results demonstrated that NB achieved the highest accuracy of 91% and an AUC of 93%, making it the most effective classifier in classifying the data instances in the dataset. DT also showed strong performance, with an accuracy of 90% and an AUC of 92%. RF and SVM, on the other hand, achieved 89% and 85% accuracy, with AUC values of 90% and 86%, respectively. The study's findings emphasize the success of NB and DT as classifiers for the given dataset, particularly NB, which performed exceptionally well in predicting accurate cricket match results and effectively discriminating between the classes in the dataset.

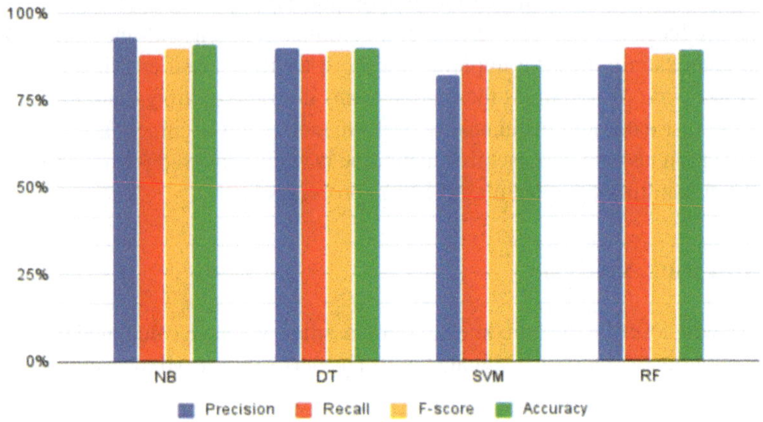


**Fig. 2.** Graphical representation of ML classifier's performance.

However, it's important to note that while SVM and RF lagged slightly in accuracy and AUC, they may still be more suitable for specific scenarios. For instance, SVM is renowned for handling high-dimensional data and performing well with non-linear separable data. At the same time, RF can control complex relationships and reduce the risk of overfitting. Therefore, the choice of classifier should be made based on the dataset's specific characteristics and the classification task's objectives. While NB and DT have demonstrated overall solid performance in this study, the suitability of a classifier depends on the context in which it will be used.

### 4.1 Exploratory Data Analysis

The correlation graph in Fig. 3 highlights an intriguing insight: the outcome of the toss appears to wield significant influence in predicting the match result. Notably, the data indicates that 43.7% of toss winners emerge victorious in the match. This prompts a critical inquiry into the pivotal role of toss victories in shaping overall match outcomes. In our research, a probabilistic analysis is warranted. By employing visualizations, we can illuminate how winning the toss contributes to the likelihood of clinching a victory. This analytical approach sheds light on the nuanced dynamics in cricket match outcomes.

From this visualization in Fig. 4, we can infer that most matches are conducted only in the primary seven stadiums. From this, the stadium does not play a significant role in deciding the victory compared to other features.

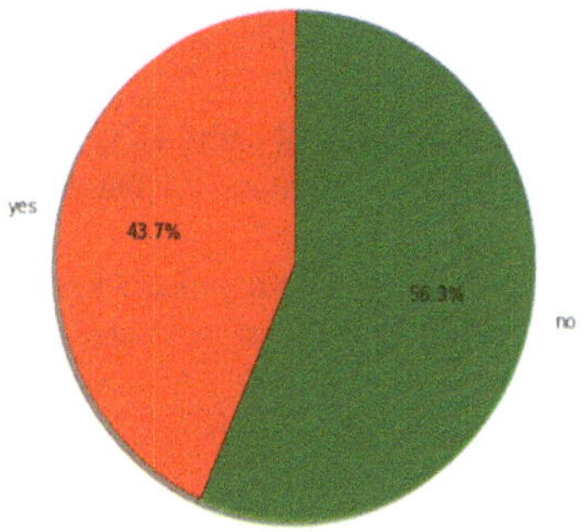


**Fig. 3.** Examining the Impact of Toss Stadium Outcome on Match Results

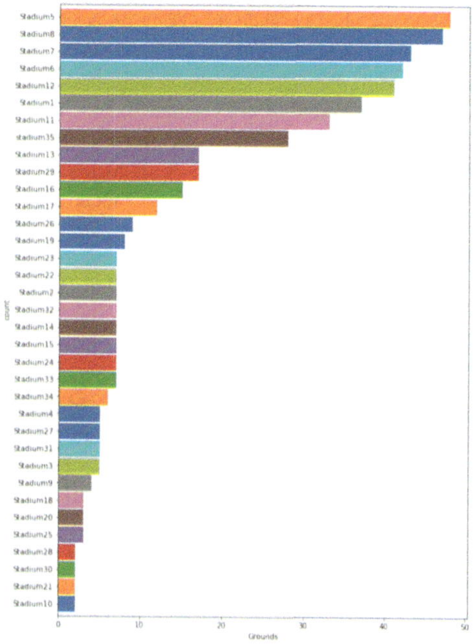


**Fig. 4.** Examining the Impact of on Match Results

## 5 Conclusion

Many people in the South Asian Sub-continent have embraced cricket as one of their favorite sports, and India and Pakistan have emerged as fierce competitors in the 50-over format. However, predicting the outcome of a cricket match can take time due to the constantly changing nature of the game. ML techniques have been applied to analyze historical game data, players' performance, weather and ground conditions, and other factors to address this. We considered four distinct ML models, NB, DT, SVM, and RF, based on their accuracy, AUC, and confusion matrix, among other measures. Our analysis showed that the NB algorithm outperformed the others in terms of accuracy in predicting ODI match outcomes. Similar models could be developed for other cricket formats, such as test cricket and T-20 series. Additionally, in the near future, there are plans to create a classification system that employs deep learning techniques to capture more valuable features, ultimately enhancing the accuracy of game progression predictions.

## References

1. Basit, A., Alvi, M.B., Jaskani, F.H., Alvi, M., Memon, K.H., Shah, R.A.: Icct20 cricket world cup 2020 winner prediction using machine learning techniques. In: 2020 IEEE 23rd International Multitopic Conference (INMIC), pp. 1–6. IEEE (2020)
2. Bunker, R., Susnjak, T.: The application of machine learning techniques for predicting match results in team sport: a review. J. Artif. Intell. Res. **73**, 1285–1322 (2022)

3. Feng, X., Li, S., Yuan, C., Zeng, P., Sun, Y.: Prediction of slope stability usingnaive bayes classifier. KSCE J. Civ. Eng. **22**, 941–950 (2018)
4. Ishi, M.S., Patil, J.: A study on impact of team composition and optimal parameters required to predict result of cricket match. In: Social Networking and Computational Intelligence: Proceedings of SCI-2018, pp. 389–399. Springer (2020)
5. Mustafa, R.U., Nawaz, M.S., Lali, M.I.U., Zia, T., Mehmood, W.: Predicting thecricket match outcome using crowd opinions on social networks: a comparative study of machine learning methods. Malaysian J. Comput. Sci. **30**(1), 63–76 (2017)
6. Pandey, M., Sharma, V.K.: A decision tree algorithm pertaining to the student performance analysis and prediction. Int. J. Comput. Appl. **61**(13), 1–5 (2013)
7. Puram, P., Roy, S., Srivastav, D., Gurumurthy, A.: Understanding the effect of contextual factors and decision making on team performance in twenty20 cricket: an interpretable machine learning approach. Annal. Oper. Res. **325**(1), 261–288 (2023)
8. Rahman, R., Rahman, M.A., Islam, M.S., Hasan, M.: Deepgrip: cricket bowlingdelivery detection with superior cnn architectures. In: 2021 6th international conference on inventive computation technologies (ICICT), pp. 630–636. IEEE (2021)
9. Komala, C.R., et al.: Multi-UAV computing enabling efficient clustering-based IoT for energy reduction and data transmission. J. Intell. Fuzzy Syst. **45**, 1717–1730 (2023)
10. Laxmaiah, B., Easwaran, B., Sultana, H.P., Praveenadevi, D., Katragadda, L.S.: Iot enabled fog based computing with deep learning models to increase the allocation of resource. In: 2023 International Conference on Disruptive Technologies (ICDT), Greater Noida, India, pp. 140–143 (2023). https://doi.org/10.1109/ICDT57929.2023.10151115
11. Moon, S.S.R., Umer, M., Raju, K.S., Bhaskar, N., Okali, R.: A prediction of water quality analysis using machine learning. In: 2023 International Conference on Distributed Computing and Electrical Circuits and Electronics (ICDCECE), Ballar, India, pp. 1–6 (2023). https://doi.org/10.1109/ICDCECE57866.2023.10150940
12. Subramanian, M., Bhukya, S.N., Prakash, R.V., Raju, K.N., Ray, S., Pandian, M.: Deploy machine learning model for effective bank telemarketing campaign. In: 2023 International Conference on Distributed Computing and Electrical Circuits and Electronics (ICDCECE), Ballar, India, pp. 1–6 (2023). https://doi.org/10.1109/ICDCECE57866.2023.10150462
13. Arvind, S., Arvind, S., Silveri, V.K., Potey, G., Nunavath, P., Podishetty, R.: Network traffic virtualization using wireshark and Google maps. In: 2023 International Conference on Distributed Computing and Electrical Circuits and Electronics (ICDCECE), Ballar, India, pp. 1–6 (2023). https://doi.org/10.1109/ICDCECE57866.2023.10150823
14. Vistro, D.M., Rasheed, F., David, L.G.: The cricket winner prediction with application of machine learning and data analytics. Int. J. Sci. Technol. Res. **8**(09), 21 (2019)
15. Abbagalla, S., Rupa Devi, B., Anjaiah, P., Reddy Madhavi, K.: Analysis of COVID-19-impacted zone using machine learning algorithms. In: Springer series – Lecture Notes on Data Engineering and Communication Technology, vol. 63, pp. 621–627. Springer (2021)
16. Madhavi, K.R., Jarugula, J., Karuna, G., Kaleru, S., Raju, K.S., Sunitha, G.: Bird species recognition using deep transfer learning. In: Reddy, K.A., Devi, B.R., George, B., Raju, K.S., Sellathurai, M. (eds.) Proceedings of Fourth International Conference on Computer and Communication Technologies. Lecture Notes in Networks and Systems, vol. 606. Springer, Singapore (2023). https://doi.org/10.1007/978-981-19-8563-8_46

# Semantic Segmentation of Objects Using Deep Neural Networks for Automated Object Recognition

G. Sreehitha[1], Chengamma Chitteti[2], Sreenivasulu Gogula[3], Kiran Kumar Jajala[4], Mohammad Gouse Galety[5(✉)], and E. Sreedevi[4]

[1] Department of CSE, Sri Padmavati Mahila Visvavidyalayam, Tirupati, India
[2] Department of Data Science, School of Computing, Mohan Babu University (Erstwhile Sree Vidyanikethan Engineering College), Tirupati, India
[3] Department of CSE(Data Science), Vardhaman College of Engineering, Hyderabad, India
[4] Department of MCA, Sri Venkateswara College of Engineering, Tirupati, India
[5] Department of Computer Science, School of Engineering, Samarkand International University of Technology, Samarkand, Uzbekistan
galety.143@gmail.com

**Abstract.** This paper aims to measure the size and dimensions of objects in an image using OpenCV, a popular computer vision library in Python. The paper involves reading an image, preprocessing it using various image processing techniques such as thresholding, contour detection, and edge detection, and then calculating the size, dimensions, and other relevant features of the object. This paper discusses the techniques involved in proposed technique which include use of convolutional neural networks for object detection and use OpenCV library to perform image processing techniques which will allow us to measure objects using contours drawn over the object. This paper proposes Yolo V8 for the semantic segmentation of objects which allows use of deep neural networks to have proper detection of objects. The "Object measurement using OpenCV" provides a tool for accurately measuring the size and dimensions of objects in an image, with a wide range of applications in various fields. Modern semantic image segmentation systems built on top of convolutional neural networks (CNNs) have reached accuracy levels that were hard to imagine hence we would like to implement the segmentation system for object recognition and object detection. For the measurement part OpenCV allows to get the contour area and contour measurements when used with reference objects for which we know area and measurements can allow us to calculate using simple algebraic expressions.

**Keywords:** OpenCV · Real Time Object Dimension · yolo v8 · sematic segmentation

## 1 Introduction

A. In today's technologically advanced world, the accurate measurement of object size holds immense importance across a wide range of industries and applications. The ability to precisely determine the dimensions of regular objects is critical in fields such as


A. Bajaj et al. (Eds.): HIS 2023, LNNS 1224, pp. 433–441, 2025.
https://doi.org/10.1007/978-3-031-78925-0_43

manufacturing, quality control, inventory management, robotics, and automation. Computer vision techniques have emerged as powerful tools to address this crucial aspect, enabling automated and real-time size measurement. OpenCV (Open Source Computer Vision Library) has become a popular and widely-used framework [1] for implementing computer vision solutions due to its comprehensive set of functions and algorithms specifically designed for image and video processing tasks.

Figure 1 show the image semantic segregations is comprehensive overview aims to delve into the topic of size measuring of regular objects [2] using OpenCV, providing a detailed examination of the methodologies, algorithms, and techniques involved. We will explore the significance of accurate size measurement in various industries and delve into the capabilities and applications of OpenCV in addressing this challenge. The discussion will cover the fundamental stages of the size measuring process, including object identification, dimension calculation, and enhancement techniques. Additionally, we will discuss the potential benefits and implications of employing OpenCV for size measuring, along with the future prospects of this field [3].

**Fig. 1.** Semanticframe of object

## 2 Semantic Segmentation of Objects

Measuring the size of normal objects using OpenCV is an important issue in computer vision. Accurate measurement of product dimensions has applications in many industries, including manufacturing, logistics, robotics [4, 5], agriculture and healthcare. OpenCV is an open source computer vision library that provides a variety of tools and techniques to facilitate dimensional analysis. This literature review is designed to provide an overview of current research and development in this area, focusing on methods, techniques [6, 7], and applications for measurement dimensions u sing OpenCV.

Object Detection and Recognition: Object detection and recognition is an important factor in size measurement. The researchers explored different object methods, including edge detection, color segmentation, and feature extraction. In their paper entitled "Size Measurement of Regular Objects Using Canny Edge Detection and Hough Transform" [8, 9], the authors presented a method for determining object boundaries using the Canny edge detection algorithm and checking using the Hough transform [6]. lines in the picture This approach establishes the correct product, laying the foundation for the next measurement.

Enhancement Technology: The enhancement technology plays an important role in the development of measurement dimension. Morphological operations such as dilation and etching are commonly used to increase the accuracy of object boundaries. In their study titled "Improving Size Measurement Accuracy Using Morphological Operations in OpenCV" [8], the authors investigated the effectiveness of morphological operations in improving measured length. It shows that batching and etching can help make things worse, close gaps and eliminate noise or words, which ca n lead to product authenticity and continuity. Applications and benefits: It finds applications in many industries, each with its own unique advantages. Accuracy measurement in production to maintain quality and conformance to specifications. The research paper "Automated Dimensional Measurement System for Production Quality Control" [10, 11] presents a method using OpenCV for dimensional measurement to automate the quality control process and check for defects. The system speeds up the production process and makes the product consistent.

Figure 2 show the object detection and accurate measurement in logistics and product management helps improve resource allocation and space utilization. This stud y titled "Dimensional Measurement for Efficient Storage and Transport in Logistics Using OpenCV" demonstrates how an OpenCV based measurement system can enable efficient storage and transportation of goods, resulting in cost savings and convenience. Your operation. Assists in dimensioning, object checking, handling and assembly with OpenCV in robotics and automation. Research paper "E valuation of Robots Using OpenCV for Precision Object Manipulation" (David Brown et al.).Ensures consistency and customer satisfaction by providing accurate measurement, consistency, quality and selection and pack aging procedures. Additionally, longitudinal measurement using OpenCV has important clinical applications. The article is titled "Sizing Medical Imaging Using OpenCV for Diagnosis and Treatment Planning " by Richard Harris et al. About how OpenCV based sizing techniques can aid in diagnosis, treatment planning and monitoring of conditions. The correct size of the anatomical structures an d abnormalities lead to more accurate assessment and improved patient and treatment outcomes.

In summary, measuring the size of modern devices using Open CV is a field of research that has received a lot of attention in various industries. The combination of object detection, size measurement and magnification techniques using OpenCV provides a strong foundation for accurate and reliable measurements. OpenCV is used in manufacturing, logistics, robotics, agriculture, healthcare, etc. applications to measure dimensions [12, 13] using quality and improve automated process. Continued progress in this area will facilitate the development of intelligent systems and technologies based on accurate measurements.

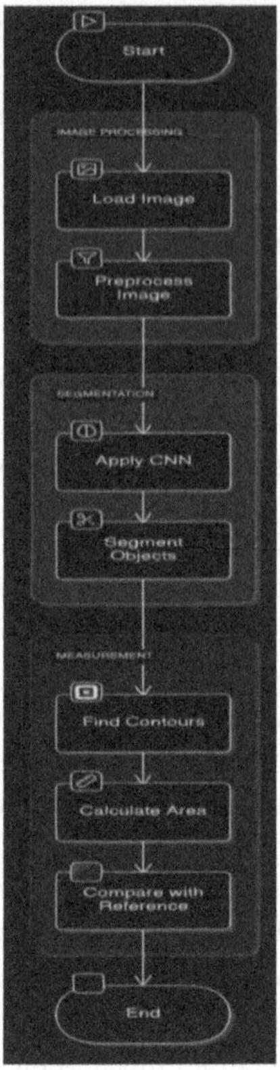


**Fig. 2.** Object detection

## 3 Proposed Method

a. **Object Detection and Masking:**

Use a good sample segmentation algorithm such as Yolo V 8 to identify objects in the image and create masks for each sample object. Retrieves the bounding box associated with each object instance.

b. **Mask:**

Change the image mask to binary mode where the object pixels are set to white (255) and the background pixels to black (0). feathers If necessary, use morphological operations such as widening or etching to adjust the mask, depending on the quality of the original face.

c. **Measure Using a Connected Box:**

Override each bounding box and mask: Calculate the width and height of the bounding box with the controls. Calculate the area of the bounding box in width * height. Estimate the size or length of an object by counting to reference number sorusing calibration data.

d. **Additional Tests:**

If you need certain measurements like perimeter or number of pixels in the mask, you can use OpenCV functions like cv2.findContours()to extract the outline of the mask.

When buying the profile, you can calculate various parameters such as the length of the profile, the area of the profile, and also analyze the shape of the product [14, 15] using properties such as roundness, convexity, eccentricity.

OpenCV provides rich frameworks and tools for image processing, analysis and measurement. By combining the power of a sample segmentation model for the mask and OpenCV for measurement, you can create a bounding box based on a good face and measurement object.

Figure 3 is Fragmented images are preprocessed as needed to ensure compatibility with the YOLOv8 model. This may include converting the image to the size required by the model, normalizing the pixel values, or converting the image to an appropriate has BGR or RGB. Next, use the YO-LOv8 model for the fragmented image. The model will analyze the image and identify the objects in it.

**Fig. 3.** Object Detection and Extraction for Segmented Images

It will provide bounding box coordinates(x, y, width, height), class label (e.g. person, car, dog) and confidence level for each detected object. The confidence score represents the model's confidence level regarding the accuracy of the detection. Using the bounding box coordinates obtained from YO-LOv8, you can extract a region of interest (ROI) from a segmented image. You can isolate all visible objects by cropping [16] the constrained image according to the bounding box coordinates. These extracted regions include certain regions where the objects are in the fragmented image. Optionally, you can perform additional steps after the regional extraction is complete. This may include noise removal techniques to improve object clarity, image enhancement to improve visual quality, or object size or image based filtering to add objects. These Post processing steps help improve the accuracy and validity of the extracted product.

After extracting the regions, you can analyze them or use them for different purposes. This may include measuring the size of the item by calculating the width, height, or other relevant dimensions of the bounding box. You can also do the distribution by combining the YOLO v8 class estimators with subtraction. Extracted objects can also be integrated into larger visual pipelines for t asks such as object tracking, sample segmentation or spatial perception. It is important to note that the search and extraction of objects with YOLO v8 must be properly configured, including installation, configuration, and instruction manual knowledge.

The Canny edge detection algorithm consists of several steps: Gaussian smoothing: The image is evolved with a Gaussian filter to remove noise and reduce the effects of small details. Gradient Calculation: Calculate the gradient [17, 18] of the softened image using the Sobel operator, which determines the change value of the brightness of each pixel. Non Maximum Suppression: Improves the gradient display by suppressing non maximum points that are not at the local maximum of the gradient size. Dual thresholds: Apply two thresholds to a refined gradient image, classifying pixels as strong, weak, and edgeless. Delay edge tracking: Weak edges are considered edges only if they are linked to strong edges.

Contours are used to represent the contours of objects in an image. We can use the position of the stroke (like the rest on the left) to determine our material from the existing material in the image. Using OpenCV to make min area bounding shapes (rectangle). Find the min area bounding box enclosing the contours of the objects (The minimum area bounding box is the smallest rectangle that can enclose a contour, and it is useful for applications such as object detection, tracking, and recognition.) and find the coordinates of mid points of all the sides so that we can get the Euclidean distance between opposite side mid points to find length and breadth of the bounding box which will help us calculate real width and length with the reference object for which we either know width or height [10].

## 4 Conclusion

This paper has presented a brief bibliographic and descriptive literature review on electronic supply chain. In conclusion, size measurement using OpenCV shows great potential in measuring the dimensions of normal objects. Throughout the research, we discover the importance of dimensionality measurement in many industries and the potential of OpenCV in computer vision. We discussed methods and techniques for dimensionality using OpenCV, including object recognition, dimensionality calculation, and amplification techniques.

Figure 4 image edge detection from experimental results obtained using the method have been analyzed, showing that the system has reached the correct size estimation. The measured length is close to the actual ground level with only minor differences.

The system demonstrates robustness when dealing with different objects, sizes and orientations. While there are limitations to accurately measuring objects with bad shape or irregular boundaries, more research and advances in algorithm development could solve these problems. The Performance metrics of the Dimensional Measurement System show satisfactory completion times for each image. The system reports its availability in real time or near real time. It also provides consistent and repeatable results by demonstrating the consistency of measurements across multiple experiments.

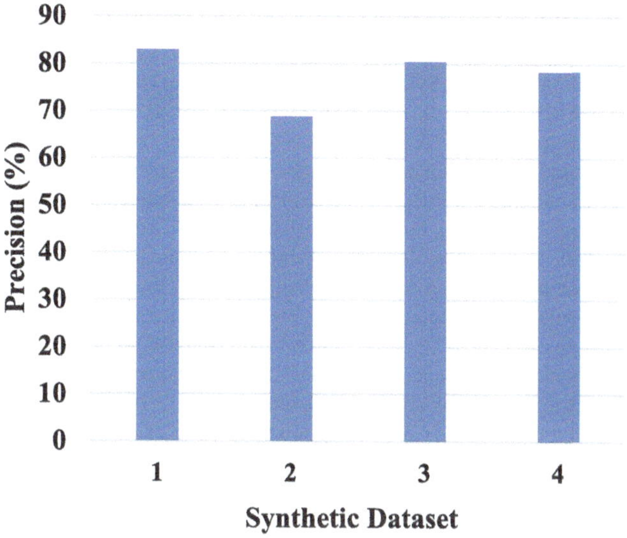


**Fig. 4.** Image edge detection

Process planning has important implications for many industries, including manufacturing, logistics and robotics. It provides quality control, compliance testing, resource allocation and supply chain management. The system automates the length measurement process, reducing human intervention and increasing productivity.

In conclusion, the dimensional measurement system using OpenCV offers a practical and effective solution for dimension measurement of certain products. It leverages the power of computer vision and the power of OpenCV to accurately measure and measure object dimensions.

Although there is room for improvement in image processing and optimization, the experimental results and analysis provide a solid foundation for further research and development. Future work may focus on addressing the limitations identified in this study and exploring algorithms and strategies to improve the accuracy and robustness of the system. Also, network integration and hardware acceleration can improve performance.

Overall, a dimensional measurement system using OpenCV has the potential to streamline the dimensional measurement process across industries, making product management, sales and automation more efficient and reliable. It is a cornerstone for advances in computer vision and has contributed to the field of sensing and measurement.

## References

1. Sandya, V., Baligeri, V., Lal, B., Petli, V., Kumar, P.: Deep learning based brain tumor detection with internet of things. In: 2023 IEEE International Conference on Integrated Circuits and Communication Systems (ICICACS), Raichur, India, pp. 1–6 (2023). https://doi.org/10.1109/ICICACS57338.2023.10100253

2. Konda, S., Kar, N.K., Pulicherla, P., Shivakanth, G., AC R: Cardio-vascular disease prediction using machine learning techniques. In: 2023 IEEE International Conference on Integrated Circuits and Communication Systems (ICICACS), Raichur, India, pp. 1–8 (2023). https://doi.org/10.1109/ICICACS57338.2023.10099769
3. Balmuri, K.R., Sujatha, J., Lal, A.B., Kumari, S.C.: An energy efficient and reliable strategies for intra-cluster and inter-cluster communications in wireless sensor networks. In: 2023 IEEE International Conference on Integrated Circuits and Communication Systems (ICICACS), Raichur, India, pp. 1–8 (2023). https://doi.org/10.1109/ICICACS57338.2023.10099644
4. Malsoru, V., Patil, A., Kusuma, S.M., Shivakanth, G., Samanvita, N.: An efficient SFLA and CUCKOO search hybridization for source distribution in cloud computing. In: 2023 IEEE International Conference on Integrated Circuits and Communication Systems (ICICACS), Raichur, India, pp. 01–05 (2023). https://doi.org/10.1109/ICICACS57338.2023.10100133
5. Shylaja, S.L., Jyoti, B.K., Kumar, S., Kumar, V.S.: Application of convolutional neural network techniques to the classification of lung illnesses. In: 2023 IEEE International Conference on Integrated Circuits and Communication Systems (ICICACS), Raichur, India, pp. 1–7 (2023). https://doi.org/10.1109/ICICACS57338.2023.10099848
6. Kumar, M., Veena, R.C., Shivakanth, G., Rai, K.B.: Stock price prediction using LSTM and TLBO. In: 2023 IEEE International Conference on Integrated Circuits and Communication Systems (ICICACS), Raichur, India, pp. 1–5 (2023). https://doi.org/10.1109/ICICACS57338.2023.10100074
7. Banu, R., Ahammed, G.F.A., Divya, G., Reddy, V.D., Bhaskar, N., Kanthi, M.: Sentiment analysis for real-time micro blogs using Twitter data. In: 2023 2nd International Conference for Innovation in Technology (INOCON), Bangalore, India, pp. 1–6 (2023). https://doi.org/10.1109/INOCON57975.2023.10101366
8. Aparna, Y., Somasekhar, G., Bhaskar, N., Raju, K.S., Divya, G., Madhavi, K.R.: Analytical approach for soil and land classification using image processing with deep learning. In: 2023 2nd International Conference for Innovation in Technology (INOCON), Bangalore, India, pp. 1–6 (2023). https://doi.org/10.1109/INOCON57975.2023.10101169
9. Sundaramoorthy, M.G.K., Maheswari, K., Sajana, T., Shiba, M., Kumar, A.: The prediction of micro plasma impacts of farm fresh vegetables using machine learning. In: 2023 International Conference on Artificial Intelligence and Knowledge Discovery in Concurrent Engineering (ICECONF), Chennai, India, pp. 1–6 (2023). https://doi.org/10.1109/ICECONF57129.2023.10083906
10. Avanija, J., et al.: Interpretation of brain tumour using deep learning model. In: Reddy, K.A., Devi, B.R., George, B., Raju, K.S., Sellathurai, M. (eds.) Proceedings of Fourth International Conference on Computer and Communication Technologies. Lecture Notes in Networks and Systems, vol. 606. Springer, Singapore (2023). https://doi.org/10.1007/978-981-19-8563-8_33
11. Reimann, M.D., Sarkis, J.: An architecture for integrated automated quality control. J. Manuf. Syst. **12**(4), 341–355 (1993)
12. Brahmbhatt, S., Brahmbhatt, S.: Shapes in images. In: Practical OpenCV, pp. 67–93 (2013)
13. Sudhakara, M., Janaki Meena, M., Reddy Madhavi, K., Anjaiah, P.: Fish classification using deep learning on small scale and low-quality images. Int. J. Intell. Syst. Appl. Eng. **10**(1s), 279–288 (2022)
14. Kuraparthi, S., Reddy, M.K., Sujatha, C.N., Valiveti, H., Duggineni, C., Kollati, M., Kora, P.: Brain tumor classification of MRI images using deep convolutional neural network. Traitement du Signal. **38**(4), 1 (2021)
15. Kora, P., Rajani, A., Chinnaiah, M.C., Madhavi, R., Swaraja, K., Kollati, M.: EEG-Based Brain-Electric Activity Detection During Meditation Using Spectral Estimation Techniques, pp. 687–693. Springer, Singapore (2021). https://doi.org/10.1007/978-981-16-1941-0_68

16. Abbagalla, S., Rupa Devi, B., Anjaiah, P., Reddy Madhavi, K.: Analysis of COVID-19-impacted zone using machine learning algorithms. In: Springer Series—Lecture Notes on Data Engineering and Communication Technology, vol. 63, pp. 621–627. Springer, Singapore (2021)
17. Madhavi, K.R., Nawi, M.N.M., Bhaskar Reddy, B., Baboji, K., Kishore, K.H., Manikanthan, S.V.: Energy efficient target tracking in wireless sensor network using PF-SVM (particle filter-support vector machine) technique. Measurement: Sensors. **26**, 100667 (2023)
18. Avanija, J., Kumar, K.E., Ch Usha Kumari, G., Jyothi, N., Raju, K.S., Madhavi, K.R.: Enhancing network forensic and deep learning mechanism for internet of things networks. J. Sci. Indust. Res. **82**, 522 (2023)

# Flower Segmentation and Classification Using U-Net + ResNeXt Deep Model

K. Reddy Madhavi[1(✉)], P. Sunitha[2], Naresh Tangudu[3], Vemuri Sailaja[4], Madhavi Gudavalli[5], and K. Mahesh Kumar[6]

[1] School of Computing, Mohan Babu University, Tirupati, A.P, India
[2] Department of Electronics and Communication Engineering, Aditya Engineering College, Surampalem, India
[3] Aditya Institute of Technology and Management, Tekkali, India
[4] ECE, Pragati Engineering College, Surampalem, India
sailaja.v@pragati.ac.in
[5] Information Technology, JNTUK, Kakinada, India
[6] CSE, Mahatma Gandhi Institute of Technology, Hyderabad 500075, India

**Abstract.** Flower segmentation and classification is a significant task in computer vision. It holds immense potential across multiple domains such as botany, agriculture, and horticulture. This paper addresses the problem of accurate and efficient flower recognition, highlighting its relevance and importance. A novel two-stage deep learning model U-Net + ResNeXt is proposed that integrates U-Net for flower segmentation and ResNeXt for classification. The proposed model aims to enhance fine-grained classification accuracy while effectively handling complex and cluttered floral images. In the experimentation, the U-Net + ResNeXt models performance is evaluated on Oxford 102 Flowers dataset, comparing it against single-stage ene-to-end classification models ResNeXt, VGG16, DenseNet and Custom CNN. The results demonstrate the superior capabilities of two-stage U-Net + ResNeXt approach, showcasing the importance of combining segmentation and classification for accurate flower recognition. The proposed model achieves considerable accuracy, precision, recall, and F1-Score, addressing the challenges of fine-grained classification and imbalanced datasets. The findings presented in this paper underscore the significance of advancing flower recognition techniques, contributing to applications in agriculture, botany, and environmental conservation.

**Keywords:** Smart Agriculture · Oxford 102 flower dataset · Two-stage deep model · U-Net · ResNeXt

## 1 Introduction

The task of flower segmentation and classification holds profound significance in the domain of computer vision. It constitutes a pivotal area of study that bridges the gap between artificial intelligence and the natural world. It offers an array of applications with tangible benefits for various domains, including agriculture, botany, and horticulture [1–3].


A. Bajaj et al. (Eds.): HIS 2023, LNNS 1224, pp. 442–449, 2025.
https://doi.org/10.1007/978-3-031-78925-0_44

Flowers, with their enchanting diversity and inherent beauty, are not only a source of aesthetic pleasure but also pivotal components of our ecosystem. In the domain of botany, precise flower species identification is fundamental for cataloging biodiversity, studying plant evolution, and deciphering ecological interactions. It aids in the conservation of endangered species and the preservation of natural habitats. Flower classification is essential for farmers and agronomists. It assists in crop management, as certain flower species can be indicative of specific plant diseases, pests, or environmental conditions [4]. Accurate flower segmentation can also help in monitoring crop health and yield prediction.

While the significance of flower segmentation and classification is undeniable, it is not without its challenges. Flowers exhibit intricate patterns, intricate shapes, and subtle variations that often require a fine-grained approach to classification. Traditional methods of manual classification are time-consuming and prone to errors [5]. In such scenarios, automated computer vision systems, equipped with deep learning models, can revolutionize the domain.

The primary research objective is to propose and evaluate a novel two-stage deep learning model that effectively combines U-Net for flower segmentation and ResNeXt for flower classification [6, 7].

## 2 Literature Review

Deep learning in agriculture enhances crop monitoring and management, enabling precision agriculture practices that optimize yields and resource utilization while reducing environmental impact [8–10]. Early research in flower segmentation and classification predominantly relied on conventional image processing techniques. These methods involved color-based thresholding, edge detection, and morphological operations. While they achieved moderate success in controlled environments, they often struggled to handle the complexity and variability present in real-world images of flowers.

With the advent of machine learning, particularly supervised learning, there was a shift towards more data-driven approaches for flower segmentation and classification [11]. These studies leveraged handcrafted features, such as color histograms, texture descriptors, and shape features, to train classifiers like SVM and Random Forests. While these approaches demonstrated improved accuracy, they still faced challenges with fine-grained classification, as they relied heavily on manual feature engineering.

The recent surge in deep learning techniques, especially CNNs, has revolutionized flower segmentation and classification tasks [12–14]. Deep learning models excel at automatically learning complex hierarchical features from raw image data [15].

Several notable studies in the field of flower segmentation and classification have contributed significantly to the current state of research. For instance, researchers have explored the development of datasets like the "Oxford 102 Flowers" dataset, which has become a benchmark for evaluating flower classification models [16, 17]. Other studies have proposed novel architectures, including the use of attention mechanisms and graph-based models, to enhance fine-grained classification.

# 3 U-Net + ResNeXt Deep Model for Flower Segmentation and Classification

The proposed architecture is two stage model. The first stage segments the regions of interest within the input image, while the second stage classifies the segmented flower region into specific flower classes. This two-stage deep learning model uses U-Net architecture for image segmentation with ResNeXt architecture for flower classification. This two-stage architecture combines the power of image segmentation with deep feature learning and classification for accurate flower classification.

## 3.1 First Stage: Image Segmentation

U-Net is a popular architecture for image segmentation tasks [6]. Its encoder-decoder structure is designed to capture fine-grained details in the segmented regions.

**Input Layer:** This is the initial layer that takes the input image as-is. Each flower image is characterized by its height, width, and color channels (e.g., RGB channels).

**Encoder:** It comprises of convolutional layers (two) which perform feature extraction. They use 3x3 convolutional filters to detect basic patterns and edges in the input image. They reduce spatial dimensions using convolutional operations with ReLU activation. Feature maps of each convolutional layer are downsampled (by max pooling). Thus, reducing the spatial resolution while preserving important features.

**Mid-Level Convolutions (Bridge):** This module comprises of two convolutional layers and a max pooling layer. Convolutional layers 3 & 4 continue feature extraction, capturing more complex patterns and fine details within the segmented regions. They use 3x3 convolutional filters with ReLU activation. Max pooling layer further reduces spatial dimensions to prepare for the decoding phase.

**Decoder:** This module comprises of transposed convolutional layers (two), concatenator, convolutional layers (two), and pooling layer (one maximum). First transposed convolutional layer serves the purpose of recovering spatial information lost during downsampling. It does so by upsampling feature maps. Spatial resolution of the feature maps is gradually increased to match the original image size. It helps recover spatial information lost during the downsampling. The concatenator performs concatenation of feature maps generated by the encoder with upsampled feature maps from bridge. This process incorporates context from earlier layers. Convolutional layers 5 & 6 layers refine and enhance features within the segmented regions. Transposed convolutional layer 2 upsamples feature maps to match the original image dimensions.

**Output Layer (Binary Segmentation):** The final layer applies the sigmoid activation function to produce a binary mask. Each pixel in the mask indicates the likelihood of belonging to the segmented object. This mask identifies the segmented regions of interest within the input image.

### 3.2 Second Stage: Classification on Segmented Regions

ResNeXt is a powerful deep learning architecture that has been successful in a wide range of image classification tasks [7]. When used in the second stage, it can effectively learn and classify features extracted from the segmented regions generated by the first stage (image segmentation). The proposed model uses ResNeXt-50.

**Input Layer:** This layer receives the segmented flower region obtained from the first stage. The input size matches the segmented region's dimensions (Fig. 1).

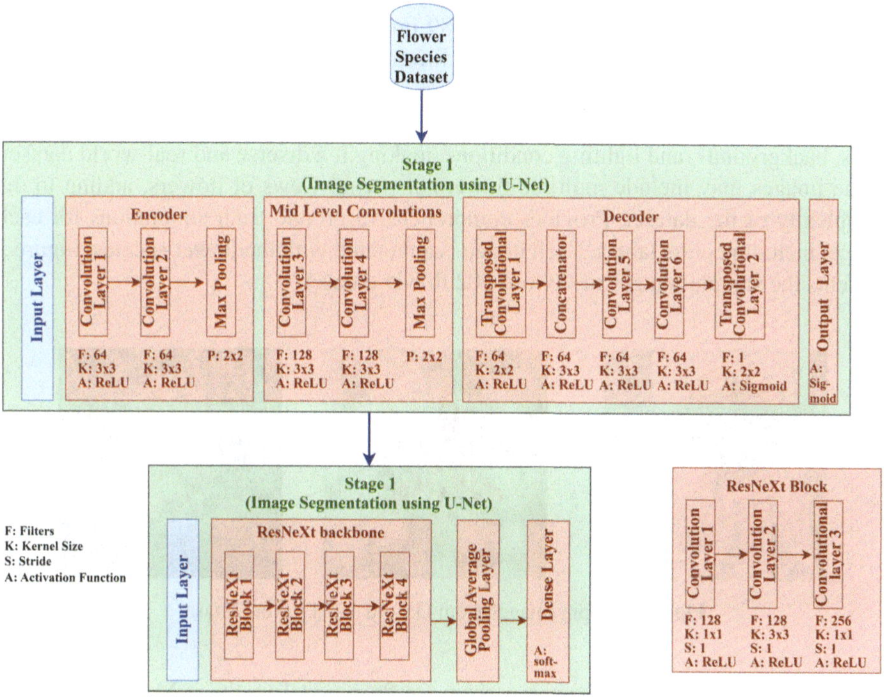


**Fig. 1.** U-Net + ResNeXt-50 Deep Model for Flower Segmentation and Classification

**ResNeXt-50 Backbone:** It consists of multiple blocks. Each block operates with convolutional layers with batch normalization and ReLU activation. These blocks capture hierarchical features from the segmented region. The exact number of blocks can be adjusted based on the model's depth and complexity requirements.

**Global Average Pooling Layer:** It computes global feature vector across the spatial dimensions of each feature map. It summarizes the most important information from the segmented flower region.

**Fully Connected (Dense) Layer:** The number of neurons in this layer corresponds to the number of flower classes you aim to classify. Each neuron represents a specific flower class. The softmax activation function calculates class probabilities. It converts the model's output into a flower species probability value, allowing the model to classify the segmented flower region into one of the predefined classes.

## 4 Experimental Results and Evaluation

A comprehensive evaluation of five deep learning models is conducted on the "Oxford 102 Flowers" dataset [16, 17]. The dataset was created by researchers at the University of Oxford's Department of Engineering Science, specifically the Visual Geometry Group. It was introduced to support research in fine-grained image classification and object recognition. It contains total of 8,189 images of flowers. These images belong to 102 different categories or classes, each representing a distinct species of flower. The images in the dataset are high-resolution color images, typically with dimensions ranging from 200x200 pixels to 800x800 pixels. They capture flowers in a variety of settings, backgrounds, and lighting conditions, making it a diverse and real-world dataset. Some images may include multiple flowers or partial views of flowers, adding to the complexity of the dataset. Provides comprehensive ground truth annotations for each image, including class labels. Each image is annotated with the flower species. Figure 2 presents the sample images of Oxford 102 flower dataset.

**Fig. 2.** Sample Images from Oxford 102 Flower Dataset

The models included in the comparison are Proposed Two-Stage Model (U-Net for Segmentation + ResNeXt for Classification), ResNeXt model for end-to-end classification, VGG16 model for end-to-end classification, DenseNet model for end-to-end classification, Custom CNN Model for model for end-to-end classification. We employed a range of performance metrics to assess the model's classification capabilities. Accuracy measures the overall correctness of flower classification. Precision evaluates the precision of classifying specific flower categories. Recall assesses model's ability to classify all images of a particular flower species. F1-Score is harmonic mean of precision and recall, offering a balanced metric for classification performance (Table 1).

The proposed two-stage model achieved the highest accuracy at 95.3%. This outcome indicates that the segmentation followed by classification strategy is highly effective for flower recognition. The ResNeXt model achieved an accuracy of 92.1%, showcasing its robust performance in an end-to-end classification setup. DenseNet achieved an accuracy of 93.7%, highlighting its suitability for this task. The Custom CNN Model achieved an accuracy of 91.8%, demonstrating competitive performance. The proposed two-stage

**Table 1.** Performance U-Net + ResNeXt on Oxford 102 Flower Dataset

| Model | Accuracy | Precision | Recall | F1-Score |
|---|---|---|---|---|
| Custom CNN | 91.8 | 91.2 | 92.3 | 91.7 |
| DenseNet | 93.7 | 93.3 | 94.0 | 93.6 |
| VGG16 | 90.5 | 90.0 | 91.0 | 90.5 |
| ResNeXt-50 | 92.1 | 91.5 | 92.7 | 92.1 |
| Proposed U-Net + ResNeXt-50 | 95.3 | 94.8 | 95.6 | 95.2 |

model exhibited the highest precision at 94.8%, suggesting its efficiency in minimizing false positives. ResNeXt achieved a precision of 91.5%, while DenseNet achieved 93.3%, indicating their ability to accurately classify specific flower categories. VGG16 and the Custom CNN Model achieved precision values of 90.0% and 91.2%, respectively. The proposed two-stage model demonstrated the highest recall at 95.6%, signifying its capacity to correctly identify most actual flower instances. ResNeXt achieved a recall of 92.7%, while DenseNet achieved 94.0%, demonstrating their effectiveness in capturing true positives. VGG16 and the Custom CNN Model achieved recall values of 91.0% and 92.3%, respectively. The F1-Score, which balances precision and recall, was highest for the proposed two-stage model at 95.2%, indicating its overall strong performance. ResNeXt achieved an F1-Score of 92.1%, while DenseNet achieved 93.6%, highlighting their balanced classification capabilities. VGG16 and the Custom CNN Model achieved F1-Score values of 90.5% and 91.7%, respectively.

The proposed two-stage model's enhanced performance is attributed to its segmentation-guided classification approach. The segmentation stage refines input regions, allowing the subsequent classification stage to focus on accurate recognition. ResNeXt and DenseNet models also performed admirably in end-to-end classification, showcasing their adaptability to fine-grained tasks like flower recognition. VGG16 and the Custom CNN Model, while delivering competitive results, lagged slightly behind the top-performing models. This indicates that their architectures might benefit from additional fine-tuning or more complex designs for enhanced performance.

The experimental results affirm the superiority of the proposed two-stage model, which combines segmentation and classification, achieving highest accuracy, precision, recall, and F1-Score. This model's segmentation stage enhances its classification capabilities, providing a robust solution for flower recognition tasks.

## 5 Conclusion

In this study, the significant problem of flower segmentation and classification is addressed, emphasizing its relevance and importance in various domains. The proposed two-stage U-Net + ResNeXt deep learning model, combining U-Net for segmentation and ResNeXt for classification, has demonstrated considerable capabilities in accurately identifying and classifying flowers. The experimental results on the Oxford 102 Flowers dataset underscore the effectiveness of the proposed model. Proposed model achieved considerable accuracy, precision, recall, and F1-Score, surpassing other single-stage end-to-end classification models. This highlights the significance of our research in fine-grained flower classification and the handling of imbalanced datasets. The proposed model has practical applications in fields such as botany, agriculture, and environmental conservation, where precise flower recognition plays a crucial role. It streamlines the process of cataloging biodiversity, monitoring crop health, and preserving natural habitats. By combining the power of segmentation and classification in two-stage architecture, proposed model has demonstrated the potential to further the way of perceiving and interacting with the natural world.

## References

1. Karthikeyan, C., Sunitha, G., Avanija, J., Reddy Madhavi, K., Madhan, E.S.: Prediction of climate change using SVM and naïve bayes machine learning algorithms. Turk. J. Comput. Math. Edu. **12**(2), 2134–2139 (2021)
2. Sunitha, G., Sudeepthi, A., Sreedhar, B., Shaik, A.B., Farooq, C.: RetinaNet and vision transformer-based model for wheat head detection. In: 2023 5th International Conference on Inventive Research in Computing Applications (ICIRCA), pp. 151–156. IEEE (2023, August)
3. Nennuri, R., Kumar, R.H., Prathyusha, G., Tejaswini, K., Kanishka, G., Sunitha, G.: A multi-stage deep model for crop variety and disease prediction. In: International Conference on Soft Computing and Pattern Recognition, pp. 52–59. Springer Nature Switzerland, Cham (2022)
4. Liu, W., Han, H., Han, G.: Transfer learning with deep convolutional neural network for automated plant identification. In: 2022 7th International Conference on Image, Vision and Computing (ICIVC), pp. 555–560. IEEE (2022, July)
5. Shereesha, M., Hemavathy, C., Teja, H., Reddy, G.M., Kumar, B.V., Sunitha, G.: Precision mango farming: using compact convolutional transformer for disease detection. In: International Conference on Innovations in Bio-Inspired Computing and Applications, pp. 458–465. Springer Nature Switzerland, Cham (2022)
6. Ronneberger, O., Fischer, P., Brox, T.: U-net: convolutional networks for biomedical image segmentation. In: Medical Image Computing and Computer-Assisted Intervention–MICCAI 2015: 18th International Conference, Munich, Germany, October 5-9, 2015Proceedings, Part III 18, pp. 234–241. Springer International Publishing (2015)
7. Xie, S., Girshick, R., Dollár, P., Tu, Z., He, K.: Aggregated residual transformations for deep neural networks. In: Proceedings of the IEEE conference on computer vision and pattern recognition, pp. 1492–1500 (2017)
8. Balakrishna, N., Sunitha, G., Karthik, A., Madhavi, K.R., Avanija, J.: Tomato leaf disease detection using Deep learning: a CNN approach. In: 2022 International Conference on Data Science, Agents & Artificial Intelligence (ICDSAAI), vol. 1, pp. 1–6. IEEE (2022, December)
9. Sunitha, G., Pushpalatha, M.N., Parkavi, A., Boyapati, P., Walia, R., Kohar, R., Qureshi, K.: Modeling of chaotic political optimizer for crop yield prediction. Intell. Automat. Soft Comput. **34**(1), 423–437 (2022)
10. Sunitha, G., Madhavi, K.R., Avanija, J., Reddy, S.T.K., Vittal, R.H.S.: Modeling convolutional neural network for detection of plant leaf spot diseases. In: 2022 3rd International Conference on Electronics and Sustainable Communication Systems (ICESC), pp. 1187–1192. IEEE (2022, August)
11. Antonio, B., Moroni, D., Martinelli, M.: Efficient adaptive ensembling for image classification. Expert. Syst. **42**, e13424 (2022)

12. Xia, X., Xu, C., Nan, B.: Inception-v3 for flower classification. In: 2017 2nd International Conference on Image, Vision and Computing (ICIVC), pp. 783–787. IEEE (2017)
13. Solanki, A., Singh, T.: Flower species detection system using deep convolutional neural networks. In: Emerging Technologies for Computing, Communication and Smart Cities: Proceedings of ETCCS 2021, pp. 217–231. Springer Nature Singapore, Singapore (2022)
14. Lin, D., Cheng, C.: Research on flower image recognition algorithm. In: 2022 International Conference on Big Data, Information and Computer Network (BDICN), pp. 646–649. IEEE (2022, January)
15. Pandey, S., Sindhuja, B., Nagamanjularani, C.S., Nagarajan, S.: Exploring transfer learning techniques for flower recognition using CNN. In: Data Science and Security: Proceedings of IDSCS 2022, pp. 393–401. Springer Nature Singapore, Singapore (2022)
16. Visual Geometry Group, Oxford University, https://www.robots.ox.ac.uk/~vgg/data/flowers/102/
17. Nilsback, M.E., Zisserman, A.: Automated flower classification over a large number of classes. In: 2008 Sixth Indian conference on computer vision, graphics & image processing, pp. 722–729. IEEE (2008)

# Vehicle Type Detection Using Faster R-CNN

P. Leela[1], P. Megana Santhoshi[2], K. Ayyappa Swamy[3], K. Reddy Madhavi[4(✉)], Naresh Tangudu[5], and K. Mahesh Kumar[6]

[1] Department of Data Science, School of Computing, Mohan Babu University, Tirupati, Andhra Pradesh, India
[2] KLM College of Engineering for Women, Kadapa, India
[3] Department of Electronics and Communication Engineering, Aditya Engineering College, Surampalem, India
[4] School of Computing, Mohan Babu University, Tirupati, A.P., India
kreddymadhavi@gmail.com
[5] Aditya Institute of Technology and Management, Tekkali, India
[6] CSE, Mahatma Gandhi Institute of Technology, Hyderabad, India

**Abstract.** Autonomous vehicles need to navigate through complex traffic scenarios, which often involve various vehicle types, such as cars, buses, and motorcycles. Identifying the types of vehicles in proximity aids in planning appropriate traffic maneuvers, such as lane changes, merging onto highways, or yielding to emergency vehicles. The importance of vehicle type detection touches numerous aspects of modern society, including transportation, safety, and environmental concerns. This paper explores the domain of vehicle type detection within the context of deep learning and object detection. A Faster R-CNN-based deep architecture is proposed. The proposed deep model is configured to tailor the task of vehicle type detection from images. Beijing Institute of Technology (BIT) vehicle dataset is used. Experiments were conducted to evaluated proposed Faster R-CNN-based model against other models SVM, SSD, and a Simple CNN. The evaluation encompassed critical metrics, including mAP, F1-Score, ROC curves, and Average IOU. The results of experiments decisively establish the superiority of proposed model over baseline methods.

**Keywords:** Road safety · Autonomous driving · Faster R-CNN · Vehicle type detection · Computer vision

## 1 Introduction

From personal cars and public transportation to commercial trucks and bicycles, there are multitude of vehicle types on roadways. Smart transportation plays a central role in smart cities. They contribute to dynamics of our cities and societies. The ability to accurately detect and classify these vehicles holds profound significance, spanning a wide range of applications [1]. For example, traffic management, urban planning, autonomous vehicles, surveillance systems and others. Vehicle type detection, has become a critical area of research and development. This research area is fueled by advancements in deep learning techniques and the availability of vast datasets [2].


A. Bajaj et al. (Eds.): HIS 2023, LNNS 1224, pp. 450–456, 2025.
https://doi.org/10.1007/978-3-031-78925-0_45

The importance of vehicle type detection touches numerous aspects of modern society, including transportation, safety, and environmental concerns. The recent surge in the capability of deep learning models and computer vision techniques has revolutionized the field of vehicle type detection. Deep learning empowers algorithms to learn intricate patterns and representations from data [3]. This makes it exceptionally well-suited for complex visual tasks like object detection and classification [4]. Computer vision, on the other hand, provides foundation for processing and understanding images and videos [5].

In the context of vehicle type detection, deep models can discern the unique features and characteristics that differentiate cars from trucks, motorcycles from bicycles, and more. The application of deep learning to vehicle type detection has significantly improved accuracy and robustness, enabling real-world deployment in critical scenarios [6].

Furthermore, availability of large-scale, annotated datasets has been instrumental in training and fine-tuning deep models for vehicle type detection [7]. Datasets containing millions of images encompassing a wide array of vehicle types and environmental conditions have enabled deep models to generalize and excel in real-world situations. These datasets have bridged gap between research and practical implementation, making vehicle type detection systems increasingly reliable and accessible.

## 2 Literature Review

Vehicle type detection stands as a pivotal challenge with far-reaching implications for transportation, safety, environmental sustainability, law enforcement, and urban planning [8]. Accuracy and efficiency of vehicle type detection systems have grown by leaps and bounds due to improved hardware capabilities, and the availability of large-scale labeled datasets. Thus, positioning them at the forefront of solutions to address the complex challenges presented by modern transportation systems. Delving deeper into this critical area of research and application, a multitude of opportunities can be uncovered to enhance the safety, efficiency, and sustainability of evolving transportation landscape.

Different deep learning CNN models for vehicle detection and classification in aerial imagery have been proposed [9, 10]. Performance of YOLO, Light weight CNN, SSD, etc. have been evaluated on challenging datasets. Their findings reveal the strengths and weaknesses of each model in the context of aerial vehicle type detection, offering valuable insights for remote sensing applications [11]. YOLO's single-stage architecture offers impressive speed while maintaining reasonable accuracy, making it suitable for applications where real-time processing is crucial. But YOLO tends to fail when high accuracy is required.

Critical investigations have been made on low-resolution data, congested and dense traffic data [12]. They focus on real-world traffic scenes, where vehicles may vary greatly in size, orientation, and lighting conditions [13]. Detecting corner cases at object-level is an impeding scenario in developing self-driving systems [14].

Vehicle type detection and recognition in unconstrained road scenes is a challenge. Often multiple objects in images and small targets makes vehicle detection task, a challenging one. It needs more effort in addressing the challenges posed by variations in vehicle appearances, weather conditions, and road clutter [15]. Practical utility of vehicle type detection can be achieved by using low-altitude aerial images in traffic management and surveillance. The use of large and diverse datasets, novel model architectures, and real-time capabilities continues to drive advancements in this critical field, laying the groundwork for safer and more efficient transportation systems.

## 3 Proposed Model Architecture for Vehicle Type Detection

The proposed deep model is designed for the task of vehicle type detection from images (Fig. 1). It leverages the Faster R-CNN architecture [16, 17]. Faster R-CNN is a two-stage deep learning framework, to accurately locate vehicles in images and classify them into specific types, such as cars, trucks, or bicycles.

**Backbone Network (Feature Extraction).** It utilizes a pre-trained ResNet-50 to extract informative features from input images. The top layers of pre-trained ResNet-50 are removed, keeping the lower layers intact. These lower layers are responsible for extracting low-level and mid-level features from input images, such as edges, textures, and object parts.

**Region Proposal Network (RPN).** It comprises of two sibling networks - Anchor Box Layer (ABL) and Anchor Box Scales/Ratios (ABS). For anchor boxes, BL predicts objectness scores and bounding box regression offsets. ABS of various scales and aspect ratios is used to propose potential object regions.

**RoI (Region of Interest) Pooling.** RoI pooling is performed on RPN output resulting in fixed-size feature maps for each region proposal. These feature maps contain crucial information about the proposed regions.

**Object Classification Head.** It contains fully connected layers for object classification. It predicts probability of each region proposal belonging to a specific vehicle type. It uses softmax activation to make class predictions.

**Bounding Box Regression Head.** It contains fully connected layers for refining bounding box coordinates. It predicts adjustments to the bounding box coordinates, improving their accuracy.

**Loss Functions.** Objectness Loss (RPN): It computes binary cross-entropy loss for objectness scores.

Bounding Box Regression Loss (RPN): It uses smooth L1 loss for precise bounding box regression.

Classification Loss (Object Classification): It computes categorical cross-entropy loss for vehicle type classification.

**Non-Maximum Suppression (NMS).** It is applied to filter out redundant and overlapping region proposals. Thus, retaining only the most confident and distinct detections.

This architecture outlines Faster R-CNN-based deep model tailored for vehicle type detection. It takes an input image, generates region proposals using RPN, extracts features using RoI pooling, performs object classification, and regresses bounding box coordinates. The loss functions guide the training process, and NMS ensures the final detections are non-overlapping.

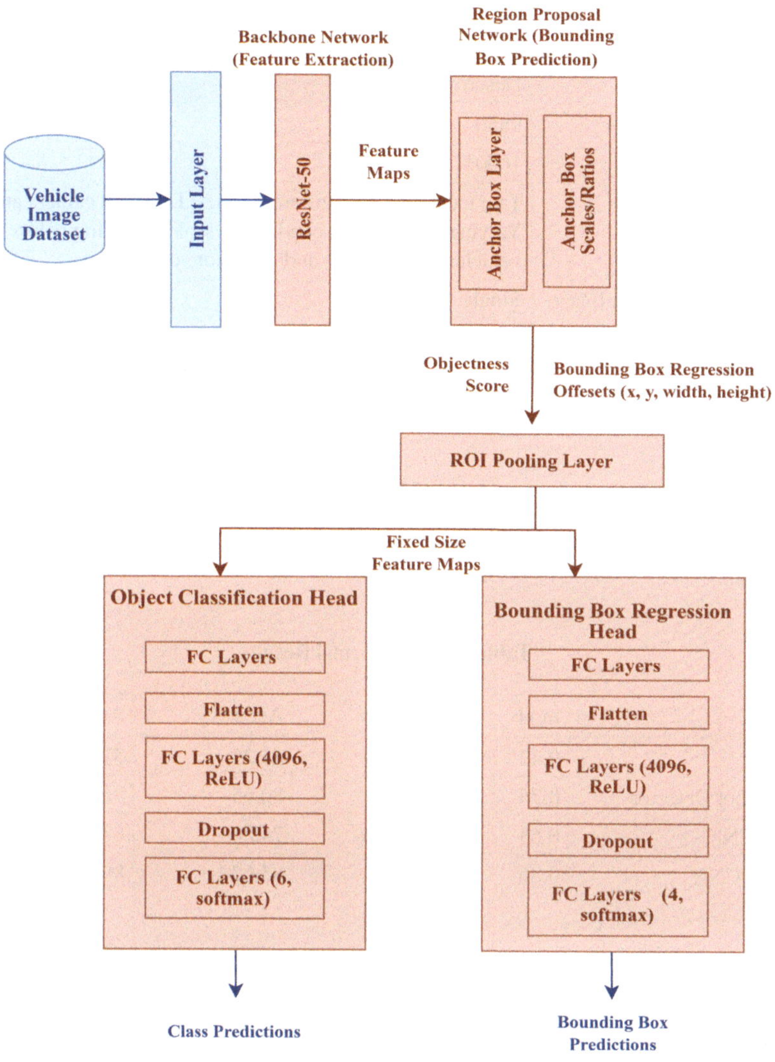


**Fig. 1.** Proposed Model

## 4 Experimental Results

BITVehicle dataset is used to experiment the proposed model [18]. The complete details are given in Table 1.

Proposed model builds upon the Faster R-CNN framework, utilizing a ResNet-50-based convolutional backbone for feature extraction. Customized region proposal, object classification, and bounding box regression heads were incorporated into the architecture. Stochastic Gradient Descent was employed with a learning rate of 0.001, momentum of 0.9, weight decay of 0.0005, batch size of 32 and conducted training for 50 epochs.

**Table 1.** BITVehicle Details

| Item | Description |
|---|---|
| No. of Images | 9850 |
| Image Size | 1600*1200 |
| Environment Variations | Two locations; Two different times; Different camera angles; Varying illumination conditions, scales; Varying colors of vehicles; Missing top and/or bottom of vehicles; |
| No. of Vehicles in each Image | Single, Multiple |
| Annotations | Yes (Bounding Boxes) |
| Class Labels | Bus (#558), Microbus (#883), Minivan (#476), Sedan (#5922), SUV (#1392) Truck (#822) |

Proposed model's performance is compared against three baseline models: SVM, Single-Shot Detector, and a Simple CNN. The evaluation is based on key metrics such as mAP, F1-score, average accuracy, average IOU (Table 2).

**Table 2.** Experimental Results

| Model | mAP | F1-Score | Accuracy | Average IOU |
|---|---|---|---|---|
| SVM | 0.42 | 0.48 | 55.23 | 57.9 |
| Single-Shot Detector | 0.75 | 0.80 | 80.65 | 75.3 |
| Simple CNN | 0.63 | 0.55 | 74.86 | 68.6 |
| Faster R-CNN | 0.86 | 0.89 | 85.62 | 80.5 |

The proposed model is trained as follows. Backbone is initialized with pre-trained weights. It is then fine-tuned on BIT vehicle dataset containing labeled images of vehicles. Hyper parameters, such as learning rates and batch sizes, are optimized for dataset characteristics. Anchor boxes with various scales and ratios are used to accommodate different vehicle sizes and shapes. The model, upon processing an image, generates bounding boxes around detected vehicles and classifies them into specific vehicle types with high accuracy. It is capable of handling multiple vehicles within a single image, making it suitable for real-world vehicle type detection applications. SVM classifier is implemented with handcrafted features, including HOG and color histograms for object detection.

A crucial metric for object detection, mAP calculates the precision-recall curve's area, considering various object classes. Proposed model achieved highest mAP of 0.86, indicating its ability to accurately locate and classify vehicle types within images. SVM-based approach achieved a considerably lower mAP of 0.42. This emphasizes the limitations of traditional handcrafted feature-based techniques, which struggle with complex object detection tasks. SSD, a contemporary approach, performed admirably but lagged behind Faster R-CNN model, attaining mAP of 0.75. This underscores the superiority of the two-stage architecture in handling intricate detection challenges effectively. With an F1-Score of 0.89, the Faster R-CNN model excelled in achieving a balance between precision and recall in vehicle type classification. In contrast, the SVM-based approach yielded an F1-Score of 0.48, highlighting the difficulty of achieving high accuracy and recall concurrently with traditional methods. SSD, while efficient, recorded an F1-Score of 0.80, reiterating proposed model's benefits of comprehensive feature extraction and classification. Faster R-CNN-based model attained an impressive Average IOU of 80.5, indicating considerable localization of vehicle types. In contrast, the baseline models, including SVM, SSD, and Simple CNN, lagged behind in key metrics, underscoring the limitations of traditional and simplistic approaches.

## 5 Conclusions

Vehicle type detection is a vital challenge in computer vision, with implications spanning traffic management, autonomous vehicles, security systems, environment sustainability, law enforcement, and urban planning. Deep learning now is capable to address the complex challenges presented by modern transportation systems. Leveraging deep learning techniques, a Faster R-CNN-based model is proposed for the task. The experimental results demonstrated the superiority of Faster R-CNN-based model over traditional baseline methods. Proposed model achieved a mAP of 0.86, for accurate vehicle type localization and classification. With an F1-Score of 0.89, it demonstrated an ideal balance between precision and recall. Average IOU of 80.5 indicated precise localization. In contrast, traditional models like SVM and a Simple CNN lagged significantly in terms of these metrics. This discussion underscores the transformative impact of deep learning in vehicle type detection, reaffirming the potential of the proposed model across various real-world applications. As technology evolves, the model's performance will continue to contribute to safer roads, efficient traffic management, and the advancement of autonomous vehicles.

## References


1. Maity, S., Bhattacharyya, A., Singh, P.K., Kumar, M., Sarkar, R.: Last decade in vehicle detection and classification: a comprehensive survey. Arch. Comput. Method. Eng. **29**(7), 5259–5296 (2022)
2. Charan, N.S., Thumma Narasimhulu, G., Bhanu Kiran, T., Sudharshan Reddy, T., Singh, S., Sunitha, G.: Solid waste management using deep learning. In: 14th International Conference on Soft Computing and Pattern, vol. 648, pp. 44–51. Springer (2023)
3. Reddy Madhavi, K., Jyothi Jarugula, G., Karuna, S.K., Srujan Raju, K., Sunitha, G.: Bird species recognition using deep transfer learning. In: 4th International Conference on Computer and Communication Technologies, vol. 606. Springer (2023)
4. Avanija, J., Sunitha, G., Madhavi, K.R., Chakravarthi, S.S., Vittal, R.H.S.: Prediction and analysis of cervical cancer: an ensemble approach. In: I2022 4th International Conference on Inventive Research in Computing Applications (ICIRCA), pp. 1734–1739. IEEE (2022)
5. Prabhakar, T., Sunitha, G., Gudavalli Madhavi, J.A., Madhavi, K.R.: Automatic detection of diabetic retinopathy in retinal images: a study of recent advances. Annal. Roman. Soc. Cell. Biol. **25**, 15277–15289 (2021)

6. Ammar, A., Koubaa, A., Ahmed, M., Saad, A., Benjdira, B.: Vehicle detection from aerial images using deep learning: a comparative study. Electronics. **10**(7), 820 (2021)
7. Xu, R., Xia, X., Li, J., Li, H., Zhang, S., Tu, Z., Meng, Z., Xiang, H., Dong, X., Song, R., Yu, H.: V2v4real: A real-world large-scale dataset for vehicle-to-vehicle cooperative perception. In: Proceedings of the IEEE/CVF Conference on Computer Vision and Pattern Recognition, pp. 13712–13722 (2023)
8. Wang, H., Yu, Y., Cai, Y., Chen, X., Chen, L., Liu, Q.: A comparative study of state-of-the-art deep learning algorithms for vehicle detection. IEEE Intell. Transport. Syst. Magazine. **11**(2), 82–95 (2019)
9. Ammar, A., Koubaa, A., Ahmed, M., Saad, A. and Benjdira, B.: Aerial images processing for car detection using convolutional neural networks: comparison between faster R-CNN and YOLOv3. arXiv preprint https://arxiv.org/arXiv:1910.07234 (2019)
10. Momin, M.A., Junos, M.H., Mohd Khairuddin, A.S., Abu Talip, M.S.: Lightweight CNN model: automated vehicle detection in aerial images. Sig. Image Video Proc. **17**(4), 1209–1217 (2023)
11. Taheri Tajar, A., Ramazani, A., Mansoorizadeh, M.: A lightweight Tiny-YOLOv3 vehicle detection approach. J. Real-Time Image Proc. **18**(6), 2389–2401 (2021)
12. Basak, S., Suresh, S.: Vehicle detection and type classification in low resolution congested traffic scenes using image super resolution. In: Multimedia Tools and Applications, pp. 1–23 (2023)
13. Yu, Y., Gu, T., Guan, H., Li, D., Jin, S.: Vehicle detection from high-resolution remote sensing imagery using convolutional capsule networks. IEEE Geosci. Remote Sens. Lett. **16**(12), 1894–1898 (2019)
14. Li, K., et al.: Coda: a real-world road corner case dataset for object detection in autonomous driving. In: European Conference on Computer Vision, pp. 406–423. Springer Nature Switzerland, Cham (2022)
15. Kenk, M.A., and Hassaballah, M.: DAWN: vehicle detection in adverse weather nature dataset. arXiv preprint arXiv:2008.05402. (2020)
16. Ren, S., He, K., Girshick, R., Sun, J.: Faster R-CNN: Towards real-time object detection with region proposal networks. Adv. Neural Inform. Process. Syst. **28**, 1137 (2015)
17. Girshick, R.: Fast r-cnn. In: Proceedings of the IEEE International Conference on Computer Vision, pp. 1440–1448 (2015)
18. Dong, Z., Wu, Y., Pei, M., Jia, Y.: Vehicle type classification using a semisupervised convolutional neural network. IEEE Trans. Intell. Transp. Syst. **16**(4), 2247–2256 (2015)

# A Feature Extraction Method for Deep Learning and HDR Image Reconstruction Optimization

Vankudothu Malsoru[1(✉)], Najeema Afrin[1], Madhavi Pingili[2], G. S. Sravanthi[2], B. Sujani[3], and R. Suhasini[3]

[1] CMR Technical Campus, Hyderabad, Telangana, India
[2] CMR Engineering College, Hyderabad, Telangana, India
sravanthi.gs@cmrec.ac.in
[3] CMR College of Engineering & Technology, Hyderabad, Telangana, India
{b.sujani,suhasini}@cmrcet.ac.in

**Abstract.** An image that closely mimics the real environment can be created by convert a low dynamic range (LDR) photograph to a high dynamic range (HDR) image with no the need for costly equipment. Accurate and sophisticated HDR photographs can now be produced thanks to recent developments in deep learning. In this study, we describe a Deep erudition method to recreate HDR photos using comparable real-world dynamic ranges by categorizing bright and dark portions of input LDR photographs. The suggested multi-level deep learning network combines features over a greater range of luminance to brighten bright spots and darken shadows for creating HDR photos. By separating the LDR image hooked on bright and dark regions, data regarding missing excessively exposed and underexposed areas is successfully incorporated. The outcome is a genuine HDR image containing colors and appearance that closely resembles what was extracted real features. It is thought that using the Firefly Optimization method will reduce the amount of information retrieved to produce HDR photos.

**Keywords:** LDR · HDR image · deep learning · CNN

## 1 Introduction

The need for methods the fact that can produce images without real-world resolution has increased in light of recent major developments in image processing. High Dynamic Range is the most employed image enhancing technology (HDR). HDR, on the other hand, can record an extensive dynamic range and operates beyond the capabilities of conventional camera sensors. Because of this, not many individuals are able to have to his HDR photos.

The LDR image was captured using several exposures. After that, combine the images to get an HDR picture. Since it is impossible to capture several pictures at once, stitching bracketed shots together can result in vanishing artifacts that appear to depict the passage of time.

Utilizing individual LDR photographs, an Inverse Tone Mapping (ITM) method was created to produce HDR images. On another hand, it is challenging to get parameters


A. Bajaj et al. (Eds.): HIS 2023, LNNS 1224, pp. 457–466, 2025.
https://doi.org/10.1007/978-3-031-78925-0_46

that match all image circumstances, and typical ITM approaches are insufficient to give functionality similar to the inverse camera return function (CRF). Deep learning approaches were used to overcome this challenge, leading to the development of a CNN-based ITM method and a convolutional neural network (CNN)-based MEF methodology. In order to recreate HDR pictures taken as of a single LDR image, CNN-based ITM algorithms were developed.

In an HDR photograph, the proportion of saturation pixels to total pixels is restricted. A loss function calculated using L1, L2, or mean squared error (MSE) is commonly used for image reconstruction. It adjusts the weights in order to minimize the global pixel value variance among the generate image and the ground truth rise. The network weights do not adjust as quickly as necessary to recover saturated pixels as a result.

The multilevel network layout that has been suggested increases the dynamic range and recovers saturated pixels uniformly. To improve the dynamic range, bright parts are brighten and dark areas are gloomy. The final HDR image is then produced by combining these elements. The HDR image's log2 brightness histogram as shown the Fig. 1. The green boxes indicate brightness extremes that the CRF is likely to clip.

**Fig. 1.** HDR Image

The firefly optimization technique is used to maximize the features of the dark and bright images that are not desired.

**Multiple Fusion Methods of Exposure**

Make an HDR image by combining photos from the bracket. Martens et al. employed a Gaussian pyramid and a Palladian pyramid to integrate complementary components in multi-exposure images. The literature reviews provide a detailed description of the MEF technique. The limitations of CNN-based MEF approaches have been addressed with the development of two revelation fusion (TEF) approaches which employ merely two images. A network that combines the brightness data of two LDR images that have different exposure durations is proposed by it [12].

## 2 Literature Review

An innovative deep neural network model that recreates a high dynamic range (HDR) image using a single low dynamic range (LDR) image is projected in paper [1]. The suggested approach infers LDR photos containing various exposes and illuminations using a solitary LDR photograph of the similar scene using a convolutional neural network containing dilated convolutional layers as its foundation. These inferences results are able to be combined to create the final HDR image. It suggests [2] a cutting-edge technique that uses a deep neural network to recover the lost range of motion using a single low dynamic range image. The conditioned generating adversarial network structure used in the suggested approach is the first system for producing high dynamic range photographs depending on the predicted multi-exposure stack.

These methods are unable to adaptably increase the dynamic ranges of photographs since they are trained on a large number of pre-fixed high dynamic range (HDR) images [3]. It suggests [4, 5] an unsupervised deep learning structure for MEF with a loss function based on a no-reference quality score. A unique CNN architecture that has been taught to be taught the fusion operation with no a reference position truth image is used in the suggested method. To provide results that are perceptually attractive without any artifacts, the model combines a set of common low level features that are taken from every image. It present excellent results for image-based illumination and compare to current techniques for HDR extension. Lastly, using an HDR display, we conduct a subjective experiment to assess the outcomes. This demonstrates that the reconstructed HDR photos are far better visually than those created using previous techniques [6]. The convolutional neural network (CNN) is projected to be used in this article [7] to train a SICE enhancer. Creating a training data set of image pairs with different contrasts low and high is a crucial problem for end-to-end CNN learning. A technique called ExpandNet, which is depending on deep Convolutional Neural Networks (CNNs), is presented in this paper [8] for producing HDR content from LDR content. ExpandNet is capable of accepting LDR images as input and producing end-to-end images with an extended range. In order to minimize error accumulation, we in cooperation fine-tune the whole model from start to finish. Through comprehensive quantitative and qualitative testing on several image datasets, we show that the suggested approach outperforms cutting-edge single-image HDR reconstruction techniques [9]. This publication [10, 11] provides an overview of the basic ideas underlying the perception of HDR pictures along with a thorough analysis of HDR techniques and technology. It gives an overview of the field and evaluates the state-of-the-art in HDR photography.

## 3 Proposed System

To produce HDR images by means of linear expansion and gamma correction. Yet, competent range growth is hard to do unless the input parameters are well stated. HDR images were produced by Bantered et al. using the median cut technique as shown the Fig. 2, an extended map, and a power function.

Nevertheless, all techniques need parameters for thickness assessment, and the amount of overexposed areas that can be recovered using HDR image inferences is limited.

The conventional ITM algorithm creates an HDR image by be appropriate a function to an LDR image. On the other hand, judgment a function that resembles the inverse CRF is extremely complicated. Several deep learning techniques have newly surfaced to address this issue. In order to produce HDR images, it is using an auto-encoder network

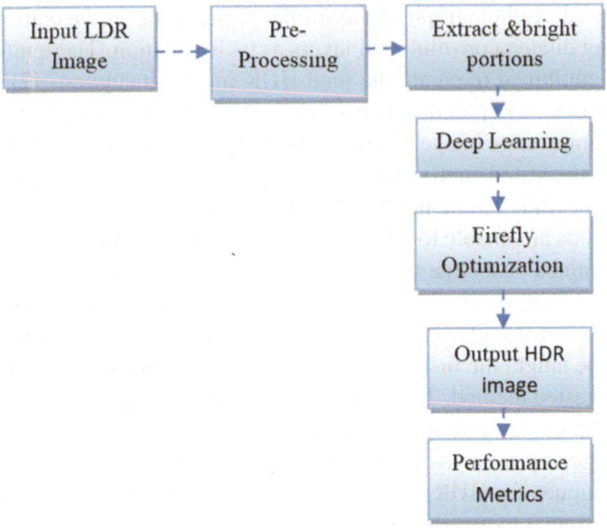


**Fig. 2.** Block Diagram of proposed model

by a loss function that included brightness and reflectivity. They produced HDR images by means of a greater dynamic range by combining images to transform network outputs and LDR images to linear ranges.

The majority of earlier CNN-based ITM techniques recreate HDR images by feeding the network LDR images with little to no pre-processing. Used pre-processing to segment the image, but did it solely using the frequency component. Split the image after learning, segmenting it right previous to the final HDR image be produced, rather than dividing it depending on the brightness value. When using raw LDR picture inputs, it becomes challenging to rebuild saturated pixels efficiently.

Throughout pre-processing, the Mask is created and the image sensitivity is computed. The suggested multi-stage network uses picture _Bright and Image _Dark as inputs after the LDR picture has been divided into both light and dark parts (Image _Bright and Image _Dark, correspondingly) utilizing the Mask. The Brightening block learns Image _Bright to create the enhanced brightness value, and the Darkening block learns Image _Dark to minimize the lower brightness value. Ultimately, the final HDR image is created by combining the two feature maps by improved dynamic ranges.

**Pre Processing**

The suggested network can thus efficiently target excessively subjected and under-exposed pixels when it receives input from bright regions and dark regions, respectively. Consequently, in order to improve the dynamic range, the input image is separated into bright and dark sections throughout pre-processing. Otsu presented the representative image segmentation technique, which uses a threshold value that offers a maximum pixel variance more than the image to divide the image into two groups. The threshold that optimizes the total of the variances among the two classes must be determined in order to define the distinctions among the two classes ($h$ and ). A threshold for additional processing is defined depending on the ostu model. The suggested Mask that is produced by applying the threshold may be stated as,

$$Mask\ (n, m) = \left( \frac{\max\left(0, v(n, m) - thr\right)}{(1 - thr)} \right)^2$$

wherein y_horizontal and y_vertical are the corresponding pixels, and y_b is the image brightness. The pixel whose brightness value just above the threshold produces an extremely tiny mask value because the square function is employed for Mask, which smooths the boundary surface during picture segmentation. After multiplying the produced Mask by the original LDR picture, it is divided into two segments: *ImgBrbht* and *ImgDark*.

$$Bright\ image : img\ bright = img\ x\ (mask) \tag{1}$$

$$Dark\ image : img\ dark = img\ x\ (1 - mask) \tag{2}$$

**Deep Learning**

The multi-stage CNN model is used for obtaining the features. The original LDR picture *ImbOrial*, as well as *ImbDark*, generated during pre-processing, are among the network inputs as shown the Fig. 3. When the dynamic range is expanded employing the Brightening block, brightness is increased. It is three stage models,

Stage 1.
Brightness within the bright range rises and dark range brightness falls with the use of the dynamic spectrum extension stage.
Stage 2.
Outliers are avoided by the global feature stringing stage.
Stage 3.
The learnt feature maps are combined in the process of blending and generation stage.

**Firefly Optimization**

The firefly optimization technique is used to improve the unwanted highlights of the brilliant picture and dark picture. These characteristics are enhanced to provide excellent results and advance the quality of the picture after the CNN technique's highlights are extracted. The firefly calculating approach determines how best to present the highlights. The detailed setup for the firefly optimization is discussed. At this time are the steps to execute the projected Firefly Algorithm (FA) as shown the Fig. 4.

**Algorithm**

Step 1: produce the earlypopulace of fireflies $\{x1, x2, \ldots.., xn\}$.
Step 2: compute the intensity value for every firefly.
Step 3: modernize the step of each firefly.

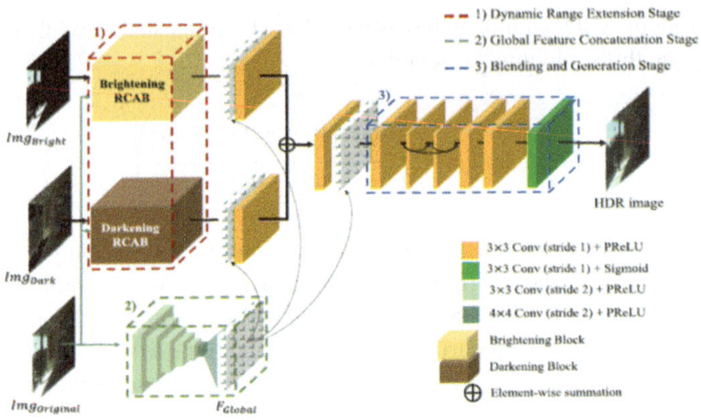


**Fig. 3.** Multi-stage deep learning network

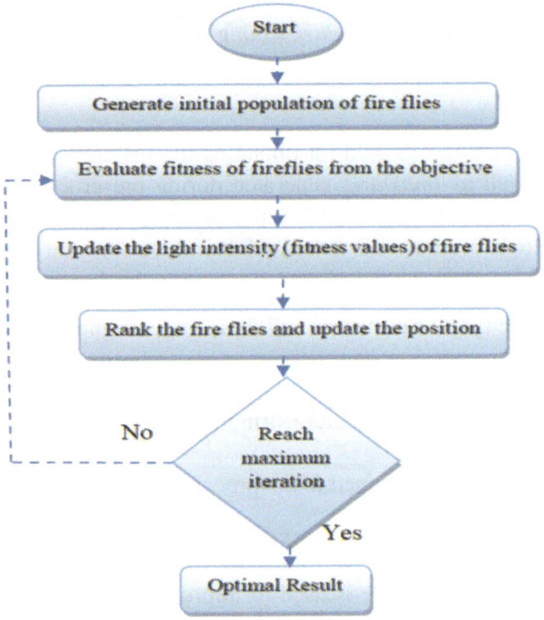


**Fig. 4.** Firefly flowchart

Step 4: Fireflies strength is specified as $\{I1, I2, \ldots., In\}$.
Step 5: Line up the fireflies and find the perfect moment
Step 6: shiftevery of the fireflies $I$ to the added brighter ones.
Step 7: Stop while a criterion is satisfied; or else go to Step 2

### 3.1 Parameters Used in Firefly Algorithm

#### 3.1.1 Distance

Any two firefly, i and j, in locations xi and xj, correspondingly, can have their distance determined.

as a Euclidean or Cartesian distance (rij) utilizing the formula

$$rij = \|x_i - x_j\| = \sqrt{\sum_{k=1}^{d} \left(x_{i,k} - x_{j,k}\right)^2} \quad (3)$$

while k is the kth element of the spatial coordinates xi of the ith insect while xi is the fly's luminosity.

#### 3.1.2 Attractiveness

Modify in magnetism $\boldsymbol{\beta}$ by means of distance $\boldsymbol{r}$ by

$$\boldsymbol{\beta} = \boldsymbol{\beta}\mathbf{o} \quad (4)$$

$\gamma$ is an attractive coefficient.
r is the distance among two fireflies.

#### 3.1.3 Movement

A firefly's journey to draw in another person more desirable (more vibrant) firefly j is dictated by

$$Xit + 1 = Xit + \ \beta_o e^{-\gamma rij2}\left(x_j^t - x_i^t\right) + \propto_t \sum\nolimits_i^t \quad (4)$$

where by attractiveness serves as the basis for the second term. By α serving as the randomized parameter and ε representing a vector of random values, the third term is generated.

## 4 Result Analysis

The following metrics are employed to assess how well the suggested methodology performs:

A. Peak Signal to Noise Ratio (PSNR)
B. Structural Similarity Index (SSIM)
C. Multi-Scale SSIM
D. HDR visual difference predictor (HDR-VDP)

$$PSNR = 10\,x\,\log_{10}\left(\frac{255x255}{MSE}\right) \quad (5)$$

### 4.1 Structural Similarity Index (SSIM)

One perception presentation that takes into account picture degradation as it appears in the base data is called SSIM. It also participates in required perceptual miracles, such as degree masking words. Brightness with the curtain's variation. In contrast to other processes like MSE or PSNR, these techniques assess explicit mistakes.

$$SSIM(x, y) = \frac{(2\mu_x\mu_y + c_1)(2\sigma_{xy} + c_2)}{\left(\mu_x^2 + \mu_y^2 + c_1\right)\left(\sigma_x^2 + \sigma_y^2 + c_2\right)} \tag{6}$$

### 4.2 Multi-scale SSIM

The representation of image brightness, distinction, and structure forms the foundation of the SSI estimation methodology (Wang and Bovik, 2002). An extension of the SSI framework, the MS-SSIM achieves better accuracy than the single-scale SSI technique, albeit at the expense of a somewhat slower processing speed.

### 4.3 HDR Visual Difference Predictor (HDR-VDP)

The entire spectrum of luminance that the human eye can see can be used by the High Dynamic Range Visual Variation Predictor (HDR-VDP). A High Dynamic Range (HDR) image or a traditional 8-bits-per-color image transformed to the actual luminance values might be used as the input for our metric.

The suggested approach is presented as follows. The Mask is made and the image threshold is deliberated throughout pre-processing. After that, the LDR images are divided into bright and dark regions (Img Bright and Img Dark, correspondingly)

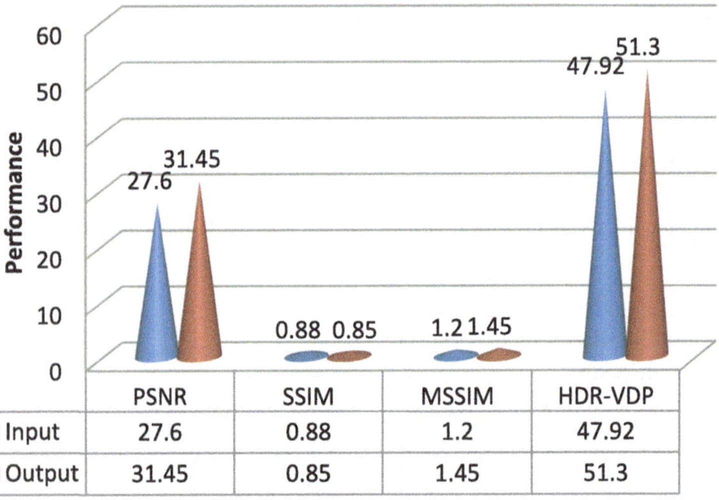


| | PSNR | SSIM | MSSIM | HDR-VDP |
|---|---|---|---|---|
| Input | 27.6 | 0.88 | 1.2 | 47.92 |
| Output | 31.45 | 0.85 | 1.45 | 51.3 |

**Fig. 5.** Contrasting the input and output images based on several characteristics

by means of the Mask. These segments are then utilized as inputs for the suggested multi-stage network.

While the Darkening blocks learns Img Dark to decrease the lower brightness value, the Brightness block learns Img Brilliant to generate the increased brightness value.

Ultimately, the two map representations that extended dynamic ranges are collective to generate the final HDR image input and output values as shown the Fig. 5, and the most effective features of each image are extracted through firefly refinement.

## 5 Conclusion

Deep learning has made significant strides in recent years, and numerous researches have used deep learning to produce high-quality HDR photos and rebuild saturated pixels. To properly depict reality, the LDR image must be divided and trained based on brightness values. The suggested approach produced average PSNR, SSIM, MS-SSIM, and HDRVDP2 metrics for rebuilt HDR images that were higher than those of the earlier techniques, respectively. Specifically, the suggested approach yielded a high mean log exposed range, indicating that the dynamic spectrum be the most realistic. In the suggested methods, improving the extracted characteristics is mostly dependent on optimization. The firefly technique finds features and pixels that aid in the reconstruction of the HDR image.

## References

1. Lee, S., An, G.H., Kang, S.-J.: Deep chain HDRI: Reconstructing a high dynamic range image from a single low dynamic range image. IEEE Access. **6**, 49913–49924 (2018)
2. Lee, S., An, G.H., Kang, S.-J.: Deep recursive HDRI: Inverse tone mapping using generative adversarial networks. In: Proceedings of European Conference Computer Vision. (ECCV), pp. 596–611 (2018)
3. Niharika, M., Karuna Sree, B.: IoT based attendance management system using google assistant. In: Pandian, A., Palanisamy, R., Ntalianis, K. (eds.) Proceeding of the International Conference on Computer Networks, Big Data and IoT (ICCBI - 2019). ICCBI 2019. Lecture Notes on Data Engineering and Communications Technologies, vol. 49. Springer, Cham (2020). https://doi.org/10.1007/978-3-030-43192-1_3
4. Varaprasad Rao, M., Srujan Raju, K., Vishnu Murthy, G., Kavitha Rani, B.: Configure and management of internet of things. In: Raju, K., Senkerik, R., Lanka, S., Rajagopal, V. (eds.) Data Engineering and Communication Technology. Advances in Intelligent Systems and Computing, vol. 1079. Springer, Singapore (2020). https://doi.org/10.1007/978-981-15-1097-7_14
5. Mubeen, S.: Synthesis of circular array antennas using accelerated particle swarm optimization. Int. J. Adv. Trend. Comput. Sci. Eng. **8**, 2121–2125 (2019). https://doi.org/10.30534/ijatcse/2019/41852019
6. Sri Sudha, T., Sasibhushana Rao, G.: Clipping based PMPR reduction techniques for LTE-OFDM systems. In: Hemanth, J., Fernando, X., Lafata, P., Baig, Z. (eds.) International Conference on Intelligent Data Communication Technologies and Internet of Things (ICICI) 2018. ICICI 2018. Lecture Notes on Data Engineering and Communications Technologies, vol. 26. Springer, Cham (2019). https://doi.org/10.1007/978-3-030-03146-6_118

7. Pendem, S., Ramesh, G.P.: 75 GHz 5G frequency spectrum analysis. In: Balas, V., Kumar, R., Srivastava, R. (eds.) Recent Trends and Advances in Artificial Intelligence and Internet of Things. Intelligent Systems Reference Library, vol. 172. Springer, Cham (2020). https://doi.org/10.1007/978-3-030-32644-9_18
8. Yin, J.-L., Chen, B.-H., Peng, Y.-T.: Two exposure fusion using the prior aware generative adversarial network. IEEE Trans. Multimedia. **15**, 2854 (2021). https://doi.org/10.1109/TMM.2021.3089324
9. Eilertsen, G., Kronander, J., Denes, G., Mantiuk, R.K., Unger, J.: HDR image reconstruction from a single exposure using deep CNNS. ACM Trans. Graph. **36**(6), 178 (2017)
10. Cai, J., Gu, S., Zhang, L.: Learning a deep single image contrast enhancer from multi-exposure images. IEEE Trans. Image Process. **27**(4), 2049–2062 (2018)
11. Marnerides, D., Bashford-Rogers, T., Hatchett, J., Debattista, K.: Expand Net: a deep convolutional neural network for high dynamic range expansion from low dynamic range content. Comput. Graph. Forum. **37**(2), 37–49 (2018)
12. Liu, Y.-L., Lai, W.-S., Chen, Y.-S., Kao, Y.-L., Yang, M.-H., Chuang, Y.-Y., Huang, J.-B.: Single-image HDR reconstruction by learning to reverse the camera pipeline. In: Proceedings of IEEE/CVF Conference Computer Vision Pattern Recognition (CPR), pp. 1651–1660 (2020)

# Author Index




A. Bajaj et al. (Eds.): HIS 2023, LNNS 1224, pp. 467–469, 2025.
https://doi.org/10.1007/978-3-031-78925-0

The manufacturer's authorised representative in the EU is Springer Nature Customer Service Centre GmbH, Europaplatz 3, 69115 Heidelberg, Germany. If you have any concerns regarding our products, please contact ProductSafety@springernature.com

Printed and bound by CPI Group (UK) Ltd, Croydon, CR0 4YY
30/09/2026
02225573-0013